Data Management

Databases and Organizations

Richard T. Watson
Department of MIS
Terry College of Business
The University of Georgia

6th edition

July 26, 2017

eGreen Press, Athens, GA

To

Clare

Cover photo: Bush, Yallingup, Western Australia

Table of Contents

Preface

This is not your traditional database textbook. It differs in three fundamental ways. *First*, it is deeper than most database books in its coverage of data modeling and SQL. The market seeks graduates who have these fundamental skills. Time and again, students who have completed my data management class have told me how these skills have been invaluable in their first and subsequent jobs. The intention is to place great emphasis on the core skills of data management. The consequence is that there is a better match between the skills students develop and those the market needs. This means that students find this text highly relevant.

Second, the treatments of data modeling and SQL are intertwined because my database teaching experience indicates that students more readily understand the intent of data modeling when they grasp the long-term goal—querying a well-designed relational database. The double helix, upward, intertwined, spiraling of data modeling and SQL is a unique pedagogical feature. Classroom testing indicates it is a superior method of teaching compared to handling data modeling and SQL separately. Students quickly understand the reason for data modeling and appreciate why it is a valuable skill. Also, rapid exposure to SQL means students gain hands-on experience that more quickly.

Third, the book is broader than most database books. Databases are one component of an expansive organizational memory. Information systems professionals need to develop a wide perspective of data management if they are to comprehend fully the organizational role of information systems.

In essence, the book is deeper where it matters—data modeling and SQL—and broader to give students a managerial outlook and an understanding of the latest technological advancements.

Information is a key resource for modern organizations. It is a critical input to managerial tasks. Because managers need high-quality information to manage change in a turbulent, global environment, many organizations have established systems for storing and retrieving data, the raw material of information. These storage and retrieval systems are an organization's memory. The organization relies on them, just as individuals rely on their personal memory, to be able to continue as a going concern.

The central concern of information systems management is to design, build, and maintain information delivery systems. Information systems management needs to discover its organization's information requirements so that it can design systems to serve these needs. It must merge a system's design and information technology to build applications that provide the organization with data in a timely manner, appropriate formats, and at convenient locations. Furthermore, it must manage applications so they evolve to meet changing needs, continue to operate under adverse conditions, and are protected from unauthorized access.

An information delivery system has two components: data and processes. This book concentrates on data, which is customarily thought of as a database. I deliberately set out to extend this horizon, however, by including all forms of organizational data stores because I believe students need to understand the role of data management that is aligned with current practice. In my view, data management is the design and maintenance of computer-based organizational memory. Thus, you will find chapters on XML and organizational intelligence technologies.

The decision to start the book with a managerial perspective arises from the belief that successful information systems practice is based on matching managerial needs, social system constraints, and technical opportunities. I want readers to appreciate the *big picture* before they become immersed in the intricacies of data modeling and SQL.

The first chapter introduces the case study, *The Expeditioner*, which is used in most of the subsequent chapters to introduce the key themes discussed. Often it sets the scene for the ensuing material by presenting a common business problem.

The second section of the book provides in-depth coverage of data modeling and SQL. Data modeling is the foundation of database quality. A solid grounding in data modeling principles and extensive practice are necessary for successful database design. In addition, this section exposes students to the full power of SQL.

I intend this book to be a long-term investment for students. There are useful reference sections for data modeling and SQL. The data modeling section details the standard structures and their relational mappings. The SQL section contains an extensive list of queries that serves as a basis for developing other SQL queries. The purpose of these sections is to facilitate *pattern matching.* For example, a student with an SQL query that is similar to a previous problem can rapidly search the SQL reference section to find the closest match. The student can then use the model answer as a guide to formulating the SQL query for the problem at hand. These reference sections are another unique teaching feature that will serve students well during the course and in their careers.

This 6th edition is a substantial revision in that it adds new chapters to provide an introduction to R, a statistics and graphics package, which provides the foundation necessary for the new chapters on data visualization, text mining, Hadoop distributed file system (HDFS) and MapReduce, and dashboards. These additions provide today's students with the skills they need to work in topical areas such as social media analytics and big data. These chapters are included in the third section, now titled Advanced Data Management, which also covers spatial and temporal data, XML, and organizational intelligence.

The fourth and final section examines the management of organizational data stores. It covers data structures and storage, data processing architectures, SQL and Java, data integrity, and data administration.

A student completing this text will

- have a broad, managerial perspective of an organization's need for a memory;
- be able to design and create a relational database;
- be able to formulate complex SQL queries;
- be able to use R to create data visualizations, mine text data, write MapReduce applications, and create dashboards;
- understand the purpose of XML and be able to design an XML schema, prepare an XML document, and write an XML stylesheet;
- have a sound understanding of database architectures and their managerial implications;
- be familiar with the full range of technologies available for organizational memory;
- be able to write a Java program to create a table from a CSV file and process a transaction;
- understand the fundamentals of data administration;

My goal is to give the reader a data management text that is innovative, relevant, and lively. I trust that you will enjoy learning about managing data in today's organization.

Supplements

Accompanying this book are an instructor's manual and an extensive Web site[1] that provides

- overhead slides in PowerPoint format;
- all relational tables in the book in electronic format;
- code for the R, Java, and XML examples in the book;

1 richardtwatson.com/dm6e/

- answers to many of the exercises;
- additional exercises.

New in the sixth edition

This edition has the following improvements and additions.

- MySQL Workbench[2] for data modeling and SQL querying
- Integration of XML and MySQL
- New chapters on R, data visualization, text mining, Hadoop distributed file system and MapReduce, and dashboards.

Acknowledgments

I thank my son, Ned, for help with the typesetting and my wife, Clare, for converting the 5th edition to Pages format and redoing the figures to support the 6th edition. I would like to thank the reviewers of this and prior editions for their many excellent suggestions and ideas for improving the quality of the content and presentation of the book.

Richard T. Watson

Athens, Georgia

2 wb.mysql.com/

Section 1: The Managerial Perspective

People only see what they are prepared to see.

Ralph Waldo Emerson, *Journals,* 1863

Organizations are accumulating vast volumes of data because of the ease with which data can be captured and distributed in digital format (e.g., smartphone cameras and tweets). The world's data are estimated to be doubling every 1-2 years, and large companies now routinely manage petabytes (10^{15} bytes) of data every day. Data management is a critical function for many organizations.

The first section of this book prepares you to see the role of data and information in an organization. The managerial perspective on data management concentrates on why enterprises design and maintain data management systems, or organizational memories. Chapter 1 examines this topic by detailing the components of organizational memory and then discussing some of its common problems. The intention is to make you aware of the scope of data management and its many facets. Chapter 2 discusses the relationship between information and organizational goals. Again, a very broad outlook is adopted in order to provide a sweeping perspective on the relationship of information to organizational success and change.

At this point, we want to give you some ***maps*** for understanding the territory you will explore. Since the terrain is possibly very new, these maps initially may be hard to read, and so you may need to consult them several times before you understand the landscape you are about to enter.

The first map is based on the Newell-Simon model[3] of the human information processing system, which shows that humans receive input, process it, and produce output. The processing is done by a processor, which is linked to a memory that stores both data and processes. The processor component retrieves both data and processes from memory.

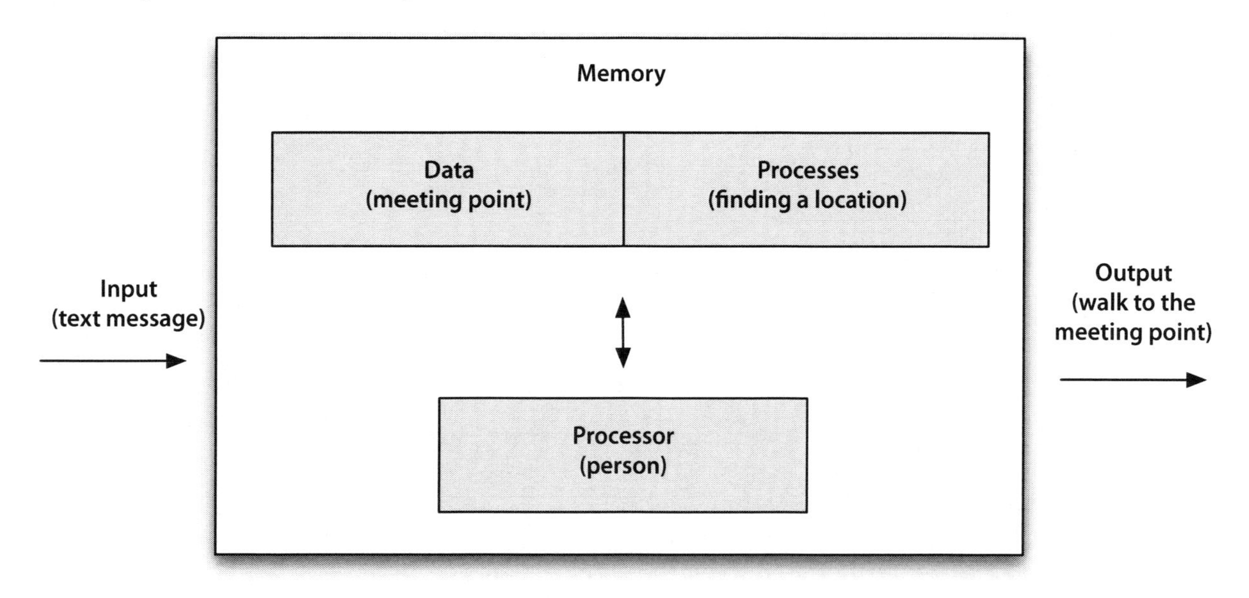

3 Newell, A., & Simon, H. A. (1972). Human problem solving. Englewood Cliffs, NJ: Prentice-Hall.

To understand this model, consider a person arriving on a flight who has agreed to meet a fellow traveler in the terminal. She receives a text message to meet her friend at B27. The message is input to her human information processing system. She retrieves the process for interpreting the message (i.e., decoding that B27 means terminal B and gate 27) and finds the terminal and gate. The person then walks to the terminal and gate, the output. Sometimes these processes are so well ingrained in our memory that we never think about retrieving them. We just do them automatically.

Human information processing systems are easily overloaded. Our memory is limited, and our ability to process data is restricted; thus we use a variety of external tools to extend and augment our capacities. A contacts app is an instance of external data memory. A recipe, a description of the process for preparing food, is an example of external process memory. Smartphones are now the external processor of choice that we use to augment our limited processing capacity.

The original model of human information processing can be extended to include external memory, for storing data and processes, and external processors, for executing processes.

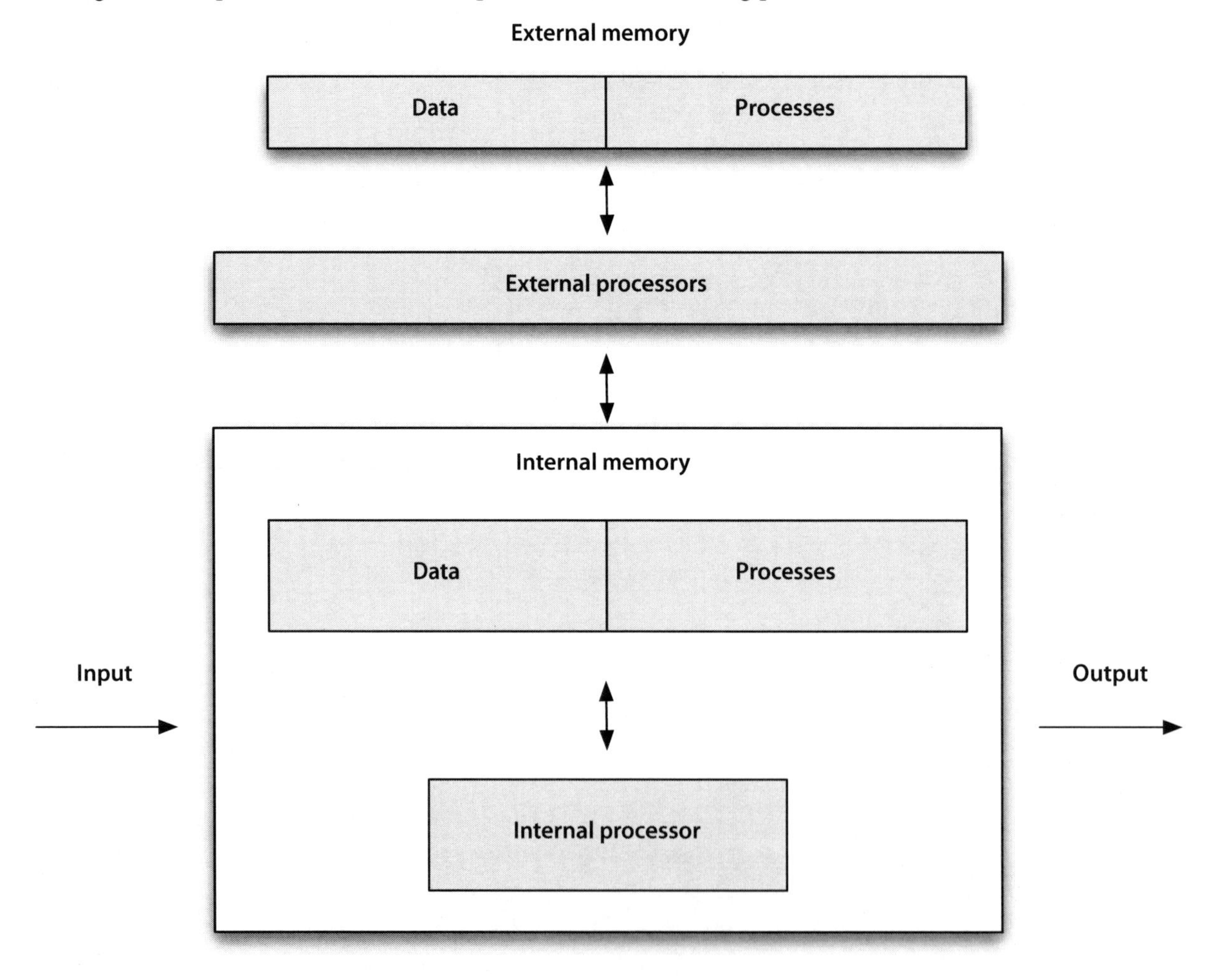

This model of augmented human information processing translates directly to an organizational setting. Organizations collect inputs from the environment: market research, customer complaints, and competitor actions. They process these data and produce outputs: sales campaigns, new products, price changes, and so on. The following figure gives an example of how an organization might process data. As a result of some market research (input), a marketing analyst (an internal processor) retrieves sales data (data) and does a sales forecast (process). The analyst also requests a marketing consultant (an external processor) to analyze (process) some demographic reports (data) before deciding to launch a new sales campaign (output).

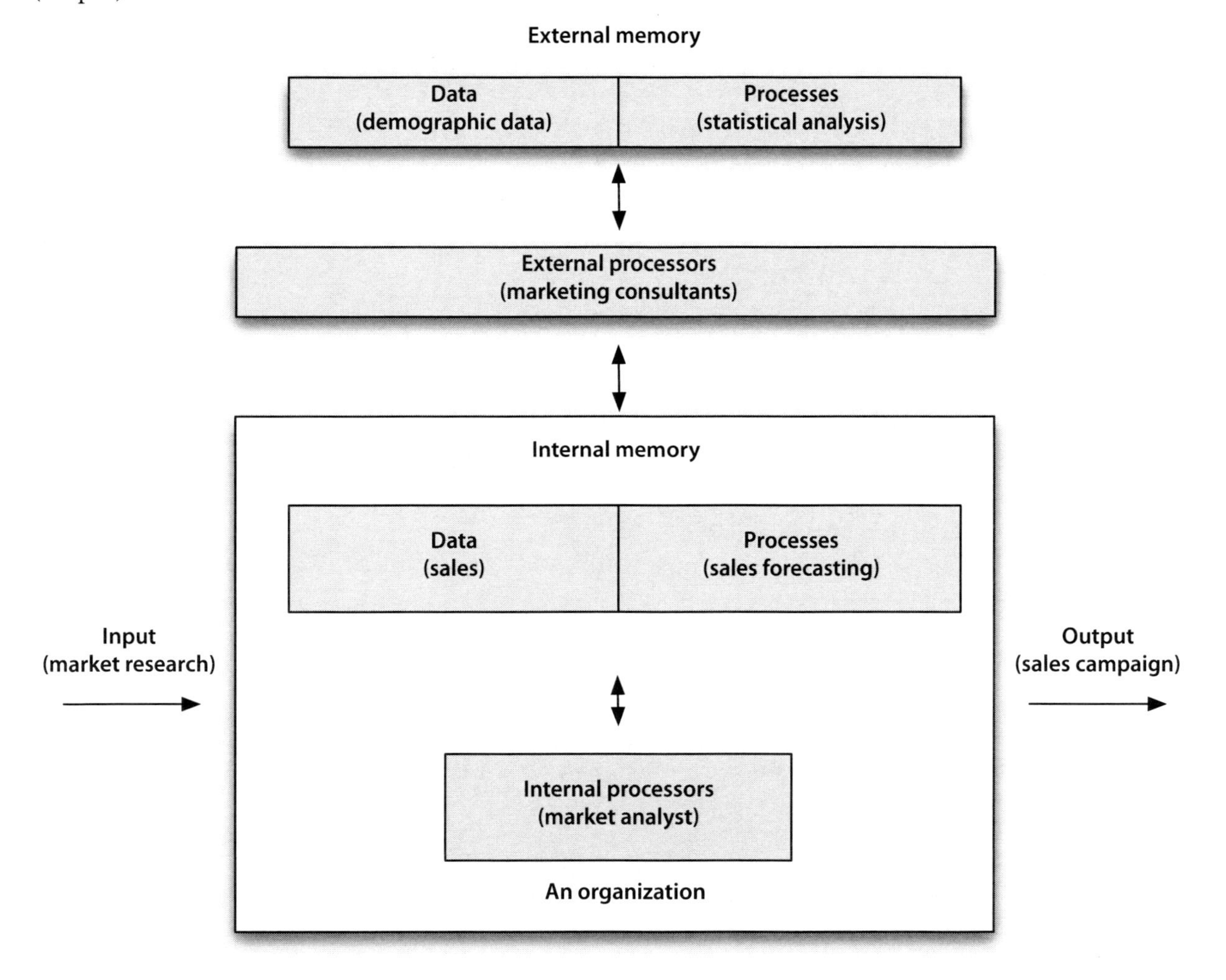

An organization's memory comes in a variety of forms, as you will see in Chapter 1. This memory also can be divided into data and processes. The data part may contain information about customers. The process portion may store details of how to handle a customer order. Organizations use a variety of processors to handle data, including people and computers. Organizations also rely on external sources to extend their information processing capacity. For example, a business may use a specialist credit agency to check a customer's creditworthiness, or an engineering firm may buy time on a supercomputer for structural analysis of a bridge. Viewed this way, the augmented human information processing model becomes a pattern for an organizational information processing system.

This book focuses on the data side of organizational memory. While it is primarily concerned with data stored within the organization, there is also coverage of data in external memory. The process side of organizational memory is typically covered in a systems analysis and design or a business process management course.

1. Managing Data

All the value of this company is in its people. If you burned down all our plants, and we just kept our people and our information files, we should soon be as strong as ever.

Thomas Watson, Jr., former chairman of IBM[4]

Learning objectives

Students completing this chapter will

- understand the key concepts of data management;
- recognize that there are many components of an organization's memory;
- understand the problems with existing data management systems;
- realize that successful data management requires an integrated understanding of organizational behavior and information technology.

Introduction

Imagine what would happen to a bank that forgot who owed it money or a magazine that lost the names and addresses of its subscribers. Both would soon be in serious difficulty, if not out of business. Organizations have data management systems to record the myriad of details necessary for transacting business and making informed decisions. Societies and organizations have always recorded data. The system may be as simple as carving a notch in a stick to keep a tally, or as intricate as modern database technology. A memory system can be as personal as a to-do list or as public as a library.

The management of organizational data, generally known as data management, requires skills in designing, using, and managing the memory systems of modern organizations. It requires multiple perspectives. Data managers need to see the organization as a social system and to understand data management technology. The integration of these views, the socio-technical perspective, is a prerequisite for successful data management.

Individuals also need to manage data. You undoubtedly are more familiar with individual memory management systems. They provide a convenient way of introducing some of the key concepts of data management.

Individual data management

As humans, we are well aware of our limited capacity to remember many things. The brain, our internal memory, can get overloaded with too much detail, and its memory decays with time. We store a few things internally: home and work telephone numbers, where we last parked our car, and faces of people we have met recently. We use external memory to keep track of those many things we would like to remember. External memory comes in a variety of forms.

On our smartphones, we have calendars to remind us of meetings and project deadlines. We have address books to record the addresses and phone numbers of those we contact frequently. We use to-do lists to remind us of the things we must do today or this week. The interesting thing about these aides-mémoire is that each has a unique way of storing data and supporting its rapid retrieval.

---

[4] Quinn, J. B. (1994). Appraising intellectual assets. *The McKinsey Quarterly*(2), 90-96.

Calendars come in many shapes and forms, but they are all based on the same organizing principle. A set amount of space is allocated for each day of the year, and the spaces are organized in date order, which supports rapid retrieval of any date. Some calendars have added features to speed up access. For example, electronic calendars usually have a button to select today's data.

A calendar

Address books also have a standard format. They typically contain predefined spaces for storing address details (e.g., name, company, phone, and email). Access is often supported by a set of buttons for each letter of the alphabet and a search engine.

An address book

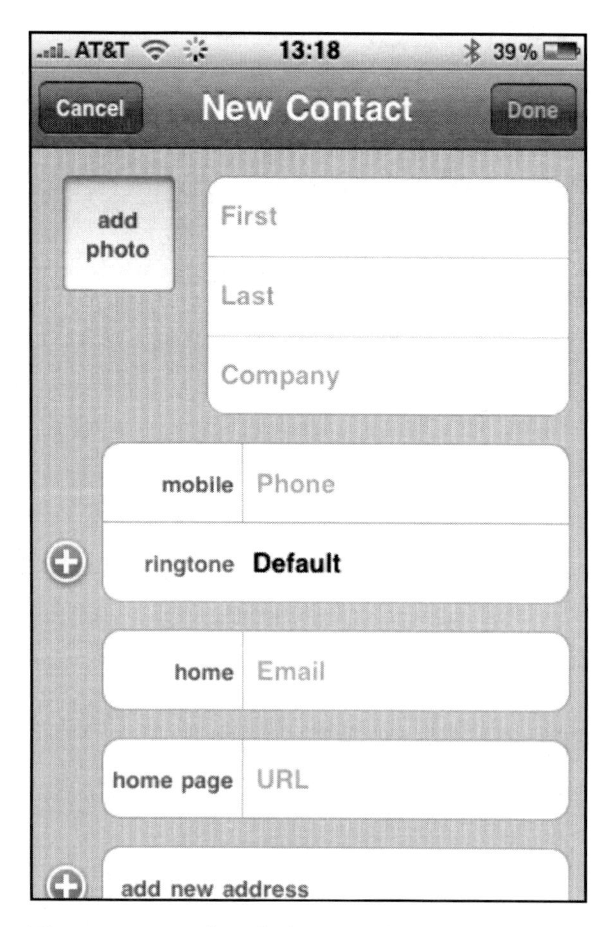

The structure of to-do lists tends to be fairly standard. They are often set up in list format with a small left-hand margin. The idea is to enter each item to be done on the right side of the screen. The left side is used to check (√) or mark those tasks that have been completed. The beauty of the check method is that you can quickly scan the left side to identify incomplete tasks.

A to-do or reminder list

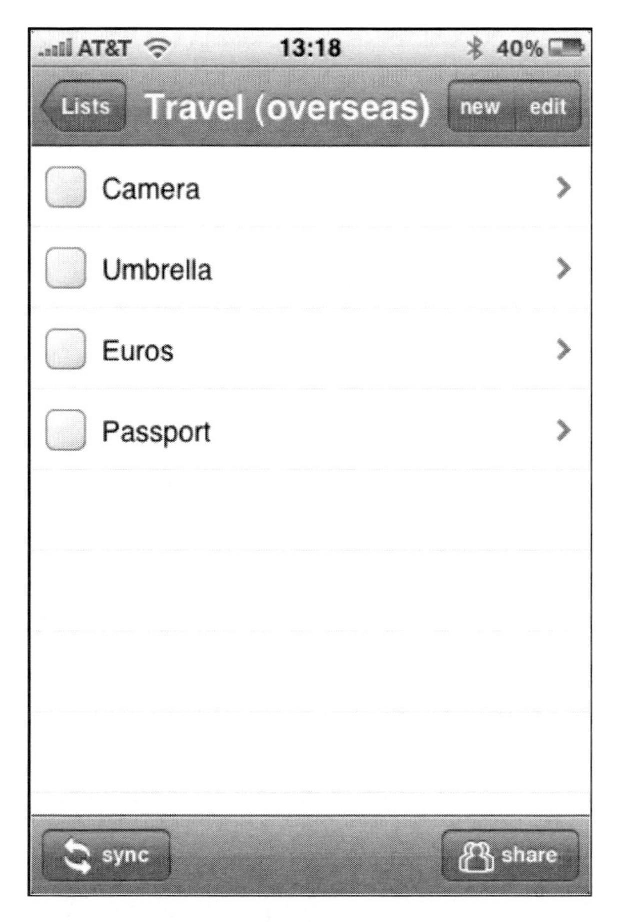

Many people use some form of the individual memory systems just described. They are frequently marketed as "time management systems" and are typically included in the suite of standard applications for a smart phone.

These three examples of individual memory systems illustrate some features common to all data management systems:

- There is a storage medium. Data are stored electronically in each case.
- There is a structure for storing data. For instance, the address book has labeled spaces for entering pertinent data.
- The storage device is organized for rapid data entry and retrieval. A calendar is stored in date and time sequence so that the data space for any appointment for a particular day can be found quickly.
- The selection of a data management system frequently requires a trade-off decision. In these examples, the trade-off is screen dimensions versus the amount of data that can be seen without scrolling. For example, you will notice the address book screen is truncated and will need to be scrolled to see full address details.

Skill builder

Smart phones have dramatically changed individual data management in the last few years. We now have calendars, address books, to-do lists, and many more apps literally in our hands. What individual data are still difficult to manage? What might be the characteristics of an app for that data?

There are differences between internal and external memories. Our internal memory is small, fast, and convenient (our brain is always with us—well, most of the time). External memory is often slower to reference and not always as convenient. The two systems are interconnected. We rely on our internal memory to access external memory. Our internal memory and our brain's processing skills manage the use of external memories. For example, we depend on our internal memory to recall how to use our smartphone and its apps. Again, we see some trade-offs. Ideally, we would like to store everything in our fast and convenient internal memory, but its limited capacity means that we are forced to use external memory for many items.

Organizational data management

Organizations, like people, need to remember many things. If you look around any office, you will see examples of the apparatus of organizational memory: people, filing cabinets, policy manuals, planning boards, and computers. The same principles found in individual memory systems apply to an organization's data management systems.

There is a storage medium. In the case of computers, the storage medium varies. Small files might be stored on a USB drive and large, archival files on a magnetic disk. In Chapter 11, we discuss electronic storage media in more detail.

A table is a common structure for storing data. For example, if we want to store details of customers, we can set up a table with each row containing individual details of a customer and each column containing data on a particular feature (e.g., customer code).

Storage devices are organized for rapid data entry and retrieval. Time is the manager's enemy: too many things to be done in too little time. Customers expect rapid responses to their questions and quick processing of their transactions. Rapid data access is a key goal of nearly all data management systems, but it always comes at a price. Fast access memories cost more, so there is nearly always a trade-off between access speed and cost.

As you will see, selecting *how* and *where* to store organizational data frequently involves a trade-off. Data managers need to know and understand what the compromises entail. They must know the key questions to ask when evaluating choices.

When we move from individual to organizational memory, some other factors come into play. To understand these factors, we need to review the different types of information systems. The automation of routine business transactions was the earliest application of information technology to business. A transaction processing system (TPS) handles common business tasks such as accounting, inventory, purchasing, and sales. The realization that the data collected by these systems could be sorted, summarized, and rearranged gave birth to the notion of a management information system (MIS). Furthermore, it was recognized that when internal data captured by a TPS is combined with appropriate external data, the raw material is available for a decision support system (DSS) or executive information system (EIS). Recently, online analytical processing (OLAP), data mining, and business intelligence (BI) have emerged as advanced data analysis techniques for data captured by business transactions and gathered from other sources (these systems are covered in detail in Chapter 14). The purpose of each of these systems is described in the following table and their interrelationship can be understood by examining the information systems cycle.

Types of information systems

	Type of information system	System's purpose
TPS	Transaction processing system	Collect and store data from routine transactions
MIS	Management information system	Convert data from a TPS into information for planning, controlling, and managing an organization
DSS	Decision support system	Support managerial decision making by providing models for processing and analyzing data
BI	Business intelligence	Gather, store, and analyze data to improve decision making
OLAP	Online analytical processing	Present a multidimensional, logical view of data
	Data mining	Use statistical analysis and artificial intelligence techniques to identify hidden relationships in data

The information systems cycle

The various systems and technologies found in an organization are linked in a cycle. The routine ongoing business of the organization is processed by TPSs, the systems that handle the present. Data collected by TPSs are stored in databases, a record of the past, the history of the organization and its interaction with those with whom it conducts business. These data are converted into information by analysts using a variety of software (e.g., a DSS). These technologies are used by the organization to prepare for the future (e.g., sales in Finland have expanded, so we will build a new service center in Helsinki). The business systems created to prepare for the future determine the transactions the company will process and the data that will be collected, and the process continues. The entire cycle is driven by people using technology (e.g., a customer booking a hotel room via a Web browser).

The information systems cycle

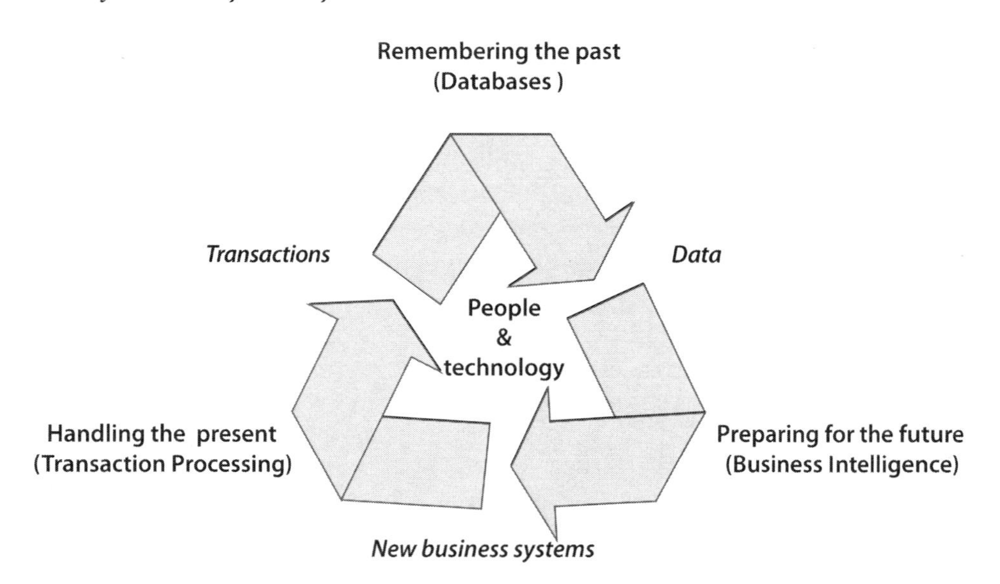

Decision Making, or preparing for the future, is the central activity of modern organizations. Today's organizations are busy turning out goods, services, and decisions. Knowledge and information workers, over half of the U.S. labor force, produce the bulk of GNP. Many of these people are decision makers. Their success, and their organization's as well, depends on the quality of their decisions.

Industrial society is a producer of goods, and the hallmark of success is product quality. Japanese manufacturers convincingly demonstrated that focusing on product quality is the key to market leadership and profitability. The methods and the philosophy of quality gurus, such as W. Edwards Deming, have been internationally recognized and adopted by many providers of goods and services. We are now in the information age as is evidenced by the hot products of the times, such as portable music players, smart phones, tablets, video recorders, digital cameras, and streaming video players. These are all information appliances, and they are supported by a host of information services. For example, consider how Apple connects together its various devices and services through iTunes. For example, an iPad owner can buy through iTunes a book to read with the iBooks app.

In the information society, which is based on innovation, knowledge, and services, the key determinant of success has shifted from product quality to decision quality. In the turbulent environment of global business, successful organizations are those able to quickly make high-quality decisions about what customers will buy, how much they will pay, and how to deliver a high-quality experience with a minimum of fuss. Companies are very dependent on information systems to create value for their customers.

Desirable attributes of data

Once we realize the critical importance of data to organizations, we can recognize some desirable attributes of data.

Desirable attributes of data

Shareable	Readily accessed by more than one person at a time
Transportable	Easily moved to a decision maker
Secure	Protected from destruction and unauthorized use
Accurate	Reliable, precise records
Timely	Current and up-to-date
Relevant	Appropriate to the decision

Shareable

Organizations contain many decision makers. There are occasions when more than one person will require access to the same data at the same time. For example, in a large bank it would not be uncommon for two customer representatives simultaneously to want data on the latest rate for a three-year certificate of deposit. As data become more volatile, shareability becomes more of a problem. Consider a restaurant. The permanent menu is printed, today's special might be displayed on a blackboard, and the waiter tells you what is no longer available.

Transportable

Data should be transportable from their storage location to the decision maker. Technologies that transport data have a long history. Homing pigeons were used to relay messages by the Egyptians and Persians 3,000 years ago. The telephone revolutionized business and social life because it rapidly transmitted voice data. Fax machines accelerated organizational correspondence because they transported both text and visual data. Computers have changed the nature of many aspects of business because they enable the transport of text, visuals, and voice data.

Today, transportability is more than just getting data to a decision maker's desk. It means getting product availability data to a salesperson in a client's office or advising a delivery driver, en route, of address details for an urgent parcel pickup. The general notion is that decision makers should have access to relevant data whenever and wherever required, although many organizations are still some way from reaching this target.

Secure

In an information society, organizations value data as a resource. As you have already learned, data support day-to-day business transactions and decision making. Because the forgetful organization will soon be out of business, organizations are very vigilant in protecting their data. There are a number of actions that organizations take to protect data against loss, sabotage, and theft. A common approach is to duplicate data and store the copy, or copies, at other locations. This technique is popular for data stored in computer systems. Access to data is often restricted through the use of physical barriers (e.g., a vault) or electronic barriers (e.g., a password). Another approach, which is popular with firms that employ knowledge workers, is a noncompete contract. For example, some software companies legally restrain computer programmers from working for a competitor for two years after they leave, hoping to prevent the transfer of valuable data, in the form of the programmer's knowledge of software, to competitors.

Accurate

You probably recall friends who excel in exams because of their good memories. Similarly, organizations with an accurate memory will do better than their less precise competitors. Organizations need to remember many details precisely. For example, an airline needs accurate data to predict the demand for each of its flights. The quality of decision making will drop dramatically if managers use a data management system riddled with errors.

Polluted data threatens a firm's profitability. One study suggests that missing, wrong, and otherwise bad data cost U.S. firms billions of dollars annually. The consequences of bad data include improper billing, cost overruns, delivery delays, and product recalls. Because data accuracy is so critical, organizations need to be watchful when capturing data—the point at which data accuracy is most vulnerable.

Timely

The value of a collection of data is often determined by its age. You can fantasize about how rich you would be if you knew tomorrow's stock prices. Although decision makers are most interested in current data, the required currency of data can vary with the task. Operational managers often want real-time data. They want to tap the pulse of the production line so that they can react quickly to machine breakdowns or quality slippages. In contrast, strategic planners might be content with data that are months old because they are more concerned with detecting long-term trends.

Relevant

Organizations must maintain data that are relevant to transaction processing and decision making. In processing a credit card application, the most relevant data might be the customer's credit history, current employment status, and income level. Hair color would be irrelevant. When assessing the success of a new product line, a marketing manager probably wants an aggregate report of sales by marketing region. A voluminous report detailing every sale would be irrelevant. Data are relevant when they pertain directly to the decision and are aggregated appropriately.

Relevance is a key concern in designing a data management system. Clients have to decide what should be stored because it is pertinent now or could have future relevance. Of course, identifying data that might be relevant in the future is difficult, and there is a tendency to accumulate too much. Relevance is also an important consideration when extracting and processing data from a data management system. Provided the pertinent data are available, query languages can be used to aggregate data appropriately.

In the final years of the twentieth century, organizations started to share much of their data, both high and low volatility, via the Web. This move has increased shareability, timeliness, and availability, and it has lowered the cost of distributing data.

In summary, a data management system for maintaining an organization's memory supports transaction processing, remembering the past, and decision making. Its contents must be shareable, secure, and accurate. Ideally, the clients of a data management system must be able to get timely and relevant data when and where required. A major challenge for data management professionals is to create data management systems that meet these criteria. Unfortunately, many existing systems fail in this regard, though we can understand some of the reasons why by reviewing the components of existing organizational memory systems.

Components of organizational memory

An organization's memory resides on a variety of media in a variety of ways. It is in people, standard operating procedures, roles, organizational culture, physical storage equipment, and electronic devices. It is scattered around the organization like pieces of a jigsaw puzzle designed by a berserk artist. The pieces don't fit together, they sometimes overlap, there are gaps, and there are no edge pieces to define the boundaries. Organizations struggle to design structures and use data management technology to link some of the pieces. To understand the complexity of this wicked puzzle, we need to examine some of the pieces. Data managers have a particular need to understand the different forms of organizational memory because their activities often influence a number of the components.

Components of organizational memory

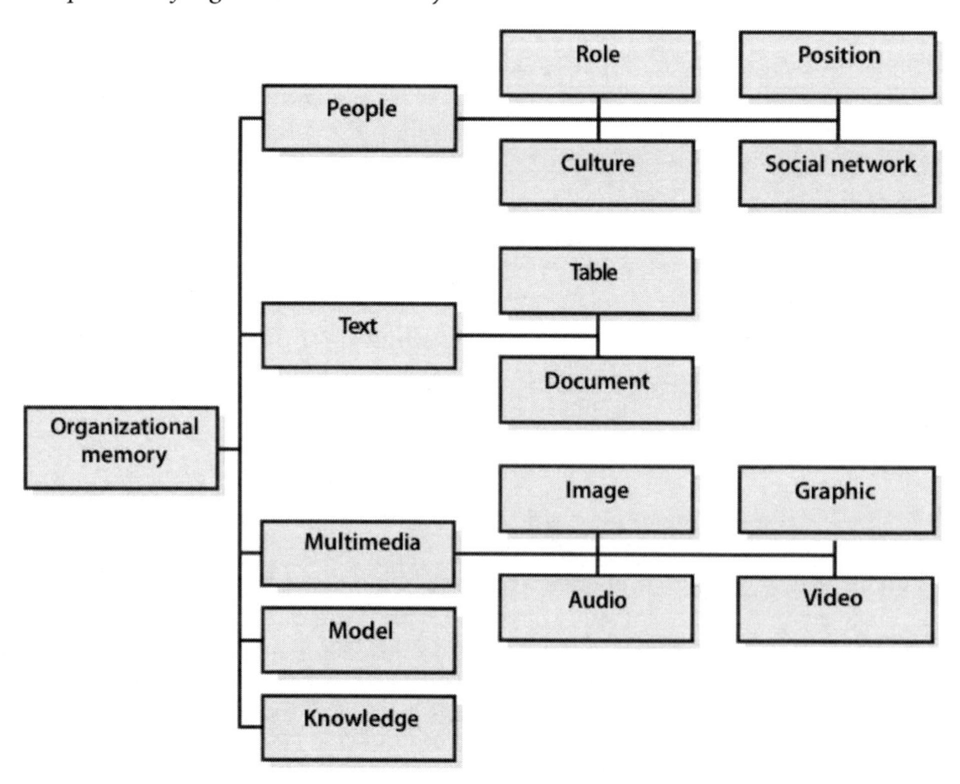

People

People are the linchpin of an organization's memory. They recall prior decisions and business actions. They create, maintain, evolve, and use data management systems. They are the major component of an organization's memory because they know how to use all the other components. People extract data from the various components of organizational memory to provide as complete a picture of a situation as possible.

Each person in an organization has a role and a position in the hierarchy. Role and position are both devices for remembering how the organization functions and how to process data. By labeling people (e.g., Chief Information Officer) and placing their names on an organizational chart, the organization creates another form of organizational memory.

Organizational culture is the shared beliefs, values, attitudes, and norms that influence the behavior and expectations of each person in an organization. As a long-lived and stable memory system, culture determines acceptable behavior and influences decision making.

People develop skills for doing their particular job—learning what to do, how to do it, and who can help them get things done. For example, they might discover someone in employee benefits who can handle personnel problems or a contact in a software company who can answer questions promptly. These social networks, which often take years to develop, are used to make things happen and to learn about the business environment. Despite their high value, they are rarely documented, at least not beyond an address book, and they are typically lost when a person leaves an organization.

Conversations are an important method for knowledge workers to create, modify, and share organizational memory and to build relationships and social networks. Discussions with customers are a key device for learning how to improve an organization's products and services and learning about competitors. The ***conversational company*** can detect change faster and react more rapidly. The telephone, instant message, e-mail, coffee machine, cocktail hour, and cafeteria are all devices for promoting conversation and creating networks. Some firms deliberately create structures for supporting dialog to make the people component of organizational memory more effective.

Standard operating procedures exist for many organizational tasks. Processing a credit application, selecting a marketing trainee, and preparing a departmental budget are typical procedures that are clearly defined by many organizations. They are described on Web pages, computer programs, and job specifications. They are the way an organization remembers how to perform routine activities.

Successful people learn how to use organizational memory. They learn what data are stored where, how to retrieve them, and how to put them together. In promoting a new product, a salesperson might send the prospect a package containing some brochures and an email of a product review in a trade journal, and supply the phone number and e-mail address of the firm's technical expert for that product. People's recall of how to use organizational memory is the core component of organizational memory. Academics call this **metamemory**; people in business call it *learning the ropes*. New employees spend a great deal of time building their metamemory so that they can use organizational memory effectively. Without this knowledge, organizational memory has little value.

Tables

A table is a common form of storing organizational data. The following table shows a price list in tabular form. Often, the first row defines the meaning of data in subsequent rows.

A price list

Product	Price
Pocket knife – Nile	4.5
Compass	10
Geopositioning system	500
Map measure	4.9

A table is a general form that describes a variety of other structures used to store data. Computer-based files are tables or can be transformed into tables; the same is true for general ledgers, worksheets, and spreadsheets. Accounting systems make frequent use of tables. As you will discover in the next section, the table is the central structure of the relational database model.

Data stored in tables typically have certain characteristics:

- Data in one column are of the same type. For example, each cell of the column headed "Price" contains a number. (Of course, the exception is the first row of each column, which contains the title of the column.)
- Data are limited by the width of the available space.

Rapid searching is one of the prime advantages of a table. For example, if the price list is sorted by product name, you can quickly find the price of any product.

Tables are a common form of storing organizational data because their structure is readily understood. People learn to read and build tables in the early years of their schooling. Also, a great deal of the data that organizations want to remember can be stored in tabular form.

Documents

The **document**—of which reports, manuals, brochures, and memos are examples—is a common medium for storing organizational data. Although documents may be subdivided into chapters, sections, paragraphs, and sentences, they lack the regularity and discipline of a table. Each row of a table has the same number of columns, but each paragraph of a document does not have the same number of sentences.

Most documents are now stored electronically. Because of the widespread use of word processing, text files are a common means of storing documents. Typically, such files are read sequentially like a book. Although there is support for limited searching of the text, such as finding the next occurrence of a specified text string, text files are usually processed linearly.

Hypertext, the familiar linking technology of the Web, supports nonlinear document processing. A hypertext document has built-in linkages between sections of text that permit the reader to jump quickly from one part to another. As a result, readers can find data they require more rapidly.

Although hypertext is certainly more reader-friendly than a flat, sequential text file, it takes time and expertise to establish the links between the various parts of the text. Someone familiar with the topic has to decide what should be linked and then establish these links. While it takes the author more time to prepare a document this way, the payoff is the speed at which readers of the document can find what they want.

Multimedia

Many Web sites display multimedia objects, such as sound and video clips. Automotive company Web sites have video clips of cars, music outlets provide sound clips of new releases, and clothing companies have online catalogs displaying photos of their latest products. Maintaining a Web site, because of the many multimedia objects that some sites contain, has become a significant data management problem for some organizations. Consider the different types of data that a news outfit such as the British Broadcasting Corporation (BBC) has to store to provide a timely, informative, and engaging Web site.

Images

Images are visual data: photographs and sketches. Image banks are maintained for several reasons. *First,* images are widely used for identification and security. Police departments keep fingerprints and mug shots. *Second,* images are used as evidence. Highly valuable items such as paintings and jewelry often are photographed for insurance records. *Third,* images are used for advertising and promotional campaigns, and organizations need to maintain records of material used in these ventures. Image archiving and retrieval are essential for mail-order companies, which often produce several photo-laden catalogs every year. *Fourth*, some organizations specialize in selling images and maintain extensive libraries of clip art and photographs.

Graphics

Maps and engineering drawings are examples of electronically stored graphics. An organization might maintain a map of sales territories and customers. Manufacturers have extensive libraries of engineering drawings that define the products they produce. Graphics often contain a high level of detail. An engineering drawing will define the dimensions of all parts and may refer to other drawings for finer detail about any components.

A graphic differs from an image in that it contains explicitly embedded data. Consider the difference between an engineering plan for a widget and a photograph of the same item. An engineering plan shows dimensional data and may describe the composition of the various components. The embedded data are used to manufacture the widget. A photograph of a widget does not have embedded data and contains insufficient data to manufacture the product. An industrial spy will receive far more for an engineering plan than for a photograph of a widget.

A geographic information systems (GIS) is a specialized graphical storage system for geographic data. The underlying structure of a GIS is a map on which data are displayed. A power company can use a GIS to store and display data about its electricity grid and the location of transformers. Using a pointing device such as a mouse, an engineer can click on a transformer's location to display a window of data about the transformer (e.g., type, capacity, installation date, and repair history). GISs have found widespread use in governments and organizations that have geographically dispersed resources.

Audio

Music stores such as iTunes enable prospective purchasers to hear audio samples. News organizations, such as National Public Radio (NPR), provide audio versions of their new stories for replay.

Some firms conduct a great deal of their business by phone. In many cases, it is important to maintain a record of the conversation between the customer and the firm's representative. The Royal Hong Kong Jockey Club, which covers horse racing gambling in Hong Kong, records all conversations between its operators and customers. Phone calls are stored on a highly specialized voice recorder, which records the time of the call and other data necessary for rapid retrieval. In the case of a customer dispute, an operator can play back the original conversation.

Video

A video clip can give a potential customer additional detail that cannot be readily conveyed by text or a still image. Consequently, some auto companies use video and virtual reality to promote their cars. On a visit to Toyota's Web site, you can view a video clip of the Prius or rotate an image to view the car from multiple angles.

Models

Organizations build mathematical models to describe their business. These models, usually placed in the broader category of DSS, are then used to analyze existing problems and forecast future business conditions. A mathematical model can often produce substantial benefits to the organization.

Knowledge

Organizations build systems to capture the knowledge of their experienced decision makers and problem solvers. This expertise is typically represented as a set of rules, semantic nets, and frames in a knowledge base, another form of organizational memory.

Decisions

Decision Making is the central activity of modern organizations. Very few organizations, however, have a formal system for recording decisions. Most keep the minutes of meetings, but these are often very brief and record only a meeting's outcome. Because they do not record details such as the objectives, criteria, assumptions, and alternatives that were considered prior to making a decision, there is no formal audit trail for decision making. As a result, most organizations rely on humans to remember the circumstances and details of prior decisions.

Specialized memories

Because of the particular nature of their business, some organizations maintain memories rarely found elsewhere. Perfume companies, for instance, maintain a library of scents, and paint manufacturers and dye makers catalog colors.

External memories

Organizations are not limited to their own memory stores. There are firms whose business is to store data for resale to other organizations. Such businesses have existed for many years and are growing as the importance of data in a postindustrial society expands. Lawyers using Mead Data Central's LEXIS® can access the laws and court decisions of all 50 American states and the U.S. federal government. Similar legal data services exist in many other nations. There is a range of other services that provide news, financial, business, scientific, and medical data.

Problems with data management systems

Successful management of data is a critical skill for nearly every organization. Yet few have gained complete mastery, and there are a variety of problems that typically afflict data management in most firms.

Problems with organizational data management systems

Redundancy	Same data are stored in different systems
Lack of data control	Data are poorly managed
Poor interface	Data are difficult to access
Delays	There are frequently delays following requests for reports
Lack of reality	Data management systems do not reflect the complexity of the real world
Lack of data integration	Data are dispersed across different systems

Redundancy

In many cases, data management systems have grown haphazardly. As a result, it is often the situation that the same data are stored in several different memories. The classic example is a customer's address, which might be stored in the sales reporting system, accounts receivable system, and the salesperson's address book. The danger is that when the customer changes address, the alteration is not recorded in all systems. Data redundancy causes additional work because the same item must be entered several times. Redundancy causes confusion when what is supposedly the same item has different values.

Lack of data control

Allied with the redundancy problem is poor data control. Although data are an important organizational resource, they frequently do not receive the same degree of management attention as other important organizational resources, such as people and money. Organizations have a personnel department to manage human resources and a treasury to handle cash. The IS department looks after data captured by the computer systems it operates, but there are many other data stores scattered around the organization. Data are stored everywhere in the organization (e.g., on personal computers and in departmental filing systems), but there is a general lack of data management. This lack is particularly surprising, since many pundits claim that data are a key competitive resource.

Poor interface

Too frequently, potential clients of data management systems have been deterred by an unfriendly interface. The computer interface for accessing a data store is sometimes difficult to remember for the occasional inquirer. People become frustrated and give up because their queries are rejected and error messages are unintelligible.

Delays

Globalization and technology have accelerated the pace of business in recent years. Managers must make more decisions more rapidly. They cannot afford to wait for programmers to write special-purpose programs to retrieve data and format reports. They expect their questions to be answered rapidly, often within a day and sometimes more quickly. Managers, or their support personnel, need query languages that provide rapid access to the data they need, in a format that they want.

Lack of reality

Organizational data stores must reflect the reality and complexity of the real world. Consider a typical bank customer who might have a personal checking account, mortgage account, credit card account, and some certificates of deposit. When a customer requests an overdraft extension, the bank officer needs full details of the customer's relationship with the bank to make an informed decision. If customer data are scattered across unrelated data stores, then these data are not easily found, and in some cases important data might be overlooked. The request for full customer details is reasonable and realistic, and the bank officer should expect to be able to enter a single, simple query to obtain it. Unfortunately, this is not always the case, because data management systems do not always reflect reality.

In this example, the reality is that the personal checking, mortgage, and credit card accounts, and certificates of deposit all belong to one customer. If the bank's data management system does not record this relationship, then it does not mimic reality. This makes it impossible to retrieve a single customer's data with a single, simple query.

A data management system must meet the decision making needs of managers, who must be able to request both routine and ad hoc reports. In order to do so effectively, a data management system must reflect the complexity of the real world. If it does not store required organizational data or record a real-world relationship between data elements, then many managerial queries cannot be answered quickly and efficiently.

Lack of data integration

There is a general lack of data integration in most organizations. Not only are data dispersed in different forms of organizational memory (e.g., files and image stores), but even within one storage format there is often a lack of integration. For example, many organizations maintain file systems that are not integrated. Appropriate files in the accounting system may not be linked to the production system.

This lack of integration will be a continuing problem for most organizations for two important reasons. ***First,*** earlier computer systems might not have been integrated because of the limitations of available technology. Organizations created simple file systems to support a particular function. Many of these systems are still in use. ***Second***, integration is a long-term goal. As new systems are developed and old ones rewritten, organizations can evolve integrated systems. It would be too costly and disruptive to try to solve the data integration problem in one step.

Many data management problems can be solved with present technology. Data modeling and relational database technology, topics covered in Section 2 of this book, help overcome many of the current problems.

A brief history of data management systems

Data management is not a new organizational concern. It is an old problem that has become more significant, important, and critical because of the emergence of data as a critical resource for effective performance in the modern economy. Organizations have always needed to manage their data so that they could remember a wide variety of facts necessary to conduct their affairs.

The recent history of computer-based data management systems is depicted in the following figure. File systems were the earliest form of data management. Limited by the sequential nature of magnetic tape technology, it was very difficult to integrate data from different files. The advent of magnetic disk technology in the mid-1950s stimulated development of integrated file systems, and the hierarchical database management system (DBMS) emerged in the 1960s, followed some years later by the network DBMS. The spatial database, or geographic information system (GIS), appeared around 1970. Until the mid-1990s, the hierarchical DBMS, mainly in the form of IBM's DL/I product, was the predominant technology for managing data. It has now been replaced by the relational DBMS, a concept first discussed by Edgar Frank Codd in an academic paper in 1970 but not commercially available until the mid-1970s. In the late 1980s, the notion of an object-oriented DBMS, primed by the ideas of object-oriented programming, emerged as a solution to situations not handled well by the relational DBMS. Also around this time, the idea of modeling a database as a graph was introduced. Towards the end of the 20th century, XML was developed for exchanging data between computers, and it can also be used as a data store as you will learn in section 3. More recently, the Hadoop distributed files system (HDFS) has emerged as a popular alternative model for data management, and it will be covered in the latter part of this book. Other recent data management systems include graph and NoSQL databases. While these are beyond the scope of an introductory data management text, if you decided to pursue a career in data management you should learn about their advantages and the applications to which they are well-suited. For example, graph databases are a good fit for the analysis of social networks.

Data management systems timeline

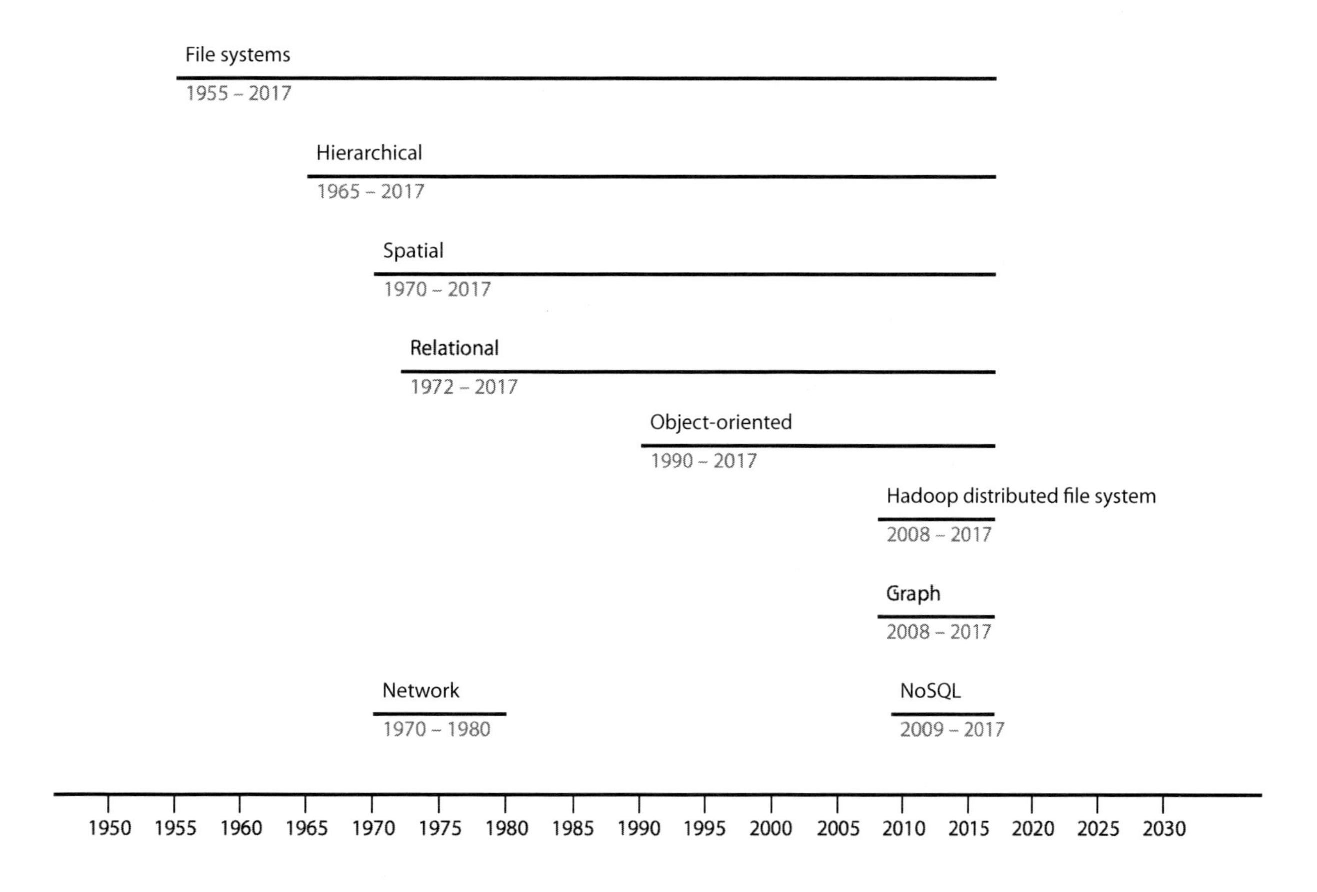

This book concentrates on the relational model, currently the most widely used data management system. As mentioned, Section 2 is devoted to the development of the necessary skills for designing and using a relational database.

Data, information, and knowledge

Often the terms ***data*** and ***information*** are used interchangeably, but they are distinctly different. Data are raw, unsummarized, and unanalyzed facts. Information is data that have been processed into a meaningful form.

A list of a supermarket's daily receipts is data, but it is not information, because it is too detailed to be very useful. A summary of the data that gives daily departmental totals is information, because the store manager can use the report to monitor store performance. The same report would be only data for a regional manager, because it is too detailed for meaningful decision making at the regional level. Information for a regional manager might be a weekly report of sales by department for each supermarket.

Data are always data, but one person's information can be another person's data. Information that is meaningful to one person can be too detailed for another. A manager's notion of information can change quickly, however. When a problem is identified, a manager might request finer levels of detail to diagnose the problem's cause. Thus, what was previously data suddenly becomes information because it helps solve the problem. When the problem is solved, the information reverts to data. There is a need for information systems that let managers customize the processing of data so that they always get information. As their needs change, they need to be able to adjust the detail of the reports they receive.

Knowledge is the capacity to use information. The education and experience that managers accumulate provide them with the expertise to make sense of the information they receive. Knowledge means that managers can interpret information and use it in decision making. In addition, knowledge is the capacity to recognize what information would be useful for making decisions. For example, a sales manager knows that requesting a report of profitability by product line is useful when she has to decide whether to employ a new product manager. Thus, when a new information system is delivered, managers need to be taught what information the system can deliver and what that information means.

The relationship between data, information, and knowledge is depicted in the following figure. A knowledgeable person requests information to support decision making. To fulfill the request, data are converted into information. Personal knowledge is then applied to interpret the requested information and reach a conclusion. Of course, the cycle can be repeated several times if more information is needed before a decision can be made. Notice how knowledge is essential for grasping what information to request and interpreting that information in terms of the required decision.

The relationship between data, information, and knowledge

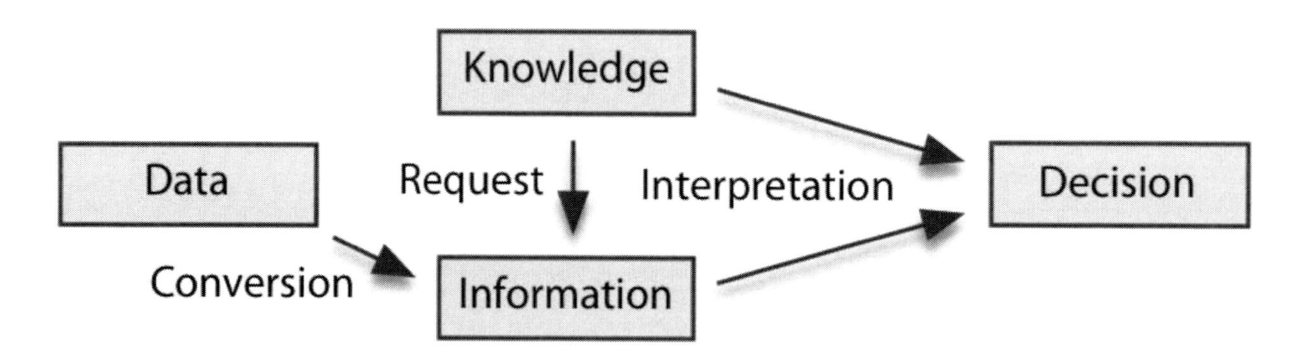

The challenge

A major challenge facing organizations is to make effective use of the data currently stored in their diverse data management systems. This challenge exists because these various systems are not integrated and many potential clients not only lack the training to access the systems but often are unaware what data exist. Before data managers can begin to address this problem, however, they must understand how organizational memories are used. In particular, they need to understand the relationship between information and managerial decision making. Data management is not a new problem. It has existed since the early days of civilization and will be an enduring problem for organizations and societies.

Alice Lindsay was enjoying the luxuries of first-class travel. It was quite a change from being an undergraduate student of business. She was enjoying her filet mignon and glass of shiraz. Good riddance to pizzas, hamburgers, and subs! Hello to fine food and gourmet restaurants! With her newly found wealth, she could travel in style and enjoy the very best restaurants. Alice, or Lady Alice to be more precise, had recently inherited a title, a valuable portfolio of stocks, and The Expeditioner. Her good fortune had coincided with the completion of her business degree. Now, instead of looking for a job, the job had found her: She was chair and managing director of The Expeditioner.

The Expeditioner is in a nineteenth-century building in Explorers Lane, opposite

Kensington Gardens and just a stone's throw from the Royal Geographical Society. Founded in the Middle Ages, The Expeditioner has a long history of equipping explorers, adventurers, and travelers. Its initial success was due to its development of a light, flexible chain mail armor. It did very well during the Middle Ages, when highway robbery was a growth business and travelers sought protection from the arrows and slings aimed at their fortunes. The branch office at the entrance to Sherwood Forest had done exceptionally well, thanks to its passing wealthy clientele. The resident company wit, one Ned Thomas, often claims that "The Expeditioner was the first mail order company."

The Expeditioner's customers included the famous and the legendary. Orders from Marco Polo, Columbus, Magellan, Cook, and Livingston can be found in the ledgers. The most long-lived customer is a Mr. Walker, who for several hundred years has had a standing yearly order of one roll of purple, premium-quality, non-rip, jungle twill. The Expeditioner's hats are very popular with a Mr. I. Jones, an American.

The nineteenth century was one long boom for The Expeditioner, the supplier de rigueur for the guardians of the European colonies. Branch offices were opened in Bombay, Sydney, Cape Town, and Singapore. For a generous fee, cad and renowned adventurer Harry Flashman assiduously promoted The Expeditioner's wares on his many trips. As a result, orders poured in from the four corners of the globe. The Expeditioner's catalog was found in polite society and clubs throughout the British Empire.

The founder of The Expeditioner was a venturesome Scot who sought his fortune in London. Carefully guarding the family business, The Expeditioner's proprietors were inclined to be closefisted and invested their massive profits wisely. Because of their close contacts with many explorers, they made some very canny investments in mining ventures in distant lands. As a result, when business contracted during the twentieth century, the investment portfolio kept the company solvent. Traditionally, ownership of the firm was inherited by the eldest child. When the most recent owner, a bachelor, died, the firm's lawyers spent months on genealogical research to identify the legal heir.

Alice had vivid memories of the day when Mr. Mainwaring phoned her to request an appointment. As she did not recognize the caller, she let the message go to voicemail. She remembered listening to it several times because she was intrigued by both the unusual accent and the message. It sounded something like, "Good afternoon, Lady Alice, I am Nigel Mannering of Chumli, Crepiny, Marchbanks, and Sinjun. If you would like to hear news to your advantage, please contact me."

Two days later, Alice met with Nigel. This also had been most memorable. The opening conversation had been confusing. He started with the very same formal introduction, "Good afternoon, Lady Alice, I am Nigel Mannering of Chumli, Crepiny, Marchbanks, and Sinjun. I am pleased to meet you," and handed Alice his card. Alice quickly inspected the card, which read, "Nigel Mainwaring, LL. B., Cholmondeley, Crespigny, Majoribanks, and St. John," but he had said nothing like that. She thought, "What planet is this guy from?"

It took fifteen minutes for Alice to discover that British English was not quite the same as American English. She quickly learned—or should that be "learnt?"—that many proper English names have a traditional pronunciation not easily inferred from the spelling. Once this names' problem had been sorted out, Nigel told Alice of her good fortune. She was the new owner of The Expeditioner. She also inherited the title that had been conferred on a previous owner by a grateful monarch who had been equipped by The Expeditioner for traveling in the Australian outback. She was now entitled to be called "Lady Alice of Bullamakanka." Nigel was in a bit of a rush. He left Alice with a first-class ticket to London and an attaché case of folders, and with an effusive "Jolly good show," disappeared.

Summary

Organizations must maintain a memory to process transactions and make decisions. Organizational data should be shareable, transportable, secure, and accurate, and provide timely, relevant information. The essential components are people (the most important), text, multimedia data, models, and knowledge. A wide variety of technologies can be used to manage data. External memories enlarge the range of data available to an organization. Data management systems often have some major shortcomings: redundancy, poor data control, poor interfaces, long lead times for query resolution, an inability to supply answers for questions posed by managers, and poor data integration. Data are raw facts; information is data processed into a meaningful form. Knowledge is the capacity to use information.

Key terms and concepts

Data
Database management system (DBMS)
Data management
Data management system
Data mining
Data security
Decision making
Decision quality
Decision support system (DSS)
Documents
Executive information system (EIS)
External memory
Geographic information system (GIS)
Graphics
Hypertext
Images
Information
Internal memory
Knowledge
Management information system (MIS)
Metamemory
Online analytical processing (OLAP)
Organizational culture
Organizational memory
Shareable data
Social networks
Standard operating procedures
Storage device
Storage medium
Storage structure
Tables
Transaction processing
Transaction processing system (TPS)
Voice data

References and additional readings

Davenport, T. H. (1998). Putting the enterprise into the enterprise system. *Harvard Business Review*, 76(4), 121-131.

Exercises

1. What are the major differences between internal and external memory?
2. What is the difference between the things you remember and the things you record on your computer?
3. What features are common to most individual memory systems?
4. What do you think organizations did before computers were invented?
5. Discuss the memory systems you use. How do they improve your performance? What are the shortcomings of your existing systems? How can you improve them?

6. Describe the most "organized" person you know. Why is that person so organized? Why haven't you adopted some of the same methods? Why do you think people differ in the extent to which they are organized?
7. Think about the last time you enrolled in a class. What data do you think were recorded for this transaction?
8. What roles do people play in organizational memory?
9. What do you think is the most important attribute of organizational memory? Justify your answer.
10. What is the difference between transaction processing and decision making?
11. When are data relevant?
12. Give some examples of specialized memories.
13. How can you measure the quality of a decision?
14. What is organizational culture? Can you name some organizations that have a distinctive culture?
15. What is hypertext? How does it differ from linear text? Why might hypertext be useful in an organizational memory system?
16. What is imaging? What are the characteristics of applications well suited for imaging?
17. What is an external memory? Why do organizations use external memories instead of building internal memories?
18. What is the common name used to refer to systems that help organizations remember knowledge?
19. What is a DSS? What is its role in organizational memory?
20. What are the major shortcomings of many data management systems? Which do you think is the most significant shortcoming?
21. What is the relationship between data, information, and knowledge?
22. Consider the following questions about The Expeditioner:
    a. What sort of data management systems would you expect to find at The Expeditioner?
    b. For each type of data management system that you identify, discuss it using the themes of data being shareable, transportable, secure, accurate, timely, and relevant.
    c. An organization's culture is the set of key values, beliefs, understandings, and norms that its members share. Discuss your ideas of the likely organizational culture of The Expeditioner.
    d. How difficult will it be to change the organizational culture of The Expeditioner? Do you anticipate that Alice will have problems making changes to the firm? If so, why?
23. Using the Web, find some stories about firms using data management systems. You might enter keywords such as "database" and "business analytics" and access the sites of publications such as Computerworld. Identify the purpose of each system. How does the system improve organizational performance? What are the attributes of the technology that make it useful? Describe any trade-offs the organization might have made. Identify other organizations in which the same technology might be applied.
24. Make a list of the organizational memory systems identified in this chapter. Interview several people working in organizations. Ask them to indicate which organizational memory systems they use. Ask which system is most important and why. Write up your findings and your conclusion.

2. Information

Effective information management must begin by thinking about how people use information—not with how people use machines.

> Davenport, T. H. (1994). Saving IT's soul: human-centered information management. *Harvard Business Review*, 72(2), 119-131.

Learning objectives

Students completing this chapter will

- understand the importance of information to society and organizations;
- be able to describe the various roles of information in organizational change;
- be able to distinguish between soft and hard information;
- know how managers use information;
- be able to describe the characteristics of common information delivery systems;
- distinguish the different types of knowledge.

Introduction

There are three characteristics of the early years of the twenty-first century: global warming, high-velocity global change, and the emerging power of information organizations. The need to create a sustainable civilization, globalization of business, and the rise of China as a major economic power are major forces contributing to mass change. Organizations are undergoing large-scale restructuring as they attempt to reposition themselves to survive the threats and exploit the opportunities presented by these changes.

In the last few years, some very powerful and highly profitable information-based organizations have emerged. Apple is the world's largest company in terms of market value. With its combination of iPod, iPhone, iPad, and iTunes, it has shown the power of a new information service to change an industry. Google has become a well-known global brand as it fulfills its mission "to organize the world's information and make it universally accessible and useful." Founded in 1995, eBay rightly calls itself the "The World's Online Marketplace®." The German software firm, SAP, markets the enterprise resource planning (ERP) software that is used by many of the world's largest organizations. Information products and information systems have become a major driver of organizational growth. We gain further insights into the value of information by considering its role in civilization.

A historical perspective

Humans have been collecting data to manage their affairs for several thousand years. In 3000 BCE, Mesopotamians recorded inventory details in cuneiform, an early script. Today's society transmits Exabytes of data every day as billions of people and millions of organizations manage their affairs. The dominant issue facing each type of economy has changed over time, and the collection and use of data has changed to reflect these concerns, as shown in the following table.

Societal focus

Economy	Subsistence	Agricultural	Industrial	Service	Sustainable
Question	How to survive?	How to farm?	How to manage resources?	How to create customers?	How to reduce environmental impact?
Dominant issue	Survival				
		Production			
				Customer service	
					Sustainability
Key information systems	Gesturing Speech	Writing Calendar Money Measuring	Accounting ERP Project management	BPM CRM Analytics	Simulation Optimization Design

Agrarian society was concerned with productive farming, and an important issue was when to plant crops. Ancient Egypt, for example, based its calendar on the flooding of the Nile. The focus shifted during the industrial era to management of resources (e.g., raw materials, labor, logistics). Accounting, ERP, and project management became key information systems for managing resources. In the current service economy, the focus has shifted to customer creation. There is an oversupply of many consumer products (e.g., cars) and companies compete to identify services and product features that will attract customers. They are concerned with determining what types of customers to recruit and finding out what they want. As a result, we have seen the rise of business analytics and customer relationship management (CRM) to address this dominant issue. As well as creating customers, firms need to serve those already recruited. High quality service often requires frequent and reliable execution of multiple processes during manifold encounters with customers by a firm's many customer facing employees (e.g., a fast food store, an airline, a hospital). Consequently, business process management (BPM) has grown in importance since the mid 1990s when there was a surge of interest in business process reengineering.

We are in transition to a new era, sustainability, where attention shifts to assessing environmental impact because, after several centuries of industrialization, atmospheric CO_2 levels have become alarmingly high. We are also reaching the limits of the planet's resources as its population now exceeds seven billion people. As a result, a new class of application is emerging, such as environmental management systems and UPS's telematics project.[5] These new systems will also include, for example, support for understanding environmental impact through simulation of energy consuming and production systems, optimization of energy systems, and design of low impact production and customer service systems. Notice that dominant issues don't disappear but rather aggregate in layers, so tomorrow's business will be concerned with survival, production, customer service, and sustainability. As a result, a firm's need for data never diminishes, and each new layer creates another set of data needs. The flood will not subside and for most firms the data flow will need to grow significantly to meet the new challenge of sustainability.

A constant across all of these economies is organizational memory, or in its larger form, social memory. Writing and paper manufacturing developed about the same time as agricultural societies. Limited writing systems appeared about 30,000 years ago. Full writing systems, which have evolved in the last 5,000 years, made possible the technological transfer that enabled humanity to move from hunting and gathering to farming. Writing enables the recording of knowledge, and information can accumulate from one generation to the next. Before writing, knowledge was confined by the limits of human memory.

5 Watson, R. T., Boudreau, M.-C., Li, S., & Levis, J. (2010). Telematics at UPS: En route to Energy Informatics. MISQ Executive, 9(1), 1-11.

There is a need for a technology that can store knowledge for extended periods and support transport of written information. The storage medium advanced from clay tablets (4000 BCE), papyrus (3500 BCE), and parchment (2000 BCE) to paper (100 CE.). Written knowledge gained great impetus from Johannes Gutenberg, whose achievement was a printing system involving movable metal type, ink, paper, and press. In less than 50 years, printing technology diffused throughout most of Europe. In the last century, a range of new storage media appeared (e.g., photographic, magnetic, and optical).

Organizational memories emerged with the development of large organizations such as governments, armies, churches, and trading companies. The growth of organizations during the industrial revolution saw a massive increase in the number and size of organizational memories. This escalation continued throughout the twentieth century.

The Web has demonstrated that we now live in a borderless world. There is a free flow of information, investment, and industry across borders. Customers ignore national boundaries to buy products and services. In the borderless information age, the old ways of creating wealth have been displaced by intelligence, marketing, global reach, and education. Excelling in the management of data, information, and knowledge has become a prerequisite to corporate and national wealth.

Wealth creation

The old	The new
Military power	Intelligence
Natural resources	Marketing
Population	Global reach
Industry	Education

This brief history shows the increasing importance of information. Civilization was facilitated by the discovery of means for recording and disseminating information. In our current society, organizations are the predominant keepers and transmitters of information.

A brief history of information systems

Information systems has three significant eras. In the first era, information work was transported to the computer. For instance, a punched card deck was physically transported to a computer center, the information was processed, and the output physically returned to the worker as a printout.

Information systems eras

Era	Focus	Period	Technology	Networks
1	Take information work to the computer	1950s – mid-1970s	Batch	Few data networks
2	Take information work to the employee	Mid-1970s – mid-1990s	Host/terminal Client/server	Spread of private networks
3	Take information work to the customer and other stakeholders	Mid-1990s – present	Browser/server	Public networks (Internet)

In the second era, private networks were used to take information work to the employee. Initially, these were time-sharing and host/terminal systems. IS departments were concerned primarily with creating systems for use by an organization's employees when interacting with customers (e.g., a hotel reservation system used by call center employees) or for the employees of another business to transact with the organization (e.g., clerks in a hospital ordering supplies).

Era 3 starts with the appearance of the Web browser in the mid-1990s. The browser, which can be used on the public and global Internet, permits organizations to take information and information work to customers and other stakeholders. Now, the customer undertakes work previously done by the organization (e.g., making an airline reservation).

The scale and complexity of era 3 is at least an order of magnitude greater than that of era 2. Nearly every company has far more customers than employees. For example, UPS, with an annual investment of more than $1 billion in information technology and 359,000 employees, is one of the world's largest employers. However, there are 11 million customers, 30 times the number of employees, who are today electronically connected to UPS.

Era 3 introduced direct electronic links between a firm and its stakeholders, such as investors and citizens. In the earlier eras, intermediaries often communicated with stakeholders on behalf of the firm (e.g., a press release to the news media). These messages could be filtered and edited, or sometimes possibly ignored, by intermediaries. Now, organizations can communicate directly with their stakeholders via the Web and e-mail. Internet technologies offer firms a chance to rethink their goals vis-à-vis each stakeholder class and to use Internet technology to pursue these goals.

This brief history leads to the conclusion that the value IS creates is determined by whom an organization can reach, how it can reach them, and where and when it can reach them.

- **Whom**. Whom an organization can contact determines whom it can influence, inform, or transfer work to. For example, if a hotel can be contacted electronically by its customers, it can promote online reservations (transfer work to customers), and reduce its costs.
- **How**. How an organization reaches a stakeholder determines the potential success of the interaction. The higher the bandwidth of the connection, the richer the message (e.g., using video instead of text), the greater the amount of information that can be conveyed, and the more information work that can be transferred.
- **Where**. Value is created when customers get information directly related to their current location (e.g., a navigation system) and what local services they want to consume (e.g., the nearest Italian restaurant).
- **When**. When a firm delivers a service to a client can greatly determine its value. Stockbrokers, for instance, who can inform clients immediately of critical corporate news or stock market movements are likely to get more business.

Information characteristics

Three useful concepts for describing information are hardness, richness, and class. Information hardness is a subjective measure of the accuracy and reliability of an item of information. Information richness describes the concept that information can be rich or lean depending on the information delivery medium. Information class groups types of information by their key features.

Information hardness

In 1812, the Austrian mineralogist Friedrich Mohs proposed a scale of hardness, in order of increasing relative hardness, based on 10 common minerals. Each mineral can scratch those with the same or a lower number, but cannot scratch higher-numbered minerals.

A similar approach can be used to describe information. Market information, such as the current price of gold, is the hardest because it is measured extremely accurately. There is no ambiguity, and its measurement is highly reliable. In contrast, the softest information, which comes from unidentified sources, is rife with uncertainty.

An information hardness scale

Minerals	Scale	Data
Talc	1	Unidentified source—rumors, gossip, and hearsay
Gypsum	2	Identified non-expert source—opinions, feelings, and ideas
Calcite	3	Identified expert source—predictions, speculations, forecasts, and estimates
Fluorite	4	Unsworn testimony—explanations, justifications, assessments, and interpretations
Apatite	5	Sworn testimony—explanations, justifications, assessments, and interpretations
Orthoclase	6	Budgets, formal plans
Quartz	7	News reports, non-financial data, industry statistics, and surveys
Topaz	8	Unaudited financial statements, and government statistics
Corundum	9	Audited financial statements
Diamond	10	Stock exchange and commodity market data

Audited financial statements are in the corundum zone. They are measured according to standard rules (known as "generally accepted accounting principles") that are promulgated by national accounting societies. External auditors monitor the application of these standards, although there is generally some leeway in their application and sometimes multiple standards for the same item. The use of different accounting principles can lead to different profit and loss statements. As a result, the information in audited financial statements has some degree of uncertainty.

There are degrees of hardness within accounting systems. The hardest data are counts, such as units sold or customers served. These are primary measures of organizational performance. Secondary measures, such as dollar sales and market share, are derived from counts. Managers vary in their preference for primary and secondary measures. Operational managers opt for counts for measuring productivity because they are uncontaminated by price changes and currency fluctuations. Senior managers, because their focus is on financial performance, select secondary measures.

The scratch test provides a convenient and reliable method of assessing the hardness of a mineral. Unfortunately, there is no scratch test for information. Managers must rely on their judgment to assess information hardness.

Although managers want hard information, there are many cases when it is not available. They compensate by seeking information from several different sources. Although this approach introduces redundancy, this is precisely what the manager seeks. Relatively consistent information from different sources is reassuring.

Information richness

Information can be described as rich or lean. Information is richest when delivered face-to-face. Conversation permits immediate feedback for verification of meaning. You can always stop the other speaker and ask, "What do you mean?" Face-to-face information delivery is rich because you see the speaker's body language, hear the tone of voice, and natural language is used. A numeric document is the leanest form of information. There is no opportunity for questions, no additional information from body movements and vocal tone. The information richness of some communication media is shown in the following table.

Information richness and communication media[6]

Richest				**Leanest**
Face-to-face	Telephone	Personal documents (letters and memos)	Impersonal written documents	Numeric documents

Managers seek rich information when they are trying to resolve equivocality or ambiguity. It means that managers cannot make sense of a situation because they arrive at multiple, conflicting interpretations of the information. An example of an equivocal situation is a class assignment where some of the instructions are missing and others are contradictory (of course, this example is an extremely rare event).

Equivocal situations cannot be resolved by collecting more information, because managers are uncertain about what questions to ask and often a clear answer is not possible. Managers reduce equivocality by sharing their interpretations of the available information and reaching a collective understanding of what the information means. By exchanging opinions and recalling their experiences, they try to make sense of an ambiguous situation.

Many of the situations that managers face each day involve a high degree of equivocality. Formal organizational memories, such as databases, are not much help, because the information they provide is lean. Many managers rely far more on talking with colleagues and using informal organizational memories such as social networks.

Data management is almost exclusively concerned with administering the formal information systems that deliver lean information. Although this is their proper role, data managers must realize that they can deliver only a portion of the data required by decision makers.

Information classes

Information can be grouped into four classes: content, form, behavior, and action. Until recently, most organizational information fell into the first category.

Information classes

Class	Description
Content	Quantity, location, and types of items
Form	Shape and composition of an object
Behavior	Simulation of a physical object
Action	Creation of action (e.g., industrial robots)

Content information records details about quantity, location, and types of items. It tends to be historical in nature and is traditionally the class of information collected and stored by organizations. The content information of a car would describe its model number, color, price, options, horsepower, and so forth. Hundreds of bytes of data may be required to record the full content information of a car. Typically, content data are captured by a TPS.

Form information describes the shape and composition of an object. For example, the form information of a car would define the dimensions and composition of every component in the car. Millions of bytes of data are needed to store the form of a car. CAD/CAM systems are used to create and store form information.

---

6 Daft, R. L., & Lengel, R. H. (1986). Organizational information requirements, media richness, and structural design. *Management Science*, 32(5), 554-571.

Behavior information is used to predict the behavior of a physical object using simulation techniques, which typically require form information as input. Massive numbers of calculations per second are required to simulate behavior. For example, simulating the flight of a new aircraft design may require trillions of computations. Behavior information is often presented visually because the vast volume of data generated cannot be easily processed by humans in other formats.

Action information enables the instantaneous creation of sophisticated action. Industrial robots take action information and manufacture parts, weld car bodies, or transport items. Antilock brakes are an example of action information in use.

A lifetime of information—every day

It is estimated that a single weekday issue of the *New York Times* contains more information than the average person in seventeenth-century England came across in a lifetime. Information, once rare, is now abundant and overflowing.

Roszak, T. 1986. The cult of information: The folklore of computers and the true art of thinking. New York, NY: Pantheon. p. 32.

Information and organizational change

Organizations are goal-directed. They undergo continual change as they use their resources, people, technology, and financial assets to reach some future desired outcome. Goals are often clearly stated, such as to make a profit of $100 million, win a football championship, or decrease the government deficit by 25 percent in five years. Goals are often not easily achieved, however, and organizations continually seek information that supports goal attainment. The information they seek falls into three categories: goal setting, gap, and change.

Organizational information categories

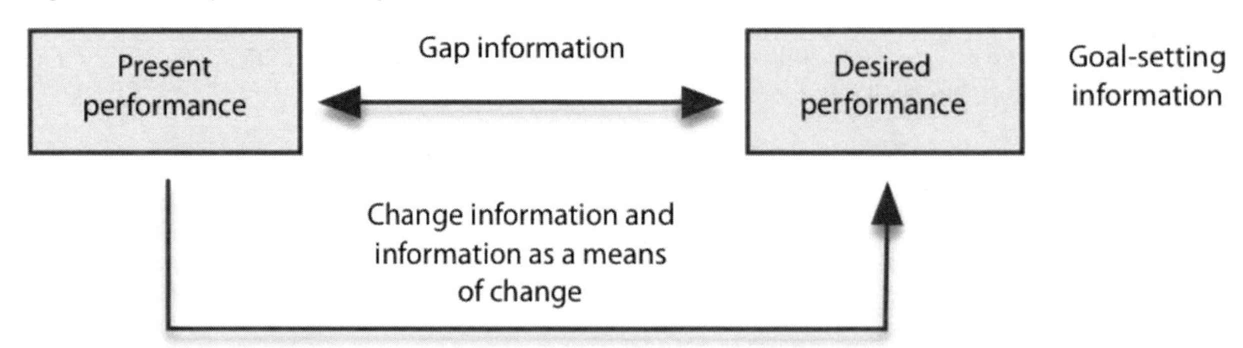

The emergence of an information society also means that information provides dual possibilities for change. Information is used to plan change, and information is a medium for change.

Goal-setting information

Organizations set goals or levels of desired performance. Managers need information to establish goals that are challenging but realistic. A common approach is to take the previous goal and stretch it. For example, a company that had a 15 percent return on investment (ROI) might set the new goal at 17 percent ROI. This technique is known as "anchoring and adjusting." Prior performance is used as a basis for setting the new performance standards. The problem with anchoring and adjusting is that it promotes incremental improvement rather than radical change because internal information is used to set performance standards. Some organizations have turned to external information and are using benchmarking as a source of information for goal setting.

Planning

Planning is an important task for senior managers. To set the direction for the company, they need information about consumers' potential demands and social, economic, technical, and political conditions. They use this information to determine the opportunities and threats facing the organization, thus permitting them to take advantage of opportunities and avoid threats.

Most of the information for long-term planning comes from sources external to the company. There are think tanks that analyze trends and publish reports on future conditions. Journal articles and books can be important sources of information about future events. There also will be a demand for some internal information to identify trends in costs and revenues. Major planning decisions, such as building a new plant, will be based on an analysis of internal data (use of existing capacity) and external data (projected customer demand).

Benchmarking

Benchmarking establishes goals based on best industry practices. It is founded on the Japanese concept of *dantotsu*, striving to be the best of the best. Benchmarking is externally directed. Information is sought on those companies, regardless of industry, that demonstrate outstanding performance in the areas to be benchmarked. Their methods are studied, documented, and used to set goals and redesign existing practices for superior performance.

Other forms of external information, such as demographic trends, economic forecasts, and competitors' actions can be used in goal setting. External information is valuable because it can force an organization to go beyond incremental goal setting.

Organizations need information to identify feasible, motivating, and challenging goals. Once these goals have been established, they need information on the extent to which these goals have been attained.

Gap information

Because goals are meant to be challenging, there is often a gap between actual and desired performance. Organizations use a number of mechanisms to detect a gap and gain some idea of its size. Problem identification and scorekeeping are two principal methods of providing gap information.

Problem identification

Business conditions are continually changing because of competitors' actions, trends in consumer demand, and government actions. Often these changes are reflected in a gap between expectations and present performance. This gap is known as a problem.

To identify problems, managers use exception reports, which are generated only when conditions vary from the established standard. Once a potential problem has been identified, managers collect additional information to confirm that a problem really exists. Problem identification information can also be delivered by printed report, computer screen, or on a tablet. Once management has been alerted, the information delivery system needs to shift into high gear. Managers will request rapid delivery of ad hoc reports from a variety of sources. The ideal organizational memory system can adapt smoothly to deliver appropriate information quickly.

Scorekeeping

Keeping track of the score provides gap information. Managers ask many questions: How many items did we make yesterday? What were the sales last week? Has our market share increased in the last year? They establish measurement systems to track variables that indicate whether organizational performance is on target. Keeping score is important; managers need to measure in order to manage. Also, measurement lets people know what is important. Once a manager starts to measure something, subordinates surmise that this variable must be important and pay more attention to the factors that influence it.

There are many aspects of the score that a manager can measure. The overwhelming variety of potentially available information is illustrated by the sales output information that a sales manager could track. Sales input information (e.g., number of service calls) can also be measured, and there is qualitative information to be considered. Because of time constraints, most managers are forced to limit their attention to 10 or fewer key variables singled out by the critical success factors (CSF) method. Scorekeeping information is usually fairly stable.

Sales output tracking information

Category	Example
Orders	Number of current customers
	Average order size
	Batting average (orders to calls)
Sales volume	Dollar sales volume
	Unit sales volume
	By customer type
	By product category
	Translated to market share
	Quota achieved
Margins	Gross margin
	Net profit
	By customer type
	By product
Customers	Number of new accounts
	Number of lost accounts
	Percentage of accounts sold
	Number of accounts overdue
	Dollar value of receivables
	Collections of receivables

Change information

Once a gap has been detected, managers take action to close it. Change information helps them determine which actions might successfully close the gap. Accurate change information is very valuable because it enables managers to predict the outcome of various actions with some certainty. Unfortunately, change information is usually not very precise, and there are many variables that can influence the effect of any planned change. Nevertheless, organizations spend a great deal of time collecting information to support problem solving and planning.

Problem solution

Managers collect information to support problem solution. Once a concern has been identified, a manager seeks to find its cause. A decrease in sales could be the result of competitors introducing a new product, an economic downturn, an ineffective advertising campaign, or many other reasons. Data can be collected to test each of these possible causes. Additional data are usually required to support analysis of each alternative. For example, if the sales decrease has been caused by an economic recession, the manager might use a decision support system (DSS) to analyze the effect of a price decrease or a range of promotional activities.

Information as a means of change

The emergence of an information society means that information can be used as a means of changing an organization's performance. Corporations are creating information-based products and services, adding information to products, and using information to enhance existing performance or gain a competitive advantage. Further insights into the use of information as a change agent are gained by examining marketing, customer service, and empowerment.

Marketing

Marketing is a key strategy for changing organizational performance by increasing sales. Information has become an important component in marketing. Airlines and retailers have established frequent flyer and buyer programs to encourage customer loyalty and gain more information about customers. These systems are heavily dependent on database technology because of the massive volume of data that must be stored. Indeed, without database technology, some marketing strategies could never be implemented.

Database technology offers the opportunity to change the very nature of communications with customers. Broadcast media have been the traditional approach to communication. Advertisements are aired on television, placed in magazines, or displayed on a Web site. Database technology can be used to address customers directly. No longer just a mailing list, today's database is a memory of customer relationships, a record of every message and response between the firm and a customer. Some companies keep track of customers' preferences and customize advertisements to their needs. Leading online retailers now send customers only those emails for products for which they estimate there is a high probability of a purchase. For example, a customer with a history of buying jazz music is sent an email promoting jazz recordings instead of one featuring classical music.

The Web has significantly enhanced the value of database technology. Many firms now use a combination of a Web site and a DBMS to market products and service customers.

Customer Service

Customer service is a key competitive issue for nearly every organization. Many American business leaders rank customer service as their most important goal for organizational success. Many of the developed economies are service economies. In the United States, services account for nearly 80 percent of the gross national product and most of the new jobs. Many companies now fight for customers by offering superior customer service. Information is frequently a key to this improved service.

Skill Builder

For many businesses, information is the key to high-quality customer service. Thus, some firms use information, and thus customer service, as a key differentiating factor, while others might compete on price. Compare the electronics component of Web sites Amazon and TigerDirect. How do they use price and information to compete? What are the implications for data management if a firm uses information to compete?

Empowerment

Empowerment means giving employees greater freedom to make decisions. More precisely, it is sharing with frontline employees

- information about the organization's performance;
- rewards based on the organization's performance;
- knowledge and information that enable employees to understand and contribute to organizational performance;
- power to make decisions that influence organizational direction and performance.

Notice that information features prominently in the process. A critical component is giving employees access to the information they need to perform their tasks with a high degree of independence and discretion. Information is empowerment. An important task for data managers is to develop and install systems that give employees ready access to an organization's memory. By linking employees to organizational memory, data managers play a pivotal role in empowering people.

Organizations believe that empowerment contributes to organizational performance by increasing the quality of products and services. Together, empowerment and information are mechanisms of planned change.

Information and managerial work

Managers are a key device for implementing organizational change. Because they frequently use data management systems in their normal activities as a source of information about change and as a means of implementing change, it is crucial for data management systems designers to understand how managers work. Failure to take account of managerial behavior can result in a system that is technically sound but rarely used because it does not fit the social system.

Studies over several decades reveal a very consistent pattern: Managerial work is very fragmented. Managers spend an average of 10 minutes on any task, and their day is organized into brief periods of concentrated attention to a variety of tasks. They work unrelentingly and are frequently interrupted by unexpected disturbances. Managers are action oriented and rely on intuition and judgment far more than contemplative analysis of information.

Managers strongly prefer oral communication. They spend a great deal of time conversing directly or by telephone. Managers use interpersonal communication to establish networks, which they later use as a source of information and a way to make things happen. The continual flow of verbal information helps them make sense of events and lets them feel the organizational pulse.

Managers rarely use formal reporting systems. They do not spend their time analyzing computer reports or querying databases but resort to formal reports to confirm impressions, should interpersonal communications suggest there is a problem. Even when managers are provided with a purpose-built, ultra-friendly executive information system, their behavior changes very little. They may access a few screens during the day, but oral communication is still their preferred method of data gathering and dissemination.

Managers' information requirements

Managers have certain requirements of the information they receive. These expectations should shape a data management system's content and how data are processed.

Managers expect to receive information that is useful for their current task under existing business conditions. Unfortunately, managerial tasks can change rapidly. The interlinked, global, economic environment is highly turbulent. Since managers' expectations are not stable, the information delivery system must be sufficiently flexible to meet changing requirements.

Managers' demands for information vary with their perception of its hardness; they require only one source that scores 10 on the information hardness scale. The Nikkei Index at the close of today's market is the same whether you read it in the *Asian Wall Street Journal* or *The Western Australian* or hear it on CNN. As perceived hardness decreases, managers demand more information, hoping to resolve uncertainty and gain a more accurate assessment of the situation. When the reliability of a source is questionable, managers seek confirmation from other sources. If a number of different sources provide consistent information, a manager gains confidence in the information's accuracy.

Relationship of perceived information hardness to volume of information requested

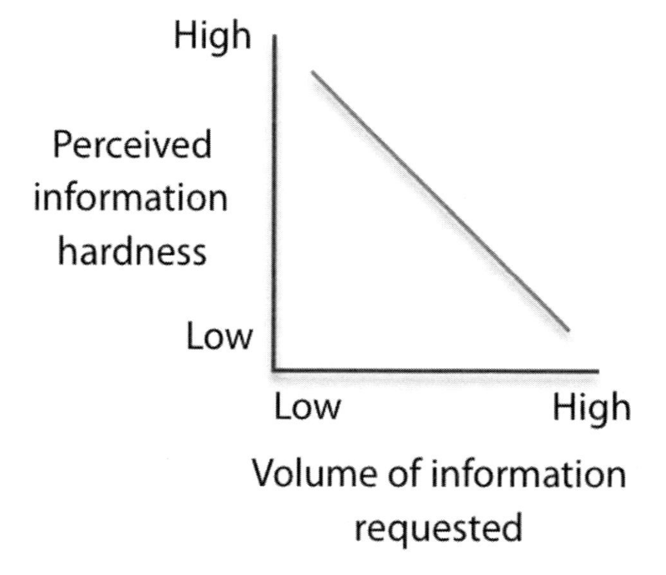

It is not unusual, therefore, to have a manager seek information from a database report, a conversation with a subordinate, and a contact in another organization. If a manager gets essentially the same information from each source, she or he has the confirmation sought. This means each data management system should be designed to minimize redundancy, but different components of organizational memory can supply overlapping information.

Managers' needs for information vary accordingly with responsibilities

Operational managers need detailed, short-term information to deal with well-defined problems in areas such as sales, service, and production. This information comes almost exclusively from internal sources that report recent performance in the manager's area. A sales manager may get weekly reports of sales by each person under that person's supervision.

As managers move up the organizational hierarchy, their information needs both expand and contract. They become responsible for a wider range of activities and are charged with planning the future of the organization, which requires information from external sources on long-term economic, demographic, political, and social trends. Despite this long-term focus, top-level managers also monitor short-term, operational performance. In this instance, they need less detail and more summary and exception reports on a small number of key indicators. To avoid information overload as new layers of information needs are added, the level of detail on the old layers naturally must decline, as illustrated in the following figure.

Management level and information need

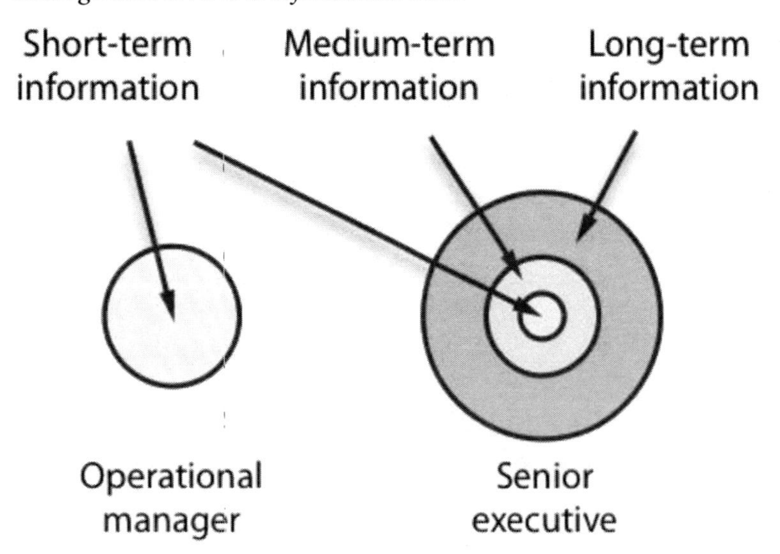

Information satisficing

Because managers face making many decisions in a short period, most do not have the time or resources to collect and interpret all the information they need to make the best decision. Consequently, they are often forced to satisfice. That is, they accept the first satisfactory decision they discover. They also satisfice in their information search, collecting only enough information to make a satisfactory decision.

Ultimately, information satisficing leads to lower-quality decision making. If, however, information systems can be used to accelerate delivery and processing of the right information, then managers should be able to move from selecting the first satisfactory decision to selecting the best of several satisfactory decisions.

Information delivery systems

Most organizations have a variety of delivery systems to provide information to managers. Developed over many years, these systems are integrated, usually via a Web site, to give managers better access to information. There are two aspects to information delivery. First, there is a need for software that accesses an organizational memory, extracts the required data, and formats it for presentation. We can use the categories of organizational memories introduced in Chapter 1 to describe the software side of information delivery systems. The second aspect of delivery is the hardware that gets information from a computer to the manager.

Information delivery systems software

Organizational memory	Delivery systems
People	Conversation
	E-mail
	Meeting
	Report
	Groupware
Files	Management information system (MIS)
Documents	Web browser

Organizational memory	Delivery systems
	E-mail attachment
Images	Image processing system (IPS)
Graphics	Computer aided design (CAD)
	Geographic information system (GIS)
Voice	Voice mail
	Voice recording system
Mathematical model	Decision support system (DSS)
Knowledge	Expert system (ES)
Decisions	Conversation
	E-mail
	Meeting
	Report
	Groupware

Software is used to move data to and from organizational memory. There is usually tight coupling between software and the format of an organizational memory. For example, a relational database management system can access tables but not decision-support models. This tight coupling is particularly frustrating for managers who often want integrated information from several organizational memories. For example, a sales manager might expect a monthly report to include details of recent sales (from a relational database) to be combined with customer comments (from email messages to a customer service system). A quick glance at the preceding table shows that there are many different information delivery systems. We will discuss each of these briefly to illustrate the lack of integration of organizational memories.

Verbal exchange

Conversations, meetings, and oral reporting are commonly used methods of information delivery. Indeed, managers show a strong preference for verbal exchange as a method for gathering information. This is not surprising because we are accustomed to oral information. This is the way we learned for thousands of years as a preliterate culture. Only recently have we learned to make decisions using spreadsheets and computer reports.

Voice mail

Voice mail is useful for people who do not want to be interrupted. It supports asynchronous communication; that is, the two parties to the conversation are not simultaneously connected by a communication line. Voice-mail systems also can store many prerecorded messages that can be selectively replayed using a phone's keypad. Organizations use this feature to support standard customer queries.

Electronic mail

Email is an important system of information delivery. It too supports asynchronous messaging, and it is less costly than voice mail for communication. Many documents are exchanged as attachments to email.

Written report

Formal, written reports have a long history in organizations. Before electronic communication, they were the main form of information delivery. They still have a role in organizations because they are an effective method of integrating information of varying hardness and from a variety of organizational memories. For example, a report can contain text, tables, graphics, and images.

Written reports are often supplemented by a verbal presentation of the key points in the report. Such a presentation enhances the information delivered, because the audience has an opportunity to ask questions and get an immediate response.

Meetings

Because managers spend between 30 and 80 percent of their time in meetings, these become a key source of information.

Groupware

Since meetings occupy so much managerial time and in many cases are poorly planned and managed, organizations are looking for improvements. Groupware is a general term applied to a range of software systems designed to improve some aspect of group work. It is excellent for tapping soft information and the informal side of organizational memory.

Management information system

Management information systems are a common method of delivering information from data management systems. A preplanned query is often used to extract the data. Managers who have developed some skills in using a query language might create their own custom reports as query languages become more powerful and easier to use.

Preplanned reports often contain too much or too detailed information, because they are written in anticipation of a manager's needs. Customized reports do not have these shortcomings, but they are often more expensive and time consuming because they need to be prepared by an analyst.

Web

Word processing, desktop publishing, or HTML editors are used for preparing documents that are disseminated by placing them on a Web server for convenient access and sharing.

Image processing system

An image processing system (IPS) captures data using a scanner to digitize an image. Images in the form of letters, reports, illustrations, and graphics have always been an important type of organizational memory. An IPS permits these forms of information to be captured electronically and disseminated.

Computer-aided design

Computer-aided design (CAD) is used extensively to create graphics. For example, engineers use CAD in product design, and architects use it for building design. These plans are a part of organizational memory for designers and manufacturers.

Geographic information system

A geographic information system (GIS) stores graphical data about a geographic region. Many cities use a GIS to record details of roads, utilities, and services. This graphical information is another form of organizational memory. Again, special-purpose software is required to access and present the information stored in this memory.

Decision support system

A decision support system (DSS) is frequently a computer-based mathematical model of a problem. DSS software, available in a range of packages, permits the model and data to be retrieved and executed. Model parameters can be varied to investigate alternatives.

Expert system

An expert system (ES) has the captured knowledge of someone who is particularly skillful at solving a certain type of problem. It is convenient to think of an ES as a set of rules. An ES is typically used interactively, as it asks the decision-maker questions and uses the responses to determine the action to recommend.

Information integration

You can think of organizational memory as a vast, disorganized data dump. A fundamental problem for most organizations is that their memory is fragmented across a wide variety of formats and technologies. Too frequently, there is a one-to-one correspondence between an organizational memory for a particular functional area and the software delivery system. For example, sales information is delivered by the sales system and production information by the production system. This is not a very desirable situation because managers want all the information they need, regardless of its source or format, amalgamated into a single report.

A response to providing access to the disorganized data dump has been the development of software, such as a Web application, that integrates information from a variety of delivery systems. An important task of this software is to integrate and present information from multiple organizational memories. Recently, some organizations have created vast integrated data stores—data warehouses— that are organized repositories of organizational data.

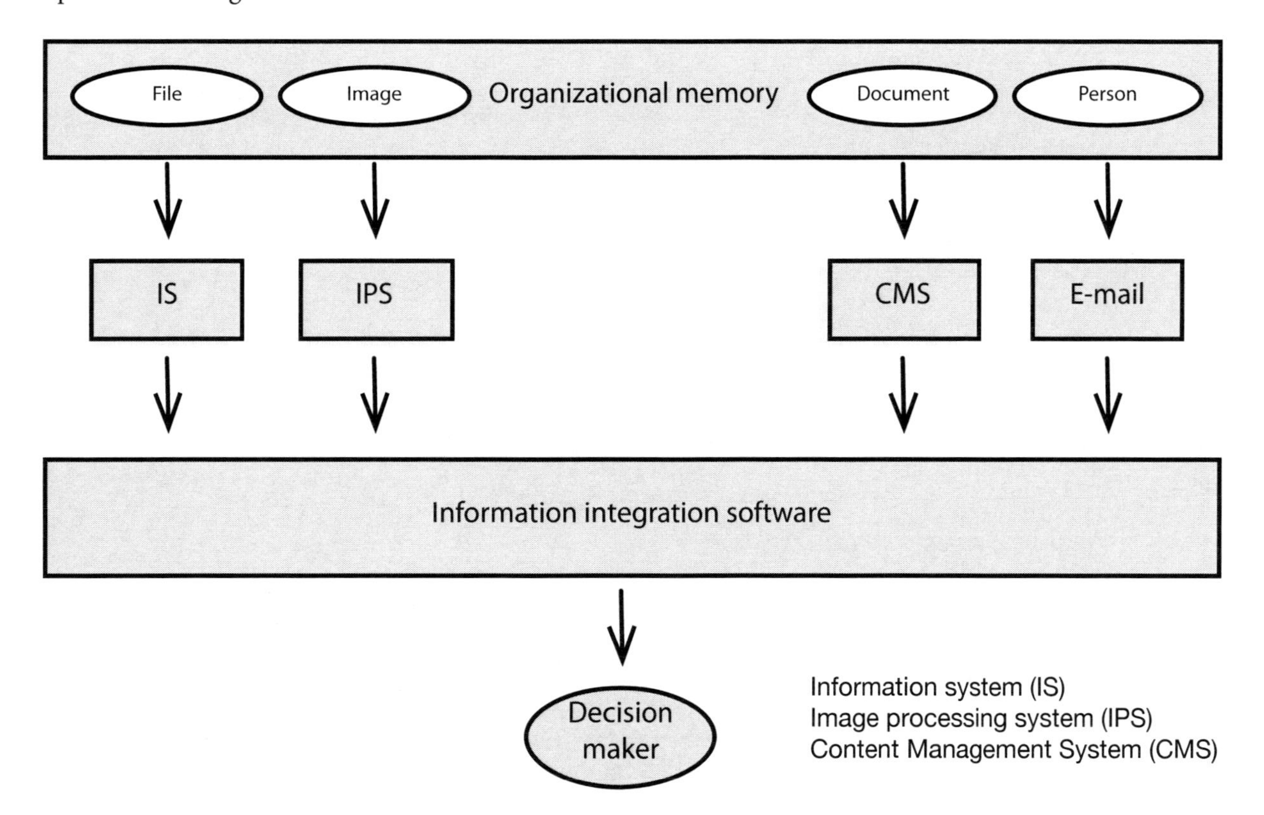

Information integration—the present situation

Information integration or mashup software is really a dirty (and not necessarily very quick) fix for a very critical need. The real requirement is to integrate organizational memory so that digital data (tables, images, graphics, and voice) can be stored together. The ideal situation is shown in the following figure.

The entire organizational memory is accessed as a single unit via one information delivery system that meets all the client's requirements.

There is a long-term trend of data integration. Early information systems were based on independent files. As these systems were rewritten, designers integrated independent files to form databases. The next level of integration is to get all forms of data together. For example, the customer database might contain file data (e.g., customer name and address) and image data (e.g., photograph supporting an insurance claim).

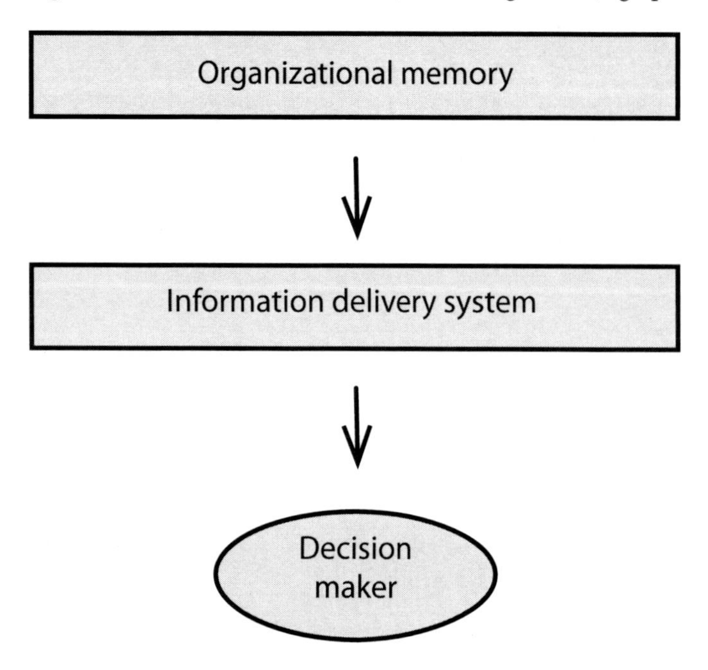

The ideal organizational memory and information delivery system

Knowledge

The performance of many organizations is determined more by their intellectual capabilities and knowledge than their physical assets. A nation's wealth is increasingly a result of the knowledge and skills of its citizens, rather than its natural resources and industrial plants. Currently, about 85 percent of all jobs in America and 80 percent of those in Europe are knowledge-based.

An organization's knowledge, in order of increasing importance, is

- cognitive knowledge (know what);
- advanced skills (know how);
- system understanding and trained intuition (know why);
- self-motivated creativity (care why).

This text illustrates the different types of knowledge. You will develop cognitive knowledge in Section 3, when you learn about data architectures and implementations. For example, knowing what storage devices can be used for archival data is cognitive knowledge. Section 2, which covers data modeling and SQL, develops advanced skills because, upon completion of that section, you will know how to model data and write SQL queries. The first two chapters are designed to expand your understanding of the influence of organizational memory on organizational performance. You need to know why you should learn data management skills. Managers know when and why to apply technology, whereas technicians know what to apply and how to apply it. Finally, you are probably an IS major, and your coursework is inculcating the values and norms of the IS profession so that you care why problems are solved using information technology.

Organizations tend to spend more on developing cognitive skills than they do on fostering creativity. This is, unfortunately, the wrong priority. Returns are likely to be much greater when higher-level knowledge skills are developed. Well-managed organizations place more attention on creating know why and care why skills because they recognize that knowledge is a key competitive weapon. Furthermore, these firms have learned that knowledge grows, often very rapidly, when shared. Knowledge is like a communication network whose potential benefit grows exponentially as the nodes within the network grow arithmetically. When knowledge is shared within the organization, or with customers and suppliers, it multiplies as each person receiving knowledge imparts it to someone else in the organization or to one of the business partners.

Skills values vs. training expenditures (Quinn et al., 1996)

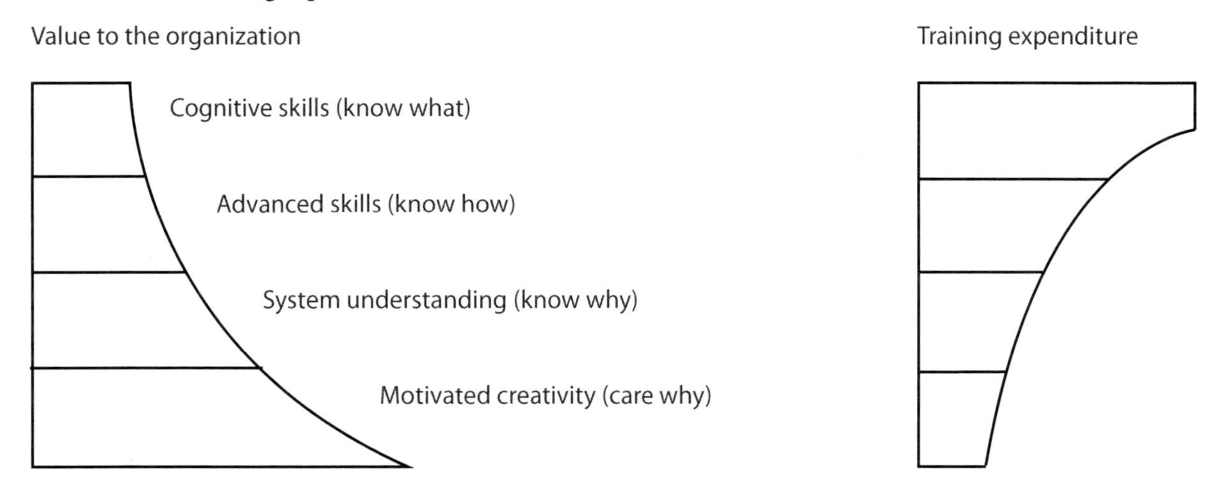

There are two types of knowledge: explicit and tacit. Explicit knowledge is codified and transferable. This textbook is an example of explicit knowledge. Knowledge about how to design databases has been formalized and communicated with the intention of transferring it to you, the reader. Tacit knowledge is personal knowledge, experience, and judgment that is difficult to codify. It is more difficult to transfer tacit knowledge because it resides in people's minds. Usually, the transfer of tacit knowledge requires the sharing of experiences. In learning to model data, the subject of the next section, you will quickly learn how to represent an entity, because this knowledge is made explicit. However, you will find it much harder to model data, because this skill comes with practice. Ideally, you should develop several models under the guidance of an experienced modeler, such as your instructor, who can pass on his or her tacit knowledge.

Summary

Information has become a key foundation for organizational growth. The information society is founded on computer and communications technology. The accumulation of knowledge requires a capacity to encode and share information. Hard information is very exact. Soft information is extremely imprecise. Rich information exchange occurs in face-to-face conversation. Numeric reports are an example of lean information. Organizations use information to set goals, determine the gap between goals and achievements, determine actions to reach goals, and create new products and services to enhance organizational performance. Managers depend more on informal communication systems than on formal reporting systems. They expect to receive information that meets their current, ever changing needs. Operational managers need short-term information. Senior executives require mainly long-term information but still have a need for both short- and medium-term information. When managers face time constraints, they collect only enough information to make a satisfactory decision. Most organizations have a wide variety of poorly integrated information delivery systems. Organizational memory should be integrated to provide managers with one interface to an organization's information stores.

Key terms and concepts

Advanced skills (know how)	Information organization
Benchmarking	Information requirements
Change information	Information richness
Cognitive knowledge (know what)	Information satisficing
Empowerment	Information society
Explicit knowledge	Knowledge
Gap information	Managerial work
Global change	Organizational change
Goal-setting information	Phases of civilization
Information as a means of change	Self-motivated creativity (care why)
Information delivery systems	Social memory
Information hardness	System understanding (know why)
Information integration	Tacit knowledge

References and additional readings

Quinn, J. B., P. Anderson, and S. Finkelstein. 1996. Leveraging intellect. *Academy of Management Executive* 10 (3):7-27.

Exercises

1. From an information perspective, what is likely to be different about the customer service era compared to the forthcoming sustainability era?
2. How does an information job differ from an industrial job?
3. Why was the development of paper and writing systems important?
4. What is the difference between soft and hard information?
5. What is the difference between rich and lean information exchange?
6. What are three major types of information connected with organizational change?
7. What is benchmarking? When might a business use benchmarking?
8. What is gap information?
9. Give some examples of how information is used as a means of change.
10. What sorts of information do senior managers want?
11. Describe the differences between the way managers handle hard and soft information.

12. What is information satisficing?
13. Describe an incident where you used information satisficing.
14. Give some examples of common information delivery systems.
15. What is a GIS? Who might use a GIS?
16. Why is information integration a problem?
17. How "hard" is an exam grade?
18. Could you develop a test for the hardness of a piece of information?
19. Is very soft information worth storing in formal organizational memory? If not, where might you draw the line?
20. If you had just failed your database exam, would you use rich or lean media to tell a parent or spouse about the result?
21. Interview a businessperson to determine his or her firm's critical success factors (CSFs). Remember, a CSF is something the firm must do right to be successful. Generally a firm has about seven CSFs. For the firm's top three CSFs, identify the information that will measure whether the CSF is being achieved.
22. If you were managing a fast-food store, what information would you want to track store performance? Classify this information as short-, medium-, or long-term.
23. Interview a manager. Identify the information that person uses to manage the company. Classify this information as short-, medium-, or long-term information. Comment on your findings.
24. Why is organizational memory like a data warehouse? What needs to be done to make good use of this data warehouse?
25. What information are you collecting to help determine your career or find a job? What problems are you having collecting this information? Is the information mainly hard or soft?
26. What type of knowledge should you gain in a university class?
27. What type of knowledge is likely to make you most valuable?

Case questions

Imagine you are the new owner of The Expeditioner.

1. What information would you request to determine the present performance of the organization?
2. What information would help you to establish goals for The Expeditioner? What goals would you set?
3. What information would you want to help you assist you in changing The Expeditioner?
4. How could you use information to achieve your goals.

Section 2: Data Modeling and SQL

It is a capital mistake to theorize before one has data.

> Sir Arthur Conan Doyle, "A Scandal in Bohemia," *The Adventures of Sherlock Holmes,* 1891

The application backlog, a large number of requests for new information systems, has been a recurring problem in many organizations for decades. The demand for new information systems and the need to maintain existing systems have usually outstripped available information systems skills. The application backlog, unfortunately, is not a new problem. In the 1970s, Codd laid out a plan for improving programmer productivity and accelerating systems development by improving the management of data. Codd's **relational model**, designed to solve many of the shortcomings of earlier systems, is currently the most popular database model.

This section develops two key skills—data modeling and query formation—that are required to take advantage of the relational model. We concentrate on the design and use of relational databases. This very abrupt change in focus is part of our plan to give you a dual understanding of data management. Section 1 is the managerial perspective, whereas this section covers technical skills development. Competent data managers are able to accommodate both views and apply whichever (or some blend of the two) is appropriate.

In Chapter 1, many forms of organizational memory were identified, and in this section we focus on files and their components. Thus, only the files branch of organizational memory is detailed in the following figure.

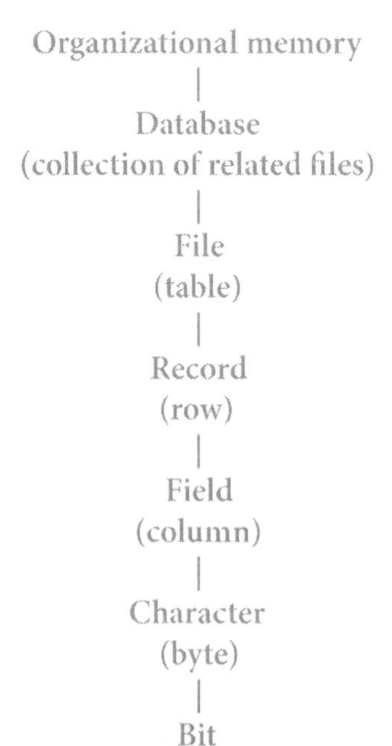

A collection of related files is a **database**. Describing the collection of files as related means that it has a common purpose (e.g., data about students). Sometimes files are also called tables, and there are synonyms for some other terms (the alternative names are shown in parentheses). Files contain **records** (or rows). Each record contains the data for one instance of the data the file stores. For example, if the file stores data about students, each record will contain data about a single student. Records have **fields** (or columns) that store the fine detail of each instance (e.g., student's first name, last name, and date of birth). Fields are composed of **characters** (a, b, c,.., 1, 2, 3,..., %, $, #,..., A, B, etc.). A **byte**, a unit of storage sufficient to store a single letter (in English) or digit, consists of a string of eight contiguous **bits** or binary digits.

The data management hierarchy stimulates three database design questions:

- What collection of files should the database contain?
- How are these files related?
- What fields should each record in the file contain?

The first objective of this section is to describe data modeling, a technique for answering the three questions. Data modeling helps you to understand the structure and meaning of data, which is necessary before a database can be created. Once a database has been designed, built, and loaded with data, the aim is to deploy it to satisfy management's requests for information. Thus, the second objective is to teach you to query a relational database. The learning of modeling and querying will be intertwined, making it easier to grasp the intent of database design and to understand why data modeling is so critical to making a database an effective tool for managerial decision making.

Chapter 3 covers modeling a single entity and querying a single-table database. This is the simplest database that can be created. As you will soon discover, a **data model** is a graphical description of the components of a database. One of these components is an entity, some feature of the real world about which data must be stored. This section also introduces the notions of a **data definition language** (DDL), which is used to describe a database, and a **data manipulation language** (DML), which is used to maintain and query a database. Subsequent chapters in this section cover advanced data modeling concepts and querying capabilities.

3. The Single Entity

I want to be alone.

Attributed to Greta Garbo

Learning objectives

Students completing this chapter will be able to

- model a single entity;
- define a single database;
- write queries for a single-table database.

The relational model

The relational model introduced by Codd in 1970 is the most popular technology for managing large collections of data. In this chapter, the major concepts of the relational model are introduced. Extensive coverage of the relational model is left until Chapter 8, by which time you will have sufficient practical experience to appreciate fully its usefulness, value, and elegance.

A **relation**, similar to the mathematical concept of a set, is a two-dimensional table arranged in rows and columns. This is a very familiar idea. You have been using tables for many years. A **relational database** is a collection of relations, where **relation** is a mathematical term for a table. One row of a table stores details of one observation, instance, or case of an item about which facts are retained—for example, one row for details of a particular student. All the rows in a table store data about the same type of item. Thus, a database might have one table for student data and another table for class data. Similarly, each column in the table contains the same type of data. For example, the first column might record a student's identification number. A key database design question is to decide what to store in each table. What should the rows and columns contain?

In a relational database, each row must be uniquely identified. There must be a **primary key**, such as student identifier, so that a particular row can be designated. The use of unique identifiers is very common. Telephone numbers and e-mail addresses are examples of unique identifiers. Selection of the primary key, or unique identifier, is another key issue of database design.

Global legal entity identifier (LEI)

There is no global standard for identifying legal entities across markets and jurisdictions. The need for such a standard was amplified by Lehman Brothers collapse in 2008. Lehman had 209 registered subsidiaries, legal entities, in 21 countries, and it was party to more than 900,000 derivatives contracts upon its collapse. Key stakeholders, such as financial regulators and Lehman's creditors, were unable to assess their exposure. Furthermore, others were unable to assess the possible ripple on them of the effects of the collapse because of the transitive nature of many investments (i.e., A owes B, B owes C, and C owes D).

The adoption of a global legal entity identifier (LEI), should improve financial system regulation and corporate risk management. Regulators will find it easier to monitor and analyze threats to financial stability and risk managers will be more able evaluate their companies' risks.

Source: www.ny.frb.org/research/epr/2014/1403flem.pdf

The tables in a relational database are **connected** or **related** by means of the data in the tables. You will learn, in the next chapter, that this connection is through a pair of values—a primary key and a foreign key. Consider a table of airlines serving a city. When examining this table, you may not recognize the code of an airline, so you then go to another table to find the name of the airline. For example, if you inspect the following table, you find that AM is an international airline serving Atlanta.

International airlines serving Atlanta

Airline
AM
JL
KX
LM
MA
OS
RG
SN
SR
LH
LY

If you don't know which airline has the abbreviation AM, then you need to look at the following table of airline codes to discover that AeroMexico, with code AM, serves Atlanta. The two tables are related by airline code. Later, you will discover which is the primary key and which is the foreign key.

A partial list of airline codes

Code	Airline
AA	American Airlines
AC	Air Canada
AD	Lone Star Airlines
AE	Mandarin Airlines
AF	Air France
AG	Interprovincial Airlines
AI	Air India
AM	AeroMexico
AQ	Aloha Airlines

When designing the relational model, Codd provided commands for processing multiple records at a time. His intention was to increase the productivity of programmers by moving beyond the record-at-a-time processing that is found in most programming languages. Consequently, the relational model supports set processing (multiple records-at-a-time), which is most frequently implemented as **Structured Query Language (SQL).**[7]

When writing applications for earlier data management systems, programmers usually had to consider the physical features of the storage device. This made programming more complex and also often meant that when data were moved from one storage device to another, software modification was necessary. The relational model separates the logical design of a database and its physical storage. This notion of **data independence** simplifies data modeling and database programming. In this section, we focus on logical database design, and now that you have had a brief introduction to the relational model, you are ready to learn data modeling.

Getting started

As with most construction projects, building a relational database must be preceded by a design phase. Data modeling, our design technique, is a method for creating a plan or blueprint of a database. The data model must accurately mirror real-world relationships if it is to support processing business transactions and managerial decision making.

Rather than getting bogged down with a *theory first, application later* approach to database design and use, we will start with application. We will get back to theory when you have some experience in data modeling and database querying. After all, you did not learn to talk by first studying sentence formation; you just started by learning and using simple words. We start with the simplest data model, a single entity, and the simplest database, a single table, as follows.

Firm code	Firm's name	Price	Quantity	Dividend	PE ratio
FC	Freedonia Copper	27.5	10,529	1.84	16
PT	Patagonian Tea	55.25	12,635	2.5	10
AR	Abyssinian Ruby	31.82	22,010	1.32	13
SLG	Sri Lankan Gold	50.37	32,868	2.68	16
ILZ	Indian Lead & Zinc	37.75	6,390	3	12
BE	Burmese Elephant	0.07	154,713	0.01	3
BS	Bolivian Sheep	12.75	231,678	1.78	11
NG	Nigerian Geese	35	12,323	1.68	10
CS	Canadian Sugar	52.78	4,716	2.5	15
ROF	Royal Ostrich Farms	33.75	1,234,923	3	6

Modeling a single-entity database

The simplest database contains information about one entity, which is some real-world thing. Some entities are physical—CUSTOMER, ORDER, and STUDENT; others are conceptual—WORK ASSIGNMENT and AUTHORSHIP. We represent an entity by a rectangle: the following figure shows a representation of the entity SHARE. The name of the entity is shown in singular form in uppercase in the top part of the rectangle.

7 Officially pronounced as "S-Q-L," but often also pronounced as "sequel."

The entity SHARE

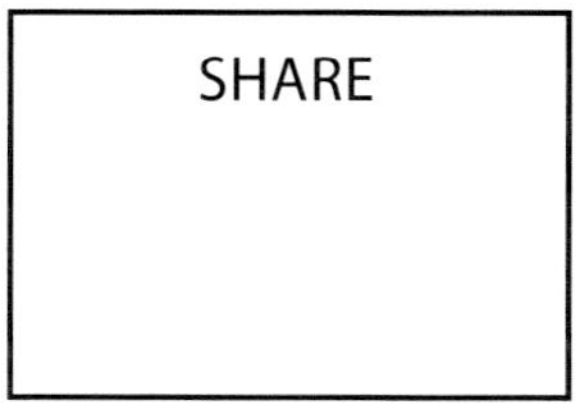

SHARE

An entity has characteristics or attributes. An **attribute** is a discrete element of data; it is not usually broken down into smaller components. Attributes describe an entity and contain the entity's data we want to store. Some attributes of the entity SHARE are *identification code, firm name, price, quantity owned, dividend*, and *price-to-earnings ratio*.[8] Attributes are shown below the entity's name. Notice that we refer to *share price*, rather than *price*, to avoid confusion if there should be another entity with an attribute called *price*. Attribute names must be carefully selected so that they are self-explanatory and unique. For example, *share dividend* is easily recognized as belonging to the entity SHARE.

The entity SHARE and its attributes

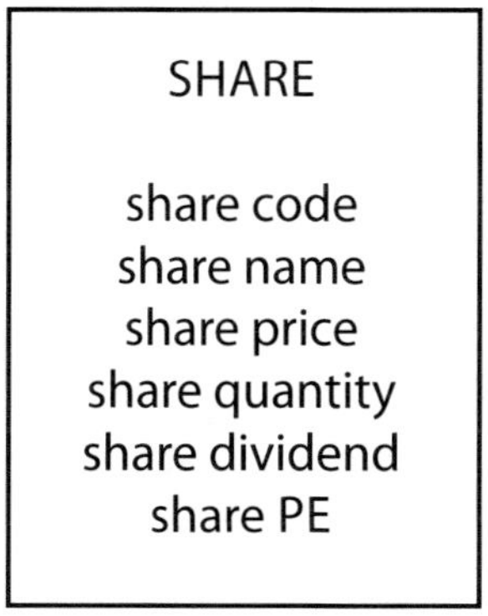

SHARE

share code
share name
share price
share quantity
share dividend
share PE

An **instance** is a particular occurrence of an entity (e.g., facts about Freedonia Copper). To avoid confusion, each instance of an entity needs to be uniquely identified. Consider the case of customer billing. In most cases, a request to bill Smith $100 cannot be accurately processed because a firm might have more than one Smith in its customer file. If a firm has carefully controlled procedures for ensuring that each customer has a unique means of identification, then a request to bill customer number 1789 $100 can be accurately processed. An attribute or collection of attributes that uniquely identifies an instance of an entity is called an **identifier**. The identifier for the entity SHARE is *share code*, a unique identifier assigned by the stock exchange.

There may be several attributes, or combinations of attributes, that are feasible identifiers for an instance of an entity. Attributes that are identifiers are prefixed by an asterisk. The following figure shows an example of a representation of an entity, its attributes, and identifier.

[8] Attributes are shown in italics.

The entity SHARE, its attributes, and identifier

SHARE
*share code share name share price share quantity share dividend share PE

Briefly, entities are things in the environment about which we wish to store information. Attributes describe an entity. An entity must have a unique identifier.

Data modeling

The modeling language used in this text is designed to record the essential details of a data model. The number of modeling symbols to learn is small, and they preserve all the fundamental concepts of data modeling. Since data modeling often occurs in a variety of settings, the symbols used have been selected so that they can be quickly drawn using pencil-and-paper, whiteboard, or a general-purpose drawing program. This also means that models can be quickly revised as parts can be readily erased and redrawn.

The symbols are distinct and visual clutter is minimized because only the absolutely essential information is recorded. This also makes the language easy for clients to learn so they can read and amend models.

Models can be rapidly translated to a set of tables for a relational database. More importantly, since this text implements the fundamental notions of all data modeling languages, you can quickly convert to another data modeling dialect. Data modeling is a high-level skill, and the emphasis needs to be on learning to think like a data modeler rather than on learning a modeling language. This text's goal is to get you off to a fast start.

Skill builder

A ship has a name, registration code, gross tonnage, and a year of construction. Ships are classified as cargo or passenger. Draw a data model for a ship.

Creating a single-table database

The next stage is to translate the data model into a relational database. The translation rules are very direct:

- Each entity becomes a table.
- The entity name becomes the table name.
- Each attribute becomes a column.
- The identifier becomes the primary key.

The American National Standards Institute's (ANSI) recommended language for relational database definition and manipulation is SQL, which is both a data definition language (DDL) (to define a database), a data manipulation language (DML) (to query and maintain a database), and a data control language (DCL) (to control access). SQL is a common standard for describing and querying databases and is available with many commercial relational database products, including DB2, Oracle, and Microsoft SQL Server, and open source products such as MySQL and PostgreSQL.

In this book, MySQL is the relational database for teaching SQL. Because SQL is a standard, it does not matter which implementation of the relational model you use as the SQL language is common across both the proprietary and open variants.[9]

SQL uses the CREATE[10] statement to define a table. It is not a particularly friendly command, and most products have friendlier interfaces for defining tables. However, it is important to learn the standard, because this is the command that SQL understands. Also, a table definition interface will generate a CREATE statement for execution by SQL. Your interface interactions ultimately translate into a standard SQL command.

It is common practice to abbreviate attribute names, as is done in the following example.

Defining a table

The CREATE command to establish a table called share[11] is as follows:

```
CREATE TABLE share (
    shrcode    CHAR(3),
    shrfirm    VARCHAR(20) NOT NULL,
    shrprice   DECIMAL(6,2),
    shrqty     DECIMAL(8),
    shrdiv     DECIMAL(5,2),
    shrpe      DECIMAL(2),
        PRIMARY KEY (shrcode));
```

The first line of the command names the table; subsequent lines describe each of the columns in it. The first component is the name of the column (e.g., shrcode). The second component is the data type (e.g., CHAR), and its length is shown in parentheses. shrfirm is a variable-length character field of length 20, which means it can store up to 20 characters, including spaces. The column shrdiv stores a decimal number that can be as large as 999.99 because its total length is 5 digits and there are 2 digits to the right of the decimal point. Some examples of allowable data types are shown in the following table. The third component (e.g., NOT NULL), which is optional, indicates any instance that cannot have null values. A column might have a null value when it is either unknown or not applicable. In the case of the share table, we specify that shrfirm must be defined for each instance in the database.

The final line of the CREATE statement defines shrcode as the primary key, the unique identifier for SHARE. When a primary key is defined, the relational database management system (RDBMS) will enforce the requirement that the primary key is unique and not null. Before any row is added to the table SHARE, the RDBMS will check that the value of shrcode is not null and that there does not already exist a duplicate value for shrcode in an existing row of share. If either of these constraints is violated, the RDBMS will not permit the new row to be inserted. This constraint, the **entity integrity rule**, ensures

9 Now would be a good time to install the MySQL Community server < www.mysql.com/downloads/mysql/ >on your computer, unless your instructor has set up a class server.

10 SQL keywords are shown in up uppercase

11 Table and column names are shown in red.

that every row has a unique, non-null primary key. Allowing the primary key to take a null value would mean there could be a row of `share` that is not uniquely identified. Note that an SQL statement is terminated by a semicolon.

SQL statements can be written in any mix of valid upper- and lowercase characters. To make it easier for you to learn the syntax, this book adopts the following conventions:

- SQL keywords are in uppercase.
- Table and column names are in lowercase.

There are more elaborate layout styles, but we will bypass those because it is more important at this stage to learn SQL. You should lay out your SQL statements so that they are easily read by you and others.

The following table shows some of the data types supported by most relational databases. Other implementations of the relational model may support some of these data types and additional ones. It is a good idea to review the available data types in your RDBMS before defining your first table.

Some allowable data types

Category	Data type	Description
Numeric	`SMALLINT`	A 15-bit signed binary value
	`INTEGER`	A 31-bit signed binary value
	`FLOAT(P)`	A scientific format number of *p* binary digits precision
	`DECIMAL(P,Q)`	A packed decimal number of *p* digits total length; *q* decimal spaces to the right of the decimal point may be specified
String	`CHAR(N)`	A fixed-length string of *n* characters
	`VARCHAR(N)`	A variable-length string of up to *n* characters
	`TEXT`	A variable-length string of up to 65,535 characters
Date/time	`DATE`	Date in the form *yyyymmdd*
	`TIME`	Time in the form *hhmmss*
	`TIMESTAMP`	A combination of date and time to the nearest microsecond
	`TIME WITH TIME ZONE`	Same as time, with the addition of an offset from universal time coordinated (UTC) of the specified time
	`TIMESTAMP WITH TIME ZONE`	Same as timestamp, with the addition of an offset from UTC of the specified time
Logical	`BOOLEAN`	A set of truth values: `TRUE`, `FALSE`, or `UNKNOWN`

The `CHAR` and `VARCHAR` data types are similar but differ in the way character strings are stored and retrieved. Both can be up to 255 characters long. The length of a `CHAR` column is fixed to the declared length. When values are stored, they are right-padded with spaces to the specified length. When `CHAR` values are retrieved, trailing spaces are removed. `VARCHAR` columns store variable-length strings and use only as many characters as are needed to store the string. Values are not padded; instead, trailing spaces are removed. In this book, we use `VARCHAR` to define most character strings, unless they are short (less than five characters is a good rule-of-thumb).

Data modeling with MySQL Workbench

MySQL Workbench is a professional quality, open source, cross-platform tool for data modeling and SQL querying.[12] In this text, you will also learn some of the features of Workbench that support data modeling and using SQL. You will find it helpful to complete the MySQL Workbench tutorial on creating a data model[13] prior to further reading, as we will assume you have such proficiency.

The entity share created with MySQL Workbench

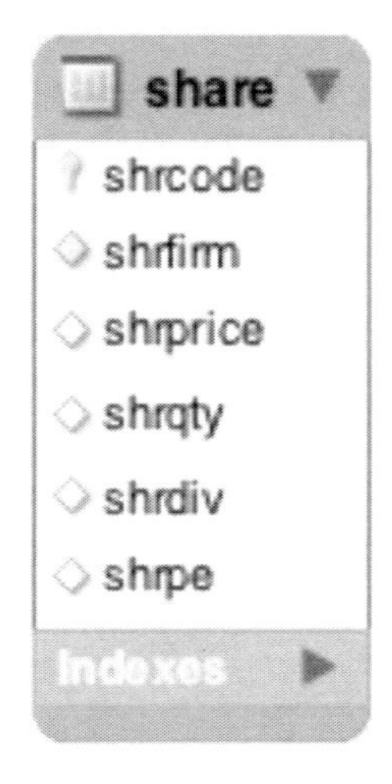

You will notice some differences from the data model we have created previously. Workbench automatically generates the SQL code to create the table, so when modeling you establish the names you want for tables and columns. A key symbol is used to indicate the identifier, which becomes the primary key. An open diamond indicates that a column can be null, whereas a blue diamond indicate the column must have a value, as with `shrfirm` in this case.

When specifying columns in Workbench you must also indicate the datatype. We opt to turn off the display of a column's datatype[14] in a model to maintain focus on the entity.

A major advantage of using a tool such as Workbench is that it will automatically generate the CREATE statement code (Database > Forward Engineer ...) and execute the code to create the database. The Workbench tutorial will have shown you how to do this, and you should try this out for yourself by creating a database with the single `share` table.

Inserting rows into a table

The rows of a table store instances of an entity. A particular shareholding (e.g., Freedonia Copper) is an example of an instance of the entity `share`. The SQL statement `INSERT` is used to add rows to a table. Although most implementations of the relational model use a spreadsheet for row insertion and you can also usually import a structured text file, the `INSERT` command is defined for completeness.

The following command adds one row to the table `share`:

---

12 wb.mysql.com/

13 dev.mysql.com/doc/workbench/en/wb-getting-started-tutorial-creating-a-model.html

14 Preferences > Diagram > Show Column Types

```
INSERT INTO share
   (shrcode,shrfirm,shrprice,shrqty,shrdiv,shrpe)
      VALUES ('FC','Freedonia Copper',27.5,10529,1.84,16);
```

There is a one-to-one correspondence between a column name in the first set of parentheses and a value in the second set of parentheses. That is, `shrcode` has the value "FC," `shrfirm` the value "Freedonia Copper," and so on. Notice that the value of a column that stores a character string (e.g., `shrfirm`) is contained within straight quotes.

The list of field names can be omitted when values are inserted in all columns of the table in the same order as that specified in the `CREATE` statement, so the preceding expression could be written

```
INSERT INTO share
   VALUES ('FC','Freedonia Copper',27.5,10529,1.84,16);
```

The data for the `share` table will be used in subsequent examples. If you have ready access to a relational database, it is a good idea to now create a table and enter the data. Then you will be able to use these data to practice querying the table.

Data for share

share					
*shrcode	shrfirm	shrprice	shrqty	shrdiv	shrpe
FC	Freedonia Copper	27.5	10,529	1.84	16
PT	Patagonian Tea	55.25	12,635	2.5	10
AR	Abyssinian Ruby	31.82	22,010	1.32	13
SLG	Sri Lankan Gold	50.37	32,868	2.68	16
ILZ	Indian Lead & Zinc	37.75	6,390	3	12
BE	Burmese Elephant	0.07	154,713	0.01	3
BS	Bolivian Sheep	12.75	231,678	1.78	11
NG	Nigerian Geese	35	12,323	1.68	10
CS	Canadian Sugar	52.78	4,716	2.5	15
ROF	Royal Ostrich Farms	33.75	1,234,923	3	6

Notice that `shrcode`, the primary key, has an asterisk prefix in the preceding table. This is the same convention as for an identifier in a data model. In the relational model, an identifier becomes a primary key, a column that guarantees that each row of the table can be uniquely addressed.

MySQL Workbench offers a spreadsheet interface for entering data, as explained in the tutorial on adding data to a database.[15]

15 dev.mysql.com/doc/workbench/en/wb-getting-started-tutorial-adding-data.html

Inserting rows with MySQL Workbench

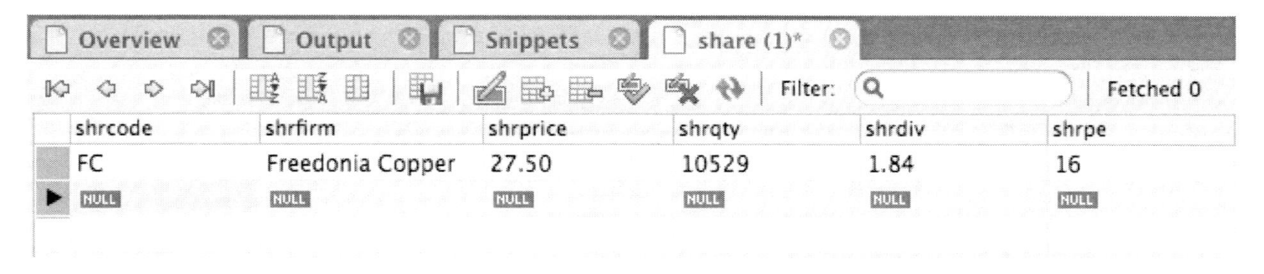

Skill builder

Create a relational database for the ship entity you modeled previously. Insert some rows.

Querying a database

The objective of developing a database is to make it easier to use the stored data to solve problems. Typically, a manager raises a question (e.g., How many shares have a PE ratio greater than 12?). A question or request for information, usually called a query, is then translated into a specific data manipulation or query language. The most widely used query language for relational databases is SQL. After the query has been executed, the resulting data are displayed. In the case of a relational database, the answer to a query is always a table.

There is also a query language called relational algebra, which describes a set of operations on tables. Sometimes it is useful to think of queries in terms of these operations. Where appropriate, we will introduce the corresponding relational algebra operation.

Generally we use a four-phase format for describing queries:

1. A brief explanation of the query's purpose
2. The query italics, prefixed by •, and some phrasing you might expect from a manager
3. The SQL version of the query
4. The results of the query.

Displaying an entire table

All the data in a table can be displayed using the SELECT statement. In SQL, the all part is indicated by an asterisk (*).

- *List all data in the share table.*

```
SELECT * FROM share;
```

shrcode	shrfirm	shrprice	shrqty	shrdiv	shrpe
FC	Freedonia Copper	27.5	10529	1.84	16
PT	Patagonian Tea	55.25	12635	2.5	10
AR	Abyssinian Ruby	31.82	22010	1.32	13
SLG	Sri Lankan Gold	50.37	32868	2.68	16
ILZ	Indian Lead & Zinc	37.75	6390	3	12

BE	Burmese Elephant	0.07	154713	0.01	3
BS	Bolivian Sheep	12.75	231678	1.78	11
NG	Nigerian Geese	35	12323	1.68	10
CS	Canadian Sugar	52.78	4716	2.5	15

Project—choosing columns

The relational algebra operation **project** creates a new table from the columns of an existing table. Project takes a vertical slice through a table by selecting all the values in specified columns. The projection of `share` on columns `shrfirm` and `shrpe` produces a new table with 10 rows and 2 columns. These columns are shaded in the table.

Projection of shrfirm and shrpe

share					
shrcode	shrfirm	shrprice	shrqty	shrdiv	shrpe
FC	Freedonia Copper	27.5	10,529	1.84	16
PT	Patagonian Tea	55.25	12,635	2.5	10
AR	Abyssinian Ruby	31.82	22,010	1.32	13
SLG	Sri Lankan Gold	50.37	32,868	2.68	16
ILZ	Indian Lead & Zinc	37.75	6,390	3	12
BE	Burmese Elephant	0.07	154,713	0.01	3
BS	Bolivian Sheep	12.75	231,678	1.78	11
NG	Nigerian Geese	35	12,323	1.68	10
CS	Canadian Sugar	52.78	4,716	2.5	15
ROF	Royal Ostrich Farms	33.75	1,234,923	3	6

The SQL syntax for the project operation simply lists the columns to be displayed.

- *Report a firm's name and price-earnings ratio.*

```
SELECT shrfirm, shrpe FROM share;
```

shrfirm	shrpe
Freedonia Copper	16
Patagonian Tea	10
Abyssinian Ruby	13
Sri Lankan Gold	16
Indian Lead & Zinc	12
Burmese Elephant	3
Bolivian Sheep	11
Nigerian Geese	10
Canadian Sugar	15

shrfirm	shrpe
Royal Ostrich Farms	6

Restrict—choosing rows

The relational algebra operation **restrict** creates a new table from the rows of an existing table. The operation restricts the new table to those rows that satisfy a specified condition. Restrict takes all columns of an existing table but only those rows that meet the specified condition. The restriction of `share` to those rows where the PE ratio is less than 12 will give a new table with five rows and six columns. These rows are shaded.

Restriction of share

share					
shrcode	shrfirm	shrprice	shrqty	shrdiv	shrpe
FC	Freedonia Copper	27.5	10,529	1.84	16
PT	Patagonian Tea	55.25	12,635	2.5	10
AR	Abyssinian Ruby	31.82	22,010	1.32	13
SLG	Sri Lankan Gold	50.37	32,868	2.68	16
ILZ	Indian Lead & Zinc	37.75	6,390	3	12
BE	Burmese Elephant	0.07	154,713	0.01	3
BS	Bolivian Sheep	12.75	231,678	1.78	11
NG	Nigerian Geese	35	12,323	1.68	10
CS	Canadian Sugar	52.78	4,716	2.5	15
ROF	Royal Ostrich Farms	33.75	1,234,923	3	6

Restrict is implemented in SQL using the `WHERE` clause to specify the condition on which rows are restricted.

- *Get all firms with a price-earnings ratio less than 12.*

```
SELECT * FROM share WHERE shrpe < 12;
```

shrcode	shrfirm	shrprice	shrqty	shrdiv	shrpe
PT	Patagonian Tea	55.25	12635	2.5	10
BE	Burmese Elephant	0.07	154713	0.01	3
BS	Bolivian Sheep	12.75	231678	1.78	11
NG	Nigerian Geese	35	12323	1.68	10
ROF	Royal Ostrich Farms	33.75	1234923	3	6

In this example, we have a less than condition for the `WHERE` clause. All permissible comparison operators are listed below.

Comparison operator	Meaning
=	Equal to

<	Less than
<=	Less than or equal to
>	Greater than
>=	Greater than or equal to
<>	Not equal to

In addition to the comparison operators, the `BETWEEN` construct is available.

a `BETWEEN` x `AND` y is equivalent to a >= x `AND` a <= y

Combining project and restrict—choosing rows and columns

SQL permits project and restrict to be combined. A single SQL `SELECT` statement can specify which columns to project and which rows to restrict.

- *List the name, price, quantity, and dividend of each firm where the share holding is at least 100,000.*

```
SELECT shrfirm, shrprice, shrqty, shrdiv FROM share
   WHERE shrqty >= 100000;
```

shrfirm	shrprice	shrqty	shrdiv
Burmese Elephant	0.07	154713	0.01
Bolivian Sheep	12.75	231678	1.78
Royal Ostrich Farms	33.75	1234923	3

More about WHERE

The `WHERE` clause can contain several conditions linked by `AND` or `OR`. A clause containing `AND` means all specified conditions must be true for a row to be selected. In the case of `OR`, at least one of the conditions must be true for a row to be selected.

- *Find all firms where the PE is 12 or higher and the share holding is less than 10,000.*

```
SELECT * FROM share
   WHERE shrpe >= 12 AND shrqty < 10000;
```

shrcode	shrfirm	shrprice	shrqty	shrdiv	shrpe
ILZ	Indian Lead & Zinc	37.75	6390	3	12
CS	Canadian Sugar	52.78	4716	2.5	15

The power of the primary key

The purpose the primary key is to guarantee that any row in a table can be uniquely addressed. In this example, we use `shrcode` to return a single row because `shrcode` is unique for each instance of `share`. The sought code (AR) must be specified in quotes because `shrcode` was defined as a character string when the table was created.

- *Report firms whose code is AR.*

```
SELECT * FROM share WHERE shrcode = 'AR';
```

shrcode	shrfirm	shrprice	shrqty	shrdiv	shrpe
AR	Abyssinian Ruby	31.82	22010	1.32	13

A query based on a non-primary-key column cannot guarantee that a single row is accessed, as the following illustrates.

- *Report firms with a dividend of 2.50.*

```
SELECT * FROM share WHERE shrdiv = 2.5;
```

shrcode	shrfirm	shrprice	shrqty	shrdiv	shrpe
PT	Patagonian Tea	55.25	12635	2.5	10
CS	Canadian Sugar	52.78	4716	2.5	15

The IN crowd

The keyword IN is used with a list to specify a set of values. IN is always paired with a column name. All rows for which a value in the specified column has a match in the list are selected. It is a simpler way of writing a series of OR statements.

- *Report data on firms with codes of FC, AR, or SLG.*

```
SELECT * FROM share WHERE shrcode IN ('FC','AR','SLG');
```

The foregoing query could have also been written as

```
SELECT * FROM share
  WHERE shrcode = 'FC' or shrcode = 'AR' or shrcode = 'SLG';
```

shrcode	shrfirm	shrprice	shrqty	shrdiv	shrpe
FC	Freedonia Copper	27.5	10529	1.84	16
AR	Abyssinian Ruby	31.82	22010	1.32	13
SLG	Sri Lankan Gold	50.37	32868	2.68	16

The NOT IN crowd

A NOT IN list is used to report instances that do not match any of the values.

- *Report all firms other than those with the code CS or PT.*

```
SELECT * FROM share WHERE shrcode NOT IN ('CS','PT');
```

is equivalent to

```
SELECT * FROM share WHERE shrcode <> 'CS' AND shrcode <> 'PT';
```

shrcode	shrfirm	shrprice	shrqty	shrdiv	shrpe
FC	Freedonia Copper	27.5	10529	1.84	16

shrcode	shrfirm	shrprice	shrqty	shrdiv	shrpe
AR	Abyssinian Ruby	31.82	22010	1.32	13
SLG	Sri Lankan Gold	50.37	32868	2.68	16
ILZ	Indian Lead & Zinc	37.75	6390	3	12
BE	Burmese Elephant	0.07	154713	0.01	3
BS	Bolivian Sheep	12.75	231678	1.78	11
NG	Nigerian Geese	35	12323	1.68	10
ROF	Royal Ostrich Farms	33.75	1234923	3	6

Skill builder

List those shares where the value of the holding exceeds one million.

Ordering columns

The order of reporting columns is identical to their order in the SQL command. For instance, compare the output of the following queries.

```
SELECT shrcode, shrfirm FROM share WHERE shrpe = 10;
```

shrcode	shrfirm
PT	Patagonian Tea
NG	Nigerian Geese

```
SELECT shrfirm, shrcode FROM share WHERE shrpe = 10;
```

shrfirm	shrcode
Patagonian Tea	PT
Nigerian Geese	NG

Ordering rows

People can generally process an ordered (e.g., sorted alphabetically) report faster than an unordered one. In SQL, the `ORDER BY` clause specifies the row order in a report. The default ordering sequence is ascending (A before B, 1 before 2). Descending is specified by adding `DESC` after the column name.

- *List all firms where PE is at least 10, and order the report in descending PE. Where PE ratios are identical, list firms in alphabetical order.*

```
SELECT * FROM share WHERE shrpe >= 10
   ORDER BY shrpe DESC, shrfirm;
```

shrcode	shrfirm	shrprice	shrqty	shrdiv	shrpe
FC	Freedonia Copper	27.5	10529	1.84	16

shrcode	shrfirm	shrprice	shrqty	shrdiv	shrpe
SLG	Sri Lankan Gold	50.37	32868	2.68	16
CS	Canadian Sugar	52.78	4716	2.5	15
AR	Abyssinian Ruby	31.82	22010	1.32	13
ILZ	Indian Lead & Zinc	37.75	6390	3	12
BS	Bolivian Sheep	12.75	231678	1.78	11
NG	Nigerian Geese	35	12323	1.68	10
PT	Patagonian Tea	55.25	12635	2.5	10

Numeric versus character sorting

Numeric data in character fields (e.g., a product code) do not always sort the way you initially expect. The difference arises from the way data are stored:

- Numeric fields are right justified and have leading zeros.
- Character fields are left justified and have trailing spaces.

For example, the value 1066 stored as `CHAR(4)` would be stored as '`1066`' and the value 45 would be stored as '`45  `'. If the column containing these data is sorted in ascending order, then '1066' precedes ' 45 ' because the leftmost character '1' is less than '4'. You can avoid this problem by always storing numeric values as numeric data types (e.g., integer or decimal) or preceding numeric values with zeros when they are stored as character data. Alternatively, start numbering at 1,000 so that all values are four digits, though the best solution is to store numeric data as numbers rather than characters.

Derived data

One of the important principles of database design is to avoid redundancy. One form of redundancy is including a column in a table when these data can be derived from other columns. For example, we do not need a column for yield because it can be calculated by dividing dividend by price and multiplying by 100 to obtain the yield as a percentage. This means that the query language does the calculation when the value is required.

- *Get firm name, price, quantity, and firm yield.*

```
SELECT shrfirm, shrprice, shrqty, shrdiv/shrprice*100 AS yield FROM share;
```

shrfirm	shrprice	shrqty	yield
Freedonia Copper	27.5	10529	6.69
Patagonian Tea	55.25	12635	4.52
Abyssinian Ruby	31.82	22010	4.15
Sri Lankan Gold	50.37	32868	5.32
Indian Lead & Zinc	37.75	6390	7.95
Burmese Elephant	0.07	154713	14.29
Bolivian Sheep	12.75	231678	13.96
Nigerian Geese	35	12323	4.8

shrfirm	shrprice	shrqty	yield
Canadian Sugar	52.78	4716	4.74
Royal Ostrich Farms	33.75	1234923	8.89

You can give the results of the calculation a column name. In this case, a good choice is yield. Note the use of AS to indicate the name of the column in which the results of the calculation are displayed.

In the preceding query, the keyword AS is introduced to specify an alias, or temporary name. The statement specifies that the result of the calculation is to be reported under the column heading yield. You can rename any column or specify a name for the results of an expression using an alias.

Aggregate functions

SQL has built-in functions to enhance its retrieval power and handle many common aggregation queries, such as computing the total value of a column. Four of these functions (AVG, SUM, MIN, and MAX) work very similarly. COUNT is a little different.

COUNT

COUNT computes the number of rows in a table. Use COUNT(*) to count all rows irrespective of their content (i.e., null or not null), and use COUNT(columnname) to count rows without a null value for columname. Count can be used with a WHERE clause to specify a condition.

- *How many firms are there in the portfolio?*

```
SELECT COUNT(shrcode) AS investments FROM share;
```

investments
10

- *How many firms have a holding greater than 50,000?*

```
SELECT COUNT(shrfirm) AS bigholdings FROM share WHERE shrqty > 50000;
```

bigholdings
3

AVG—averaging

AVG computes the average of the values in a column of numeric data. Null values in the column are not included in the calculation.

- *Find the average dividend.*

```
SELECT AVG(shrdiv) AS avgdiv FROM share;
```

avgdiv
2.03

- *What is the average yield for the portfolio?*

```
SELECT AVG(shrdiv/shrprice*100) AS avgyield FROM share;
```

avgyield
7.53

SUM, MIN, and MAX

SUM, MIN, and MAX differ in the statistic they calculate but are used similarly to AVG. As with AVG, null values in a column are not included in the calculation. SUM computes the sum of a column of values. MIN finds the smallest value in a column; MAX finds the largest.

Subqueries

Sometimes we need the answer to another query before we can write the query of ultimate interest. For example, to list all shares with a PE ratio greater than the portfolio average, you first must find the average PE ratio for the portfolio. You could do the query in two stages:

```
a. SELECT AVG(shrpe) FROM share;
b. SELECT shrfirm, shrpe FROM share WHERE shrpe > x;
```

where x is the value returned from the first query.

Unfortunately, the two-stage method introduces the possibility of errors. You might forget the value returned by the first query or enter it incorrectly. It also takes longer to get the results of the query. We can solve these problems by using parentheses to indicate the first query is nested within the second one. As a result, the value returned by the inner or nested subquery, the one in parentheses, is used in the outer query. In the following example, the nested query returns 11.20, which is then automatically substituted in the outer query.

- *Report all firms with a PE ratio greater than the average for the portfolio.*

```
SELECT shrfirm, shrpe FROM share
   WHERE shrpe > (SELECT AVG(shrpe) FROM share);
```

shrfirm	shrpe
Freedonia Copper	16
Abyssinian Ruby	13
Sri Lankan Gold	16
Indian Lead & Zinc	12
Canadian Sugar	15

> Warning: The preceding query is often mistakenly written as
>
> ```
> SELECT shrfirm, shrpe FROM SHARE
>   WHERE shrpe > AVG(shrpe);
> ```

Skill builder

Find the name of the firm for which the value of the holding is greatest.

Regular expression—pattern matching

Regular expression is a concise and powerful method for searching for a specified pattern in a nominated column. Regular expression processing is supported by programming languages such as Java and PHP. In this chapter, we introduce a few typical regular expressions and will continue developing your knowledge of this feature in the next chapter.

Search for a string

- *List all firms containing 'Ruby' in their name.*

```
SELECT shrfirm FROM share WHERE shrfirm REGEXP 'Ruby';
```

shrfirm
Abyssinian Ruby

Search for alternative strings

In some situations you want to find columns that contain more than one string. In this case, we use the alternation symbol '|' to indicate the alternatives being sought. For example, a|b finds 'a' or 'b'.

- *List the firms containing gold or zinc in their name, irrespective of case.*

```
SELECT shrfirm FROM share WHERE shrfirm REGEXP 'gold|zinc|Gold|Zinc';
```

shrfirm
Sri Lankan Gold
Indian Lead & Zinc

Search for a beginning string

If you are interested in finding a value in a column that starts with a particular character string, then use ^ to indicate this option.

- *List the firms whose name begins with Sri.*

```
SELECT shrfirm FROM share WHERE shrfirm REGEXP '^Sri';
```

shrfirm
Sri Lankan Gold

Search for an ending string

If you are interested in finding if a value in a column ends with a particular character string, then use $ to indicate this option.

- *List the firms whose name ends with Geese.*

```
SELECT shrfirm FROM share WHERE shrfirm REGEXP 'Geese$';
```

shrfirm
Nigerian Geese

Skill builder

List the firms containing "ian" in their name.

DISTINCT—eliminating duplicate rows

The DISTINCT clause is used to eliminate duplicate rows. It can be used with column functions or before a column name. When used with a column function, it ignores duplicate values.

- *Report the different values of the PE ratio.*

```
SELECT DISTINCT shrpe FROM share;
```

shrpe
3
10
11
12
13
15
16

- *Find the number of different PE ratios.*

```
SELECT COUNT(DISTINCT shrpe) as 'Different PEs' FROM share;
```

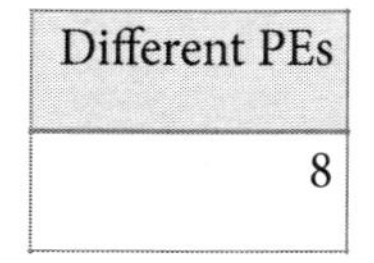

Different PEs
8

When used before a column name, DISTINCT prevents the selection of duplicate rows. Notice a slightly different use of the keyword AS. In this case, because the alias includes a space, the entire alias is enclosed in straight quotes.

DELETE

Rows in a table can be deleted using the DELETE clause in an SQL statement. DELETE is typically used with a WHERE clause to specify the rows to be deleted. If there is no WHERE clause, all rows are deleted.

- *Erase the data for Burmese Elephant. All the shares have been sold.*

```
DELETE FROM share WHERE shrfirm = 'Burmese Elephant';
```

In the preceding statement, shrfirm is used to indicate the row to be deleted.

UPDATE

Rows can be modified using SQL's UPDATE clause, which is used with a WHERE clause to specify the rows to be updated.

- *Change the share price of FC to 31.50.*

```
UPDATE share SET shrprice = 31.50 WHERE shrcode = 'FC';
```

- *Increase the total number of shares for Nigerian Geese by 10% because of the recent bonus issue.*

```
UPDATE share SET shrqty = shrqty*1.1 WHERE shrfirm = 'Nigerian Geese';
```

Quotes

There are three types of quotes that you can typically use with SQL. Double and single quotes are equivalent and can be used interchangeably. Note that single and double quotes must be straight rather than curly, and the back quote is to the left of the 1 key.

Type of quote	Representation
Single	'
Double	"
Back	`

The following SQL illustrates the use of three types of quotes to find a person with a last name of O'Hara and where the column names are person first and person last.

```
SELECT `person first` FROM person WHERE `person last` = "O'Hara";
```

Debriefing

Now that you have learned how to model a single entity, create a table, and specify queries, you are on the way to mastering the fundamental skills of database design, implementation, and use. Remember, planning occurs before action. A data model is a plan for a database. The action side of a database is inserting rows and running queries.

Summary

The relational database model is an effective means of representing real-world relationships. Data modeling is used to determine what data must be stored and how data are related. An entity is something in the environment. An entity has attributes, which describe it, and an identifier, which uniquely distinguishes an instance of an entity. Every entity must have a unique identifier. A relational database consists of tables with rows and columns. A data model is readily translated into a relational database. The SQL statement CREATE is used to define a table. Rows are added to a table using INSERT. In SQL, queries are written using the SELECT statement. Project (choosing columns) and restrict (choosing rows) are common table operations. The WHERE clause is used to specify row selection criteria. WHERE can be combined with `IN` and `NOT IN`, which specify values for a single column. The rows of a report are sorted using the `ORDER BY` clause. Arithmetic expressions can appear in SQL statements, and SQL has built-in functions for common arithmetic operations. A subquery is a query within a query. Regular expressions are used to find string patterns within character strings. Duplicate rows are eliminated with the `DISTINCT` clause. Rows can be erased using `DELETE` or modified with `UPDATE`.

Key terms and concepts

Alias
AS
Attribute
AVG
Column
COUNT
CREATE
Data modeling
Data type
Database
DELETE
DISTINCT
Entity
Entity integrity rule
Identifier
IN
INSERT
Instance
MAX
MIN
NOT IN
ORDER BY
Primary key
Project
Relational database
Restrict
Row
SELECT
SQL
Subquery
SUM
Table
UPDATE
WHERE

Exercises

1. Draw data models for the following entities. In each case, make certain that you show the attributes and identifiers. Also, create a relational database and insert some rows for each of the models.

    a. Aircraft: An aircraft has a manufacturer, model number, call sign (e.g., N123D), payload, and a year of construction. Aircraft are classified as civilian or military.

    b. Car: A car has a manufacturer, range name, and style code (e.g., a Honda Accord DX, where Honda is the manufacturer, Accord is the range, and DX is the style). A car also has a vehicle identification code, registration code, and color.

    c. Restaurant: A restaurant has an address, seating capacity, phone number, and style of food (e.g., French, Russian, Chinese).

    d. Cow: A dairy cow has a name, date of birth, breed (e.g., Holstein), and a numbered plastic ear tag.

2. Do the following queries using SQL:
   a. List a share's name and its code.
   b. List full details for all shares with a price less than $1.
   c. List the names and prices of all shares with a price of at least $10.
   d. Create a report showing firm name, share price, share holding, and total value of shares held. (Value of shares held is price times quantity.)
   e. List the names of all shares with a yield exceeding 5 percent.
   f. Report the total dividend payment of Patagonian Tea. (The total dividend payment is dividend times quantity.)
   g. Find all shares where the price is less than 20 times the dividend.
   h. Find the share(s) with the minimum yield.
   i. Find the total value of all shares with a PE ratio > 10.
   j. Find the share(s) with the maximum total dividend payment.
   k. Find the value of the holdings in Abyssinian Ruby and Sri Lankan Gold.
   l. Find the yield of all firms except Bolivian Sheep and Canadian Sugar.
   m. Find the total value of the portfolio.
   n. List firm name and value in descending order of value.
   o. List shares with a firm name containing "Gold."
   p. Find shares with a code starting with "B."
3. Run the following queries and explain the differences in output. Write each query as a manager might state it.
   a. `SELECT shrfirm FROM share WHERE shrfirm NOT REGEXP 's';`
   b. `SELECT shrfirm FROM share WHERE shrfirm NOT REGEXP 'S';`
   c. `SELECT shrfirm FROM share WHERE shrfirm NOT REGEXP 's|S';`
   d. `SELECT shrfirm FROM share WHERE shrfirm NOT REGEXP '^S';`
   e. `SELECT shrfirm FROM share WHERE shrfirm NOT REGEXP 's$';`
4. A weekly newspaper, sold at supermarket checkouts, frequently reports stories of aliens visiting Earth and taking humans on short trips. Sometimes a captured human sees Elvis commanding the spaceship. To keep track of all these reports, the newspaper has created the following data model.

ALIEN
*al#
alname
alheads
alcolor
alsmell

The paper has also supplied some data for the last few sightings and asked you to create the database, add details of these aliens, and answer the following queries:

a. What's the average number of heads of an alien?

b. Which alien has the most heads?

c. Are there any aliens with a double o in their names?

d. How many aliens are chartreuse?

e. Report details of all aliens sorted by smell and color.

5. Eduardo, a bibliophile, has a collection of several hundred books. Being a little disorganized, he has his books scattered around his den. They are piled on the floor, some are in bookcases, and others sit on top of his desk. Because he has so many books, he finds it difficult to remember what he has, and sometimes he cannot find the book he wants. Eduardo has a simple personal computer file system that is fine for a single entity or file. He has decided that he would like to list each book by author(s)' name and type of book (e.g., literature, travel, reference). Draw a data model for this problem, create a single entity table, and write some SQL queries.

a. How do you identify each instance of a book? (It might help to look at a few books.)

b. How should Eduardo physically organize his books to permit fast retrieval of a particular one?

c. Are there any shortcomings with the data model you have created?

6. What is an identifier? Why does a data model have an identifier?

7. What are entities?

8. What is the entity integrity rule?

CD library case

Ajay, a DJ in the student-operated radio station at his university, is impressed by the station's large collection of CDs but is frustrated by the time he spends searching for a particular piece of music. Recently, when preparing his weekly jazz session, he spent more than half an hour searching for a recording of Duke Ellington's "Creole Blues." An IS major, Ajay had recently commenced a data management class. He quickly realized that a relational database, storing details of all the CDs owned by the radio station, would enable him to find quickly any piece of music.

Having just learned how to model a single entity, Ajay decided to start building a CD database. He took one of his favorite CDs, John Coltrane's Giant Steps, and drew a model to record details of the tracks on this CD.

CD library V1.0

TRACK
*trkid trknum trktitle trklength

Ajay pondered using track number as the identifier since each track on a CD is uniquely numbered, but he quickly realized that track number was not suitable because it uniquely identifies only those tracks on a specific CD. So, he introduced `trackid` to identify uniquely each track irrespective of the CD on which it is found.

He also recorded the name of the track and its length. He had originally thought that he would record the track's length in minutes and seconds (e.g., 4:43) but soon discovered that the TIME data type of his DBMS is designed to store time of day rather than the length of time of an event. Thus, he concluded that he should store the length of a track in minutes as a decimal (i.e., 4:43 is stored as 4.72).

*trkid	trknum	trktitle	trklength
1	1	Giant Steps	4.72
2	2	Cousin Mary	5.75
3	3	Countdown	2.35
4	4	Spiral	5.93
5	5	Syeeda's Song Flute	7
6	6	Naima	4.35
7	7	Mr. P.C.	6.95
8	8	Giant Steps	3.67
9	9	Naima	4.45
10	10	Cousin Mary	5.9
11	11	Countdown	4.55
12	12	Syeeda's Song Flute	7.03

Skill builder

Create a single table database using the data in the preceding table.

Justify your choice of data type for each of the attributes.

Ajay ran a few queries on his single table database.

- *On what track is "Giant Steps"?*

```
SELECT trknum, trktitle FROM track
   WHERE trktitle = 'Giant Steps';
```

trknum	trktitle
1	Giant Steps

8	Giant Steps

Observe that there are two tracks with the title "Giant Steps."

- *Report all recordings longer than 5 minutes.*

```
SELECT trknum, trktitle, trklength FROM track
   WHERE trklength > 5;
```

trknum	trktitle	trklength
2	Cousin Mary	5.75
4	Spiral	5.93
5	Syeeda's Song Flute	7
7	Mr. P.C.	6.95
10	Cousin Mary	5.9
12	Syeeda's song Flute	7.03

- *What is the total length of recordings on the CD?*

This query makes use of the function SUM to total the length of tracks on the CD.

```
SELECT SUM(trklength) AS sumtrklen FROM track;
```

sumtrklen
62.65

When you run the preceding query, the value reported may not be exactly 62.65. Why?

- *Find the longest track on the CD.*

```
SELECT trknum, trktitle, trklength FROM track
   WHERE trklength = (SELECT MAX(trklength) FROM track);
```

This query follows the model described previously in this chapter. First determine the longest track (i.e., 7.03 minutes) and then find the track, or tracks, that are this length.

trknum	trktitle	trklength
12	Syeeda's Song Flute	7.03

- *How many tracks are there on the CD?*

```
SELECT COUNT(*) AS tracks FROM track;
```

tracks
12

- *Sort the different titles on the CD by their name.*

```
SELECT DISTINCT(trktitle) FROM track
```

```
    ORDER BY trktitle;
```

trktitle
Countdown
Cousin Mary
Giant Steps
Mr. P.C.
Naima
Spiral
Syeeda's Song Flute

- *List the shortest track containing "song" in its title.*

```
SELECT trknum, trktitle, trklength FROM track
    WHERE trktitle REGEXP 'song'
    AND trklength = (SELECT MIN(trklength) FROM track
        WHERE trktitle REGEXP 'song');
```

The subquery determines the length of the shortest track with "song" in its title. Then any track with a length equal to the minimum and also containing "song" in its title (just in case there is another without "song" in its title that is the same length) is reported.

trknum	trktitle	trklength
5	Syeeda's Song Flute	7

- *Report details of tracks 1 and 5.*

```
SELECT trknum, trktitle, trklength FROM track
    WHERE trknum IN (1,5);
```

trknum	trktitle	trklength
1	Giant Steps	4.72
5	Syeeda's Song Flute	7

Skill builder

1. Write SQL commands for the following queries:
    a. Report all tracks between 4 and 5 minutes long.
    b. On what track is "Naima"?
    c. Sort the tracks by descending length.
    d. What tracks are longer than the average length of tracks on the CD?
    e. How many tracks are less than 5 minutes?
    f. What tracks start with "Cou"?
2. Add the 13 tracks from The Manhattan Transfer's Swing CD in the following table to the track table:

Track	Title	Length
1	Stomp of King Porter	3.2
2	Sing a Study in Brown	2.85
3	Sing Moten's Swing	3.6
4	A-tisket, A-tasket	2.95
5	I Know Why	3.57
6	Sing You Sinners	2.75
7	Java Jive	2.85
8	Down South Camp Meetin'	3.25
9	Topsy	3.23
10	Clouds	7.2
11	Skyliner	3.18
12	It's Good Enough to Keep	3.18
13	Choo Choo Ch' Boogie	3

3. Write SQL to answer the following queries:
    a. What is the longest track on Swing?
    b. What tracks on Swing include the word Java?
4. What is the data model missing?

4. The One-to-Many Relationship

Cow of many—well milked and badly fed.

Spanish proverb

Learning objectives

Students completing this chapter will be able to

- model a one-to-many relationship between two entities;
- define a database with a one-to-many relationship;
- write queries for a database with a one-to-many relationship.

Alice sat with a self-satisfied smirk on her face. Her previous foray into her attaché case had revealed the extent of her considerable stock holdings. She was wealthy beyond her dreams. What else was there in this marvelous, magic attaché case? She rummaged further into the case and retrieved a folder labeled "Stocks—foreign." More shares!

On the inside cover of the folder was a note stating that the stocks in this folder were not listed in the United Kingdom. The folder contained three sheets of paper. Each was headed by the name of a country and followed by a list of stocks and the number of shares. The smirk became the cattiest of Cheshire grins. Alice ordered another bottle of champagne and turned to her iPad to check the latest prices on the Google Finance page. She wanted to calculate the current value of each stock and then use current exchange rates to convert the values into British currency.

Relationships

Entities are not isolated; they are related to other entities. When we move beyond the single entity, we need to identify the relationships between entities to accurately represent the real world. Once we recognize that all stocks in our case study are not listed in the United Kingdom, we need to introduce an entity called NATION. We now have two entities, STOCK and NATION. Consider the relationship between them. A NATION can have many listed stocks. A stock, in this case, is listed in only one nation. There is a 1:m (one-to-many) relationship between NATION and STOCK.

A 1:m relationship between two entities is depicted by a line connecting the two with a crow's foot at the many end of the relationship. The following figure shows the 1:m relationship between NATION and STOCK. This can be read as: "a nation can have many stocks, but a stock belongs to only one nation." The entity NATION is identified by *nation code* and has attributes *nation name* and *exchange rate.*

A 1:m relationship between NATION and STOCK

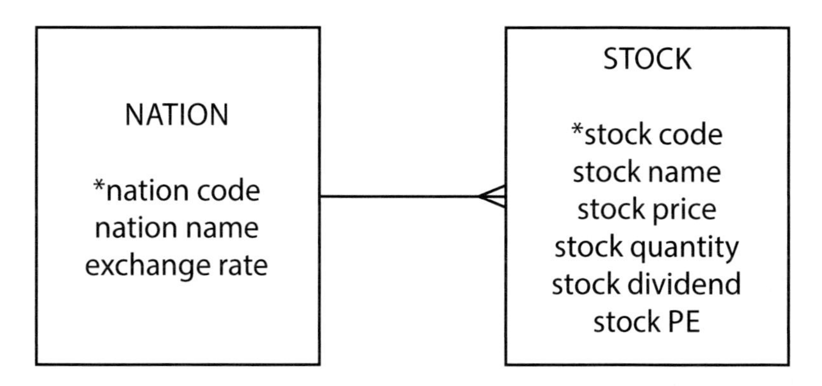

The 1:m relationship occurs frequently in business situations. Sometimes it occurs in a tree or hierarchical fashion. Consider a very hierarchical firm. It has many divisions, but a division belongs to only one firm. A division has many departments, but a department belongs to only one division. A department has many sections, but a section belongs to only one department.

A series of 1:m relationships

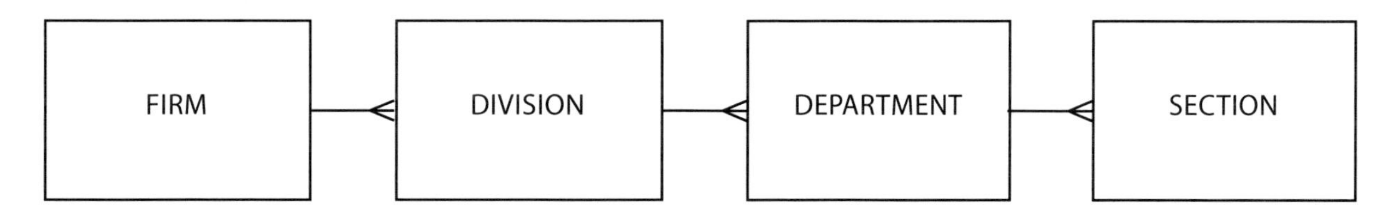

Why did we create an additional entity?

Another approach to adding data about listing nation and exchange rate is to add two attributes to STOCK: *nation name* and *exchange rate*. At first glance, this seems a very workable solution; however, this will introduce considerable redundancy, as the following table illustrates.

The table stock with additional columns

stock							
*stkcode	stkfirm	stkprice	stkqty	stkdiv	stkpe	natname	exchrate
FC	Freedonia Copper	27.5	10529	1.84	16	United Kingdom	1
PT	Patagonian Tea	55.25	12635	2.5	10	United Kingdom	1
AR	Abyssinian Ruby	31.82	22010	1.32	13	United Kingdom	1
SLG	Sri Lankan Gold	50.37	32868	2.68	16	United Kingdom	1
ILZ	Indian Lead & Zinc	37.75	6390	3	12	United Kingdom	1
BE	Burmese Elephant	0.07	154713	0.01	3	United Kingdom	1
BS	Bolivian Sheep	12.75	231678	1.78	11	United Kingdom	1
NG	Nigerian Geese	35	12323	1.68	10	United Kingdom	1
CS	Canadian Sugar	52.78	4716	2.5	15	United Kingdom	1
ROF	Royal Ostrich Farms	33.75	1234923	3	6	United Kingdom	1
MG	Minnesota Gold	53.87	816122	1	25	USA	0.67

stock							
*stkcode	stkfirm	stkprice	stkqty	stkdiv	stkpe	natname	exchrate
GP	Georgia Peach	2.35	387333	0.2	5	USA	0.67
NE	Narembeen Emu	12.34	45619	1	8	Australia	0.46
QD	Queensland Diamond	6.73	89251	0.5	7	Australia	0.46
IR	Indooroopilly Ruby	15.92	56147	0.5	20	Australia	0.46
BD	Bombay Duck	25.55	167382	1	12	India	0.0228

The exact same `nation name` and `exchange rate` data pair occurs 10 times for stocks listed in the United Kingdom. Redundancy presents problems when we want to insert, delete, or update data. These problems, generally known as **update anomalies**, occur with these three basic operations.

Insert anomalies
We cannot insert a fact about a nation's exchange rate unless we first buy a stock that is listed in that nation. Consider the case where we want to keep a record of France's exchange rate and we have no French stocks. We cannot skirt this problem by putting in a null entry for stock details because `stkcode`, the primary key, would be null, and this is not allowed. If we have a separate table for facts about a nation, then we can easily add new nations without having to buy stocks. This is particularly useful when other parts of the organization, say International Trading, also need access to exchange rates for many nations.

Delete anomalies
If we delete data about a particular stock, we might also lose a fact about exchange rates. For example, if we delete details of Bombay Duck, we also erase the Indian exchange rate.

Update anomalies
Exchange rates are volatile. Most companies need to update them every day. What happens when the Australian exchange rate changes? Every row in `stock` with `nation = 'Australia'` will have to be updated. In a large portfolio, many rows will be changed. There is also the danger of someone forgetting to update all the instances of the nation and exchange rate pair. As a result, there could be two exchange rates for the one nation. If exchange rate is stored in `nation`, however, only one change is necessary, there is no redundancy, and there is no danger of inconsistent exchange rates.

Creating a database with a 1:m relationship

As before, each entity becomes a table in a relational database, the entity name becomes the table name, each attribute becomes a column, and each identifier becomes a primary key. The 1:m relationship is mapped by adding a column to the entity at the many end of the relationship. The additional column contains the identifier of the one end of the relationship.t

Consider the relationship between the entities STOCK and NATION. The database has two tables: `stock` and `nation`. The table `stock` has an additional column, `natcode`, which contains the primary key of `nation`. If `natcode` is not stored in `stock`, then there is no way of knowing the identity of the nation where the stock is listed.

A relational database with tables nation and stock

nation		
*natcode	natname	exchrate
AUS	Australia	0.46
IND	India	0.0228
UK	United Kingdom	1
USA	United States	0.67 →

stock						
*stkcode	stkfirm	stkprice	stkqty	stkdiv	stkpe	*natcode*
FC	Freedonia Copper	27.5	10,529	1.84	16	UK
PT	Patagonian Tea	55.25	12,635	2.5	10	UK
AR	Abyssinian Ruby	31.82	22,010	1.32	13	UK
SLG	Sri Lankan Gold	50.37	32,868	2.68	16	UK
ILZ	Indian Lead & Zinc	37.75	6,390	3	12	UK
BE	Burmese Elephant	0.07	154,713	0.01	3	UK
BS	Bolivian Sheep	12.75	231,678	1.78	11	UK
NG	Nigerian Geese	35	12,323	1.68	10	UK
CS	Canadian Sugar	52.78	4,716	2.5	15	UK
ROF	Royal Ostrich Farms	33.75	1,234,923	3	6	UK
MG	Minnesota Gold	53.87	816,122	1	25	USA ←
GP	Georgia Peach	2.35	387,333	0.2	5	USA ←
NE	Narembeen Emu	12.34	45,619	1	8	AUS
QD	Queensland Diamond	6.73	89,251	0.5	7	AUS
IR	Indooroopilly Ruby	15.92	56,147	0.5	20	AUS
BD	Bombay Duck	25.55	167,382	1	12	IND

Notice that `natcode` appears in both the `stock` and `nation` tables. In `nation`, `natcode` is the primary key; it is unique for each instance of nation. In table `stock`, `natcode` is a **foreign key** because it is the primary key of `nation`, the one end of the 1:m relationship. The column `natcode` is a foreign key in `stock` because it is a primary key in `nation`. A matched primary key–foreign key pair is the method for recording the 1:m relationship between the two tables. This method of representing a relationship is illustrated using dark blue shading and arrows for the two USA stocks. In the `stock` table, `natcode` is italicized to indicate that it is a foreign key. This method, like an asterisk prefix for a primary key, is a useful reminder.

Although the same name has been used for the primary key and the foreign key in this example, it is not mandatory. The two columns can have different names, and in some cases you are forced to use different names. When possible, we find it convenient to use identical column names to help us remember that the tables are related. To distinguish between columns with identical names, they must by **qualified** by prefixing the table name. In this case, use nation.natcode and stock.natcode. Thus, nation.natcode refers to the natcode column in the table nation.

Although a nation can have many stocks, it is not mandatory to have any. That is, in data modeling terminology, many can be zero, one, or more, but it is mandatory to have a value for natcode in nation for every value of natcode in stock. This requirement, known as the **referential integrity constraint**, maintains the accuracy of a database. Its application means that every foreign key in a table has an identical primary key in that same table or another table. In this example, it means that for every value of natcode in stock, there is a corresponding entry in nation. As a result, a primary key row must be created before its corresponding foreign key row. In other words, details for a nation must be added before any data about its listed stocks are entered.

Every foreign key must have a matching primary key (referential integrity rule), and every primary key must be non-null (entity integrity rule). A foreign key cannot be null when a relationship is mandatory, as in the case where a stock must belong to a nation. If a relationship is optional (a person can have a boss), then a foreign key can be null (i.e., a person is the head of the organization, and thus has no boss). The ideas of mandatory and optional will be discussed later in this book.

Why is the foreign key in the table at the "many" end of the relationship? Because each instance of stock is associated with exactly one instance of nation. The rule is that a stock must be listed in one, and only one, nation. Thus, the foreign key field is single-valued when it is at the "many" end of a relationship. The foreign key is not at the "one" end of the relationship because each instance of nation can be associated with more than one instance of stock, and this implies a multivalued foreign key. The relational model does not support multivalued fields because of the processing problems they can cause.

Using SQL, the two tables are defined in a similar manner to the way we created a single table in Chapter 3. Here are the SQL statements:

```
CREATE TABLE nation (
   natcode        CHAR(3),
   natname        VARCHAR(20),
   exchrate       DECIMAL(9,5),
   PRIMARY KEY(natcode));

CREATE TABLE stock (
   stkcode        CHAR(3),
   stkfirm        VARCHAR(20),
   stkprice       DECIMAL(6,2),
   stkqty         DECIMAL(8),
   stkdiv         DECIMAL(5,2),
   stkpe          DECIMAL(5),
   natcode        CHAR(3),
   PRIMARY KEY(stkcode),
   CONSTRAINT fk_has_nation FOREIGN KEY(natcode)
      REFERENCES nation(natcode) ON DELETE RESTRICT);
```

Notice that the definition of stock includes an additional phrase to specify the foreign key and the referential integrity constraint. The CONSTRAINT clause defines the column or columns in the table being created that constitute the foreign key. A referential integrity constraint can be named, and in this case, the constraint's name is fk_has_nation. The foreign key is the column natcode in STOCK, and it references the primary key of nation, which is natcode.

The ON DELETE clause specifies what processing should occur if an attempt is made to delete a row in nation with a primary key that is a foreign key in stock. In this case, the ON DELETE clause specifies that it is not permissible (the meaning of RESTRICT) to delete a primary key row in nation while a corresponding foreign key in stock exists. In other words, the system will not execute the delete. You must first delete all corresponding rows in stock before attempting to delete the row containing the primary key. ON DELETE is the default clause for most RDBMSs, so we will dispense with specifying it for future foreign key constraints.

Observe that both the primary and foreign keys are defined as CHAR(3). The relational model requires that a primary key–foreign key pair have the same data type and are the same length.

Skill Builder

The university architect has asked you to develop a data model to record details of the campus buildings. A building can have many rooms, but a room can be in only one building. Buildings have names, and rooms have a size and purpose (e.g., lecture, laboratory, seminar). Draw a data model for this situation and create the matching relational database.

MySQL Workbench

In Workbench, a 1:m relationship is represented in a similar manner to the method you have just learned. Also, note that the foreign key is shown in the entity at the many end with a red diamond. We omit the foreign key when data modeling because it can be inferred. You will observe some additional symbols on the line between the two entities, and these will be explained later, but take note of the crow's foot indicating the 1:m relationship between nation and stock. Because Workbench can generate automatically the SQL to create the tables,[16] we use lowercase table names and abbreviated column names.

16 Database > Forward Engineer...

Specifying a 1:m relationship in MySQL Workbench

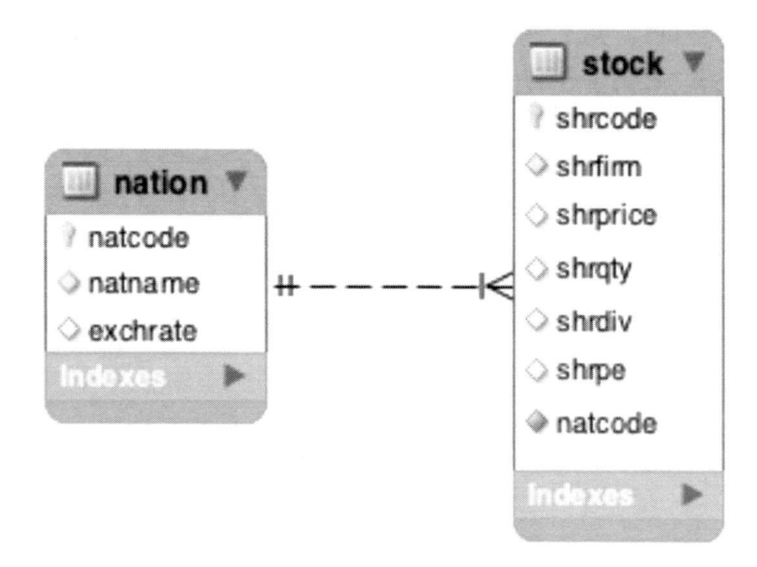

Querying a two-table database

A two-table database offers the opportunity to learn more SQL and another relational algebra operation: join.

Join

Join creates a new table from two existing tables by matching on a column common to both tables. Usually, the common column is a primary key–foreign key pair: The primary key of one table is matched with the foreign key of another table. Join is frequently used to get the data for a query into a single row. Consider the tables `nation` and `stock`. If we want to calculate the value—in British pounds—of a stock, we multiply stock price by stock quantity and then exchange rate. To find the appropriate exchange rate for a stock, get its `natcode` from `stock` and then find the exchange rate in the matching row in `nation`, the one with the same value for `natcode`. For example, to calculate the value of Georgia Peach, which has `natcode = 'US'`, find the row in `nation` that also has `natcode = 'US'`. In this case, the stock's value is 2.35*387333*0.67 = £609,855.81.

Calculation of stock value is very easy once a join is used to get the three values in one row. The SQL command for joining the two tables is:

```
SELECT * FROM stock JOIN nation
   ON stock.natcode = nation.natcode;
```

The join of stock and nation

stkcode	stkfirm	stkprice	stkqty	stkdiv	stkpe	natcode	natcode	natname	exchrate
NE	Narembeen Emu	12.34	45619	1	8	AUS	AUS	Australia	0.4600
IR	Indooroopilly Ruby	15.92	56147	0.5	20	AUS	AUS	Australia	0.4600
QD	Queensland Diamond	6.73	89251	0.5	7	AUS	AUS	Australia	0.4600
BD	Bombay Duck	25.55	167382	1	12	IND	IND	India	0.0228

stkcode	stkfirm	stkprice	stkqty	stkdiv	stkpe	natcode	natcode	natname	exchrate
ROF	Royal Ostrich Farms	33.75	1234923	3	6	UK	UK	United Kingdom	1.0000
CS	Canadian Sugar	52.78	4716	2.5	15	UK	UK	United Kingdom	1.0000
FC	Freedonia Copper	27.5	10529	1.84	16	UK	UK	United Kingdom	1.0000
BS	Bolivian Sheep	12.75	231678	1.78	11	UK	UK	United Kingdom	1.0000
BE	Burmese Elephant	0.07	154713	0.01	3	UK	UK	United Kingdom	1.0000
ILZ	Indian Lead & Zinc	37.75	6390	3	12	UK	UK	United Kingdom	1.0000
SLG	Sri Lankan Gold	50.37	32868	2.68	16	UK	UK	United Kingdom	1.0000
AR	Abyssinian Ruby	31.82	22010	1.32	13	UK	UK	United Kingdom	1.0000
PT	Patagonian Tea	55.25	12635	2.5	10	UK	UK	United Kingdom	1.0000
NG	Nigerian Geese	35	12323	1.68	10	UK	UK	United Kingdom	1.0000
MG	Minnesota Gold	53.87	816122	1	25	US	US	United States	0.6700
GP	Georgia Peach	2.35	387333	0.2	5	US	US	United States	0.6700

The columns `stkprice` and `stkdiv` record values in the country's currency. Thus, the price of Bombay Duck is 25.55 Indian rupees. To find the value in U.K. pounds (GPB), multiply the price by 0.0228, because one rupee is worth 0.0228 GPB. The value of one share of Bombay Duck in U.S. dollars (USD) is 25.55*0.0228/0.67 because one USD is worth 0.67 GBP.

There are several things to notice about the SQL command and the result:

- To avoid confusion because `natcode` is a column name in both stock and nation, it needs to be qualified. If `natcode` is not qualified, the system will reject the query because it cannot distinguish between the two columns titled `natcode`.
- The new table has the `natcode` column replicated. Both are called `natcode`. The naming convention for the replicated column varies with the RDBMS. The columns, for example, should be labeled `stock.natcode` and `nation.natcode`.
- The SQL command specifies the names of the tables to be joined, the columns to be used for matching, and the condition for the match (equality in this case).
- The number of columns in the new table is the sum of the columns in the two tables.
- The stock value calculation is now easily specified in an SQL command because all the data are in one row.

Remember that during data modeling we created two entities, STOCK and NATION, and defined the relationship between them. We showed that if the data were stored in one table, there could be updating problems. Now, with a join, we have combined these data. So why separate the data only to put them back together later? There are two reasons. First, we want to avoid update anomalies. Second, as you will discover, we do not join the same tables every time.

Join comes in several flavors. The matching condition can be =, <>, <=, <, >=, and >. This generalized version is called a **theta join**. Generally, when people refer to a join, however, they mean an equijoin, when the matching condition is equality.

In an alphabetical list of employees, how many appear before Clare?

```
SELECT count(*) FROM emp A JOIN emp B
  ON A.empfname > B.empfname
  WHERE A.empfname = "Clare"
```

A join can be combined with other SQL commands.

- *Report the value of each stockholding in UK pounds. Sort the report by nation and firm.*

```
SELECT natname, stkfirm, stkprice, stkqty, exchrate,
   stkprice*stkqty*exchrate as stkvalue
      FROM stock JOIN nation
         ON stock.natcode = nation.natcode
            ORDER BY natname, stkfirm;
```

natname	stkfirm	stkprice	stkqty	exchrate	stkvalue
Australia	Indooroopilly Ruby	15.92	56147	0.4600	411176
Australia	Narembeen Emu	12.34	45619	0.4600	258952
Australia	Queensland Diamond	6.73	89251	0.4600	276303
India	Bombay Duck	25.55	167382	0.0228	97507
United Kingdom	Abyssinian Ruby	31.82	22010	1.0000	700358
United Kingdom	Bolivian Sheep	12.75	231678	1.0000	2953895
United Kingdom	Burmese Elephant	0.07	154713	1.0000	10830
United Kingdom	Canadian Sugar	52.78	4716	1.0000	248910
United Kingdom	Freedonia Copper	27.5	10529	1.0000	289548
United Kingdom	Indian Lead & Zinc	37.75	6390	1.0000	241223
United Kingdom	Nigerian Geese	35	12323	1.0000	431305
United Kingdom	Patagonian Tea	55.25	12635	1.0000	698084
United Kingdom	Royal Ostrich Farms	33.75	1234923	1.0000	41678651
United Kingdom	Sri Lankan Gold	50.37	32868	1.0000	1655561
United States	Georgia Peach	2.35	387333	0.6700	609856
United States	Minnesota Gold	53.87	816122	0.6700	29456210

Control break reporting

The purpose of a join is to collect the necessary data for a report. When two tables in a 1:m relationship are joined, the report will contain repetitive data. If you re-examine the report from the join, you will see that `nation` and `exchrate` are often repeated because the same values apply to many stocks. A more appropriate format is shown in the following figure, an example of a **control break report**.

```
Nation              Exchange rate
   Firm                               Price      Quantity           Value
Australia                0.4600
   Indooroopilly Ruby                 15.92        56,147      411,175.71
   Narembeen Emu                      12.34        45,619      258,951.69
   Queensland Diamond                  6.73        89,251      276,303.25
India                    0.0228
   Bombay Duck                        25.55       167,382       97,506.71
United Kingdom           1.0000
   Abyssinian Ruby                    31.82        22,010      700,358.20
   Bolivian Sheep                     12.75       231,678    2,953,894.50
   Burmese Elephant                    0.07       154,713       10,829.91
   Canadian Sugar                     52.78         4,716      248,910.48
   Freedonia Copper                   27.50        10,529      289,547.50
   Indian Lead & Zinc                 37.75         6,390      241,222.50
   Nigerian Geese                     35.00        12,323      431,305.00
   Patagonian Tea                     55.25        12,635      698,083.75
   Royal Ostrich Farms                33.75     1,234,923   41,678,651.25
   Sri Lankan Gold                    50.37        32,868    1,655,561.16
United States            0.6700
   Georgia Peach                       2.35       387,333      609,855.81
   Minnesota Gold                     53.87       816,122   29,456,209.73
```

A control break report recognizes that the values in a particular column or columns seldom change. In this case, natname and exchrate are often the same from one row to the next, so it makes sense to report these data only when they change. The report is also easier to read. The column natname is known as a *control field.* Notice that there are four groups of data, because natname has four different values.

Many RDBMS packages have report-writing languages to facilitate creating a control break report. These languages typically support summary reporting for each group of rows having the same value for the control field(s). A table must usually be sorted on the control break field(s) before the report is created.

GROUP BY—reporting by groups

The GROUP BY clause is an elementary form of control break reporting. It permits grouping of rows that have the same value for a specified column or columns, and it produces one row for each different value of the grouping column(s).

- *Report by nation the total value of stockholdings.*

```
SELECT natname, sum(stkprice*stkqty*exchrate) as stkvalue
   FROM stock JOIN nation ON stock.natcode = nation.natcode
      GROUP BY natname;
```

natname	stkvalue
Australia	946431
India	97507
United Kingdom	48908364
United States	30066066

SQL's built-in functions (COUNT, SUM, AVERAGE, MIN, and MAX) can be used with the GROUP BY clause. They are applied to a group of rows having the same value for a specified column. You can specify more than one function in a SELECT statement. For example, we can compute total value and number of different stocks and group by nation using:

- *Report the number of stocks and their total value by nation.*

```
SELECT natname, COUNT(*), SUM(stkprice*stkqty*exchrate) AS stkvalue
   FROM stock JOIN nation ON stock.natcode = nation.natcode
      GROUP BY natname;
```

natname	count	stkvalue
Australia	3	946431
India	1	97507
United Kingdom	10	48908364
United States	2	30066066

You can group by more than one column name; however, all column names appearing in the SELECT clause must be associated with a built-in function or be in a GROUP BY clause.

- *List stocks by nation, and for each nation show the number of stocks for each PE ratio and the total value of those stock holdings in UK pounds.*

```
SELECT natname,stkpe,COUNT(*),
   SUM(stkprice*stkqty*exchrate) AS stkvalue
      FROM stock JOIN nation ON stock.natcode = nation.natcode
         GROUP BY natname, stkpe;
```

natname	stkpe	count	stkvalue
Australia	7	1	276303
Australia	8	1	258952
Australia	20	1	411176
India	12	1	97507
United Kingdom	3	1	10830
United Kingdom	6	1	41678651
United Kingdom	10	2	1129389
United Kingdom	11	1	2953895
United Kingdom	12	1	241223
United Kingdom	13	1	700358
United Kingdom	15	1	248910
United Kingdom	16	2	1945109
United States	5	1	609856
United States	25	1	29456210

In this example, stocks are grouped by both natname and stkpe. In most cases, there is only one stock for each pair of natname and stkpe; however, there are two situations (U.K. stocks with PEs of 10 and 16) where details of multiple stocks are grouped into one report line. Examining the values in the COUNT column helps you to identify these stocks.

HAVING—the WHERE clause of groups

HAVING does in the GROUP BY what the WHERE clause does in a SELECT. It restricts the number of groups reported, whereas WHERE restricts the number of rows reported. Used with built-in functions,

HAVING is always preceded by GROUP BY and is always followed by a function (SUM, AVG, MAX, MIN, OR COUNT).

- *Report the total value of stocks for nations with two or more listed stocks.*

```
SELECT natname, SUM(stkprice*stkqty*exchrate) AS stkvalue
   FROM stock JOIN nation ON stock.natcode = nation.natcode
      GROUP BY natname
         HAVING COUNT(*) >= 2;
```

natname	stkvalue
Australia	946431
United Kingdom	48908364
United States	30066066

Skill Builder

Report by nation the total value of dividends.

Regular expression—pattern matching

Regular expression was introduced in the previous chapter, and we will now continue to learn some more of its features.

Search for a string not containing specified characters

The ^ (carat) is the symbol for NOT. It is used when we want to find a string not containing a character in one or more specified strings. For example, [^a-f] means any character not in the set containing a, b, c, d, e, or f.

- *List the names of nations with non-alphabetic characters in their names*

```
SELECT natname FROM nation WHERE natname REGEXP '[^a-z|A-Z]'
```

natname
United Kingdom
United States

Notice that the nations reported have a space in their name, which is a character not in the range a-z and not in A-Z.

Search for string containing a repeated pattern or repetition

A pair of curly brackets is used to denote the repetition factor for a pattern. For example, {n} means repeat a specified pattern n times.

- *List the names of firms with a double 'e'.*

```
SELECT stkfirm FROM stock WHERE stkfirm REGEXP '[e]{2}';
```

stkfirm
Bolivian Sheep
Freedonia Copper
Narembeen Emu

stkfirm
Nigerian Geese

Search combining alternation and repetition

Regular expressions becomes very powerful when you combine several of the basic capabilities into a single search expression.

- *List the names of firms with a double 's' or a double 'n'.*

```
SELECT stkfirm FROM stock WHERE stkfirm REGEXP '[s]{2}|[n]{2}';
```

stkfirm
Abyssinian Ruby
Minnesota Gold

Search for multiple versions of a string

If you are interested in find a string containing several specified string, you can use the square brackets to indicate the sought strings. For example, [ea] means any character from the set containing e and a.

- *List the names of firms with names that include 'inia' or 'onia'.*

```
SELECT stkfirm FROM stock WHERE stkfirm REGEXP '[io]nia';
```

shrfirm
Abyssinian Ruby
Freedonia Copper
Patagonian Tea

Search for a string in a particular position

Sometimes you might be interested in identifying a string with a character in a particular position.

- *Find firms with 't' as the third letter of their name.*

```
SELECT stkfirm FROM stock WHERE stkfirm REGEXP '^(.){2}t';
```

shrfirm
Patagonian Tea

The regular expression has three elements:

^ indicates start searching at the beginning of the string;

(.){2} specifies that anything is acceptable for the next two characters;

t indicates what the next character, the third, must be.

You have seen a few of the features of a very powerful tool. To learn more about regular expressions, see regexlib.com,[17] which contains a library of regular expressions and a feature for finding expressions to solve specific problems. Check out the regular expression for checking whether a character string is a valid email address.

17 http://regexlib.com

Subqueries

A subquery, or nested SELECT, is a SELECT nested within another SELECT. A subquery can be used to return a list of values subsequently searched with an IN clause.

- *Report the names of all Australian stocks.*

```
SELECT stkfirm FROM stock
    WHERE natcode IN
        (SELECT natcode FROM nation
            WHERE natname = 'Australia');
```

stkfirm
Narembeen Emu
Queensland Diamond
Indooroopilly Ruby

Conceptually, the subquery is evaluated first. It returns a list of values for natcode ('AUS') so that the query then is the same as:

```
SELECT stkfirm FROM stock
    WHERE natcode IN ('AUS');
```

When discussing subqueries, sometimes a subquery is also called an **inner query**. The term **outer query** is applied to the SQL preceding the inner query. In this case, the outer and inner queries are:

Outer query	SELECT stkfirm FROM stock WHERE natcode IN
Inner query	(SELECT natcode FROM nation WHERE natname = 'AUSTRALIA');

Note that in this case we do not have to qualify natcode. There is no identity crisis, because natcode in the inner query is implicitly qualified as nation.natcode and natcode in the outer query is understood to be stock.natcode.

This query also can be run as a join by writing:

```
SELECT stkfirm FROM stock JOIN nation
    ON stock.natcode = nation.natcode
    AND natname = 'Australia';
```

Correlated subquery

In a correlated subquery, the subquery cannot be evaluated independently of the outer query. It depends on the outer query for the values it needs to resolve the inner query. The subquery is evaluated for each value passed to it by the outer query. An example illustrates when you might use a correlated subquery and how it operates.

- *Find those stocks where the quantity is greater than the average for that country.*

An approach to this query is to examine the rows of stock one a time, and each time compare the quantity of stock to the average for that country. This means that for each row, the subquery must receive the outer query's country code so it can compute the average for that country.

```
SELECT natname, stkfirm, stkqty FROM stock JOIN nation
   ON stock.natcode = nation.natcode
   WHERE stkqty >
      (SELECT avg(stkqty) FROM stock
         WHERE stock.natcode = nation.natcode);
```

natname	stkfirm	stkqty
Australia	Queensland Diamond	89251
United Kingdom	Bolivian Sheep	231678
United Kingdom	Royal Ostrich Farms	1234923
United States	Minnesota Gold	816122

Conceptually, think of this query as stepping through the join of `stock` and `nation` one row at a time and executing the subquery each time. The first row has `natcode = 'AUS'` so the subquery becomes

```
SELECT AVG(stkqty) FROM stock
   WHERE stock.natcode = 'AUS';
```

Since the average stock quantity for Australian stocks is 63,672.33, the first row in the join, Narembeen Emu, is not reported. Neither is the second row reported, but the third is.

The term **correlated subquery** is used because the inner query's execution depends on receiving a value for a variable (`nation.natcode` in this instance) from the outer query. Thus, the inner query of correlated subquery cannot be evaluated once and for all. It must be evaluated repeatedly—once for each value of the variable received from the outer query. In this respect, a correlated subquery is different from a subquery, where the inner query needs to be evaluated only once. The requirement to compare each row of a table against a function (e.g., average or count) for some rows of a column is usually a clue that you need to write a correlated subquery.

Skill builder

Why are no Indian stocks reported in the correlated subquery example? How would you change the query to report an Indian stock?

Report only the three stocks with the largest quantities (i.e., do the query without using `ORDER BY`).

Views—virtual tables

You might have noticed that in these examples we repeated the join and stock value calculation for each query. Ideally, we should do this once, store the result, and be able to use it with other queries. We can do so if we create a ***view***, a virtual table. A view does not physically exist as stored data; it is an imaginary table constructed from existing tables as required. You can treat a view as if it were a table and write SQL to query it.

A view contains selected columns from one or more tables. The selected columns can be renamed and rearranged. New columns based on arithmetic expressions can be created. `GROUP BY` can also be used to create a view. Remember, a view contains no actual data. It is a virtual table.

This SQL command does the join, calculates stock value, and saves the result as a view:

```
CREATE VIEW stkvalue
   (nation, firm, price, qty, exchrate, value)
   AS SELECT natname, stkfirm, stkprice, stkqty, exchrate,
      stkprice*stkqty*exchrate
```

```
        FROM stock JOIN nation
        ON stock.natcode = nation.natcode;
```

There are several things to notice about creating a view:

- The six names enclosed by parentheses are the column names for the view.
- There is a one-to-one correspondence between the names in parentheses and the names or expressions in the SELECT clause. Thus the view column named value contains the result of the arithmetic expression stkprice*stkqty*exchrate.

A view can be used in a query, such as:

- *Find stocks with a value greater than £100,000.*

```
SELECT nation, firm, value FROM stkvalue WHERE value > 100000;
```

nation	firm	value
United Kingdom	Freedonia Copper	289548
United Kingdom	Patagonian Tea	698084
United Kingdom	Abyssinian Ruby	700358
United Kingdom	Sri Lankan Gold	1655561
United Kingdom	Indian Lead & Zinc	241223
United Kingdom	Bolivian Sheep	2953895
United Kingdom	Nigerian Geese	431305
United Kingdom	Canadian Sugar	248910
United Kingdom	Royal Ostrich Farms	41678651
United States	Minnesota Gold	29456210
United States	Georgia Peach	609856
Australia	Narembeen Emu	258952
Australia	Queensland Diamond	276303
Australia	Indooroopilly Ruby	411176

There are two main reasons for creating a view. *First,* as we have seen, query writing can be simplified. If you find that you are frequently writing the same section of code for a variety of queries, then isolate the common section and put it in a view. This means that you will usually create a view when a fact, such as stock value, is derived from other facts in the table.

The *second* reason is to restrict access to certain columns or rows. For example, the person who updates stock could be given a view that excludes stkqty. In this case, changes in stock prices could be updated without revealing confidential information, such as the value of the stock portfolio.

Skill builder

How could you use a view to solve the following query that was used when discussing the correlated subquery?

Find those stocks where the quantity is greater than the average for that country.

Summary

Entities are related to other entities by relationships. The 1:m (one-to-many) relationship occurs frequently in data models. An additional entity is required to represent a 1:m relationship to avoid update anomalies. In a relational database, a 1:m relationship is represented by an additional column, the foreign key, in the table at the many end of the relationship. The referential integrity constraint insists that a foreign key must always exist as a primary key in a table. A foreign key constraint is specified in a `CREATE` statement.

Join creates a new table from two existing tables by matching on a column common to both tables. Often the common column is a primary key–foreign key combination. A theta-join can have matching conditions of =, <>, <=, <, >=, and >. An equijoin describes the situation where the matching condition is equality. The `GROUP BY` clause is used to create an elementary control break report. The `HAVING` clause of `GROUP BY` is like the `WHERE` clause of `SELECT`. A subquery, which has a `SELECT` statement within another `SELECT` statement, causes two `SELECT` statements to be executed—one for the inner query and one for the outer query. A correlated subquery is executed as many times as there are rows selected by the outer query. A view is a virtual table that is created when required. Views can simplify report writing and restrict access to specified columns or rows.

Key terms and concepts

Constraint

Control break reporting

Correlated subquery

Delete anomalies

Equijoin

Foreign key

GROUP BY

HAVING

Insert anomalies

JOIN

One-to-many (1:m) relationship

Referential integrity

Relationship

Theta-join

Update anomalies

Views

Virtual table

Exercises

1. Draw data models for the following situations. In each case, make certain that you show the attributes and feasible identifiers:
    a. A farmer can have many cows, but a cow belongs to only one farmer.
    b. A university has many students, and a student can attend at most one university.
    c. An aircraft can have many passengers, but a passenger can be on only one flight at a time.
    d. A nation can have many states and a state many cities.
    e. An art researcher has asked you to design a database to record details of artists and the museums in which their paintings are displayed. For each painting, the researcher wants to know the size of the canvas, year painted, title, and style. The nationality, date of birth, and death of each artist must be recorded. For each museum, record details of its location and specialty, if it has one.
2. Report all values in British pounds:

a. Report the value of stocks listed in Australia.
b. Report the dividend payment of all stocks.
c. Report the total dividend payment by nation.
d. Create a view containing nation, firm, price, quantity, exchange rate, value, and yield.
e. Report the average yield by nation.
f. Report the minimum and maximum yield for each nation.
g. Report the nations where the average yield of stocks exceeds the average yield of all stocks.

3. How would you change the queries in exercise 4-2 if you were required to report the values in American dollars, Australian dollars, or Indian rupees?
4. What is a foreign key and what role does it serve?
5. What is the referential integrity constraint? Why should it be enforced?
6. Kisha, against the advice of her friends, is simultaneously studying data management and Shakespearean drama. She thought the two subjects would be an interesting contrast. However, the classes are very demanding and often enter her midsummer dreams. Last night, she dreamed that William Shakespeare wanted her to draw a data model. He explained, before she woke up in a cold sweat, that a play had many characters but the same character never appeared in more than one play. "Methinks," he said, "the same name may have appeareth more than the once, but 'twas always a person of a different ilk." He then, she hazily recollects, went on to spout about the quality of data dropping like the gentle rain. Draw a data model to keep old Bill quiet and help Kisha get some sleep.
7. An orchestra has four broad classes of instruments (strings, woodwinds, brass, and percussion). Each class contains musicians who play different instruments. For example, the strings section of a full symphony orchestra contains 2 harps, 16 to 18 first violins, 14 to 16 second violins, 12 violas, 10 cellos, and 8 double basses. A city has asked you to develop a database to store details of the musicians in its three orchestras. All the musicians are specialists and play only one instrument for one orchestra.
8. Answer the following queries based on the following database for a car dealer:

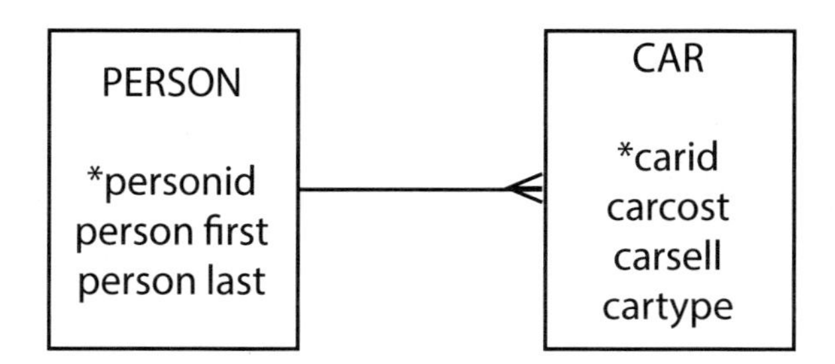

a. What is the personid of Sheila O'Hara?
b. List sales personnel sorted by last name and within last name, first name.
c. List details of the sales made by Bruce Bush.
d. List details of all sales showing the gross profit (selling price minus cost price).
e. Report the number of cars sold of each type.
f. What is the average selling price of cars sold by Sue Lim?

g. Report details of all sales where the gross profit is less than the average.

h. What was the maximum selling price of any car?

i. What is the total gross profit?

j. Report the gross profit made by each salesperson who sold at least three cars.

k. Create a view containing all the details in the CAR table and the gross profit

9. Find stocks where the third or fourth letter in their name is an 'm'.

10. An electricity supply company needs a database to record details of solar panels installed on its customers' homes so it can estimate how much solar energy will be generated based on the forecast level of solar radiation for each house's location. A solar panel has an area, measured in square meters, and an efficiency expressed as a percentage (e.g., 22% efficiency means that 22% of the incident solar energy is converted into electrical energy). Create a data model. How will you identify each customer and each panel?

CD library case

Ajay soon realized that a CD contains far more data that just a list of track titles and their length. Now that he had learned about the 1:m relationship, he was ready to add some more detail. He recognized that a CD contains many tracks, and a label (e.g., Atlantic) releases many CDs. He revised his data model to include these additional entities (CD and LABEL) and their relationships.

CD library V2.0

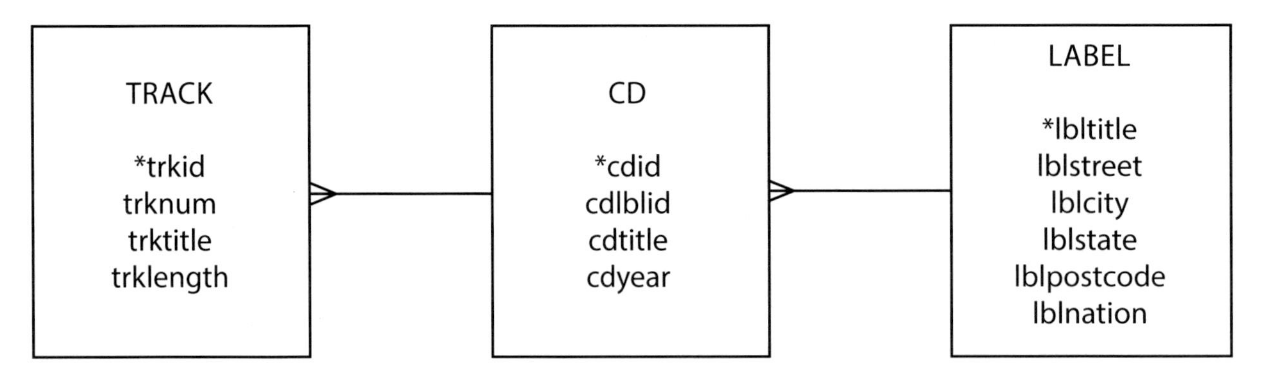

Giant Steps and *Swing* are both on the Atlantic label; the data are

*lbltitle	lblstreet	lblcity	lblstate	lblpostcode	lblnation
Atlantic	75 Rockefeller Plaza	New York	NY	10019	USA

The CD data, where `lbltitle` is the foreign key, are

*cdid	cdlblid	cdtitle	cdyear	*lbltitle*
1	A2 1311	Giant Steps	1960	Atlantic
2	83012-2	Swing	1977	Atlantic

The TRACK data, where *cdid* is the foreign key, are

trkid	trknum	trktitle	trklength	cdid
1	1	Giant Steps	4.72	1

trkid	trknum	trktitle	trklength	cdid
2	2	Cousin Mary	5.75	1
3	3	Countdown	2.35	1
4	4	Spiral	5.93	1
5	5	Syeeda's Song Flute	7	1
6	6	Naima	4.35	1
7	7	Mr. P.C.	6.95	1
8	8	Giant Steps	3.67	1
9	9	Naima	4.45	1
10	10	Cousin Mary	5.9	1
11	11	Countdown	4.55	1
12	12	Syeeda's Song Flute	7.03	1
13	1	Stomp of King Porter	3.2	2
14	2	Sing a Study in Brown	2.85	2
15	3	Sing Moten's Swing	3.6	2
16	4	A-tisket, A-tasket	2.95	2
17	5	I Know Why	3.57	2
18	6	Sing You Sinners	2.75	2
19	7	Java Jive	2.85	2
20	8	Down South Camp Meetin'	3.25	2
21	9	Topsy	3.23	2
22	10	Clouds	7.2	2
23	11	Skyliner	3.18	2
24	12	It's Good Enough to Keep	3.18	2
25	13	Choo Choo Ch' Boogie	3	2

Skill builder

Create tables to store the label and CD data. Add a column to `track` to store the foreign key. Then, either enter the new data or download it from the Web site. Define `lbltitle` as a foreign key in `cd` and `cdid` as a foreign key in `track`.

Ajay ran a few queries on his revised database.

- *What are the tracks on Swing?*

This query requires joining `track` and `cd` because the name of a CD (*Swing* in this case) is stored in `cd` and track data are stored in `track`. The foreign key of `track` is matched to the primary key of `cd` (i.e., `track.cdid = cd.cdid`).

```
SELECT trknum, trktitle, trklength FROM track JOIN cd
   ON track.cdid = cd.cdid
   WHERE cdtitle = 'Swing';
```

trknum	trktitle	trklength
1	Stomp of King Porter	3.2
2	Sing a Study in Brown	2.85

3	Sing Moten's Swing	3.6
4	A-tisket, A-tasket	2.95
5	I Know Why	3.57
6	Sing You Sinners	2.75
7	Java Jive	2.85
8	Down South Camp Meetin'	3.25
9	Topsy	3.23
10	Clouds	7.2
11	Skyliner	3.18
12	It's Good Enough to Keep	3.18
13	Choo Choo Ch' Boogie	3

- *What is the longest track on Swing?*

Like the prior query, this one requires joining `track` and `cd`. The inner query isolates the tracks on *Swing* and then selects the longest of these using the `MAX` function. The outer query also isolates the tracks on *Swing* and selects the track equal to the maximum length. The outer query must restrict attention to tracks on *Swing* in case there are other tracks in the `track` table with a time equal to the value returned by the inner query.

```
SELECT trknum, trktitle, trklength FROM track JOIN cd
   ON track.cdid = cd.cdid
   WHERE cdtitle = 'Swing'
   AND trklength =
      (select max(trklength) FROM track, cd
         WHERE track.cdid = cd.cdid
         AND cdtitle = 'Swing');
```

trknum	trktitle	trklength
10	Clouds	7.2

- *What are the titles of CDs containing some tracks less than 3 minutes long and on U.S.-based labels? List details of these tracks as well.*

This is a three-table join since resolution of the query requires data from `cd` (`cdtitle`), `track` (`trktitle`), and `label` (`lblnation`).

```
SELECT cdtitle, trktitle, trklength FROM track JOIN cd
   ON track.cdid = cd.cdid
   JOIN label
   ON cd.lbltitle = label.lbltitle
   WHERE trklength < 3
   AND lblnation = 'USA';
```

cdtitle	trktitle	trklength
Giant Steps	Countdown	2.35
Swing	Sing a Study in Brown	2.85
Swing	A-tisket, A-tasket	2.95
Swing	Sing You Sinners	2.75
Swing	Java Jive	2.85

- *Report the number of tracks on each CD.*

This query requires GROUP BY because the number of tracks on each CD is totaled by using the aggregate function COUNT.

```
SELECT cdtitle, count(*) AS tracks FROM track JOIN cd
   ON track.cdid = cd.cdid
      GROUP BY cdtitle;
```

cdtitle	tracks
Giant Steps	12
Swing	13

- *Report, in ascending order, the total length of tracks on each CD.*

This is another example of GROUP BY. In this case, SUM is used to accumulate track length, and ORDER BY is used for sorting in ascending order, the default sort order.

```
SELECT cdtitle, SUM(trklength) AS sumtrklen FROM track JOIN cd
   ON track.cdid = cd.cdid
      GROUP BY cdtitle
         ORDER BY SUM(trklength);
```

cdtitle	sumtrklen
Swing	44.81
Giant Steps	62.65

- *Does either CD have more than 60 minutes of music?*

This query uses the HAVING clause to limit the CDs reported based on their total length.

```
SELECT cdtitle, SUM(trklength) AS sumtrklen FROM track JOIN cd
   ON track.cdid = cd.cdid
      GROUP BY cdtitle HAVING SUM(trklength) > 60;
```

cdtitle	sumtrklen
Giant Steps	62.65

Skill builder

1. Write SQL for the following queries:
   a. List the tracks by CD in order of track length.
   b. What is the longest track on each CD?
2. What is wrong with the current data model?
3. Could `cdlblid` be used as an identifier for CD?

5. The Many-to-Many Relationship

Fearful concatenation of circumstances.

Daniel Webster

Learning objectives

Students completing this chapter will be able to

- model a many-to-many relationship between two entities;
- define a database with a many-to-many relationship;
- write queries for a database with a many-to-many relationship.

Alice was deeply involved in uncovering the workings of her firm. She spent hours talking to the staff, wanting to know everything about the products sold. She engaged many a customer in conversation to find out why they shopped at The Expeditioner, what they were looking for, and how they thought the business could be improved. She examined all the products herself and pestered the staff to tell her who bought them, how they were used, how many were sold, who supplied them, and a host of other questions. She plowed (or should that be "ploughed") through accounting journals, sales reports, and market forecasts. She was more than a new broom; she was a giant vacuum cleaner sucking up data about the firm so that she would be prepared to manage it successfully.

Ned, the marketing manager, was a Jekyll and Hyde character. By day he was a charming, savvy, marketing executive. Walking to work in his three-piece suit, bowler hat, and furled umbrella, he was the epitome of the conservative English businessman. By night he became a computer nerd with ragged jeans, a T-shirt, black-framed glasses, and unruly hair. Ned had been desperate to introduce computers to The Expeditioner for years, but he knew that he could never reveal his second nature. The other staff at The Expeditioner just would not accept such a radical change in technology. Why, they had not even switched to fountain pens until their personal stock of quill pens was finally depleted. Ned was just not prepared to face the indignant silence and contempt that would greet his mere mention of computers. It was better to continue a secret double life than admit to being a cyber punk.

But times were changing, and Ned saw the opportunity. He furtively suggested to Lady Alice that perhaps she needed a database of sales facts. Her response was instantaneous: "Yes, and do it as soon as possible." Ned was ecstatic. This was truly a wonderful woman.

The many-to-many relationship

Consider the case when items are sold. We can immediately identify two entities: SALE and ITEM. A sale can contain many items, and an item can appear in many sales. We are not saying the same item can be sold many times, but the particular type of item (e.g., a compass) can be sold many times; thus we have a many-to-many (m:m) relationship between SALE and ITEM. When we have an m:m relationship, we create a third entity to link the entities through two 1:m relationships. Usually, it is fairly easy to name this third entity. In this case, this third entity, typically known as an **associative entity**, is called LINE ITEM. A typical sales form lists the items purchased by a customer. Each of the lines appearing on the order form is generally known in retailing as a line item, which links an item and a sale.

A sales form

The Expeditioner Sale of Goods					
Sale#			Date:		
	Item#	Description	Quantity	Unit price	Total
1					
2					
3					
4					
5					
6					
				Grand total	

The representation of this m:m relationship is shown. We say many-to-many because there are two relationships—an ITEM is related to many SALEs, and a SALE is related to many ITEMs. This data model can also be read as: "a sale has many line items, but a line item refers to only one sale. Similarly, an item can appear as many line items, but a line item references only one item."

An m:m relationship between SALE and ITEM

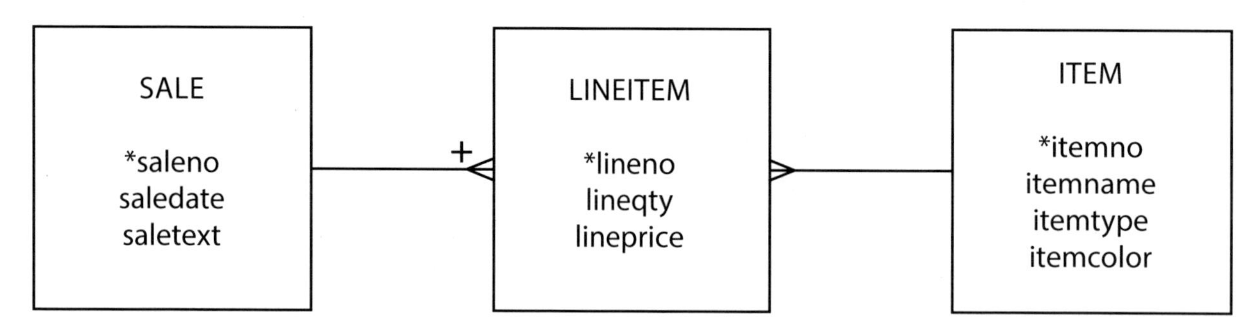

The entity SALE is identified by *saleno* and has the attributes *saledate* and *saletext* (a brief comment on the customer—soft information). LINEITEM is partially identified by *lineno* and has attributes *lineqty* (the number of units sold) and *lineprice* (the unit selling price for this sale). ITEM is identified by *itemno* and has attributes *itemname*, *itemtype* (e.g., clothing, equipment, navigation aids, furniture, and so on), and *itemcolor*.

If you look carefully at the m:m relationship figure, you will notice that there is a plus sign (+) above the crow's foot at the "many" end of the 1:m relationship between SALE and LINEITEM. This plus sign provides information about the identifier of LINEITEM. As you know, every entity must have a unique identifier. A sales order is a series of rows or lines, and *lineno* is unique only within a particular order. If we just use *lineno* as the identifier, we cannot guarantee that every instance of LINEITEM is unique. If we use *saleno* and *lineno* together, however, we have a unique identifier for every instance of LINEITEM. Identifier *saleno* is unique for every sale, and *lineno* is unique within any sale. The plus indicates that LINEITEM's unique identifier is the concatenation of *saleno* and *lineno*. The order of concatenation does not matter.

LINEITEM is termed a **weak entity** because it relies on another entity for its existence and identification.

MySQL Workbench

Workbench automatically creates an associative entity for an m:m relationship and populates it with a composite primary key based on concatenating the primary keys of the two entities forming the m:m relationship. *First*, draw the two tables and enter their respective primary keys and columns. *Second*, select the m:m symbol and connect the two tables through clicking on one and dragging to the second and releasing. You can then modify the associative entity as required, such as changing its primary key. The capability to automatically create an associative entity for an m:m relationship is a very useful Workbench feature.

An m:m relationship with Workbench

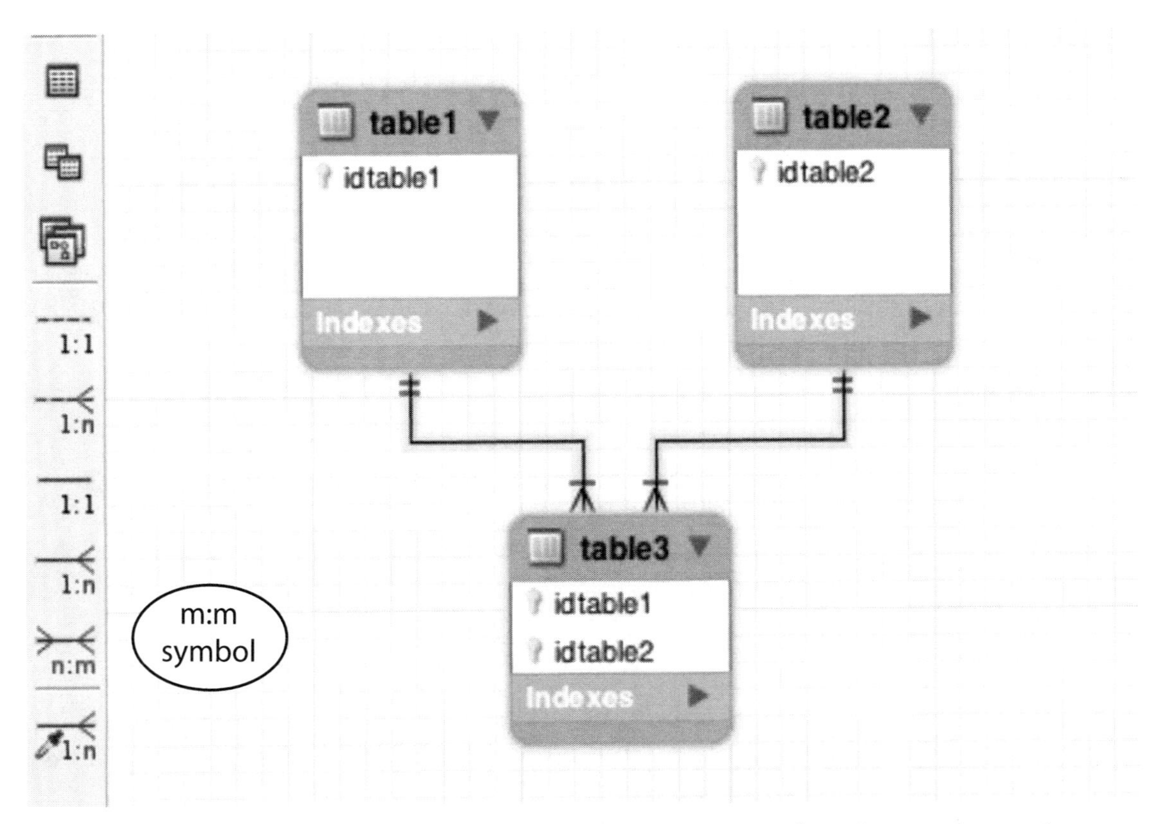

Workbench distinguishes between two types of relationships. An **identifying relationship**, shown by a solid line, is used when the entity at the many end of the relationship is a weak entity and needs the identifier of the one end of the relationship to uniquely identify an instance of the relationship, as in LINEITEM. An identifying relationship corresponds to the + sign associated with a crow's foot. The other type of relationship, shown by a dashed line, is known as a *non-identifying* relationship. The mapping between the type of relationship and the representation (i.e., dashed or solid line) is arbitrary and thus not always easily recalled. We think that using a + on the crow's foot is a better way of denoting weak entities.

When the relationship between SALE and ITEM is drawn in Workbench, as shown in the following figure, there are two things to notice. *First*, the table, `LINEITEM`, maps the associative entity generated for the m:m relationship. *Second*, `LINEITEM` has an identifying relationship with `SALE` and a non-identifying relationship with `ITEM`.

An m:m relationship between SALE and ITEM in MySQL Workbench

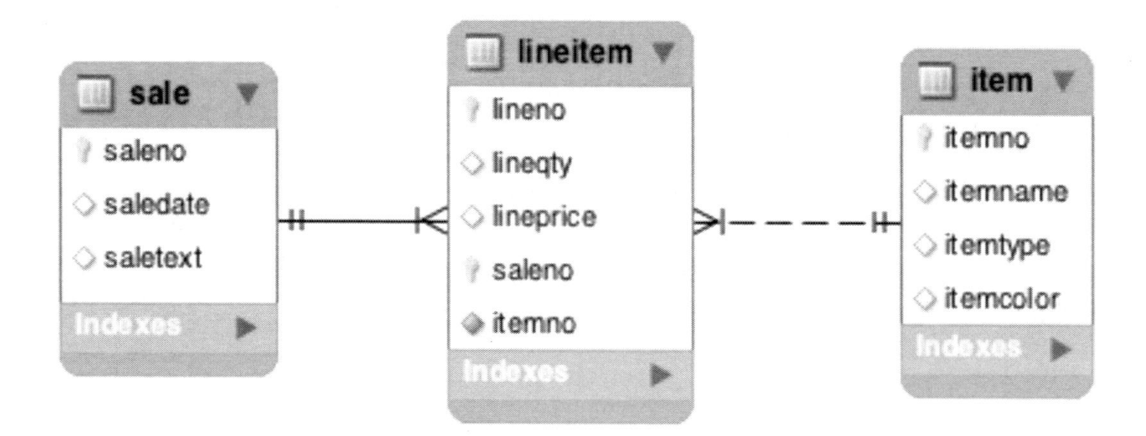

Why did we create a third entity?

When we have an m:m relationship, we create an associative entity to store data about the relationship. In this case, we have to store data about the items sold. We cannot store the data with SALE because a sale can be many items, and an instance of an entity stores only single-value facts. Similarly, we cannot store data with ITEM because an item can appear in many sales. Since we cannot store data in SALE or ITEM, we must create another entity to store data about the m:m relationship.

You might find it useful to think of the m:m relationship as two 1: m relationships. An item can appear on many line item listings, and a line item entry refers to only one item. A sale has many line items, and each line item entry refers to only one sale.

Social Security number is not unique!

Two girls named Sarah Lee Ferguson were born on May 3, 1959. The U.S. government considered them one and the same and issued both the same Social Security number (SSN), a nine-digit identifier of U.S. residents. Now Sarah Lee Ferguson Boles and Sarah Lee Ferguson Johnson share the same SSN.[18]

Mrs. Boles became aware of her SSN twin in 1987 when the Internal Revenue Service claimed there was a discrepancy in her reported income. Because SSN is widely used as an attribute or identifier in many computer systems, Mrs. Boles encountered other incidents of mistaken identity. Some of Mrs. Johnson's purchases appeared on Mrs. Boles' credit reports.

In late 1989, the Social Security Administration notified Mrs. Boles that her original number was given to her in error and she had to provide evidence of her age, identity, and citizenship to get a new number. When Mrs. Boles got her new SSN, it is likely she had to also get a new driver's license and establish a new credit history.

Creating a relational database with an m:m relationship

As before, each entity becomes a table in a relational database, the entity name becomes the table name, each attribute becomes a column, and each identifier becomes a primary key. Remember, a 1:m relationship is mapped by adding a column to the entity of the many end of the relationship. The new column contains the identifier of the one end of the relationship.

18 "Two women share a name, birthday, and S.S. number!" *Athens [Georgia] Daily News*, January 29 1990, 7A. Also, see wnyt.com/article/stories/S1589778.shtml?cat=10114

Conversion of the foregoing data model results in the three tables following. Note the one-to-one correspondence between attributes and columns for `sale` and `item`. Observe the `lineitem` has two additional columns, `saleno` and `itemno`. Both of these columns are foreign keys in `lineitem` (remember the use of italics to signify foreign keys). Two foreign keys are required to record the two 1:m relationships. Notice in `lineitem` that `saleno` is both part of the primary key and a foreign key.

Tables sale, lineitem, and item

sale		
*saleno	**saledate**	**saletext**
1	2003-01-15	Scruffy Australian—called himself Bruce.
2	2003-01-15	Man. Rather fond of hats.
3	2003-01-15	Woman. Planning to row Atlantic—lengthwise!
4	2003-01-15	Man. Trip to New York—thinks NY is a jungle!
5	2003-01-16	Expedition leader for African safari.

lineitem				
*lineno	lineqty	lineprice	**saleno*	*itemno*
1	1	4.5	1	2
1	1	25	2	6
2	1	20	2	16
3	1	25	2	19
4	1	2.25	2	2
1	1	500	3	4
2	1	2.25	3	2
1	1	500	4	4
2	1	65	4	9
3	1	60	4	13
4	1	75	4	14
5	1	10	4	3
6	1	2.25	4	2
1	50	36	5	10
2	50	40.5	5	11
3	8	153	5	12
4	1	60	5	13
5	1	0	5	2

item			
*itemno	itemname	itemtype	itemcolor
1	Pocket knife—Nile	E	Brown
2	Pocket knife—Avon	E	Brown
3	Compass	N	—
4	Geopositioning system	N	—
5	Map measure	N	—
6	Hat—Polar Explorer	C	Red
7	Hat—Polar Explorer	C	White
8	Boots—snake proof	C	Green
9	Boots—snake proof	C	Black
10	Safari chair	F	Khaki
11	Hammock	F	Khaki
12	Tent—8 person	F	Khaki
13	Tent—2 person	F	Khaki
14	Safari cooking kit	E	—
15	Pith helmet	C	Khaki
16	Pith helmet	C	White
17	Map case	N	Brown
18	Sextant	N	—
19	Stetson	C	Black
20	Stetson	C	Brown

The SQL commands to create the three tables are as follows:

```
CREATE TABLE sale (
   saleno     INTEGER,
   saledate      DATE NOT NULL,
   saletext      VARCHAR(50),
      PRIMARY KEY(saleno));
CREATE TABLE item (
   itemno     INTEGER,
   itemname      VARCHAR(30) NOT NULL,
   itemtype      CHAR(1) NOT NULL,
   itemcolor     VARCHAR(10),
      PRIMARY KEY(itemno));
CREATE TABLE lineitem (
   lineno     INTEGER,
   lineqty       INTEGER NOT NULL,
   lineprice     DECIMAL(7,2) NOT NULL,
   saleno     INTEGER,
   itemno     INTEGER,
      PRIMARY KEY(lineno,saleno),
      CONSTRAINT fk_has_sale FOREIGN KEY(saleno)
         REFERENCES sale(saleno),
      CONSTRAINT fk_has_item FOREIGN KEY(itemno)
         REFERENCES item(itemno));
```

Although the sale and item tables are created in a similar fashion to previous examples, there are two things to note about the definition of lineitem. *First,* the primary key is a composite of lineno and saleno, because together they uniquely identify an instance of lineitem. *Second,* there are two foreign keys, because lineno is at the "many" end of two 1: m relationships.

Skill builder

A hamburger shop makes several types of hamburgers, and the same type of ingredient can be used with several types of hamburgers. This does not literally mean the same piece of lettuce is used many times, but lettuce is used with several types of hamburgers. Draw the data model for this situation. What is a good name for the associative entity?

Querying an m:m relationship

A three-table join

The join operation can be easily extended from two tables to three or more merely by specifying the tables to be joined and the matching conditions. For example:

```
SELECT * FROM sale JOIN lineitem
    ON sale.saleno = lineitem.saleno
    JOIN item
    ON item.itemno = lineitem.itemno;
```

There are two matching conditions: one for sale and lineitem (sale.saleno = lineitem.saleno) and one for the ITEM and lineitem tables (item.itemno = lineitem.itemno). The table lineitem is the link between sale and item and must be referenced in both matching conditions.

You can tailor the join to be more precise and report some columns rather than all.

- *List the name, quantity, price, and value of items sold on January 16, 2011.*

```
SELECT itemname, lineqty, lineprice, lineqty*lineprice AS total
    FROM sale, lineitem, item
        WHERE lineitem.saleno = sale.saleno
        AND item.itemno = lineitem.itemno
        AND saledate = '2011-01-16';
```

itemname	lineqty	lineprice	total
Pocket knife—Avon	1	0	0
Safari chair	50	36	1800
Hammock	50	40.5	2025
Tent— 8 person	8	153	1224
Tent— 2 person	1	60	60

EXISTS—does a value exist

EXISTS is used in a WHERE clause to test whether a table contains at least one row satisfying a specified condition. It returns the value **true** if and only if some row satisfies the condition; otherwise it returns **false**. EXISTS represents the **existential quantifier** of formal logic. The best way to get a feel for EXISTS is to examine a query.

- *Report all clothing items (type "C") for which a sale is recorded.*

```
SELECT itemname, itemcolor FROM item
   WHERE itemtype = 'C'
   AND EXISTS (SELECT * FROM lineitem
      WHERE lineitem.itemno = item.itemno);
```

itemname	itemcolor
Hat—Polar Explorer	Red
Boots—snake proof	Black
Pith helmet	White
Stetson	Black

Conceptually, we can think of this query as evaluating the subquery for each row of item. The first item with itemtype = 'C', Hat–Polar Explorer (red), in item has itemno = 6. Thus, the query becomes

```
SELECT itemname, itemcolor FROM item
   WHERE itemtype = 'C'
   AND EXISTS (SELECT * FROM lineitem
      WHERE lineitem.itemno = 6);
```

Because there is at least one row in lineitem with itemno = 6, the subquery returns *true*. The item has been sold and should be reported. The second clothing item, Hat–Polar Explorer (white), in item has itemno = 7. There are no rows in lineitem with itemno = 7, so the subquery returns *false*. That item has not been sold and should not be reported.

You can also think of the query as, "Select clothing items for which a sale exists." Remember, for EXISTS to return *true*, there needs to be only one row for which the condition is *true*.

NOT EXISTS—select a value if it does not exist

NOT EXISTS is the negative of EXISTS. It is used in a WHERE clause to test whether all rows in a table fail to satisfy a specified condition. It returns the value *true* if there are no rows satisfying the condition; otherwise it returns *false*.

- *Report all clothing items that have not been sold.*

```
SELECT itemname, itemcolor FROM item
   WHERE itemtype = 'C'
   AND NOT EXISTS
      (SELECT * FROM lineitem
         WHERE item.itemno = lineitem.itemno);
```

If we consider this query as the opposite of that used to illustrate EXISTS, it seems logical to use NOT EXISTS. Conceptually, we can also think of this query as evaluating the subquery for each row of ITEM. The first item with itemtype = 'C', Hat–Polar Explorer (red), in item has itemno = 6. Thus, the query becomes

```
SELECT itemname, itemcolor FROM item
   WHERE itemtype = 'C'
      AND NOT EXISTS
         (SELECT * FROM lineitem
            WHERE lineitem.itemno = 6);
```

There is at least one row in lineitem with itemno = 6, so the subquery returns *true*. The NOT before EXISTS then negates the *true* to give *false*; the item will not be reported because it has been sold.

The second item with itemtype = 'C', Hat–Polar Explorer (white), in item has itemno = 7. The query becomes

```
SELECT itemname, itemcolor FROM item
   WHERE itemtype = 'C'
   AND NOT EXISTS
      (SELECT * FROM lineitem
         WHERE lineitem.itemno = 7);
```

Because there are no rows in lineitem with itemno = 7, the subquery returns *false*, and this is negated by the NOT before EXISTS to give *true*. The item has not been sold and should be reported.

itemname	itemcolor
Hat — Polar Explorer	White
Boots — snake proof	Green
Pith helmet	Khaki
Stetson	Brown

You can also think of the query as, "Select clothing items for which no sales exist." Also remember, for NOT EXISTS to return *true*, no rows should satisfy the condition.

Skill builder

Report all red items that have not been sold. Write the query twice, once using EXISTS and once without EXISTS.

Divide (and be conquered)

In addition to the existential quantifier that you have already encountered, formal logic has a **universal quantifier** known as **forall** that is necessary for queries such as

- *Find the items that have appeared in all sales.*

If a universal quantifier were supported by SQL, this query could be phrased as, "Select item names where *forall* sales there *exists* a lineitem row recording that this item was sold." A quick inspection of the first set of tables shows that one item satisfies this condition (itemno = 2).

While SQL does not directly support the universal quantifier, formal logic shows that *forall* can be expressed using EXISTS. The query becomes, "Find items such that there does not exist a sale in which this item does not appear." The equivalent SQL expression is

```
SELECT itemno, itemname FROM item
   WHERE NOT EXISTS
      (SELECT * FROM sale
         WHERE NOT EXISTS
            (SELECT * FROM lineitem
               WHERE lineitem.itemno = item.itemno
               AND lineitem.saleno = sale.saleno));
```

itemname
Pocket knife – Avon

If you are interested in learning the inner workings of the preceding SQL for divide, see the additional material for Chapter 5 on the book's Web site.

Relational algebra (Chapter 9) has the divide operation, which makes divide queries easy to write. Be careful: Not all queries containing the word *all* are divides. With experience, you will learn to recognize and conquer divide.

To save the tedium of formulating this query from scratch, we have developed a template for dealing with these sorts of queries. Divide queries typically occur with m:m relationships.

A template for divide

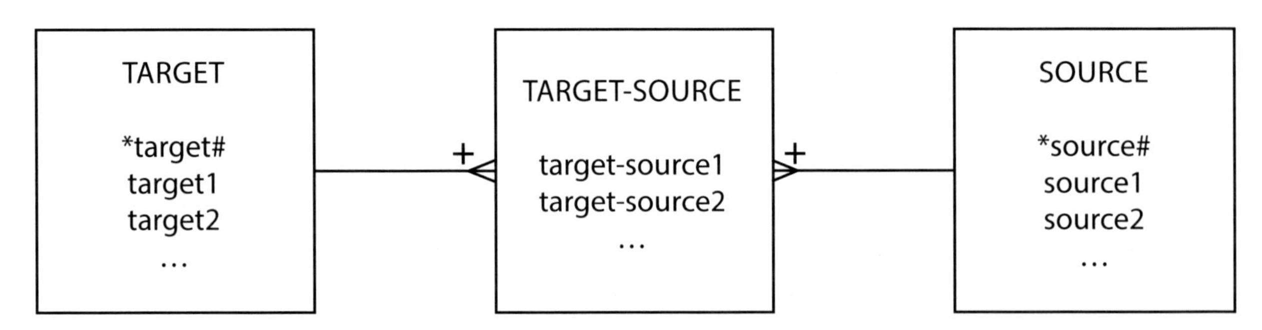

An appropriate generic query and template SQL command are

- *Find the target1 that have appeared in all sources.*

```
SELECT target1 FROM target
   WHERE NOT EXISTS
      (SELECT * FROM source
         WHERE NOT EXISTS
            (SELECT * FROM target-source
               WHERE target-source.target# = target.target#
               AND target-source.source# = source.source#));
```

Skill builder

Find the brown items that have appeared in all sales.

Beyond the great divide

Divide proves troublesome to most people because of the double negative—we just don't think that way. If divide sends your neurons into knots, then try the following approach.

The query "Find the items that have appeared in all sales" can be rephrased as "Find all the items for which the number of sales that include this item is equal to the total number of sales." This is an easier query to write than "Find items such that there does not exist a sale in which this item does not appear." The rephrased query has two parts. *First,* determine the total number of sales. Here we mean distinct sales (i.e., the number of rows with a distinct value for `saleno`). The SQL is

```
SELECT COUNT (DISTINCT saleno) FROM sale;
```

Second, group the items sold by `itemno` and `itemname` and use a `HAVING` clause with `COUNT` to calculate the number of sales in which the item has occurred. Forcing the count in the `HAVING` clause to equal the result of the first query, which becomes an inner query, results in a list of items appearing in all sales.

```
SELECT item.itemno, item.itemname
   FROM item JOIN lineitem
      ON item.itemno = lineitem.itemno
         GROUP BY item.itemno, item.itemname
            HAVING COUNT(DISTINCT saleno)
               = (SELECT COUNT(DISTINCT saleno) FROM sale);
```

Set operations

Set operators are useful for combining the values derived from two or more SQL queries. The `UNION` operation is equivalent to *or*, and `INTERSECT` is equivalent to *and.*

- *List items that were sold on January 16, 2011, or are brown.*

Resolution of this query requires two tables: one to report items sold on January 16, 2011, and one to report the brown items. `UNION` (i.e., or) then combines the results of the tables, including *any* rows in both tables and excluding duplicate rows.

```
SELECT itemname FROM item JOIN lineitem
   ON item.itemno = lineitem.itemno
   JOIN sale
   ON lineitem.saleno = sale.saleno
   WHERE saledate = '2011-01-16'
UNION
   SELECT itemname FROM item WHERE itemcolor = 'Brown';
```

itemname
Hammock
Map case
Pocket knife—Avon
Pocket knife—Nile
Safari chair
Stetson

itemname
Tent—2 person
Tent—8 person

- *List items that were sold on January 16, 2011, and are brown.*

This query uses the same two tables as the previous query. In this case, INTERSECT (i.e., and) then combines the results of the tables including ***only*** rows in both tables and excluding duplicates.[19]

```
SELECT itemname FROM item JOIN lineitem
   ON item.itemno = lineitem.itemno
   JOIN sale
   ON lineitem.saleno = sale.saleno
   WHERE saledate = '2011-01-16'
INTERSECT
   SELECT itemname FROM item WHERE itemcolor = 'Brown';
```

itemname
Pocket knife—Avon

Skill builder

List the items that contain the words “Hat”, “Helmet”, or “Stetson” in their names.

Summary

There can be a many-to-many (m:m) relationship between entities, which is represented by creating an associative entity and two 1:m relationships. An associative entity stores data about an m: m relationship. The join operation can be extended from two tables to three or more tables. EXISTS tests whether a table has at least one row that meets a specified condition. NOT EXISTS tests whether all rows in a table do not satisfy a specified condition. Both EXISTS and NOT EXISTS can return *true* or *false*. The relational operation divide, also known as *forall*, can be translated into a double negative. It is represented in SQL by a query containing two NOT EXISTS statements. Set operations enable the results of queries to be combined.

Key terms and concepts

Associative entity

Divide

Existential quantifier

EXISTS

INTERSECT

Many-to-many (m:m) relationship

NOT EXISTS

UNION

Universal quantifier

Exercises

19 MySQL does not support INTERSECT. Use another AND in the WHERE statement.

1. Draw data models for the following situations. In each case, think about the names you give each entity:

   a. Farmers can own cows or share cows with other farmers.

   b. A track and field meet can have many competitors, and a competitor can participate in more than one event.

   c. A patient can have many physicians, and a physician can have many patients.

   d. A student can attend more than one class, and the same class can have many students.

   e. *The Marathoner*, a monthly magazine, regularly reports the performance of professional marathon runners. It has asked you to design a database to record the details of all major marathons (e.g., Boston, London, and Paris). Professional marathon runners compete in several races each year. A race may have thousands of competitors, but only about 200 or so are professional runners, the ones *The Marathoner* tracks. For each race, the magazine reports a runner's time and finishing position and some personal details such as name, gender, and age.

2. The data model shown was designed by a golf statistician. Write SQL statements to create the corresponding relational database.

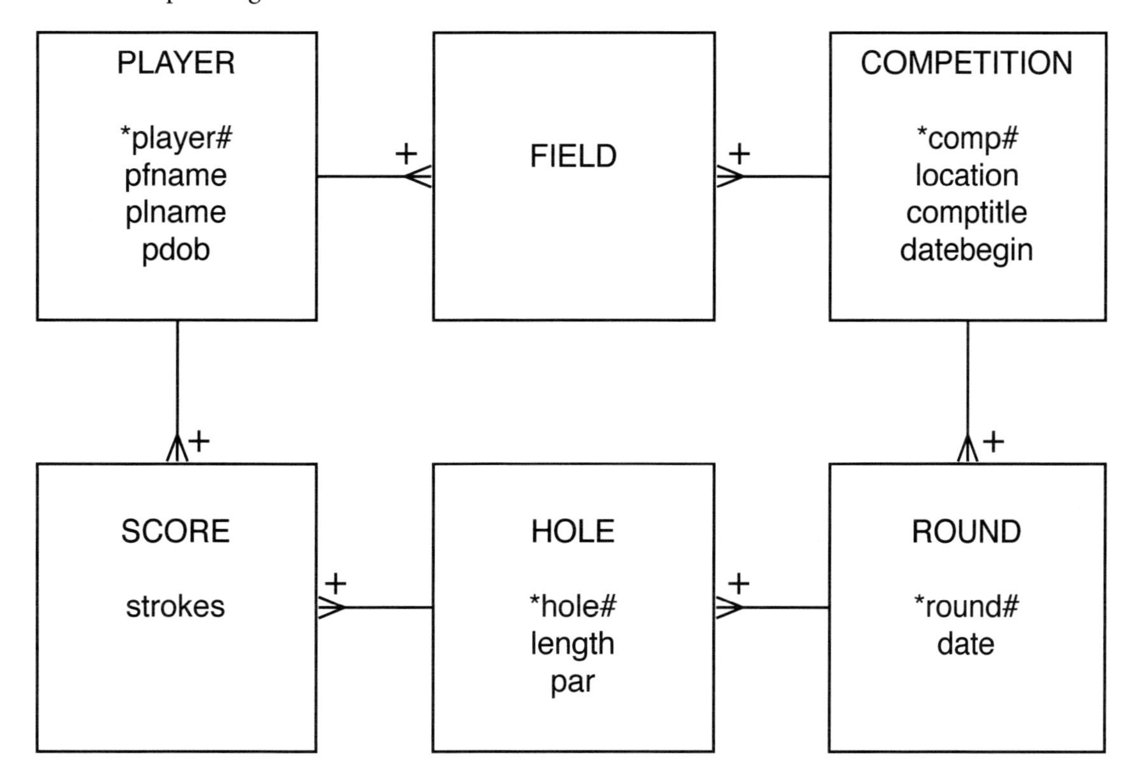

3. Write the following SQL queries for the database described in this chapter:

   a. List the names of items for which the quantity sold is greater than one for any sale.

   b. Compute the total value of sales for each item by date.

   c. Report all items of type "F" that have been sold.

   d. List all items of type "F" that have not been sold.

   e. Compute the total value of each sale.

4. Why do you have to create a third entity when you have an m:m relationship?

5. What does a plus sign near a relationship arc mean?
6. How does `EXISTS` differ from other clauses in an SQL statement?
7. Answer the following queries based on the described relational database.

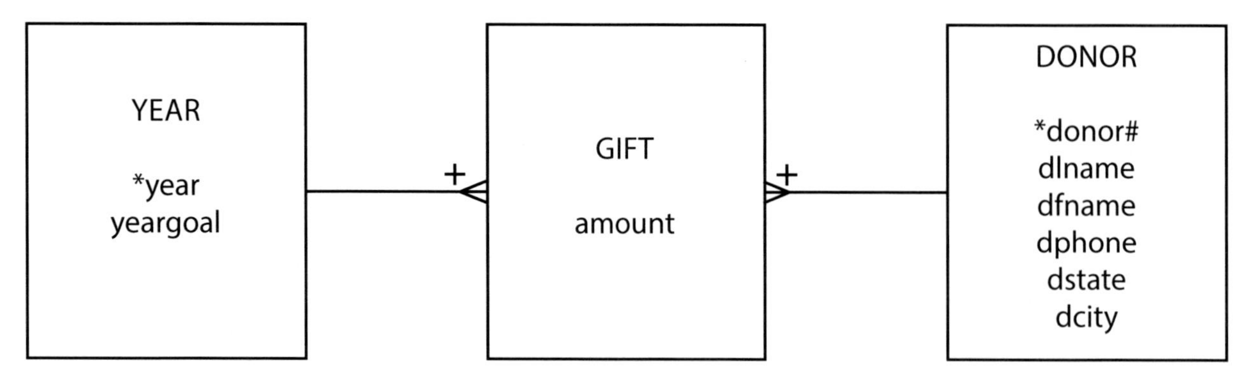

   a. List the phone numbers of donors Hays and Jefts.
   b. How many donors are there in the donor table?
   c. How many people made donations in 1999?
   d. What is the name of the person who made the largest donation in 1999?
   e. What was the total amount donated in 2000?
   f. List the donors who have made a donation every year.
   g. List the donors whose average donation is more than twice the average donation of all donors.
   h. List the total amount given by each person across all years; sort the report by the donor's name.
   i. Report the total donations in 2001 by state.
   j. In which years did the total donated exceed the goal for the year?
8. The following table records data found on the side of a breakfast cereal carton. Use these data as a guide to develop a data model to record nutrition facts for a meal. In this case, a meal is a cup of cereal and 1/2 cup of skim milk.

Nutrition facts

Serving size 1 cup (30g)

Servings per container about 17

Amount per serving	Cereal	with 1/2 cup of skim milk
Calories	110	150
Calories from Fat	10	10
		% Daily Value
Total Fat 1g	1%	2%
Saturated Fat 0g	0%	0%
Polyunsaturated Fat 0g		
Monounsaturated Fat 0g		
Cholesterol 0mg	0%	1%
Sodium 220mg	9%	12%
Potassium 105 mg	3%	9%
Total Carbohydrate 24g	8%	10%
Dietary Fiber 3g	13%	13%
Sugars 4g		
Other Carbohydrate 17g		
Protein 3g		
Vitamin A	10%	15%
Vitamin C	10%	10%
Calcium	2%	15%
Iron	45%	45%
Vitamin D	10%	25%
Thiamin	50%	50%
Riboflavin	50%	50%
Niacin	50%	50%
Vitamin B12	50%	60%
Phosphorus	10%	20%
Magnesium	8%	10%
Zinc	50%	50%
Copper	4%	4%

CD library case

After learning how to model m:m relationships, Ajay realized he could now model a number of situations that had bothered him. He had been puzzling over some relationships that he had recognized but was unsure how to model:

- A person can appear on many tracks, and a track can have many persons (e.g., *The Manhattan Transfer*, a quartet, has recorded many tracks).
- A person can compose many songs, and a song can have multiple composers (e.g., Rodgers and Hammerstein collaborated many times with Rodgers composing the music and Hammerstein writing the lyrics).
- An artist can release many CDs, and a CD can feature several artists (e.g., John Coltrane, who has many CDs, has a CD, *Miles & Coltrane,* with Miles Davis).

Ajay revised his data model to include the m:m relationships he had recognized. He wrestled with what to call the entity that stored details of people. Sometimes these people are musicians; other times singers, composers, and so forth. He decided that the most general approach was to call the entity PERSON. The extended data model is shown in the following figure.

CD library V3.0

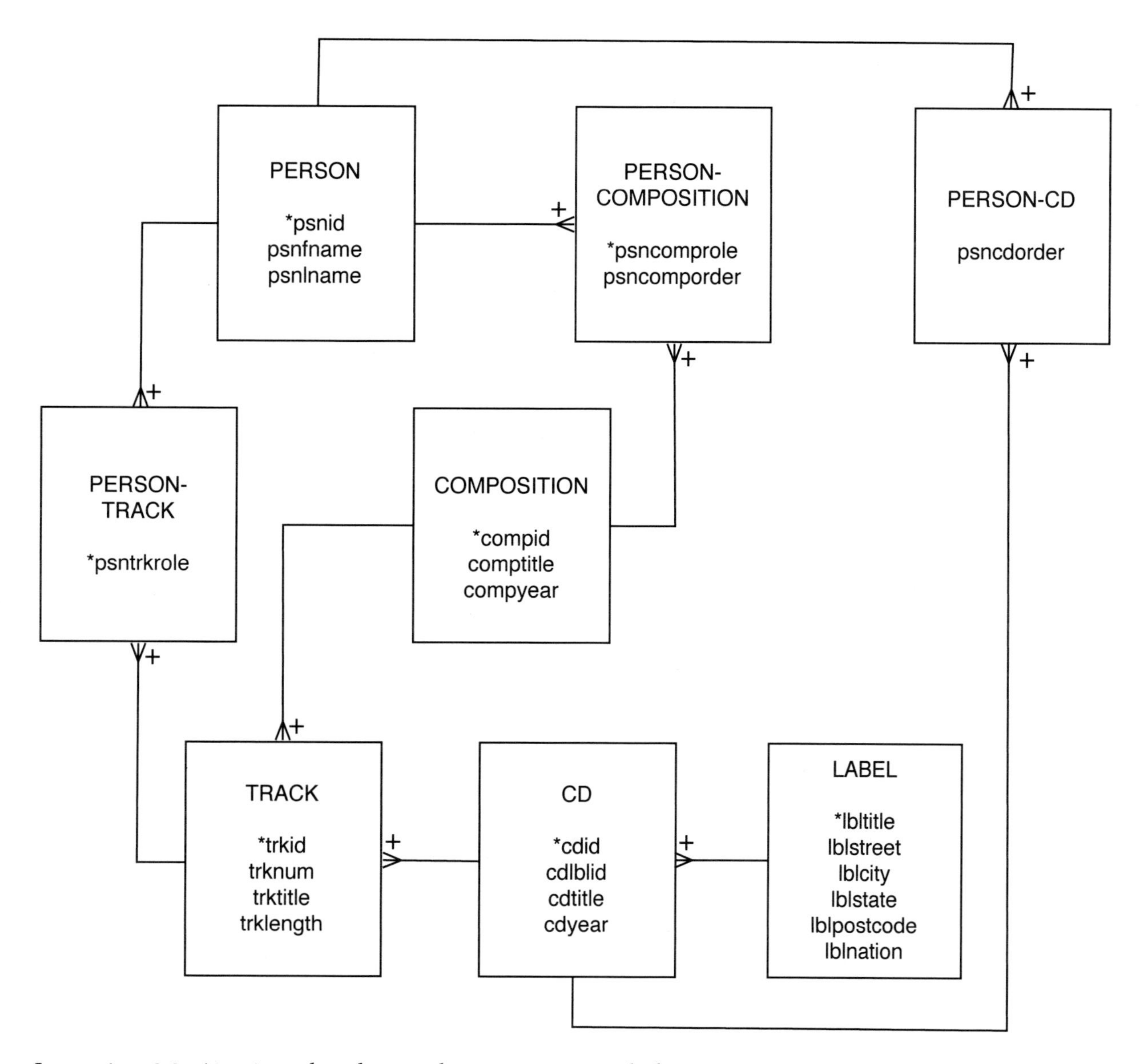

In version 3.0, Ajay introduced several new entities, including COMPOSITION to record details of a composition. He initially linked COMPOSITION to TRACK with a 1:m relationship. In other words, a composition can have many tracks, but a track is of one composition. This did not seem right.

After further thought, he realized there was a missing entity—RECORDING. A composition has many recordings (e.g., multiple recordings of John Coltrane's composition "Giant Steps"), and a recording can appear as a track on different CDs (typically those whose title begins with *The Best of …* feature tracks on earlier CDs). Also, because a recording is a performance of a composition, it is a composition that has a title, not recording or track. So, he revised version 3.0 of the data model to produce version 4.0.

CD library V4.0

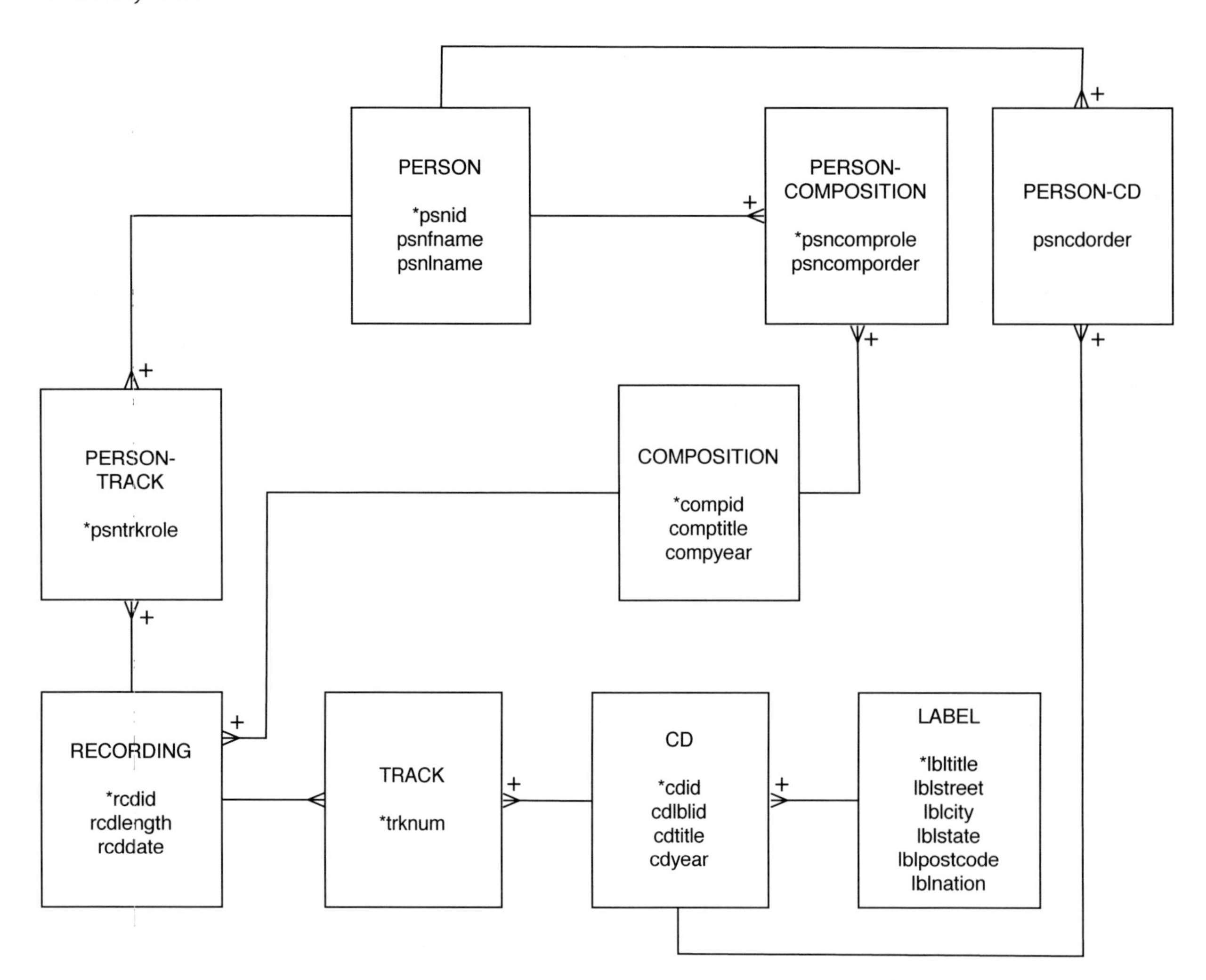

Version 4.0 of the model captures:

- the m:m between PERSON and RECORDING with the associative entity PERSON-RECORDING. The identifier for PERSON-RECORDING is a composite key (*psnid, rcdid, psnrcdrole*). The identifiers of PERSON and RECORDING are required to identify uniquely an instance of PERSON-RECORDING, which is why there are plus signs on the crow's feet attached to PERSON-RECORDING. Also, because a person can have multiple roles on a recording, (e.g., play the piano and sing), then *psnrcdrole* is required to uniquely identify these situations.
- the m:m between PERSON and COMPOSITION with the associative entity PERSON-COMPOSITION. The identifier is a composite key (*psnid, compid, psncomprole*). The attribute *psncomporder* is for remembering the correct sequence in those cases where there are multiple people involved in a composition. That is, the database needs to record that it is Rodgers and Hammerstein, not Hammerstein and Rodgers.
- the m:m between PERSON and CD with the associative entity PERSON-CD. Again the identifier is a composite key (*psnid, cdid*) of the identifiers of entities in the m:m relationship, and there is an attribute (*psncdorder*) to record the order of the people featured on the CD.
- the 1:m between RECORDING and TRACK. A recording can appear on many tracks, but a track has only one recording.

- the 1:m between CD and TRACK. A CD can have many tracks, but a track appears on only one CD. You can also think of TRACK as the name for the associative entity of the m:m relationship between RECORDING and CD. A recording can appear on many CDs, and a CD has many recordings.

After creating the new tables, Ajay was ready to add some data. He decided first to correct the errors introduced by his initial, incorrect modeling of `track` by inserting data for `recording` and revising `track`. He realized that because `recording` contains a foreign key to record the 1:m relationship with `composition`, he needed to start by entering the data for `composition`. Note that some data are missing for the year of composition.

The table composition

*compid	comptitle	compyear
1	Giant Steps	
2	Cousin Mary	
3	Countdown	
4	Spiral	
5	Syeeda's Song Flute	
6	Naima	
7	Mr. P.C.	
8	Stomp of King Porter	1924
9	Sing a Study in Brown	1937
10	Sing Moten's Swing	1997
11	A-tisket, A-tasket	1938
12	I Know Why	1941
13	Sing You Sinners	1930
14	Java Jive	1940
15	Down South Camp Meetin'	1997
16	Topsy	1936
17	Clouds	
18	Skyliner	1944
19	It's Good Enough to Keep	1997
20	Choo Choo Ch' Boogie	1945

Once `composition` was entered, Ajay moved on to the `recording` table. Again, some data are missing.

The table recording

*rcdid	*compid	rcdlength	rcddate
1	1	4.72	1959-May-04
2	2	5.75	1959-May-04
3	3	2.35	1959-May-04
4	4	5.93	1959-May-04
5	5	7	1959-May-04
6	6	4.35	1959-Dec-02
7	7	2.95	1959-May-04
8	1	5.93	1959-Apr-01
9	6	7	1959-Apr-01

*rcdid	*compid	rcdlength	rcddate
10	2	6.95	1959-May-04
11	3	3.67	1959-May-04
12	2	4.45	1959-May-04
13	8	3.2	
14	9	2.85	
15	10	3.6	
16	11	2.95	
17	12	3.57	
18	13	2.75	
19	14	2.85	
20	15	3.25	
21	16	3.23	
22	17	7.2	
23	18	3.18	
24	19	3.18	
25	20	3	

The last row of the `recording` table indicates a recording of the composition "Choo Choo Ch' Boogie," which has `compid = 20`.

Next, the data for `track` could be entered.

The table track

*cdid	*trknum	*rcdid
1	1	1
1	2	2
1	3	3
1	4	4
1	5	5
1	6	6
1	7	7
1	8	1
1	9	6
1	10	2
1	11	3
1	12	5
2	1	13
2	2	14
2	3	15
2	4	16
2	5	17
2	6	18
2	7	19
2	8	20
2	9	21
2	10	22

*cdid	*trknum	*rcdid
2	11	23
2	12	24
2	13	25

The last row of the `track` data indicates that track 13 of *Swing* (`cdid = 2`) is a recording of "Choo Choo Ch' Boogie" (`rcdid = 25 and compid = 20`).

The table `PERSON-RECORDING` stores details of the people involved in each recording. Referring to the accompanying notes for the CD *Giant Steps*, Ajay learned that the composition "Giants Steps," recorded on May 4, 1959, included the following personnel: John Coltrane on tenor sax, Tommy Flanagan on piano, Paul Chambers on bass, and Art Taylor on drums. Using the following data, he inserted four rows in the `PERSON` table.

The table person

*psnid	psnfname	psnlname
1	John	Coltrane
2	Tommy	Flanagan
3	Paul	Chamber
4	Art	Taylor

Then, he linked these people to the particular recording of "Giant Steps" by entering the data in the table, `PERSON-RECORDING`.

The table person-recording

*psnid	*rcdid	*psncdrole
1	1	tenor sax
2	1	piano
3	1	bass
4	1	drums

Skill builder

1. What data will you enter in `person-cd` to relate John Coltrane to the *Giant Steps* CD?
2. Enter appropriate values in `person-composition` to relate John Coltrane to compositions with `compid` 1 through 7. You should indicate he wrote the music by entering a value of "music" for the person's role in the composition.
3. Use SQL to solve the following queries:
   a. List the tracks on "Swing."
   b. Who composed the music for "Spiral"?
   c. Who played which instruments for the May 4, 1959 recording of "Giant Steps"?
   d. List the composers who write music and play the tenor sax.
4. What is the data model missing?

6. One-to-One and Recursive Relationships

Self-reflection is the school of wisdom.

Baltasar Gracián, *The Art of Worldly Wisdom,* 1647

Learning objectives

Students completing this chapter will be able to

- model one-to-one and recursive relationships;
- define a database with one-to-one and recursive relationships;
- write queries for a database with one-to-one and recursive relationships.

Alice was convinced that a database of business facts would speed up decision making. Her initial experience with the sales database had reinforced this conviction. She decided that employee data should be computerized next. When she arrived at The Expeditioner, she found the company lacked an organization chart and job descriptions for employees. One of her first tasks had been to draw an organization chart.

The Expeditioner's organization chart

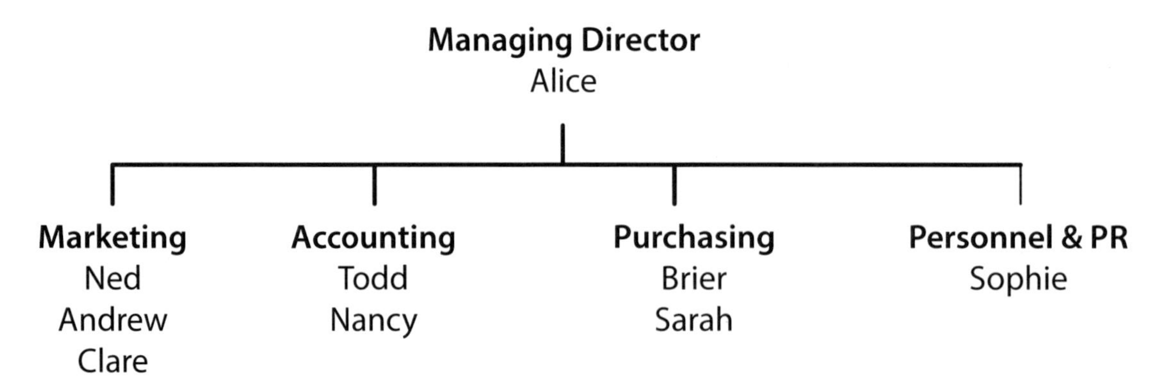

She divided the firm into four departments and appointed a boss for each department. The first person listed in each department was its boss; of course, Alice was paramount.

Ned had finished the sales database, so the time was right to create an employee database. Ned seemed to be enjoying his new role, though his dress standard had slipped. Maybe Alice would have to move him out of Marketing. In fact, he sometimes reminded her of fellows who used to hang around the computer lab all day playing weird games. She wasn't quite ready to sound a "nerd alert," but it was getting close.

Modeling a one-to-one relationship

Initially, the organization chart appears to record two relationships. *First,* a department has one or more employees, and an employee belongs to one department. *Second,* a department has one boss, and a person is boss of only one department. That is, boss is a 1:1 relationship between DEPT and EMP. The data model for this situation is shown.

A data model illustrating a 1:1 relationship

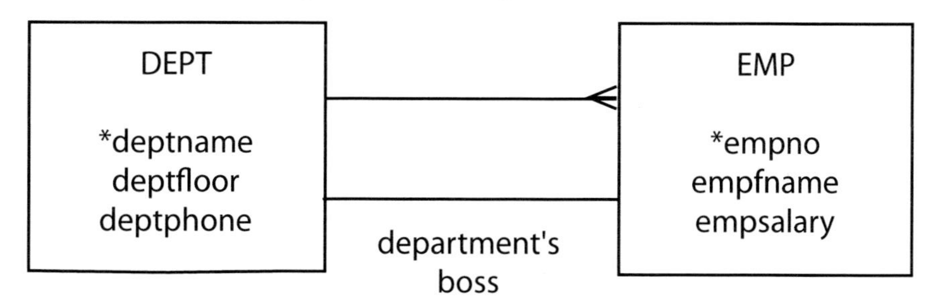

MySQL Workbench version of a 1:1 relationship

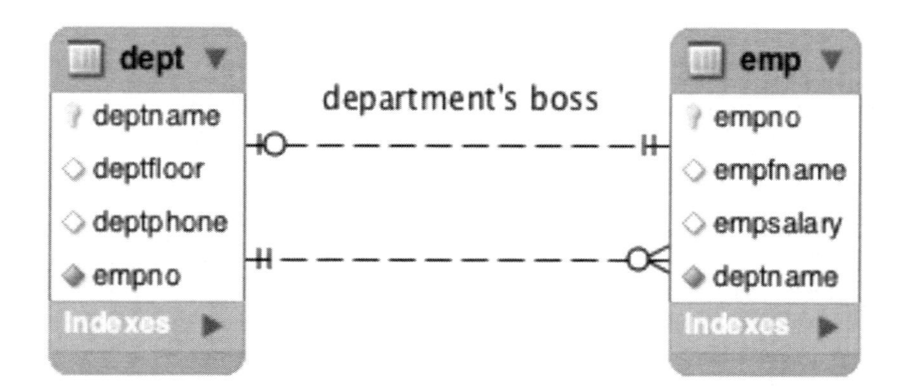

As a general rule, the 1:1 relationship is labeled to avoid confusion because the meaning of such a relationship cannot always be inferred. This label is called a **relationship descriptor**. The 1:m relationship between DEPT and EMP is not labeled because its meaning is readily understood by reading the model. Use a relationship descriptor when there is more than one relationship between entities or when the meaning of the relationship is not readily inferred from the model.

If we think about this problem, we realize there is more to boss than just a department. People also have a boss. Thus, Alice is the boss of all the other employees. In this case, we are mainly interested in who directly bosses someone else. So, Alice is the direct boss of Ned, Todd, Brier, and Sophie. We need to record the person - boss relationship as well as the department - boss relationship.

The person - boss relationship is a **recursive 1:m relationship** because it is a relationship between employees — an employee has one boss and a boss can have many employees. The data model is shown in the following figure.

A data model illustrating a recursive 1:m relationship

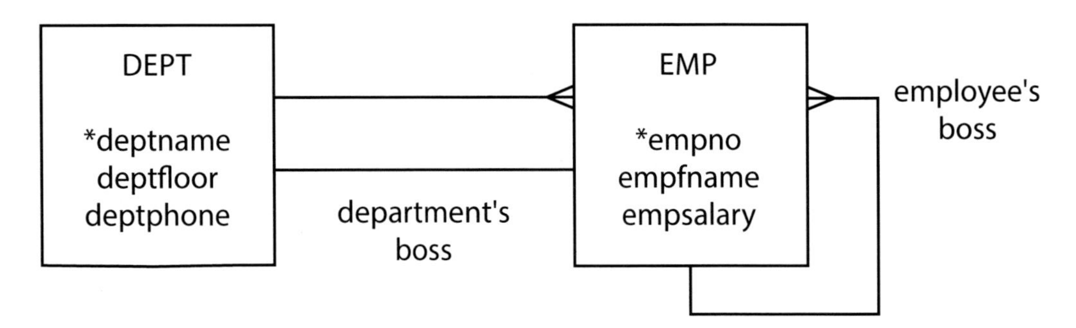

MySQL Workbench version of a recursive 1:m relationship

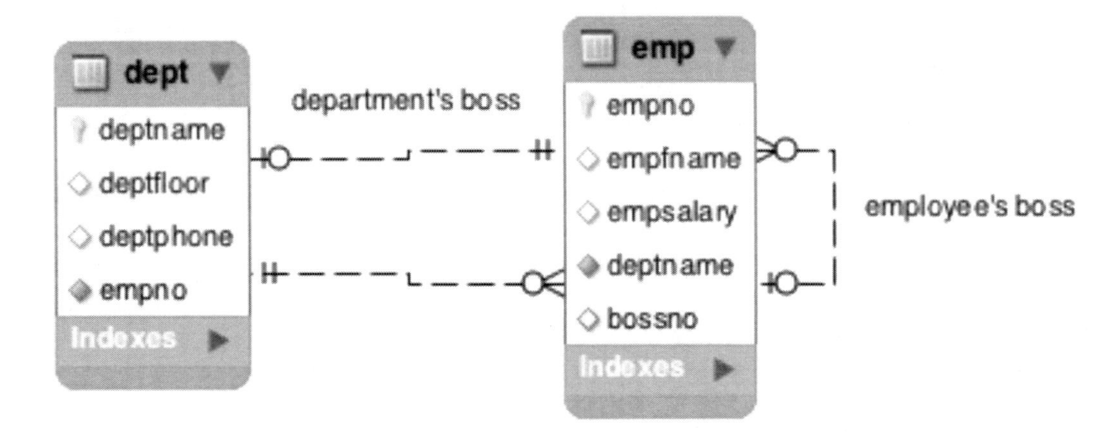

It is a good idea to label the recursive relationship, because its meaning is often not obvious from the data model.

Mapping a one-to-one relationship

Since mapping a 1:1 relationship follows the same rules as for any other data model, the major consideration is where to place the foreign key(s). There are three alternatives:

1. Put the foreign key in `dept`.

   Doing so means that every instance of `dept` will record the `empno` of the employee who is boss. Because all departments in this case have a boss, the foreign key will always be non-null.

2. Put the foreign key in `emp`.

   Choosing this alternative means that every instance of `emp` should record `deptname` of the department this employee bosses. Since many employees are not bosses, the value of the foreign key column will generally be null.

3. Put a foreign key in both `dept` and `emp`.
   The consequence of putting a foreign key in both tables in the 1:1 relationship is the combination of points 1 and 2.

A sound approach is to select the entity that results in the fewest nulls, since this tends to be less confusing to clients. In this case, the simplest approach is to put the foreign key in `dept`.

Mapping a recursive one-to-many relationship

A recursive 1:m relationship is mapped like a standard 1:m relationship. An additional column, for the foreign key, is created for the entity at the "many" end of the relationship. Of course, in this case the "one" and "many" ends are the same entity, so an additional column is added to emp. This column contains the key empno of the "one" end of the relationship. Since empno is already used as a column name, a different name needs to be selected. In this case, it makes sense to call the foreign key column BOSSNO because it stores the boss's employee number.

The mapping of the data model is shown in the following table. Note that deptname becomes a column in emp, the "many" end of the 1:m relationship, and empno becomes a foreign key in dept, an end of the 1:1 relationship.

The tables dept and emp

dept

*deptname	deptfloor	deptphone	*empno*
Management	5	2001	1
Marketing	1	2002	2
Accounting	4	2003	5
Purchasing	4	2004	7
Personnel & PR	1	2005	9

emp

*empno	empfname	empsalary	deptname	*bossno*
1	*Alice*	75000	*Management*	
2	*Ned*	45000	*Marketing*	1
3	*Andrew*	25000	*Marketing*	2
4	*Clare*	22000	*Marketing*	2
5	*Todd*	38000	*Accounting*	1
6	*Nancy*	22000	*Accounting*	5
7	*Brier*	43000	*Purchasing*	1
8	*Sarah*	56000	*Purchasing*	7
9	*Sophie*	35000	*Personnel & PR*	1

If you examine emp, you will see that the boss of the employee with empno = 2 (Ned) has bossno = 1. You can then look up the row in emp with empno = 1 to find that Ned's boss is Alice. This "double lookup" is frequently used when manually interrogating a table that represents a recursive relationship. Soon you will discover how this is handled with SQL.

Here is the SQL to create the two tables:

```
CREATE TABLE dept (
  deptname      VARCHAR(15),
  deptfloor     SMALLINT NOT NULL,
  deptphone     SMALLINT NOT NULL,
  empno         SMALLINT NOT NULL,
  PRIMARY KEY(deptname));
```

```
CREATE TABLE emp (
   empno         SMALLINT,
   empfname      VARCHAR(10),
   empsalary     DECIMAL(7,0),
   deptname      VARCHAR(15),
   bossno        SMALLINT,
   PRIMARY KEY(empno),
   CONSTRAINT fk_belong_dept FOREIGN KEY(deptname)
      REFERENCES dept(deptname),
   CONSTRAINT fk_has_boss FOREIGN KEY(bossno)
      REFERENCES emp(empno));
```

You will notice that there is no foreign key definition for `empno` in `dept` (the 1:1 department's boss relationship). Why? Observe that `deptname` is a foreign key in `emp`. If we make `empno` a foreign key in `dept`, then we have a *deadly embrace*. A new department cannot be added to the `dept` table until there is a boss for that department (i.e., there is a person in the `emp` table with the `empno` of the boss); however, the other constraint states that an employee cannot be added to the `emp` table unless there is a department to which that person is assigned. If we have both foreign key constraints, we cannot add a new department until we have added a boss, and we cannot add a boss until we have added a department for that person. Nothing, under these circumstances, can happen if both foreign key constraints are in place. Thus, only one of them is specified.

In the case of the recursive employee relationship, we can create a constraint to ensure that `bossno` exists for each employee, except of course the person, Alice, who is top of the pyramid. This form of constraint is known as a **self-referential** foreign key. However, we must make certain that the first person inserted into `emp` is Alice. The following statements illustrate that we must always insert a person's boss before we insert the person.

```
INSERT INTO emp (empno, empfname, empsalary, deptname)
   VALUES (1,'Alice',75000,'Management');
INSERT INTO emp VALUES (2,'Ned',45000,'Marketing',1);
INSERT INTO emp VALUES (3,'Andrew',25000,'Marketing',2);
INSERT INTO emp VALUES (4,'Clare',22000,'Marketing',2);
INSERT INTO emp VALUES (5,'Todd',38000,'Accounting',1);
INSERT INTO emp VALUES (6,'Nancy',22000,'Accounting',5);
INSERT INTO emp VALUES (7,'Brier',43000,'Purchasing',1);
INSERT INTO emp VALUES (8,'Sarah',56000,'Purchasing',7);
INSERT INTO emp VALUES (9,'Sophie',35000,'Personnel',1);
```

In more complex modeling situations, such as when there are multiple relationships between a pair of entities, use of a `FOREIGN KEY` clause may result in a deadlock. Always consider the consequences of using a `FOREIGN KEY` clause before applying it.

Skill builder

A consulting company has assigned each of its employees to a specialist group (e.g., database management). Each specialist group has a team leader. When employees join the company, they are assigned a mentor for the first year. One person might mentor several employees, but an employee has at most one mentor.

Querying a one-to-one relationship

Querying a 1:1 relationship presents no special difficulties but does allow us to see additional SQL features.

- *List the salary of each department's boss.*

```
SELECT empfname, deptname, empsalary FROM emp
   WHERE empno IN (SELECT empno FROM dept);
```

or

```
SELECT empfname, dept.deptname, empsalary
   FROM emp JOIN dept
      ON dept.empno = emp.empno;
```

empfname	deptname	empsalary
Alice	Management	75000
Ned	Marketing	45000
Todd	Accounting	38000
Brier	Purchasing	43000
Sophie	Personnel & PR	35000

Querying a recursive 1:m relationship

Querying a recursive relationship is puzzling until you realize that you can join a table to itself by creating two copies of the table. In SQL, you create a temporary copy, a **table alias**, by following the table's name with the alias (e.g., `emp wrk` creates a temporary copy, called `wrk`, of the permanent table `emp`). Table aliases are always required so that SQL can distinguish the copy of the table being referenced. To demonstrate:

- *Find the salary of Nancy's boss.*

```
SELECT wrk.empfname, wrk.empsalary,boss.empfname, boss.empsalary
   FROM emp wrk JOIN emp boss
      ON wrk.empfname = 'Nancy'
      WHERE wrk.bossno = boss.empno;
```

Many queries are solved by getting all the data you need to answer the request in one row. In this case, the query is easy to answer once the data for Nancy and her boss are in the one row. Thus, think of this query as joining two copies of the table `emp` to get the worker and her boss's data in one row. Notice that there is a qualifier (`wrk` and `boss`) for each copy of the table to distinguish between them. It helps to use a qualifier that makes sense. In this case, the `wrk` and `boss` qualifiers can be thought of as referring to the worker and boss tables, respectively. You can understand how the query works by examining the following table illustrating the self-join.

wrk					boss				
empno	empfname	empsalary	deptname	bossno	empno	empfname	empsalary	deptname	bossno
2	Ned	45000	Marketing	1	1	Alice	75000	Management	
3	Andrew	25000	Marketing	2	2	Ned	45000	Marketing	1
4	Clare	22000	Marketing	2	2	Ned	45000	Marketing	1
5	Todd	38000	Accounting	1	1	Alice	75000	Management	
6	Nancy	22000	Accounting	5	5	Todd	38000	Accounting	1
7	Brier	43000	Purchasing	1	1	Alice	75000	Management	
8	Sarah	56000	Purchasing	7	7	Brier	43000	Purchasing	1
9	Sophie	35000	Personnel & PR	1	1	Alice	75000	Management	

The result of the SQL query is now quite clear once we apply the WHERE clause (see the highlighted row in the preceding table):

wrk.empfname	wrk.empsalary	boss.empfname	boss.empsalary
Nancy	22000	Todd	38000

- *Find the names of employees who earn more than their boss.*

```
SELECT wrk.empfname
  FROM emp wrk JOIN emp boss
     ON wrk.bossno = boss.empno
     WHERE wrk.empsalary > boss.empsalary;
```

This would be very easy if the employee and boss data were in the same row. We could simply compare the salaries of the two people. To get the data in the one row, we repeat the self-join with a different WHERE condition. The result is as follows:

empfname
Sarah

Skill builder

Find the name of Sophie's boss.

Alice has found several histories of The Expeditioner from a variety of eras. Because many expeditions they outfitted were conducted under royal patronage, it was not uncommon for these histories to refer to British monarchs. Alice could remember very little about British history, let alone when various kings and queens reigned. This sounded like another database problem. She would ask Ned to create a database that recorded details of each monarch. She thought it also would be useful to record details of royal succession.

Modeling a recursive one-to-one relationship

The British monarchy can be represented by a simple one-entity model. A monarch has one direct successor and one direct predecessor. The sequencing of monarchs can be modeled by a recursive 1:1 relationship, shown in the following figure.

A data model illustrating a recursive 1:1 relationship

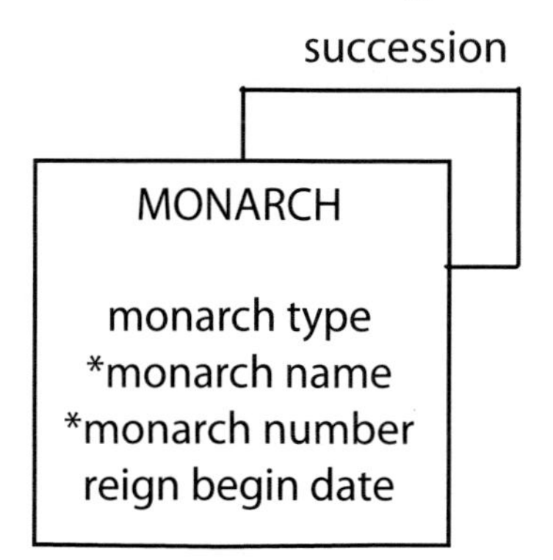

MySQL Workbench version of a recursive 1:1 relationship

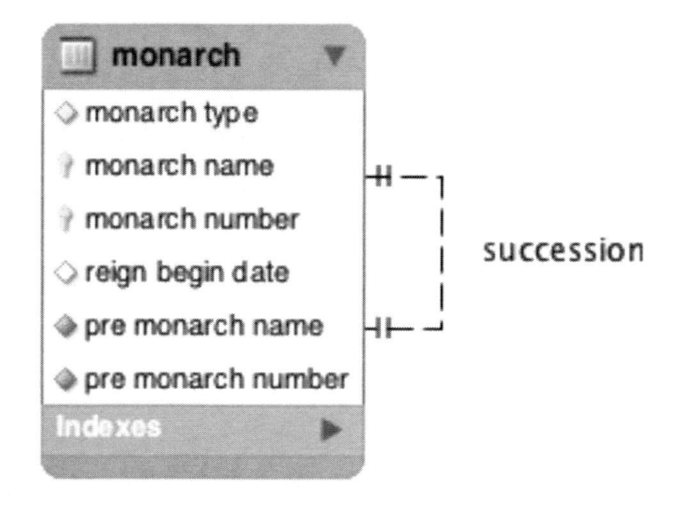

Mapping a recursive one-to-one relationship

The recursive 1:1 relationship is mapped by adding a foreign key to `monarch`. You can add a foreign key to represent either the successor or predecessor relationship. In this case, for no particular reason, the preceding relationship is selected. Because each instance of a monarch is identified by a composite key, two columns are added to `monarch` for the foreign key. Data for recent monarchs are shown in the following table.

The table `MONARCH`

monarch					
montype	monname	monnum	rgnbeg	*premonname*	*premonnum*
King	William	IV	1830/6/26		
Queen	Victoria	I	1837/6/20	William	IV

monarch					
montype	monname	monnum	rgnbeg	*premonname*	*premonnum*
King	Edward	VII	1901/1/22	Victoria	I
King	George	V	1910/5/6	Edward	VII
King	Edward	VIII	1936/1/20	George	V
King	George	VI	1936/12/11	Edward	VIII
Queen	Elizabeth	II	1952/02/06	George	VI

The SQL statements to create the table are very straightforward.

```
CREATE TABLE monarch (
   montype       CHAR(5) NOT NULL,
   monname       VARCHAR(15),
   monnum        VARCHAR(5),
   rgnbeg        DATE,
   premonname       VARCHAR(15),
   premonnum     VARCHAR(5),
   PRIMARY KEY(monname,monnum),
   CONSTRAINT fk_monarch FOREIGN KEY (premonname, premonnum)
      REFERENCES monarch(monname, monnum);
```

Because the 1:1 relationship is recursive, you cannot insert Queen Victoria without first inserting King William IV. What you can do is first insert King William, without any reference to the preceding monarch (i.e., a null foreign key). The following code illustrates the order of record insertion so that the referential integrity constraint is obeyed.

```
INSERT INTO MONARCH (montype,monname, monnum,rgnbeg)
   VALUES ('King','William','IV','1830-06-26');
INSERT INTO MONARCH
   VALUES ('Queen','Victoria','I','1837-06-20','William','IV');
INSERT INTO MONARCH
   VALUES ('King','Edward','VII','1901-01-22','Victoria','I');
INSERT INTO MONARCH
   VALUES ('King','George','V','1910-05-06','Edward','VII');
INSERT INTO MONARCH
   VALUES ('King','Edward','VIII','1936-01-20','George','V');
INSERT INTO MONARCH
   VALUES('King','George','VI','1936-12-11','Edward','VIII');
INSERT INTO MONARCH
   VALUES('Queen','Elizabeth','II','1952-02-06','George','VI');
```

Skill builder

In a competitive bridge competition, the same pair of players play together for the entire tournament. Draw a data model to record details of all the players and the pairs of players.

Querying a recursive one-to-one relationship

Some queries on the `monarch` table demonstrate querying a recursive 1:1 relationship.

- *Who preceded Elizabeth II?*

```
SELECT premonname, premonnum FROM monarch
   WHERE monname = 'Elizabeth' and monnum = 'II';
```

premonname	premonnum
George	VI

This is simple because all the data are in one row. A more complex query is:

- *Was Elizabeth II's predecessor a king or queen?*

```
SELECT pre.montype FROM monarch cur JOIN monarch pre
   ON cur.premonname = pre.monname AND cur.premonnum = pre.monnum
   WHERE cur.monname = 'Elizabeth'
   AND cur.monnum = 'II';
```

Notice in the preceding query how to specify the ON clause when you have a composite key.

montype
King

This is very similar to the query to find the salary of Nancy's boss. The `monarch` table is joined with itself to create a row that contains all the details to answer the query.

- *List the kings and queens of England in ascending chronological order.*

```
SELECT montype, monname, monnum, rgnbeg
   FROM monarch ORDER BY rgnbeg;
```

montype	monname	monnum	rgnbeg
Queen	Victoria	I	1837-06-20
King	Edward	VII	1901-01-22
King	George	V	1910-05-06
King	Edward	VIII	1936-01-20
King	George	VI	1936-12-11
Queen	Elizabeth	II	1952-02-06

This is a simple query because `rgnbeg` is like a ranking column. It would not be enough to store just the year in `rgnbeg`, because two kings started their reigns in 1936; hence, the full date is required.

Skill builder

Who succeeded Queen Victoria?

Of course, Alice soon had another project for Ned. The Expeditioner keeps a wide range of products that are sometimes assembled into kits to make other products. For example, the animal photography kit is

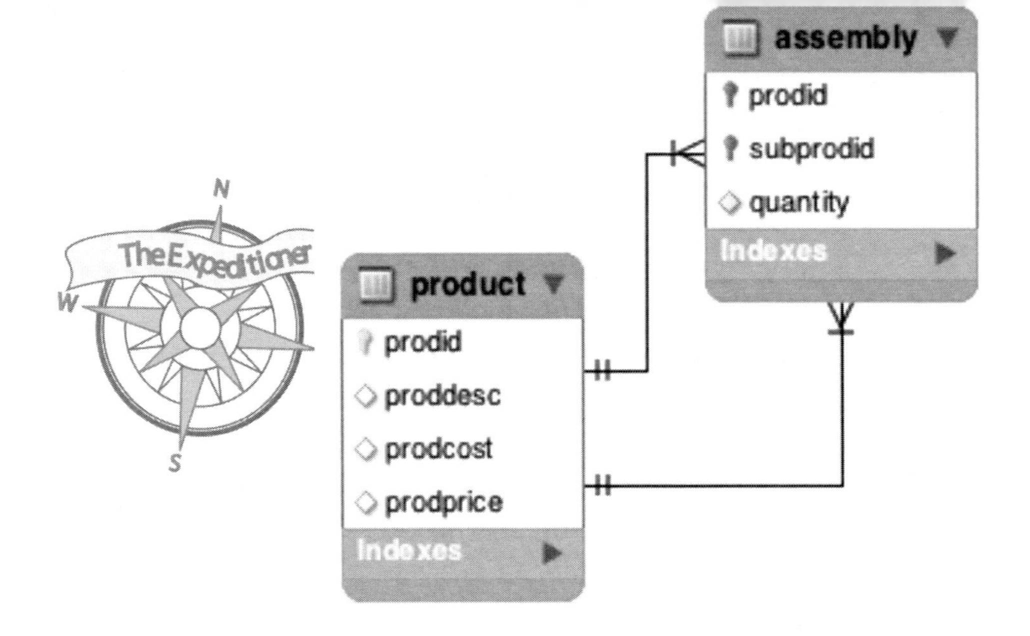

made up of eight items that The Expeditioner also sells separately. In addition, some kits became part of much larger kits. The animal photography kit is included as one of 45 items in the East African Safari package. All of the various items are considered products, and each has its own product code. Ned was now required to create a product database that would keep track of all the items in The Expeditioner's stock.

Modeling a recursive many-to-many relationship

The assembly of products to create other products is very common in business. Manufacturing even has a special term to describe it: a **bill of materials**. The data model is relatively simple once you realize that a product can appear as part of many other products and can be composed of many other products; that is, we have a **recursive many-to-many (m:m) relationship** for product. As usual, we turn an m:m relationship into two one-to-many (1:m) relationships. Thus, we get the data model displayed in the following figure.

A data model illustrating a recursive m:m relationship

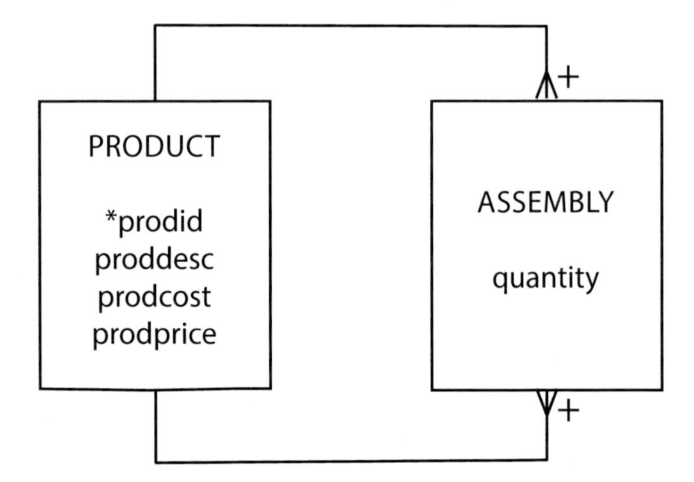

MySQL Workbench version of a recursive m:m relationship

Tables product and assembly

product			
prodid	proddesc	prodcost	prodprice
1000	Animal photography kit		725
101	Camera	150	300
102	Camera case	10	15
103	70-210 zoom lens	125	200
104	28-85 zoom lens	115	185
105	Photographer's vest	25	40
106	Lens cleaning cloth	1	1.25
107	Tripod	35	45
108	16 GB SDHC memory card	30	37

assembly		
quantity	*prodid*	*subprodid*
1	1000	101
1	1000	102
1	1000	103
1	1000	104
1	1000	105
2	1000	106
1	1000	107
10	1000	108

Mapping a recursive many-to-many relationship

Mapping follows the same procedure described previously, producing the two tables shown below. The SQL statements to create the tables are shown next. Observe that `assembly` has a composite key, and there are two foreign key constraints.

```
CREATE TABLE product (
   prodid         INTEGER,
   proddesc       VARCHAR(30),
   prodcost       DECIMAL(9,2),
   prodprice      DECIMAL(9,2),
      PRIMARY KEY(prodid));
CREATE TABLE assembly (
   quantity       INTEGER NOT NULL,
   prodid         INTEGER,
   subprodid      INTEGER,
      PRIMARY KEY(prodid, subprodid),
      CONSTRAINT fk_assembly_product FOREIGN KEY(prodid)
         REFERENCES product(prodid),
      CONSTRAINT fk_assembly_subproduct FOREIGN KEY(subprodid)
```

```
        REFERENCES product(prodid));
```

Skill builder

An army is broken up into many administrative units (e.g., army, brigade, platoon). A unit can contain many other units (e.g., a regiment contains two or more battalions), and a unit can be part of a larger unit (e.g., a squad is a member of a platoon). Draw a data model for this situation.

Querying a recursive many-to-many relationship

- *List the product identifier of each component of the animal photography kit.*

```
SELECT subprodid FROM product JOIN assembly
   ON product.prodid = assembly.prodid
   WHERE proddesc = 'Animal photography kit';
```

subprodid
101
106
107
105
104
103
102
108

Why are the values for subprodid listed in no apparent order? Remember, there is no implied ordering of rows in a table, and it is quite possible, as this example illustrates, for the rows to have what appears to be an unusual ordering. If you want to order rows, use the ORDER BY clause.

- *List the product description and cost of each component of the animal photography kit.*

```
SELECT proddesc, prodcost FROM product
   WHERE prodid IN
      (SELECT subprodid FROM product JOIN assembly
            ON product.prodid = assembly.prodid
            WHERE proddesc = 'Animal photography kit');
```

In this case, first determine the prodid of those products in the animal photography kit (the inner query), and then report the description of these products. Alternatively, a three-way join can be done using two copies of product.

```
SELECT b.proddesc, b.prodcost FROM product a JOIN assembly
      ON a.prodid = assembly.prodid
      JOIN product b
      ON assembly.subprodid = b.prodid
      WHERE a.proddesc = 'Animal photography kit'
```

proddesc	prodcost
Camera	150
Camera case	10
70–210 zoom lens	125
28–85 zoom lens	115
Photographer's vest	25
Lens cleaning cloth	1
Tripod	35
16 GB SDHC memory card	0.85

Skill builder

How many lens cleaning cloths are there in the animal photography kit?

Summary

Relationships can be one-to-one and recursive. A recursive relationship is within a single entity rather than between entities. Recursive relationships are mapped to the relational model in the same way as other relationships. A self-referential foreign key constraint permits a foreign key reference to a key within the same table. Resolution of queries involving recursive relationships often requires a table to be joined with itself. Recursive many-to-many relationships occur in business in the form of a bill of materials.

Key terms and concepts

One-to-one (1:1) relationship

Recursive relationship

Recursive m:m relationship

Recursive 1:m relationship

Recursive 1:1 relationship

Relationship descriptor

Self-join

Self-referential foreign key

Exercises

1. Draw data models for the following two problems:

   a. (i) A dairy farmer, who is also a part-time cartoonist, has several herds of cows. He has assigned each cow to a particular herd. In each herd, the farmer has one cow that is his favorite—often that cow is featured in a cartoon.
   (ii) A few malcontents in each herd, mainly those who feel they should have appeared in the cartoon, disagree with the farmer's choice of a favorite cow, whom they disparagingly refer to as the sacred cow. As a result, each herd now has elected a herd leader.

   b. The originator of a pyramid marketing scheme has a system for selling ethnic jewelry. The pyramid has three levels—gold, silver, and bronze. New associates join the pyramid at the bronze level. They contribute 30 percent of the revenue of their sales of jewelry to the silver chief in charge of their clan. In turn, silver chiefs contribute 30 percent of what they receive from bronze associates to the gold master in command of their tribe. Finally, gold masters pass on 30 percent of what they receive to the originator of the scheme.

   c. The legion, the basic combat unit of the ancient Roman army, contained 3,000 to 6,000 men, consisting primarily of heavy infantry (hoplites), supported by light infantry (velites), and sometimes by cavalry. The hoplites were drawn up in three lines. The hastati (youngest men) were in the first, the principes (seasoned troops) in the second, and the triarii (oldest men) behind them, reinforced by velites. Each line was divided into 10 maniples, consisting of two centuries (60 to 80 men per century) each. Each legion had a commander, and a century was commanded by a centurion. Julius Caesar, through one of his Californian channelers, has asked you to design a database to maintain details of soldiers. Of course, Julius is a little forgetful at times, and he has not supplied the titles of the officers who command maniples, lines, and hoplites, but he expects that you can handle this lack of fine detail.

d. A travel agency is frequently asked questions about tourist destinations. For example, customers want to know details of the climate for a particular month, the population of the city, and other geographic facts. Sometimes they request the flying time and distance between two cities. The manager has asked you to create a database to maintain these facts.

e. The Center for the Study of World Trade keeps track of trade treaties between nations. For each treaty, it records details of the countries signing the treaty and where and when it was signed.

f. Design a database to store details about U.S. presidents and their terms in office. Also, record details of their date and place of birth, gender, and political party affiliation (e.g., Caluthumpian Progress Party). You are required to record the sequence of presidents so that the predecessor and successor of any president can be identified. How will you model the case of Grover Cleveland, who served nonconsecutive terms as president? Is it feasible that political party affiliation may change? If so, how will you handle it?

g. The IS department of a large organization makes extensive use of software modules. New applications are built, where possible, from existing modules. Software modules can also contain other modules. The IS manager realizes that she now needs a database to keep track of which modules are used in which applications or other modules. (Hint: It is helpful to think of an application as a module.)

h. Data modeling is finally getting to you. Last night you dreamed you were asked by Noah to design a database to store data about the animals on the ark. All you can remember from Sunday school is the bit about the animals entering the ark two-by-two, so you thought you should check the real thing.
Take with you seven pairs of every kind of clean animal, a male and its mate, and two of every kind of unclean animal, a male and its mate, and also seven pair of every kind of bird, male and female. Genesis 7:2
Next time Noah disturbs your sleep, you want to be ready. So, draw a data model and make certain you record the two-by-two relationship.

2. Write SQL to answer the following queries using the DEPT and EMP tables described in this chapter:

   a. Find the departments where all the employees earn less than their boss.

   b. Find the names of employees who are in the same department as their boss (as an employee).

   c. List the departments having an average salary greater than $25,000.

   d. List the departments where the average salary of the employees, excluding the boss, is greater than $25,000.

   e. List the names and manager of the employees of the Marketing department who have a salary greater than $25,000.

   f. List the names of the employees who earn more than any employee in the Marketing department.

3. Write SQL to answer the following queries using the `MONARCH` table described in this chapter:
    a. Who succeeded Victoria I?
    b. How many days did Victoria I reign?
    c. How many kings are there in the table?
    d. Which monarch had the shortest reign?
4. Write SQL to answer the following queries using the `PRODUCT` and `ASSEMBLY` tables:
    a. How many different items are there in the animal photography kit?
    b. What is the most expensive item in the animal photography kit?
    c. What is the total cost of the components of the animal photography kit?
    d. Compute the total quantity for each of the items required to assemble 15 animal photography kits.

CD Library case

Ajay had some more time to work on this CD Library database. He picked up The Manhattan Transfer's *Swing* CD to enter its data. Very quickly he recognized the need for yet another entity, if he were going to store details of the people in a group. A group can contain many people (there are four artists in The Manhattan Transfer), and also it is feasible that over time a person could be a member of multiple groups. So, there is an m:m between PERSON and GROUP. Building on his understanding of the relationship between PERSON and RECORDING and CD, Ajay realized there are two other required relationships: an m:m between GROUP and RECORDING, and between GROUP and CD. This led to version 5.0, shown in the following figure.

CD library V5.0

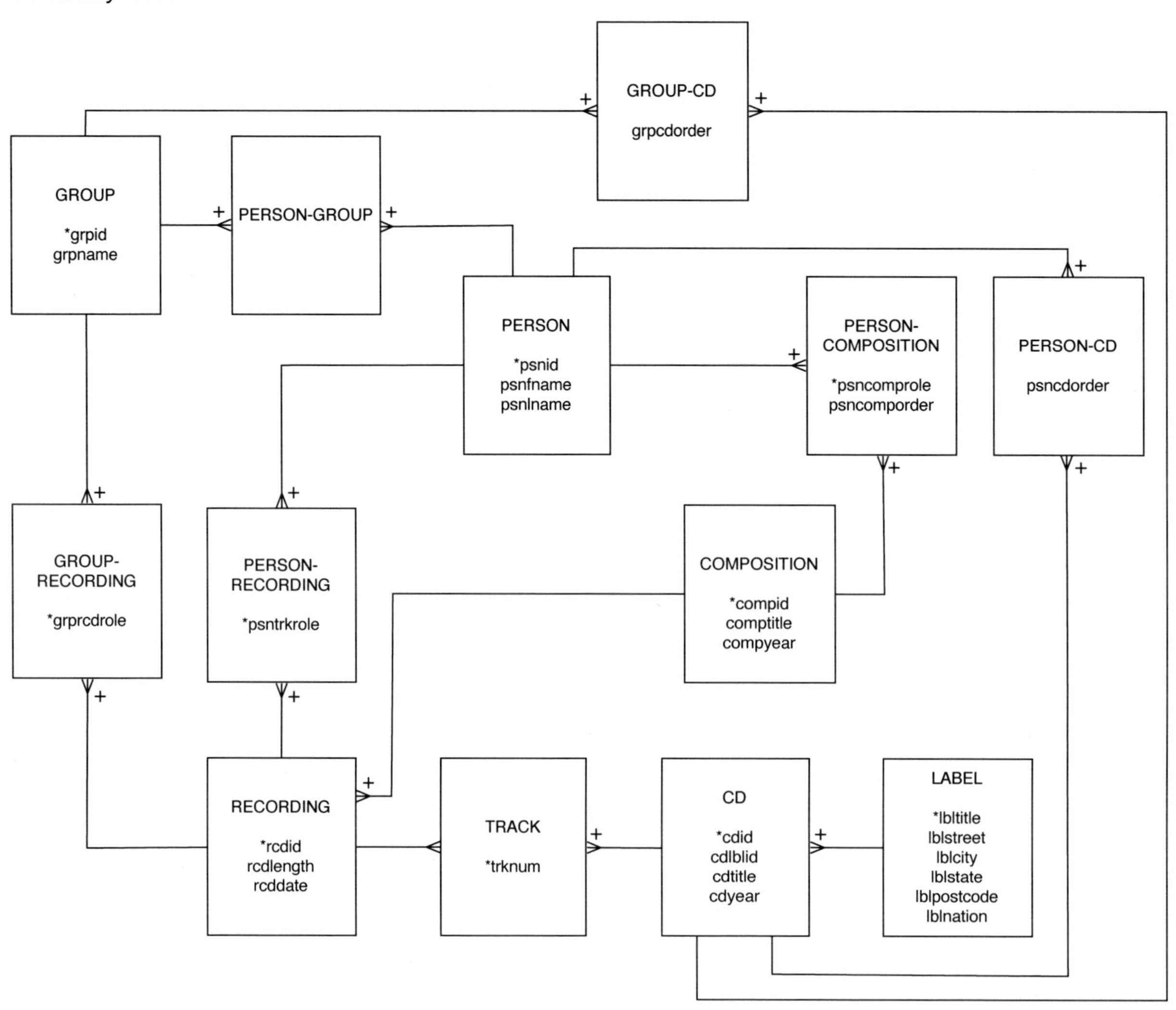

Skill builder

1. The group known as *Asleep at the Wheel* is featured on tracks 3, 4, and 7 of *Swing*. Record these data and write SQL to answer the following:

   a. What are the titles of the recordings on which *Asleep at the Wheel* appear?

   b. List all CDs that have any tracks featuring *Asleep at the Wheel.*

   The members of *The Manhattan Transfer* are Cheryl Bentyne, Janis Siegel, Tim Hauser, and Alan Paul. Update the database with this information, and write SQL to answer the following:

   Who are the members of *The Manhattan Transfer*?

   What CDs have been released by *The Manhattan Transfer*?

   What CDs feature Cheryl Bentyne as an individual or member of a group?

2. Record the following facts. The music of "*Sing a Song of Brown,*" composed in 1937, is by Count Basie and Larry Clinton and lyrics by Jon Hendricks. "*Sing Moten's Swing,*" composed in 1932, features music by Buster and Benny Moten, and lyrics by Jon Hendricks. Write SQL to answer the following:

a. For what songs has Jon Hendricks written the lyrics?

b. Report all compositions and their composers where more than one person was involved in composing the music. Make certain you report them in the correct order.

3. Test your SQL skills with the following:

   a. List the tracks on which Alan Paul appears as an individual or as a member of The Manhattan Transfer.

   b. List all CDs featuring a group, and report the names of the group members.

   c. Report all tracks that feature more than one group.

   d. List the composers appearing on each CD.

4. How might you extend the data model?

7. Data Modeling

Man is a knot, a web, a mesh into which relationships are tied. Only those relationships matter.

Antoine de Saint-Exupéry in *Flight to Arras*

Learning objectives

Students completing this chapter will be able to create a well-formed, high-fidelity data model.

Modeling

Modeling is widely used within business to learn about organizational problems and design solutions. To understand where data modeling fits within the broader context, it is useful to review the full range of modeling activities. The types of modeling methods follow the 5W-H model of journalism.[20]

		Scope	**Model**	**Technology**
Motivation	Why	Goals	Business plan canvas	Groupware
People	Who	Business units	Organization chart	System interface
Time	When	Key events	PERT chart	Scheduling
Data	What	Key entities	Data model	Relational database
Network	Where	Locations	Logistics network	System architecture
Function	How	Key processes	Process model	Application software

A broad perspective on modeling

Modeling occurs at multiple levels. At the highest level, an organization needs to determine the scope of its business by identifying the major elements of its environment, such as its goals, business units, where it operates, and critical events, entities, and processes. At the top level, textual models are frequently used. For example, an organization might list its major goals and business processes. A map will be typically used to display where the business operates.

Once senior managers have clarified the scope of a business, models can be constructed for each of the major elements. Goals will be converted into a business plan, business units will be shown on an organizational chart, and so on. The key elements, from an IS perspective, are data and process modeling, which are typically covered in the core courses of an IS program.

Technology, the final stage, converts models into operational systems to support the organization. Many organizations rely extensively on e-mail and Web technology for communicating business decisions. People connect to systems through an interface, such as a Web browser. Ensuring that events occur at the right time is managed by scheduling software. This could be implemented with operating system procedures that schedule the execution of applications. In some database management systems (DBMSs), triggers can be established. A **trigger** is a database procedure that is automatically executed when some event is recognized. For example, U.S. banks are required to report all deposits exceeding USD 10,000, and a trigger could be coded for this event.

[20] Roberto Franzosi, "On Quantitative Narrative Analysis," in Varieties of Narrative Analysis, ed. James A. Holstein and Jaber F. Gubrium (SAGE Publications, Inc., 2012), 75–96.

Data models are typically converted into relational databases, and process models become computer programs. Thus, you can see that data modeling is one element of a comprehensive modeling activity that is often required to design business systems. When a business undergoes a major change, such as a reengineering project, many dimensions can be altered, and it may be appropriate to rethink many elements of the business, starting with its goals. Because such major change is very disruptive and costly, it occurs less frequently. It is more likely that data modeling is conducted as a stand-alone activity or part of process modeling to create a new business application.

Data modeling

You were introduced to the basic building blocks of data modeling in the earlier chapters of this section. Now it is time to learn how to assemble blocks to build a data model. Data modeling is a method for determining what data and relationships should be stored in the database. It is also a way of communicating a database design.

The goal of data modeling is to identify the facts that must be stored in a database. A data model is not concerned with how the data will be stored. This is the concern of those who implement the database. A data model is not concerned with how the data will be processed. This is the province of process modeling. The goal is to create a data model that is an accurate representation of data needs and real-world data relationships.

Building a data model is a partnership between a client, a representative of the eventual owners of the database, and a designer. Of course, there can be a team of clients and designers. For simplicity, we assume there is one client and one designer.

Drawing a data model is an iterative process of trial and revision. A data model is a working document that will change as you learn more about the client's needs and world. Your early versions are likely to be quite different from the final product. Use a product such as MySQL Workbench for drawing and revising a data model.

The building blocks

The purpose of a database is to store data about things, which can include facts (e.g., an exchange rate), plans (e.g., scheduled production of two-person tents for June), estimates (e.g., forecast of demand for geopositioning systems), and a variety of other data. A data model describes these things and their relationships with other things using four components: entity, attribute, relationship, and identifier.

Entity

The entity is the basic building block of a data model. an entity is a thing about which data should be stored, something we need to describe. Each entity in a data model has a unique name that we write in singular form. Why singular? Because we want to emphasize that an entity describes an instance of a thing. Thus, we previously used the word SHARE to define an entity because it describes each instance of a share rather than shares in general.

We have already introduced the convention that an entity is represented by a rectangle, and the name of the entity is shown in uppercase letters, as follows.

The entity SHARE

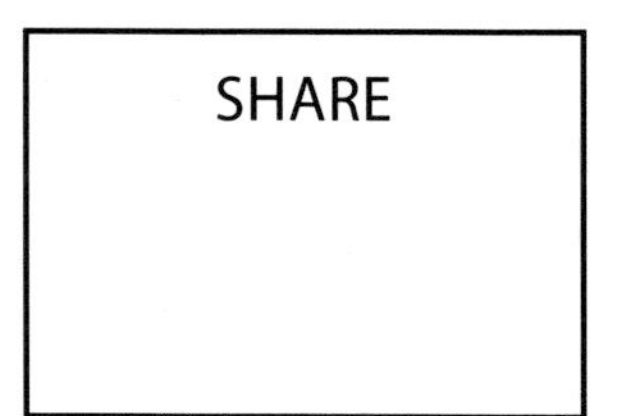

A large data model might contain over a 1,000 entities. A database can easily contain hundreds of millions of instances (rows) for any one entity (table). Imagine the number of instances in a national tax department's database.

How do you begin to identify entities? One approach is to underline any *nouns* in the system proposal or provided documentation. Most nouns are possible entities, and underlining ensures that you do not overlook any potential entities. Start by selecting an entity that seems central to the problem. If you were designing a student database, you might start with the student. Once you have picked a central entity, describe it. Then move to the others.

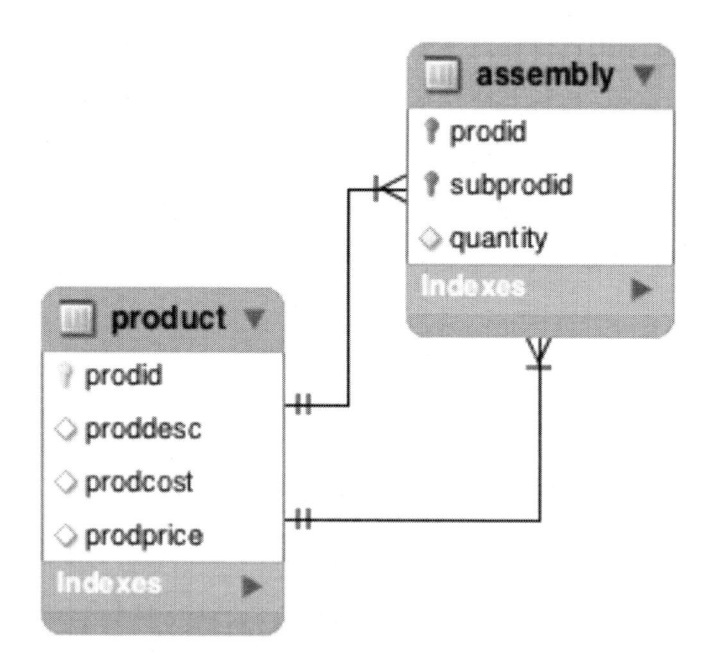

Attribute

An attribute describes an entity. When an entity has been identified, the next step is to determine its attributes, that is, the data that should be kept to describe fully the entity. An attribute name is singular and unique within the data model. You may need to use a modifier to make an attribute name unique (e.g., "hire date" and "sale date" rather than just "date").

Our convention is that the name of an attribute is recorded in lowercase letters within the entity rectangle. The following figure illustrates that SHARE has attributes *share code, share name, share price, share quantity, share dividend,* and *share PE.* Notice the frequent use of the modifier "share". It is possible that other entities in the database might also have a price and quantity, so we use a modifier to create unique attribute names. Note also that *share code* is the identifier.

The entity SHARE with its attributes

SHARE

*share code
share name
share price
share quantity
share dividend
share PE

Defining attributes generally takes considerable discussion with the client. You should include any attribute that is likely to be required for present or future decision making, but don't get carried away. Avoid storing unnecessary data. For example, to describe the entity STUDENT you usually record date of birth, but it is unlikely that you would store height. If you were describing an entity PATIENT, on the other hand, it might be necessary to record height, because that is sometimes relevant to medical decision making.

An attribute has a single value, which may be null. Multiple values are not allowed. If you need to store multiple values, it is a signal that you have a one-to-many (1:m) relationship and need to define another entity.

Relationship

Entities are related to other entities. If there were no relationships between entities, there would be no need for a data model and no need for a relational database. A simple, flat file (a single-entity database) would be sufficient. A relationship is binary. It describes a linkage between two entities and is represented by a line between them.

Because a relationship is binary, strictly speaking it has two relationship descriptors, one for each entity. Each relationship descriptor has a degree stating how many instances of the other entity may be related to each instance of the described entity. Consider the entities STOCK and NATION and their 1:m relationship (see the following figure). We have two relationship descriptors: *stocks of nation* for NATION and *nation of stock* for STOCK. The relationship descriptor *stocks of nation* has a degree of m because a nation may have zero or more listed stocks. The relationship descriptor *nation of stock* has a degree of 1 because a stock is listed in at most one nation. To reduce data model clutter, where necessary for clarification, we will use a single label for the pair of relationship descriptors. Experience shows this approach captures the meaning of the relationship and improves readability.

A 1:m relationship between STOCK and NATION

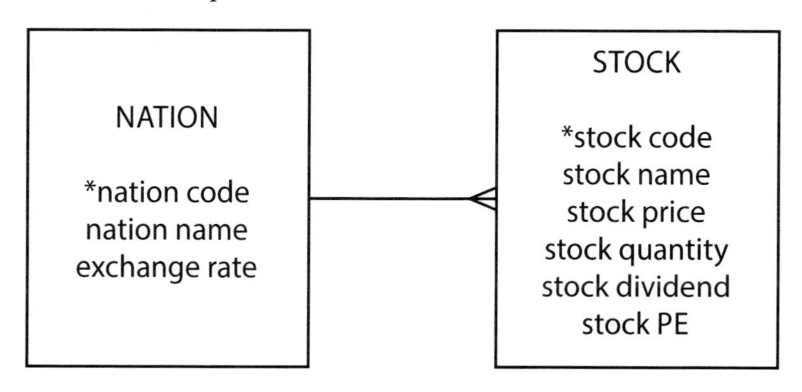

Because the meaning of relationships frequently can be inferred, there is no need to label every one; however, each additional relationship between two entities should be labeled to clarify meaning. Also, it is a good idea to label one-to-one (1:1) relationships, because the meaning of the relationship is not always obvious.

Consider the fragment in the following figure. The descriptors of the 1:m relationship can be inferred as a firm has employees and an employee belongs to a firm. the 1:1 relationship is clarified by labeling the 1:1 relationship as firm's boss.

Relationship labeling

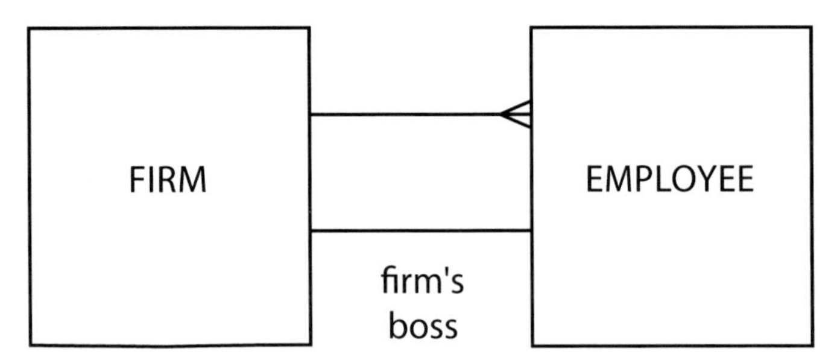

Identifier

An identifier uniquely distinguishes an instance of an entity. An identifier can be one or more attributes and may include the identifier of a related entity. When a related entity's identifier is part of an identifier, a plus sign is placed on the line closest to the entity being identified. In the following figure, an instance of the entity LINEITEM is identified by the composite of *lineno* and *saleno*, the identifier of SALE. The value *lineno* does not uniquely identify an instance of LINEITEM, because it is simply a number that appears on the sales form. Thus, the identifier must include *saleno* (the identifier of SALE). So, any instance of LINEITEM is uniquely identified by the composite *saleno* and *lineno*.

A related entity's identifier as part of the identifier

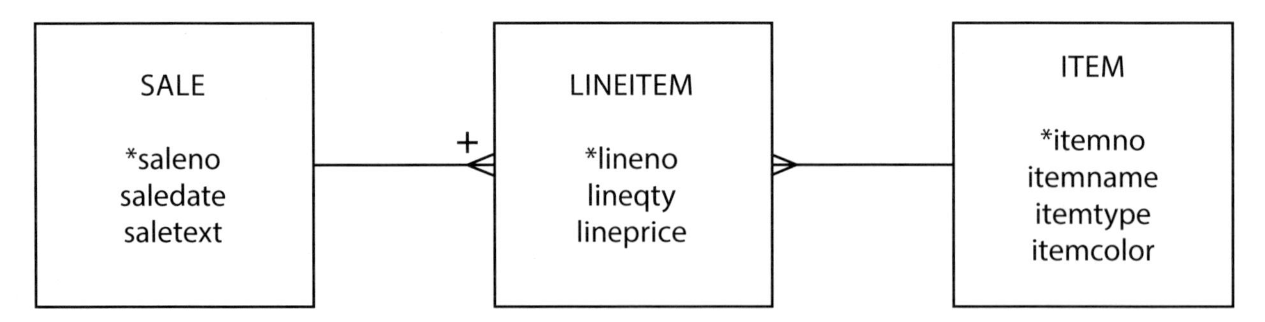

Occasionally, there will be several possible identifiers, and these are each given a different symbol. Our convention is to prefix an identifier with an asterisk (*). If there are multiple identifiers, use other symbols such as #, !, or &. Be careful: If you have too many possible identifiers, your model will look like comic book swearing. In most cases, you will have only one identifier.

No part of an identifier can be null. If this were permitted, there would be no guarantee that the other parts of the identifier were sufficiently unique to distinguish all instances in the database.

Data model quality

There are two criteria for judging the quality of a data model. It must be well-formed and have high fidelity.

A well-formed data model

A well-formed data model clearly communicates information to the client. Being well-formed means the construction rules have been obeyed. There is no ambiguity; all entities are named, and all entities have identifiers. If identifiers are missing, the client may make incorrect inferences about the data model. All relationships are recorded using the proper notation and labeled whenever there is a possibility of confusion.

Characteristics of a well-formed data model

All construction rules are obeyed.
There is no ambiguity.
All entities are named.
Every entity has an identifier.
All relationships are represented, using the correct notation.
Relationships are labeled to avoid misunderstanding.
All attributes of each entity are listed.
All attribute names are meaningful and unique.

All the attributes of an entity are listed because missing attributes create two types of problems. *First,* it is unclear what data will be stored about each instance. *Second*, the data model may be missing some relationships. Attributes that can have multiple values become entities. It is only by listing all of an entity's attributes during data modeling that these additional entities are recognized.

In a well-formed data model, all attribute names are meaningful and unique. The names of entities, identifiers, attributes, and relationships must be meaningful to the client because they are meant to describe the client's world. Indeed, in nearly all cases they are the client's everyday names. Take care in selecting words because they are critical to communicating meaning. The acid test for comprehension is to get the client to read the data model to other potential users. Names need to be unique to avoid confusion.

A high-fidelity image

Music lovers aspire to own a high-fidelity stereo system—one that faithfully reproduces the original performance with minimal or no distortion. A data model is a high-fidelity image when it faithfully describes the world it is supposed to represent. All relationships are recorded and are of the correct degree. There are no compromises or distortions. If the real-world relationship is many-to-many (m:m), then so is the relationship shown in the data model. a well-formed, high-fidelity data model is complete, understandable, accurate, and syntactically correct.

Quality improvement

A data model is an evolving representation. Each change should be an incremental improvement in quality. Occasionally, you will find a major quality problem at the data model's core and have to change the model dramatically.

Detail and context

The quality of a data model can be determined only by understanding the context in which it will be used. Consider the data model (see the following figure) used in Chapter 4 to discuss the 1:m relationship. This fragment says a nation has many stocks. Is that really what we want to represent? Stocks are listed on a stock exchange, and a nation may have several stock exchanges. For example, the U.S. has the New York Stock Exchange and NASDAQ. Furthermore, some foreign stocks are listed on the New York Stock

Exchange. If this is the world we have to describe, the data model is likely to differ from that shown in the figure. Try drawing the revised data model.

A 1:m relationship between STOCK and NATION

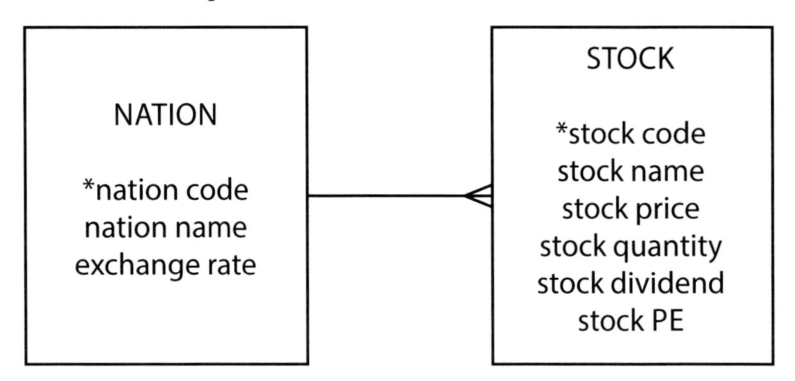

As you can see, the data model in the following figure is quite different from the initial data model of preceding figure. Which one is better? They both can be valid models; it just depends on the world you are trying to represent and possible future queries. In the first case, the purpose was to determine the value of a portfolio. There was no interest in on which exchange stocks were listed or their home exchange. So the first model has high fidelity for the described situation.

The second data model would be appropriate if the client needed to know the home exchange of stocks and their price on the various exchanges where they were listed. The second data model is an improvement on the first if it incorporates the additional facts and relationships required. If it does not, then the additional detail is not worth the extra cost.

Revised NATION-STOCK data model

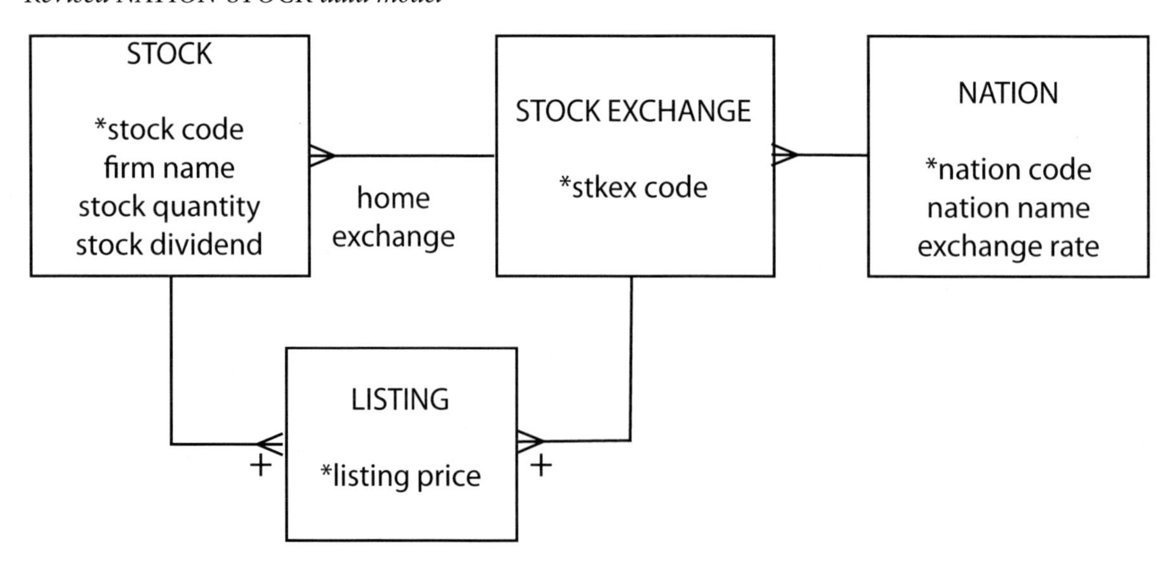

A lesson in pure geography

A data model must be an accurate representation of the world you want to model. The data model must account for all the exceptions—there should be no impurities. Let's say you want to establish a database to store details of the world's cities. You might start with the data model depicted in the following figure.

A world's cities data model

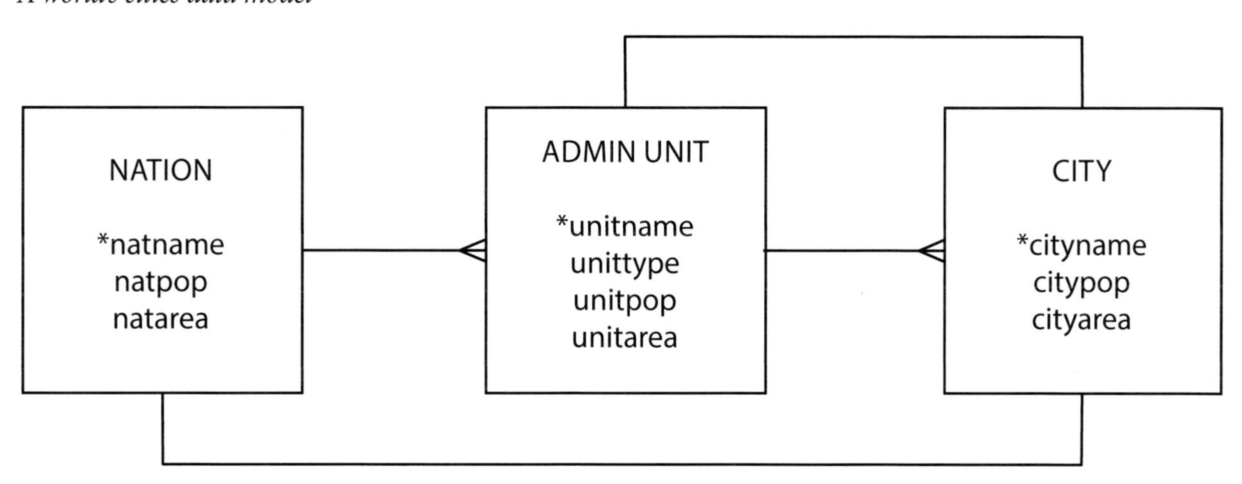

Before looking more closely at the data model, let's clarify the meaning of *unittype*. Because most countries are divided into administrative units that are variously called states, provinces, territories, and so forth, *unittype* indicates the type of administrative unit (e.g., state). How many errors can you find in the initial data model?

Problems and solutions for the initial world's cities data model

	Problem	**Solution**
1	City names are not unique. There is an Athens in Greece and the U.S. has an Athens in Georgia, Alabama, Ohio, Pennsylvania, Tennessee, and Texas.	To identify a city uniquely, you need to specify its administrative unit and country. Add a plus sign to the crow's feet between CITY and ADMIN UNIT and between ADMIN UNIT and NATION.
2	Administrative unit names are not necessarily unique. There used to be a Georgia in the old U.S.S.R., and there is a Georgia in the U.S. There is no guarantee that administrative unit names will always be unique.	To identify an administrative unit uniquely, you also need to know the country in which it is found. Add a plus sign to the crow's foot between ADMIN UNIT and NATION.
3	There is an unlabeled 1:1 relationship between `CITY` and `ADMIN UNIT` and CITY and NATION. What do these mean? They are supposed to indicate that an administrative unit has a capital, and a nation has a capital.	Label relationships (e.g., national capital city).
4	The assumption is that a nation has only one capital, but there are exceptions. South Africa has three capitals: Cape Town (legislative), Pretoria (administrative), and Bloemfontein (judicial). You need only one exception to lower the fidelity of a data model significantly.	Change the relationship between NATION and CITY to 1:m. Add an attribute *natcaptype* to CITY to distinguish between types of capitals.

	Problem	Solution
5	The assumption is that an administrative unit has only one capital, but there are exceptions. Chandigarh is the capital of two Indian states, Punjab and Haryana. The Indian state of Jammu & Kashmir has two capitals: Jammu (summer) and Srinagar (winter).	Change the relationship between ADMIN UNIT and CITY to m:m by creating an associative entity and include a distinguishing attribute of *unitcaptype*.
6	Some values can be derived. National population, *natpop*, and area, *natarea*, are the sum of the regional populations, *regpop*, and areas, *regarea*, respectively. The same rule does not apply to regions and cities because not everyone lives in a city.	Remove the attributes *natpop* and *natarea* from NATION.

A revised world's cities data model

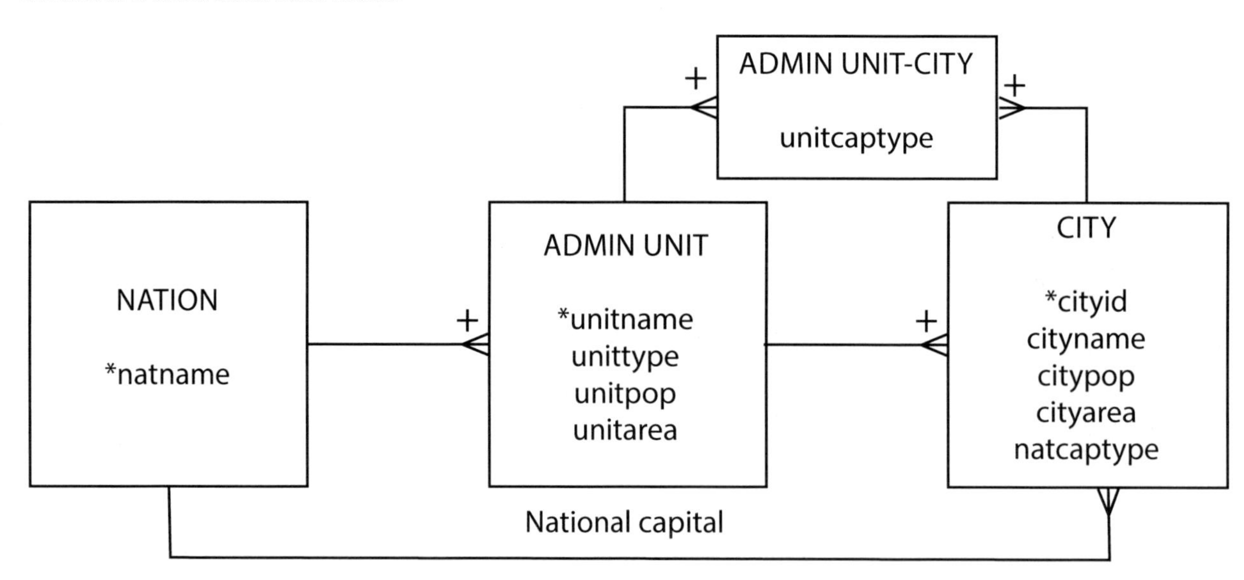

This geography lesson demonstrates how you often start with a simple model of low fidelity. By additional thinking about relationships and consideration of exceptions, you gradually create a high-fidelity data model.

Skill Builder

Write SQL to determine which administrative units have two capitals and which have a shared capital.

Family matters

Families can be very complicated. It can be tricky to keep track of all those relations—maybe you need a relational database (this is the worst joke in the book; they improve after this). We start with a very limited view of marriage and gradually ease the restrictions to demonstrate how any and all aspects of a relationship can be modeled. An initial fragment of the data model is shown in the following figure. There are several things to notice about it.

A man-woman data model

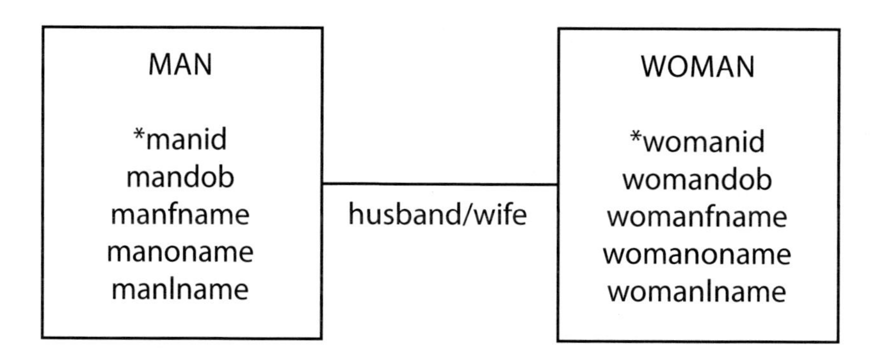

1. There is a 1:1 relationship between man and woman. The labels indicate that this relationship is marriage, but there is no indication of the marriage date. We are left to infer that the data model records the current marriage.
2. MAN and WOMAN have the same attributes. A prefix is used to make them unique. Both have an identifier (*id*, in the United States this would probably be a Social Security number), date of birth (*dob*), first name (*fname*), other names (*oname*), and last name (*name*).
3. Other names (*oname*) looks like a multivalued attribute, which is not allowed. Should *oname* be single or multivalue? This is a tricky decision. It depends on how these data will be used. If queries such as "find all men whose other names include Herbert" are likely, then a person's other names—kept as a separate entity that has a 1:m relationship for MAN and WOMAN—must be established. That is, a man or woman can have one or more other names. However, if you just want to store the data for completeness and retrieve it in its entirety, then *oname* is fine. It is just a text string. The key question to ask is, "What is the lowest level of detail possibly required for future queries?" Also, remember that SQL's REGEXP clause can be used to search for values within a text string.

The use of a prefix to distinguish attribute names in different entities suggests you should consider combining the entities. In this case, we could combine the entities MAN and WOMAN to create an entity called PERSON. Usually when entities are combined, you have to create a new attribute to distinguish between the different types. In this example, the attribute *gender* is added. We can also generalize the relationship label to *spouse*. The revised data model appears in the following figure.

A PERSON data model

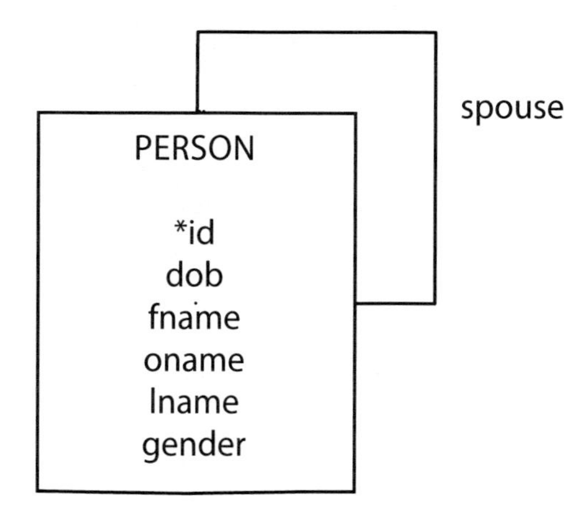

Now for a bit of marriage counseling. Marriage normally is a relationship between two people. Is that so? Well, some societies permit polygyny (one man can have many wives), others allow polyandry (one woman can have many husbands), and some permit same-sex marriages. So marriage, if we consider all the possibilities, is an m:m relationship between people, and partner might be a better choice than spouse for labeling the relationship Also, a person can be married more than once. To distinguish between different marriages, we really need to record the start and end date of each relationship and who was involved.

A marriage data model

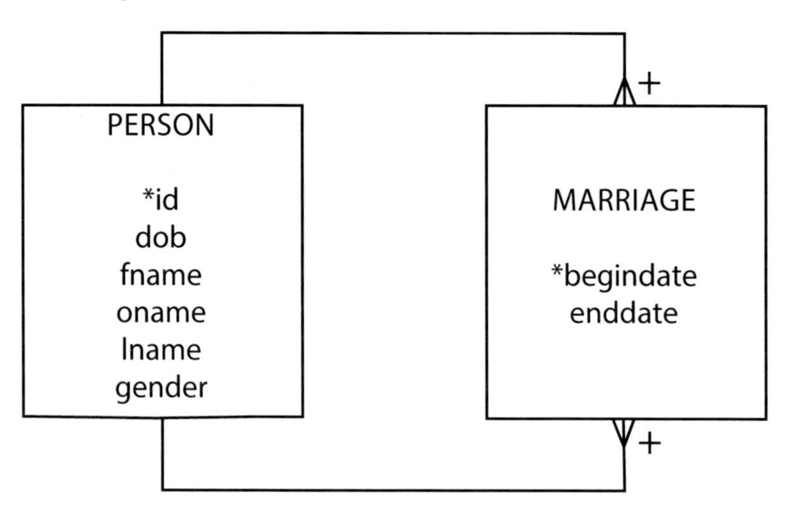

Marriage is an m:m relationship between two persons. It has attributes *begindate* and *enddate*. An instance of marriage is uniquely identified by a composite identifier: the two spouse identifiers and *begindate*. This means any marriage is uniquely identified by the composite of two person identifiers and the beginning date of the marriage. We need *begindate* as part of the identifier because the same couple might have more than one marriage (e.g., get divorced and remarry each other later). Furthermore, we can safely assume that it is impossible for a couple to get married, divorced, and remarried all on the one day. *Begindate* and *enddate* can be used to determine the current state of a marriage. If *enddate* is null, the marriage is current; otherwise, the couple has divorced.

This data model assumes a couple goes through some formal process to get married or divorced, and there is an official date for both of these events. What happens if they just gradually drift into cohabitation, and there is no official beginning date? Think about it. (The data model problem, that is—not cohabitation!) Many countries recognize this situation as a common-law marriage, so the data model needs to recognize it. The present data model cannot handle this situation because *begindate* cannot be null—it is an identifier. Instead, a new identifier is needed, and *begindate* should become an attribute.

Two new attributes can handle a common-law marriage. *Marriageno* can count the number of times a couple has been married to each other. In the majority of cases, *marriageno* will be 1. *Marriagestatus* can record whether a marriage is current or ended. Now we have a data model that can also handle common-law marriages. This is also a high-quality data model in that the client does not have to remember to examine *enddate* to determine a marriage's current status. It is easier to remember to examine *marriagestatus* to check status. Also, we can allow a couple to be married, divorced, and remarried as many times as they like on the one day—which means we can now use the database in Las Vegas.

A revised marriage data model

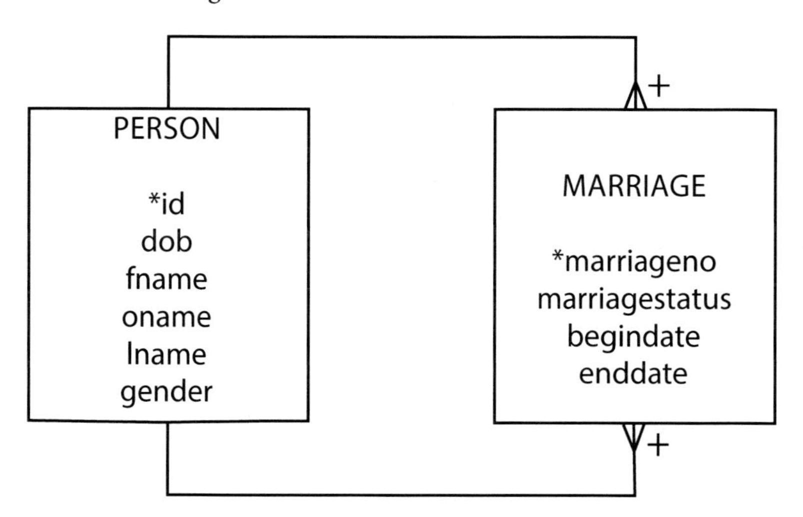

All right, now that we have the couple successfully married, we need to start thinking about children. A marriage has zero or more children, and let's start with the assumption a child belongs to only one marriage. Therefore, we have a 1:m relationship between marriage and person to represent the children of a marriage.

A marriage with children data model

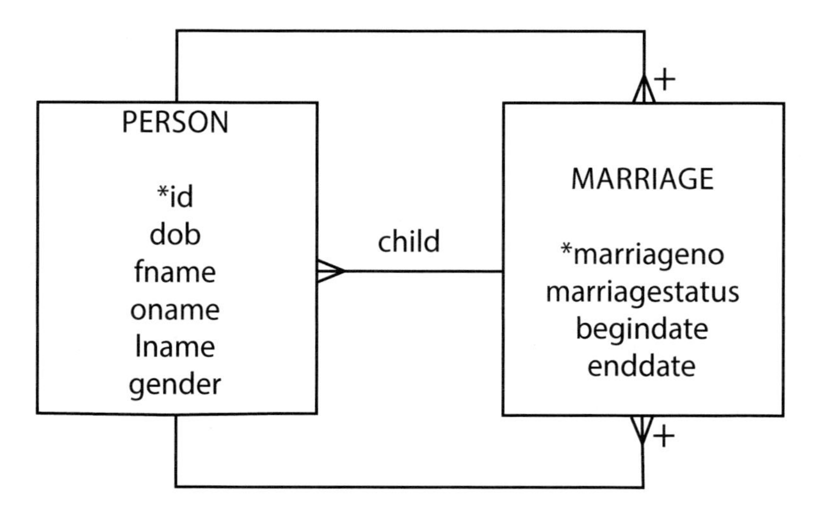

You might want to consider how the model would change to handle single-parent families, adopted children, and other aspects of human relationships.

Skill builder

The International Commission for Border Resolution Disputes requires a database to record details of which countries have common borders. Design the database. Incidentally, which country borders the most other countries?

When's a book not a book?

Sometimes we have to rethink our ideas of physical objects. Consider a data model for a library.

A library data model fragment

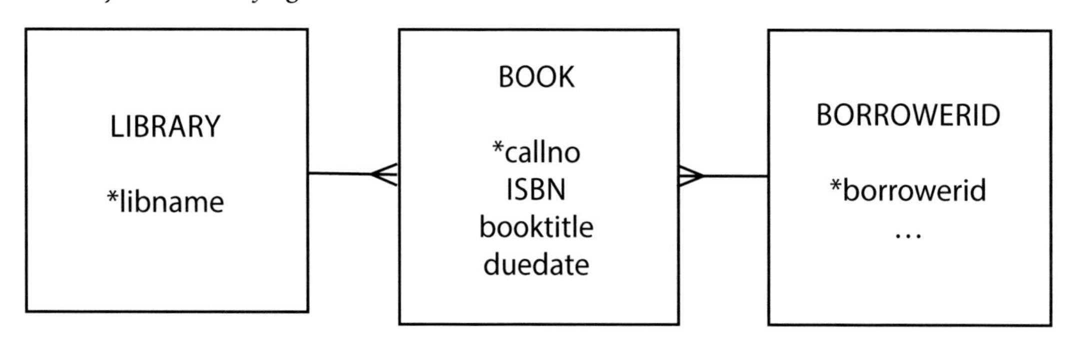

The preceding fragment assumes that a person borrows a book. What happens if the library has two copies of the book? Do we add an attribute to BOOK called *copy number*? No, because we would introduce redundancy by repeating the same information for each book. What you need to recognize is that in the realm of data modeling, a book is not really a physical thing, but the copy is, because it is what you borrow.

A revised library data model fragment

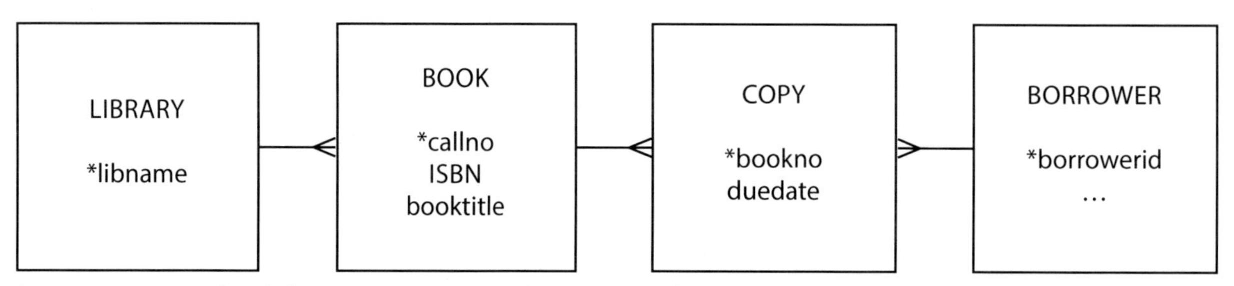

As you can see, a book has many copies, and a copy is of one book. A book can be identified by *callno*, which is usually a Library of Congress number or Dewey number. A copy is identified by a *bookno*. This is a unique number allocated by the library to the copy of the book. If you look in a library book, you will generally find it pasted on the inside back cover and shown in numeric and bar code format. Notice that it is called *book number* despite the fact that it really identifies a copy of a book. This is because most people, including librarians, think of the copy as a book.

The International Standard Book Number (ISBN) uniquely identifies any instance of a book (not copy). Although it sounds like a potential identifier for BOOK, it is not. ISBNs were introduced in the second half of the twentieth century, and books published before then do not have an ISBN.

A history lesson

Many organizations maintain historical data (e.g., a person's job history or a student's enrollment record). A data model can depict historical data relationships just as readily as current data relationships.

Consider the case of an employee who works for a firm that consists of divisions (e.g., production) and departments (e.g., quality control). The firm contains many departments, but a department belongs to only one division. At any one time, an employee belongs to only one department.

Employment history—take 1

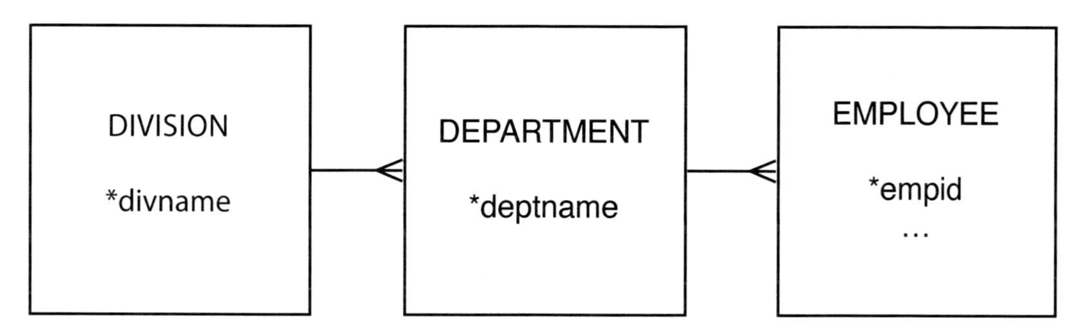

The fragment in the figure can be amended to keep track of the divisions and departments in which a person works. While employed with a firm, a person can work in more than one department. Since a department can have many employees, we have an m:m relationship between DEPARTMENT and EMPLOYEE. We might call the resulting associative entity POSITION.

Employment history—take 2

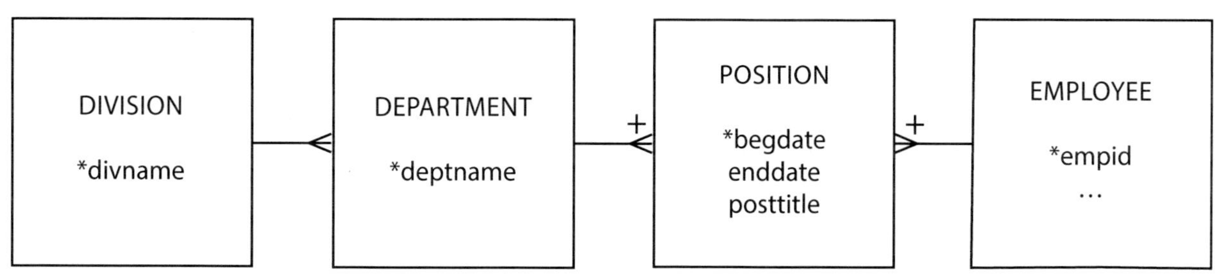

The revised fragment records employee work history. Note that any instance of POSITION is identified by *begdate* and the identifiers for EMPLOYEE and DEPARTMENT. The fragment is typical of what happens when you move from just keeping current data to recording history. A 1:m becomes an m:m relationship to record history.

People who work get paid. How do we keep track of an employee's pay data? An employee has many pay slips, but a pay slip belongs to one employee. When you look at a pay slip, you will find it contains many items: gross pay and a series of deductions for tax, medical insurance, and so on. Think of a pay slip as containing many lines, analogous to a sales form. Now look at the revised data model.

Employment history—take 3

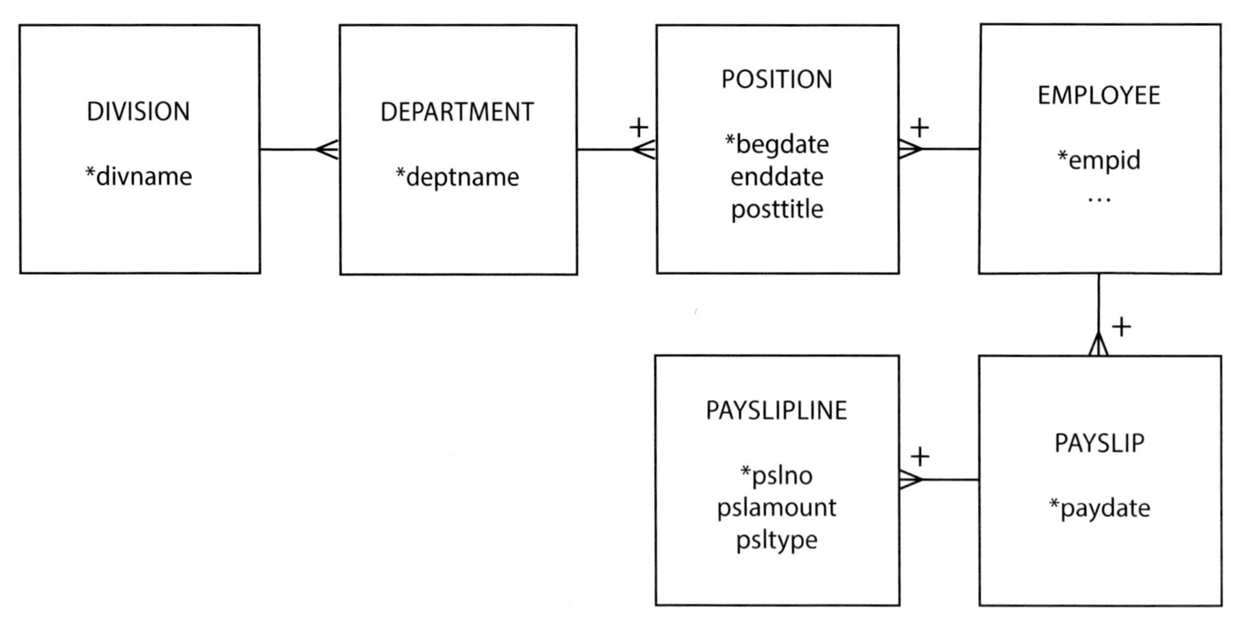

Typically, an amount shown on a pay slip is identified by a short text field (e.g., gross pay). PAYSLIPLINE contains an attribute *psltype* to identify the text that should accompany an amount, but where is the text? When dealing with codes like *psltype* and their associated text, the fidelity of a data model is improved by creating a separate entity, PSLTEXT, for the code and its text.

Employment history—take 4

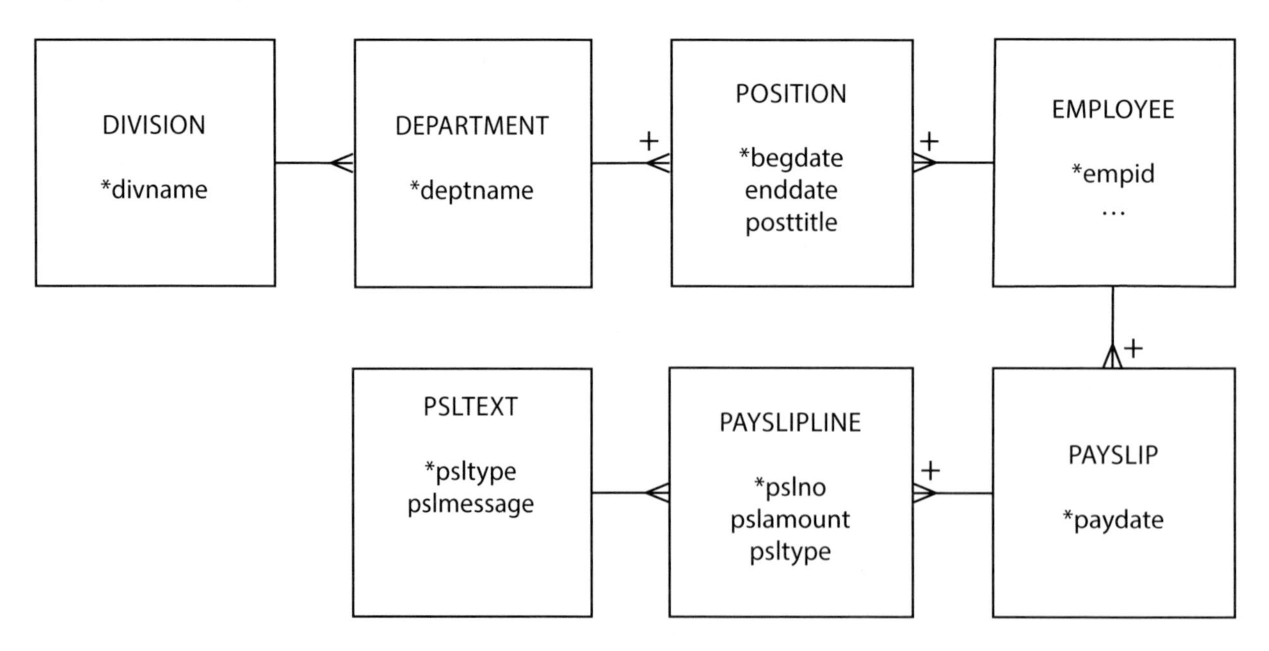

A ménage à trois for entities

Consider aircraft leasing. A plane is leased to an airline for a specific period. When a lease expires, the plane can be leased to another airline. So, an aircraft can be leased many times, and an airline can lease many aircraft. Furthermore, there is an agent responsible for handling each deal. An agent can lease many aircraft and deal with many airlines. Over time, an airline will deal with many agents. When an agent reaches a deal with an airline to lease a particular aircraft, you have a transaction. If you analyze the aircraft leasing business, you discover there are three m:m relationships. You might think of these as three separate relationships.

An aircraft-airline-agent data model

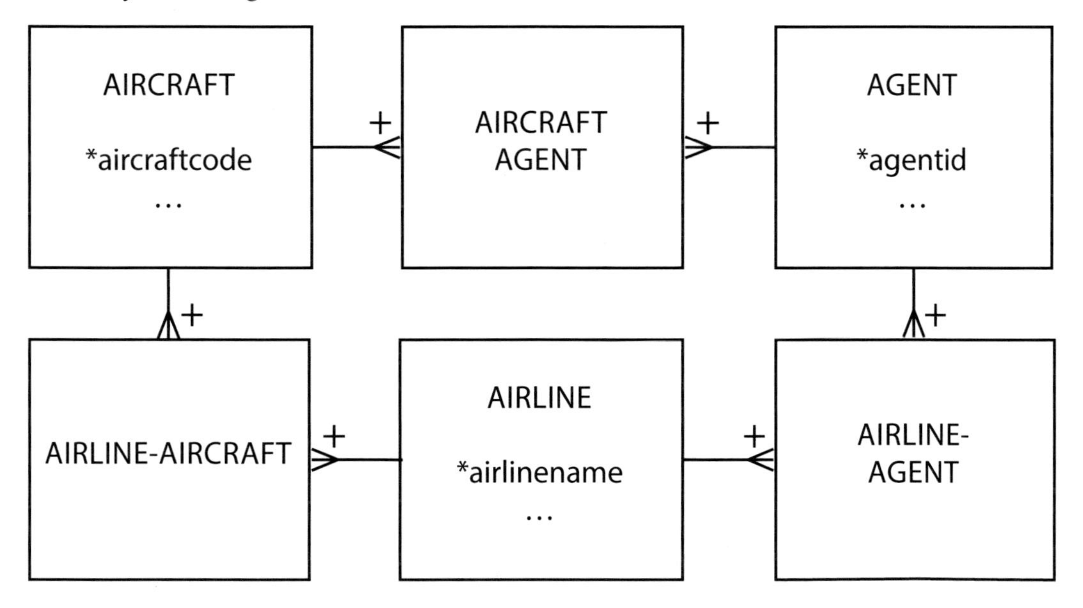

The problem with three separate m:m relationships is that it is unclear where to store data about the lease. Is it stored in AIRLINE-AIRCRAFT? If you store the information there, what do you do about recording the agent who closed the deal? After you read the fine print and do some more thinking, you discover that a lease is the association of these three entities in an m:m relationship.

A revised aircraft-airline-agent data model

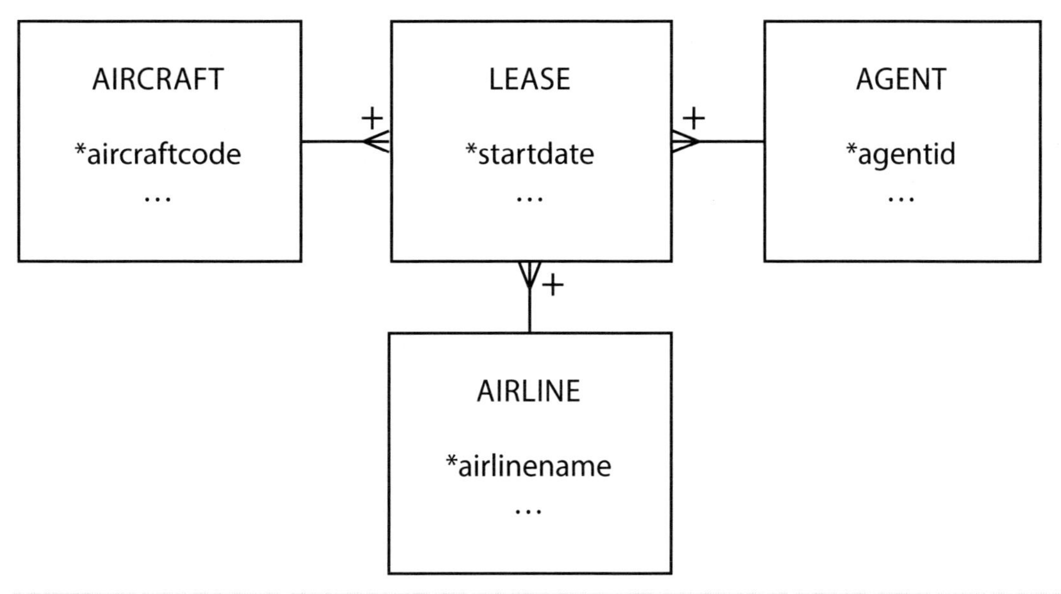

Skill builder

Horse racing is a popular sport in some parts of the world. A horse competes in at most one race on a course at a particular date. Over time, a horse can compete in many races on many courses. A horse's rider is called a jockey, and a jockey can ride many horses and a horse can have many jockeys. Of course, there is only ever one jockey riding a horse at a particular time. Courses vary in their features, such as the length of the course and the type of surface (e.g., dirt or grass). Design a database to keep track of the results of horse races.

Project management—planning and doing

Project management involves both planned and actual data. A project is divided into a number of activities that use resources. Planning includes estimation of the resources to be consumed. When a project is being executed, managers keep track of the resources used to monitor progress and keep the project on budget. Planning data may not be as detailed as actual data and is typically fairly broad, such as an estimate of the number of hours that an activity will take. Actual data will be more detailed because they are usually collected by getting those assigned to the project to log a daily account of how they spent their time. Also, resources used on a project are allocated to a particular activity. For the purposes of this data model, we will focus only on recording time and nothing else.

A project management data model

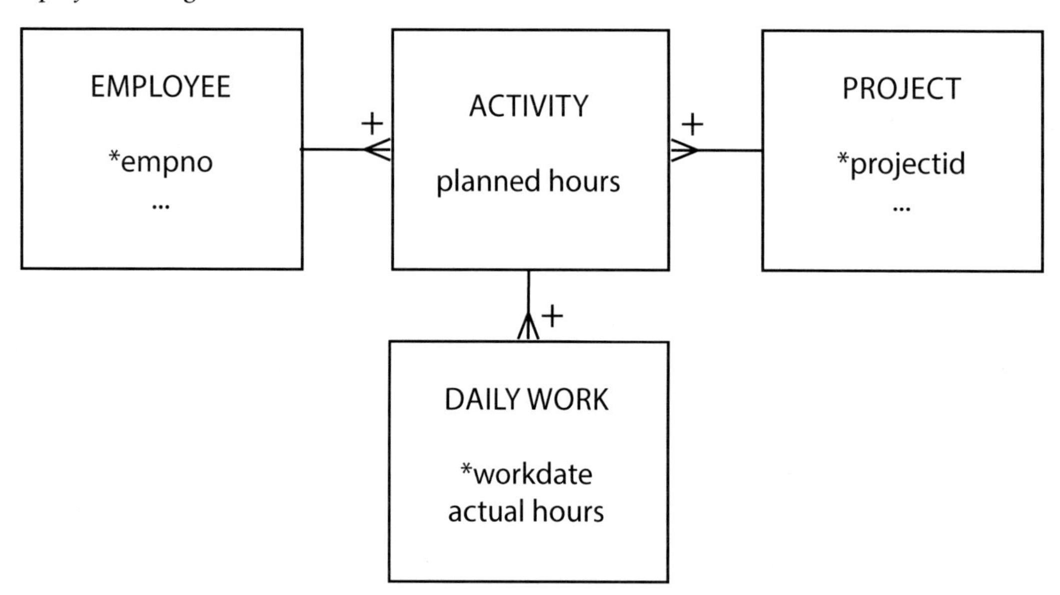

Notice that *planned hours* is an attribute of ACTIVITY, and *actual hours* is an attribute of DAILY WORK. The hours spent on an activity are derived by summing *actual hours* in the associated DAILY WORK entity. This is a high-fidelity data model if planning is done at the activity level and employees submit daily worksheets. Planning can be done in greater detail, however, if planners indicate how many hours of each day each employee should spend on a project.

A revised project management data model

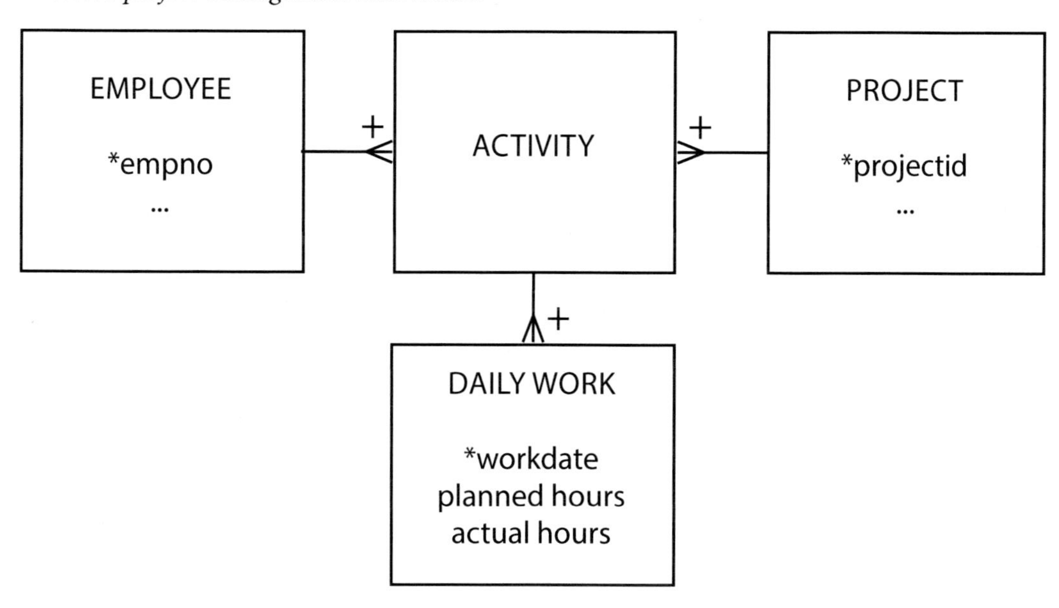

Now you see that *planned hours* and *actual hours* are both attributes of DAILY WORK. The message of this example, therefore, is not to assume that planned and actual data have the same level of detail, but do not be surprised if they do.

Cardinality

Some data modeling languages are quite precise in specifying the cardinality, or multiplicity, of a relationship. Thus, in a 1:m relationship, you might see additional information on the diagram. For example, the "many" end of the relationship might contain notation (such as 0,n) to indicate zero or more instances of the entity at the "many" end of the relationship.

Cardinality and modality options

Cardinality	Modality	Meaning
0,1	Optional	There can be zero or one instance of the entity relative to the other entity.
0,n		There can be zero or many instances of the entity relative to the other entity.
1,1	Mandatory	There is exactly one instance of the entity relative to the other entity.
1,n		The entity must have at least one and can have many instances relative to the other entity.

The data modeling method of this text, as you now realize, has taken a broad approach to cardinality (is it 1:1 or 1:m?). We have not been concerned with greater precision (is it 0,1 or 1,1?), because the focus has been on acquiring basic data modeling skills. Once you know how to model, you can add more detail. When learning to model, too much detail and too much terminology can get in the way of mastering this difficult skill. Also, it is far easier to sketch models on a whiteboard or paper when you have less to draw. Once you switch to a data modeling tool, then adding cardinality precision is easier and appropriate.

Modality

Modality, also known as optionality, specifies whether an instance of an entity must participate in a relationship. Cardinality indicates the range of instances of an entity participating in a relationship, while modality defines the minimum number of instances. Cardinality and modality are linked, as shown in the table above. If an entity is optional, the minimum cardinality will be 0, and if mandatory, the minimum cardinality is 1.

By asking the question, "Does an occurrence of this entity require an occurrence of the other entity?" you can assess whether an entity is mandatory. The usual practice is to use "O" for optional and place it near the entity that is optional. Mandatory relationships are often indicated by a short line (or bar) at right angles to the relationship between the entities.

In the following figure, it is mandatory for an instance of STOCK to have an instance of NATION, and optional for an instance of NATION to have an instance of STOCK, which means we can record details of a nation for which stocks have not yet been purchased. During conversion of the data model to a relational database, the mandatory requirement (i.e., each stock must have a nation) is enforced by the foreign key constraint.

1:m relationship showing modality

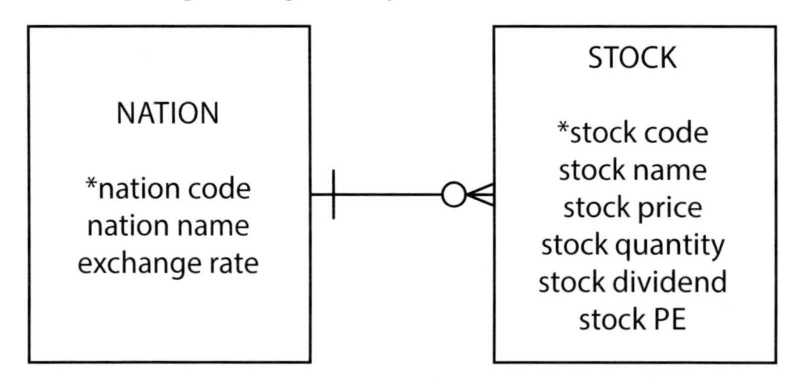

Now consider an example of an m:m relationship. A SALE can have many LINEITEMs, and a LINEITEM has only one SALE. An instance of LINEITEM must be related to an instance of SALE, otherwise it can't exist. Similarly, it is mandatory for an instance of SALE to have an instance of LINEITEM, because it makes no sense to have a sale without any items being sold.

In a relational database, the foreign key constraint handles the mandatory relationship between instances of LINEITEM and SALE, The mandatory requirement between SALE and LINEITEM has to be built into the processing logic of the application that processes a sales transaction. Basic business logic requires that a sale include some items, so a sales transaction creates one instance of SALE and multiple instances of LINEITEM. It does not make sense to have a sale that does not sell something.

An m:m relationship showing modality

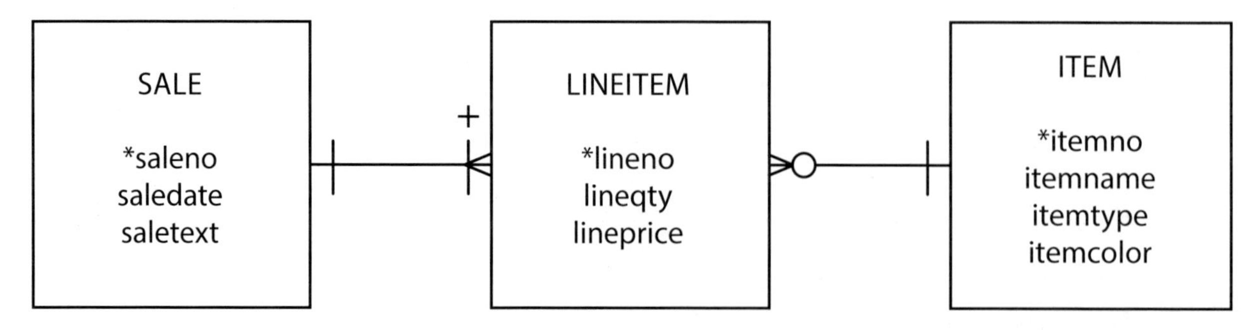

The relationship between ITEM and LINEITEM is similar in structure to that of NATION and STOCK, which we saw in the previous example. An instance of a LINEITEM has a mandatory requirement for an instance of ITEM. An instance of an ITEM can exist without the need for an instance of a LINEITEM (i.e., the items that have not been sold).

We gain further understanding of modality by considering a 1:1 relationship. We focus attention on the 1:1 because the 1:m relationship has been covered. The 1:1 implies that it is mandatory for an instance of DEPT to be related to one instance of EMP (i.e., a department must have a person who is the boss), and an instance of EMP is optionally related to one instance of DEPT (i.e., an employee can be a boss, but not all employees are departmental bosses).

A 1:1 relationship showing modality

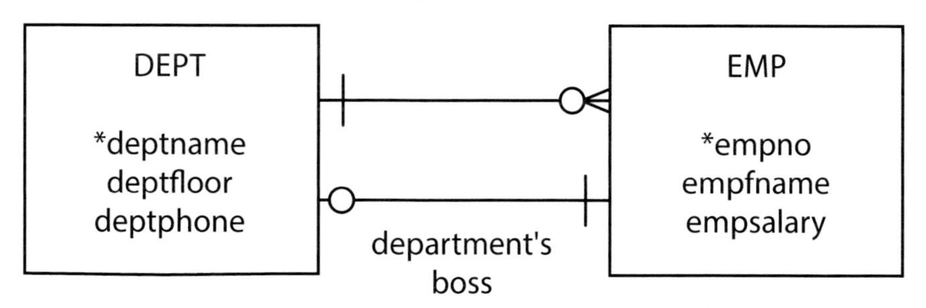

Earlier, when deciding where to store the foreign key for the 1:1 relationship, we settled on placing the foreign key in DEPT, because all departments have a boss. Now we have a rule. The foreign key goes with the entity for which the relationship is mandatory. When specifying the create statements, you will need to be aware of the chicken-and-egg connection between DEPT and EMP. We cannot insert an employee in EMP unless we know that person's department, and we cannot insert a department unless we have already inserted the boss for that department in EMP. We can get around this problem by using a **deferrable foreign key constraint**, but at this point we don't know enough about transaction processing to discuss this approach.

Let's now consider recursive relationships. The following figure shows a recursive 1:m between employees, representing that one employee can be the boss of many other employees and a person has one boss. It is optional that a person is a boss, and optional that everyone has a boss, mainly because there has to be one person who is the boss of everyone. However, we can get around this one exception by inserting employees in a particular order. By inserting the biggest boss first, we effectively make it mandatory that all employees have a boss. Nevertheless, data models cover every situation and exception, so we still show the relationship as optional.

1:m recursive with modality

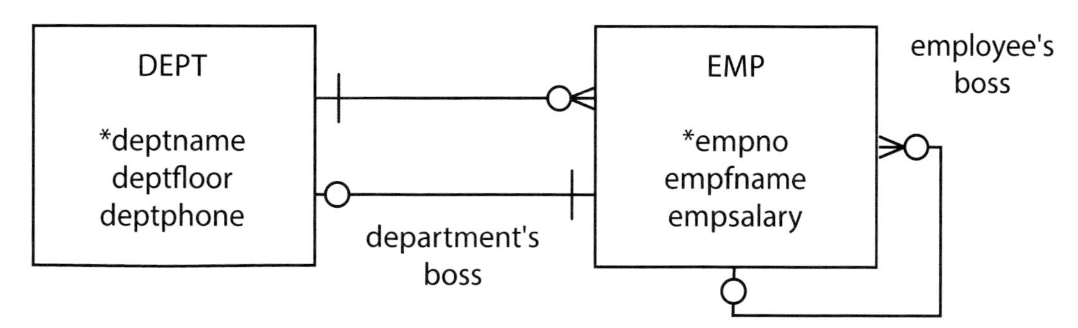

We rely again on the British monarchy. This time we use it to illustrate modality with a 1:1 recursive relationship. As the following figure shows, it is optional for a monarch to have a successor. We are not talking about the end of a monarchy, but rather how we address the data modeling issue of not knowing who succeeds the current monarch. Similarly, there must be a monarch who did not succeed someone (i.e., the first king or queen). Thus, both ends of the succession relationship have optional modality.

1:1 recursive with modality

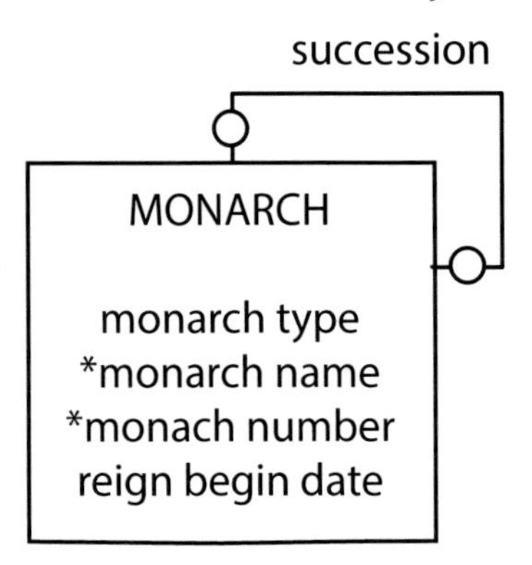

Finally, in our exploration of modality, we examine the m:m recursive. The model indicates that it is optional for a product to have components and optional for a product to be a component in other products. However, every assembly must have associated products.

m:m recursive with modality

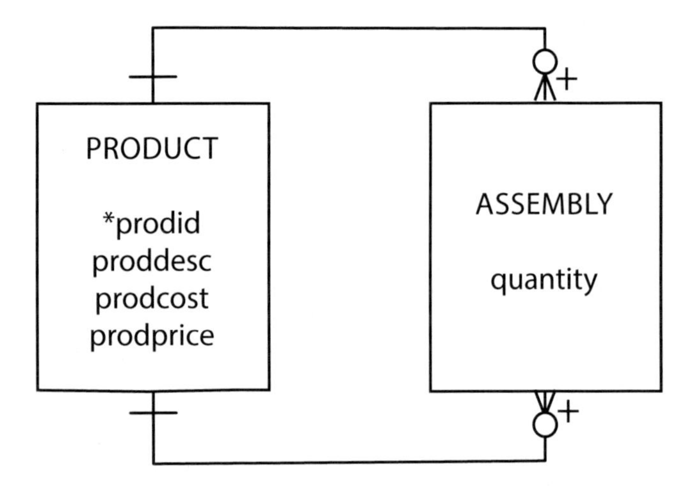

Summary

Modality adds additional information to a data model, and once you have grasped the fundamental ideas of entities and relationships, it is quite straightforward to consider whether a relationship is optional or mandatory. When a relationship is mandatory, you need to consider what constraint you can add to a table's definition to enforce the relationship. As we have seen, in some cases the foreign key constraint handles mandatory relationships. In other cases, the logic for enforcing a requirement will be built into the application.

Entity types

A data model contains different kinds of entities, which are distinguished by the format of their identifiers. Labeling each of these entities by type will help you to determine what questions to ask.

Independent entity

An independent entity is often central to a data model and foremost in the client's mind. Independent entities are frequently the starting points of a data model. Remember that in the investment database, the independent entities are STOCK and NATION. Independent entities typically have clearly distinguishable names because they occur so frequently in the client's world. In addition, they usually have a single identifier, such as *stock code* or *nation code.*

Independent entities are often connected to other independent entities in a 1:m or m:m relationship. The data model in the following figure shows two independent entities linked in a 1:m relationship.

NATION and STOCK—independent entities

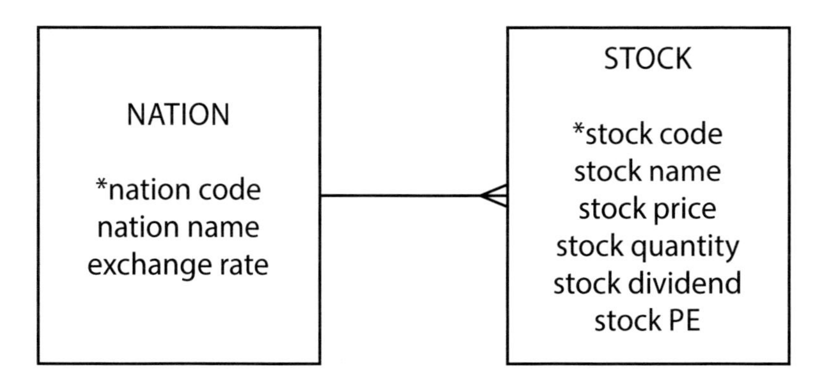

Weak or dependent entity

A weak entity (also known as a dependent entity) relies on another entity for its existence and identification. It is recognized by a plus on the weak entity's end of the relationship. CITY cannot exist without REGION. A CITY is uniquely identified by *cityname* and *regname.* If the composite identifier becomes unwieldy, creating an arbitrary identifier (e.g., cityno) will change the weak entity into an independent one.

CITY—a weak entity

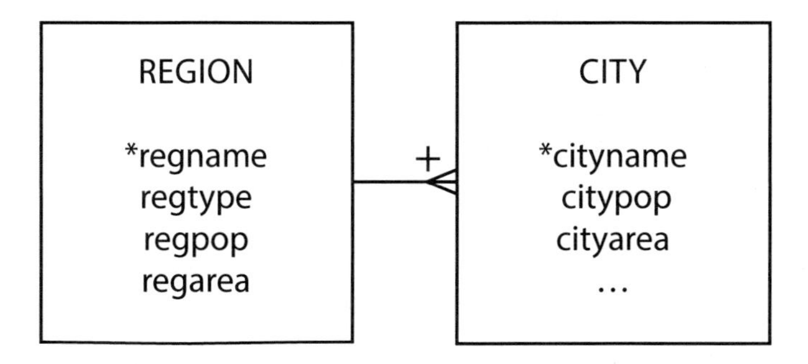

Associative entity

Associative entities are by-products of m:m relationships. They are typically found between independent entities. Associative entities sometimes have obvious names because they occur in the real world. For instance, the associative entity for the m:m relationship between DEPARTMENT and EMPLOYEE is usually called POSITION. If the associative entity does not have a common name, the two entity names are generally hyphenated (e.g., DEPARTMENT-EMPLOYEE). Always search for the appropriate name, because it will improve the quality of the data model. Hyphenated names are a last resort.

POSITION is an associative entity

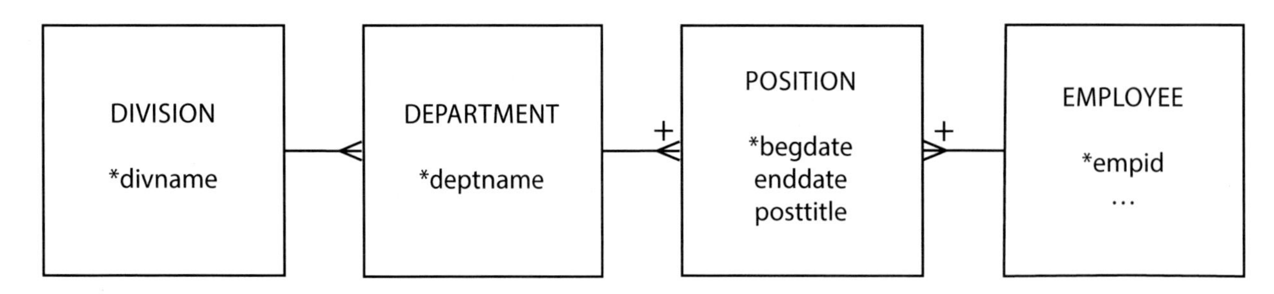

Associative entities can show either the current or the historic relationship between two entities. If an associative entity's only identifiers are the two related entities' identifiers, then it records the current relationship between the entities. If the associative entity has some time measure (e.g., date or hour) as a partial identifier, then it records the history of the relationship. Whenever you find an associative entity, ask the client whether the history of the relationship should be recorded.

Creating a single, arbitrary identifier for an associative entity will change it to an independent one. This is likely to happen if the associative entity becomes central to an application. For example, if a personnel department does a lot of work with POSITION, it may find it more expedient to give it a separate identifier (e.g., *position number*).

Aggregate entity

An aggregate entity is created when several different entities have similar attributes that are distinguished by a preceding or following modifier to keep their names unique. For example, because components of an address might occur in several entities (e.g., CUSTOMER and SUPPLIER), an aggregate address entity can be created to store details of all addresses. Aggregate entities usually become independent entities. In this case, we could use *address number* to identify an instance of address uniquely.

Subordinate entity

A subordinate entity stores data about an entity that can vary among instances. A subordinate entity is useful when an entity consists of mutually exclusive classes that have different descriptions. The farm animal database shown in the following figure indicates that the farmer requires different data for sheep and horses. Notice that each of the subordinate entities is identified by the related entity's identifier. You could avoid subordinate entities by placing all the attributes in ANIMAL, but then you make it incumbent on the client to remember which attributes apply to horses and which to sheep. This becomes an important issue for null fields. For example, is the attribute *hay consumption* null because it does not apply to sheep, or is it null because the value is unknown?

SHEEP and HORSE are subordinate entities

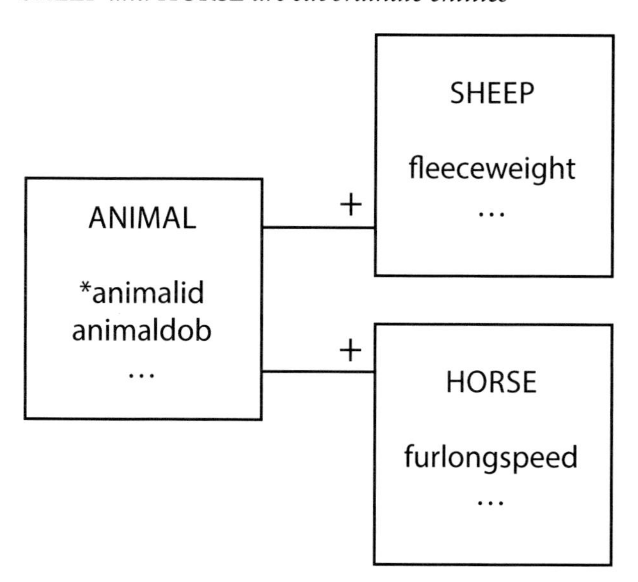

If a subordinate entity becomes important, it is likely to evolve into an independent entity. The framing of a problem often determines whether subordinate entities are created. Stating the problem as, "a farmer has many animals, and these animals can be horses or sheep," might lead to the creation of subordinate entities. Alternatively, saying that "a farmer has many sheep and horses" might lead to setting up SHEEP and HORSE as independent entities.

Generalization and aggregation

Generalization and aggregation are common ideas in many modeling languages, particularly object-oriented modeling languages such as the Unified Modeling Language (UML), which we will use in this section.

Generalization

A generalization is a relationship between a more general element and a more specific element. In the following figure, the general element is *animal*, and the specific elements are *sheep* and *horse*. A horse is a subtype of animal. A generalization is often called an "is a" relationship because the subtype element is a member of the generalization.

A generalization is directly mapped to the relational model with one table for each entity. For each of the subtype entities (i.e., SHEEP and HORSE), the primary key is that of the supertype entity (i.e., ANIMAL). You must also make this column a foreign key so that a subtype cannot be inserted without the presence of the matching supertype. A generalization is represented by a series of 1:1 relationships to weak entities.

A generalization hierarchy in UML

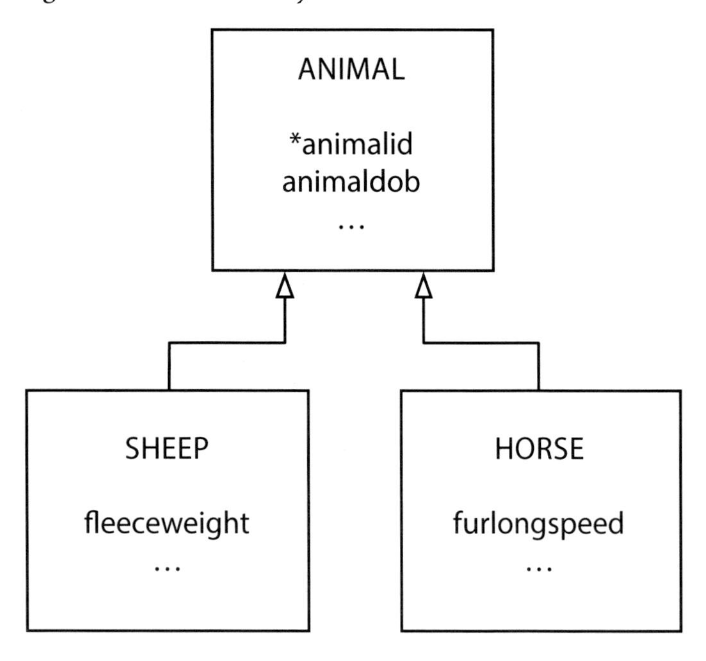

Aggregation

Aggregation is a part-whole relationship between two entities. Common phrases used to describe an aggregation are "consists of," "contains," and "is part of." The following figure shows an aggregation where a herd consists of many cows. Cows can be removed or added, but it is still a herd. An open diamond next to the whole denotes an aggregation. In data modeling terms, there is a 1:m relationship between herd and cow.

An aggregation in UML and data modeling

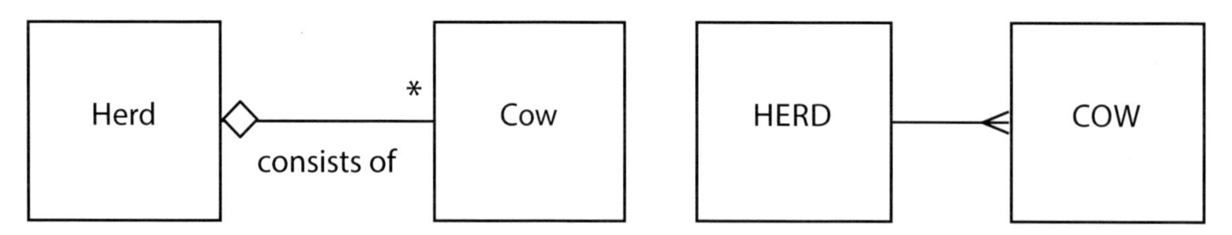

There are two special types of aggregation: shared and composition. With **shared aggregation,** one entity owns another entity, but other entities can own that entity as well. A sample shared aggregation is displayed in the following figure, which shows that a class has many students, and a student can be a member of many classes. Students, the parts in this case, can be part of many classes. An open diamond next to the whole denotes a shared aggregation, which we represent in data modeling as an m:m relationship.

Shared aggregation in UML and data modeling

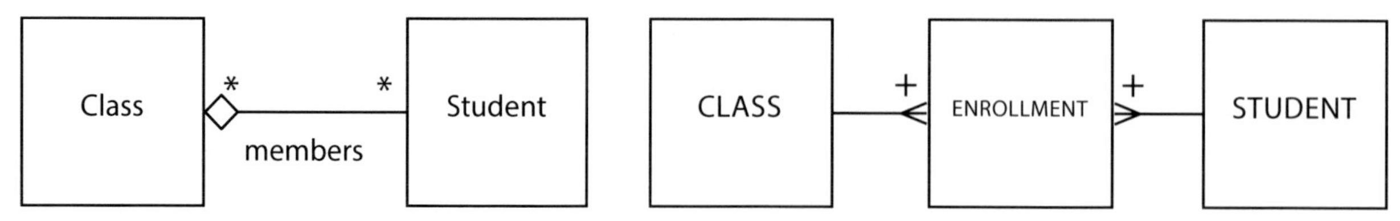

In a **composition aggregation,** one entity exclusively owns the other entity. A solid diamond at the whole end of the relationship denotes composition aggregation. The appropriate data model is a weak entity, as shown in the following figure.

Composition aggregation in UML and data modeling

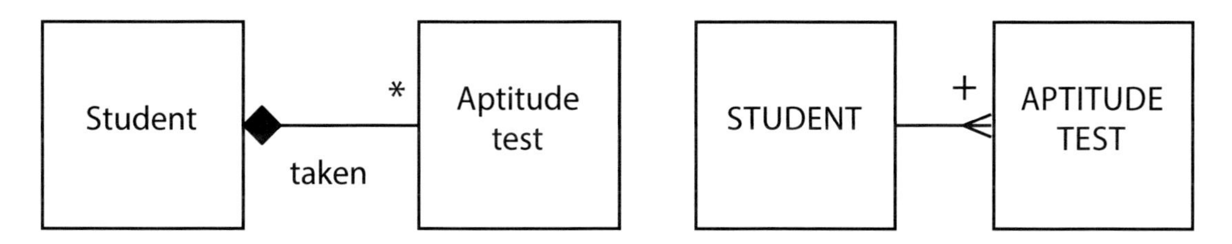

Data modeling hints

A high-fidelity data model takes time to develop. The client must explain the problem fully so that the database designer can understand the client's requirements and translate them into a data model. Data modeling is like prototyping; the database designer gradually creates a model of the database as a result of interaction with the client. Some issues will frequently arise in this progressive creation, and the following hints should help you resolve many of the common problems you will encounter.

The rise and fall of a data model

Expect your data model to both expand and contract. Your initial data model will expand as you extend the boundaries of the application. You will discover some attributes will evolve into entities, and some 1:m relationships will become m:m relationships. It will grow because you will add entities, attributes, and relationships to allow for exceptions. Your data model will grow because you are representing the complexity of the real world. Do not try to constrain growth. Let the data model be as large as is necessary. As a rough rule, expect your final data model to grow to about two to three times the number of entities of your initial data model.

Expect your data model to contract as you generalize structures. The following fragment is typical of the models produced by novice data modelers. It can be reduced by recognizing that PAINTING, SCULPTURE, and CERAMIC are all types of art. A more general entity, ART, could be used to represent the same data. Of course, it would need an additional attribute to distinguish the different types of art recordings.

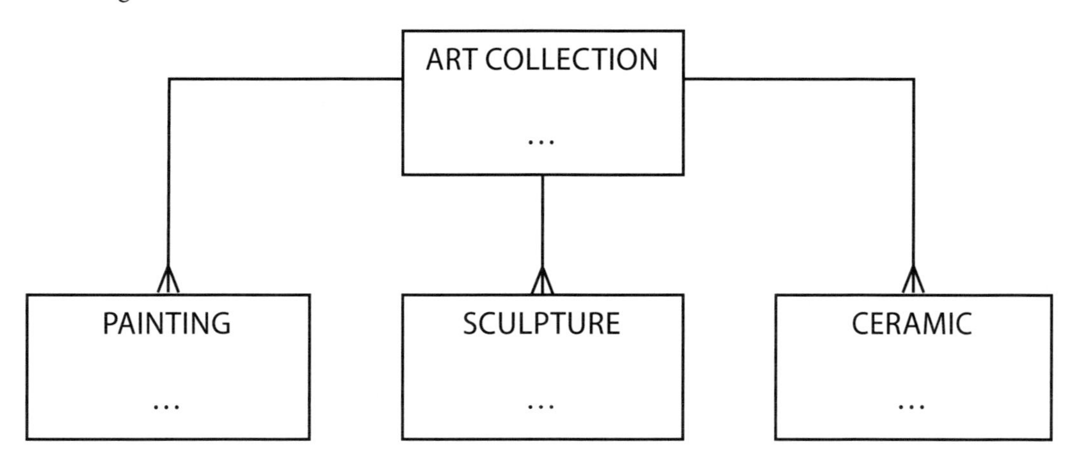

The shrunken data model

ART
...
art type

As you discover new entities, your data model will grow. As you generalize, your data model will shrink.

Identifier

If there is no obvious simple identifier, invent one. The simplest is a meaningless code (e.g., *person number* or *order number*). If you create an identifier, you can also guarantee its uniqueness.

Don't overwork an identifier. The worst case we have seen is a 22-character product code used by a plastics company that supposedly not only uniquely identified an item but told you its color, its manufacturing process, and type of plastic used to make it! Color, manufacturing process, and type of plastic are all attributes. This product code was unwieldy. Data entry error rates were extremely high, and very few people could remember how to decipher the code.

An identifier has to do only one thing: uniquely identify every instance of the entity. Sometimes trying to make it do double, or even triple, duty only creates more work for the client.

Position and order

There is no ordering in a data model. Entities can appear anywhere. You will usually find a central entity near the middle of the data model only because you tend to start with prominent entities (e.g., starting with STUDENT when modeling a student information system). What really matters is that you identify all relevant entities.

Attributes are in no order. You find that you will tend to list them as they are identified. For readability, it is a good idea to list some attributes sequentially. For example, first name, other names, and last name are usually together. This is not necessary, but it speeds up verification of a data model's completeness.

Instances are also assumed to have no ordering. There is no first instance, next instance, or last instance. If you must recognize a particular order, create an attribute to record the order. For example, if ranking of potential investment projects must be recorded, include an attribute (*projrank*) to store this data. This does not mean the instances will be stored in *projrank* order, but it does allow you to use the ORDER BY clause of SQL to report the projects in rank order.

If you need to store details of a precedence relationship (i.e., there is an ordering of instances and a need to know the successor and predecessor of any instance), then use a 1:1 recursive relationship.

Using an attribute to record an ordering of instances

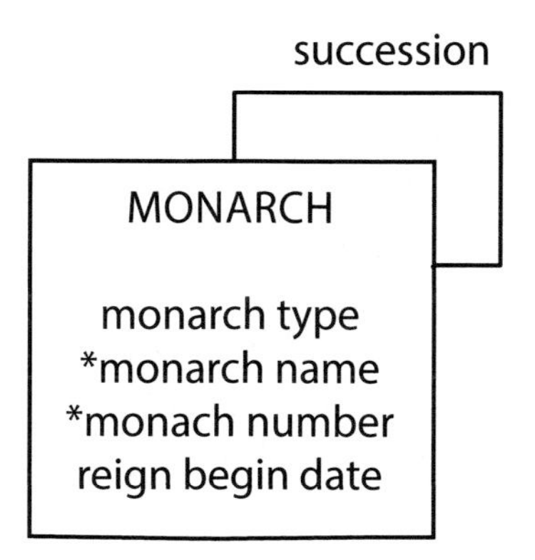

Attributes and consistency

Attributes must be consistent, retaining the same meaning for every instance. An example of an inconsistent attribute would be an attribute *stock info* that contains a stock's PE ratio or its ROI (return on investment). Another attribute, say *stock info code*, is required to decipher which meaning applies to any particular instance. The code could be "1" for PE and "2" for ROI. If one attribute determines the meaning of another, you have inconsistency. Writing SQL queries will be extremely challenging because inconsistent data increases query complexity. It is better to create separate attributes for PE and ROI.

Names and addresses

An attribute is the smallest piece of data that will conceivably form part of a query. If an attribute has segments (e.g., a person's name), determine whether these could form part of a query. If so, then make them separate attributes and reapply the query test. When you apply the query test to *person name*, you will usually decide to create three attributes: *first name*, *other names*, and *last name*.

What about titles? There are two sorts of modifier titles: preceding titles (e.g., Mr., Mrs., Ms., and Dr.) and following titles (e.g., Jr. and III). These should be separate attributes, especially if you have divided *name* into separate attributes.

Addresses seem to cause more concern than names because they are more variant. Business addresses tend to be the most complicated, and foreign addresses add a few more twists. The data model fragment in the following figure works in most cases.

Handling addresses

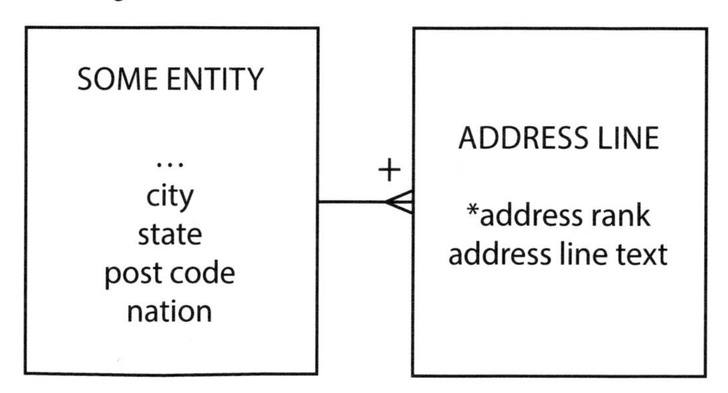

Common features of every address are city, state (province in Canada, county in Eire), postal code (ZIP code in the United States), and nation. Some of these can be null. For example, Singapore does not have any units corresponding to states. In the United States, city and state can be derived from the zip code, but this is not necessarily true for other countries. Furthermore, even if this were true for every nation, you would need a different derivation rule for each country, which is a complexity you might want to avoid. If there is a lifetime guarantee that every address in the database will be for one country, then examine the possibility of reducing redundancy by just storing post code.

Notice that the problem of multiple address lines is represented by a 1:m relationship. An address line is a text string that appears as one line of an address. There is often a set sequence in which these are displayed. The attribute *address rank* records this order.

How do you address students? First names may be all right for class, but it is not adequate enough for student records. Students typically have multiple addresses: a home address, a school address, and maybe a summer address. The following data model fragment shows how to handle this situation. Any instance of ADDRESS LINE is identified by the composite of *studentid, addresstype*, and *address rank.*

Students with multiple addresses

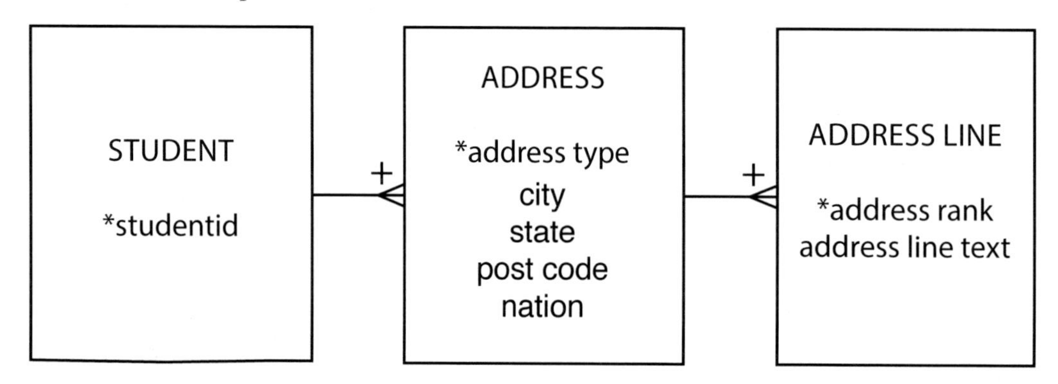

The identifier *address type* is used to distinguish among the different types of addresses. The same data model fragment works in situations where a business has different mailing and shipping addresses.

When data modeling, can you take a shortcut with names and addresses? It is time consuming to write out all the components of name and address. It creates clutter and does not add much fidelity. Our practical advice is to do two things. *First,* create a policy for all names and addresses (e.g., all names will be stored as three parts: first name, other names, last name). *Second,* use shorthand forms of attributes for names and addresses when they obey the policy. So from now on, we will sometimes use name and address as attributes with the understanding that when the database is created, the parts will become separate columns.

Single-instance entities

Do not be afraid of creating an entity with a single instance. Consider the data model fragment in the following figure which describes a single fast-food chain. Because the data model describes only one firm, the entity FIRM will have only one instance. The inclusion of this single instance entity permits facts about the firm to be maintained. Furthermore, it provides flexibility for expansion. If two fast-food chains combine, the data model needs no amendment.

FIRM—a single-instance entity

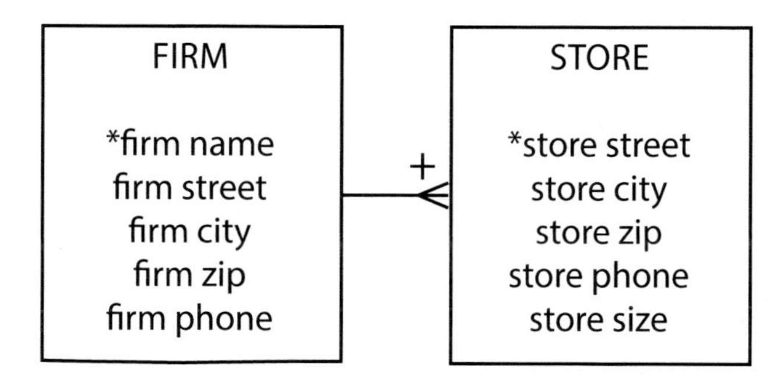

Picking words

Words are very important. They are all we have to make ourselves understood. Let the client choose the words, because the data model belongs to the client and describes the client's world. If you try to impose your words on the client, you are likely to create a misunderstanding.

Synonyms

Synonyms are words that have the same meaning. For example, task, assignment, and project might all refer to the same real-world object. One of these terms, maybe the one in most common use, needs to be selected for naming the entity. Clients need to be told the official name for the entity and encouraged to adopt this term for describing it. Alternatively, create views that enable clients to use their own term. Synonyms are not a technical problem. It is a social problem of getting clients to agree on one word.

Homonyms

Homonyms are words that sound the same but have different meanings. Homonyms can create real confusion. *Sale date* is a classic example. In business it can have several meanings. To the salespeople, it might be the day that the customer shakes hands and says, "We have a deal." For the lawyers, it is could be the day when the contract is signed, and for production, it might be the day the customer takes delivery. In this case, the solution is reasonably obvious; redefine *sale date* to be separate terms for each area (e.g., *sales date*, *contract date*, and *delivery date* for sales, legal, and production departments, respectively).

Homonyms cause real confusion when clients do not realize that they are using the same word for different entities or attributes. Hunt down homonyms by asking lots of questions and querying a range of clients.

Exception hunting

Go hunting for exceptions. Keep asking the client questions such as

- Is it always like this?
- Would there be any situations where this could be an m:m relationship?
- Have there ever been any exceptions?
- Are things likely to change in the future?

Always probe for exceptions and look for them in the examples the client uses. Redesigning the data model to handle exceptions increases fidelity.

Relationship labeling

Relationship labeling clutters a data model. In most cases, labels can be correctly inferred, so use labels only when there is a possibility of ambiguity.

Keeping the data model in shape

As you add to a data model, maintain fidelity. Do not add entities without also completing details of the identifier and attributes. By keeping the data model well formed, you avoid ambiguity, which frequently leads to miscommunication.

Used entities

Would you buy a used data model? Certainly, because a used data model is more likely to have higher fidelity than a brand new one. A used data model has been subjected to much scrutiny and revision and should be a more accurate representation of the real world.

Meaningful identifiers

An identifier is said to be meaningful when some attributes of the entity can be inferred from the identifier's value (e.g., a code of B78 for an item informs a clerk that the item is black). The word "meaningful" in everyday speech usually connotes something desirable, but many data managers agree that meaningful identifiers or codes create problems. While avoiding meaningful identifiers is generally accepted as good data management practice, they are still widely used. After a short description of some examples of meaningful identifiers, this section discusses the pros and cons of meaningful identifiers.

The invoice number "12dec0001" is a meaningful identifier. The first five positions indicate in which period the invoice was generated. The last four digits of the invoice number have a meaning. The 0001 indicates it was the first invoice sent in December. In addition, e-mail addresses, like mvdpas@bizzo.nl, have a meaning.

When coding a person's gender, some systems use the codes "m" and "f" and others "1" and "2." The pair (m, f) is meaningful, but (1, 2) is not and puts the onus on the user to remember the meaning. The codes "m" and "f"" are derived from the reality that they represent. They are the first letters of the words male and female. The "1" and "2" are not abstracted from reality, and therefore they do not have a meaning. This observation is reiterated by the following quote from the International Organization for Standardization (ISO) information interchange standard for the representation of human sexes (ISO 5218):

No significance is to be placed on the fact that "Male" is coded "1" and "Female" is coded "2." This standard was developed based upon predominant practices of the countries involved and does not convey any meaning of importance, ranking or any other basis that could imply discrimination.

In determining whether an identifier is meaningful, one can look at the way new values of that identifier are generated. If the creation takes place on the basis of characteristics that appear in reality, the identifier is meaningful. That is why the identifier of the invoice "12dec0002," which is the identifier assigned to the second invoice generated in December 2012, is meaningful. It is also why the e-mail address rwatson@terry.uga.edu is meaningful.

Meaningful identifiers clearly have some advantages and disadvantages, and these are now considered. Advantages and disadvantages of meaningful identifiers

Advantages	Disadvantages
Recognizable and rememberable	Identifier exhaustion
Administrative simplicity	Reality changes
	Loss of meaningfulness

Advantages of meaningful identifiers

Recognizable and rememberable

People can often recognize and remember the significance of meaningful identifiers. If, for example, someone receives an e-mail from rwatson@terry.uga.edu, this person can probably deduce the sender's name and workplace. If a manufacturer, for instance, uses the last two digits of its furniture code to indicate the color (e.g., 08 = beech), employees can then quickly determine the color of a packed piece of furniture. Of course, the firm could also show the color of the furniture on the label, in which case customers, who are most unlikely to know the coding scheme, can quickly determine the color of the enclosed product.

Administrative simplicity

By using meaningful identifiers, administration can be relatively easily decentralized. Administering e-mail addresses, for instance, can be handled at the domain level. This way, one can avoid the process of synchronizing with other domains whenever a new e-mail address is created.

A second example is the EAN (European article number), the bar code on European products. Issuing EANs can be administered per country, since each country has a unique number in the EAN. When issuing a new number, a country need only issue a code that is not yet in use in that country. The method of contacting all countries to ask whether an EAN has already been issued is time consuming and open to error. On the other hand, creating a central database of issued EAN codes is an option. If every issuer of EAN codes can access such a database, alignment among countries is created and duplicates are prevented.

Disadvantages of meaningful identifiers

Identifier exhaustion

Suppose a company with a six-digit item identifier decides to use the first three digits to specify the product group (e.g., stationery) uniquely. Within this group, the remaining three digits are used to define the item (e.g., yellow lined pad). As a consequence, only one thousand items can be specified per product group. This problem remains even when some numbers in a specific product group, or even when entire product groups, are not used.

Consider the case when product group "010" is exhausted and is supplemented by the still unused product group code "940." What seems to be a simple quick fix causes problems, because a particular product group is not uniquely identified by a single identifier. Clients have to remember there is an exception.

A real-world example of identifier exhaustion is a holding company that identifies its subsidiaries by using a code, where the first character is the first letter of the subsidiary's name. When the holding company acquired its seventh subsidiary, the meaningfulness of the coding system failed. The newly acquired subsidiary name started with the same letter as one of the existing subsidiaries. The system had already run out of identifiers.

When existing identifiers have to be converted because of exhaustion, printed codes on labels, packages, shelves, and catalogs will have to be redone. Recoding is an expensive process and should be avoided.

Exhaustion is not only a problem with meaningful identifiers. It can happen with non-meaningful identifiers. The problem is that meaningful identifier systems tend to exhaust sooner. To avoid identifier exhaustion and maintain meaningful identifiers, some designers increase identifier lengths (e.g., four digits for product group).

Reality changes

The second problem of meaningful identifiers is that the reality they record changes. For example, a company changes its name. The College of Business at the University of Georgia was renamed the Terry College of Business. E-mail addresses changed (e.g., rwatson@cba.uga.edu became rwatson@terry.uga.edu). An identifier can remain meaningful over a long period only if the meaningful part rarely or never changes.

Non-meaningful identifiers avoid reality changes. Consider the case of most telephone billing systems, which use an account ID to identify a customer rather than a telephone number. This means a customer can change telephone numbers without causing any coding problems because telephone number is not the unique identifier of each customer.

A meaningful identifier can lose its meaningfulness

Consider the problems that can occur with the identifier that records both the color and the material of an article. Chipboard with a beech wood veneer is coded BW. A black product, on the other hand, is coded BL. Problems arise whenever there is an overlap between these characteristics. What happens with a product that is finished with a black veneer on beech wood? Should the code be BW or BL? What about a black-painted product that is made out of beech wood? Maybe a new code, BB for black beech wood, is required. There will always be employees and customers who do not know the meaning of these not-so-meaningful identifiers and attribute the wrong meaning to them.

The solution—non-meaningful identifiers

Most people are initially inclined to try to make identifiers meaningful. There is a sense that something is lost by using a simple numeric to identify a product or customer. Nothing, however, is lost and much is gained. Non-meaningful identifiers serve their sole purpose well—to identify an entity uniquely. Attributes are used to describe the characteristics of the entity (e.g., color, type of wood). A clear distinction between the role of identifiers and attributes creates fewer data management problems now and in the future.

Vehicle identification

In most countries, every road vehicle is uniquely identified. The Vehicle Identification Number (VIN) was originally described in ISO Standard 3779 in February 1977 and last revised in 1983. It is designed to identify motor vehicles, trailers, motorcycles, and mopeds. A VIN is 17 characters, A through Z and 0 through 9.

The European Union and the United States have different implementations of the standard. The U.S. VIN is divided into four parts

- World Manufacturer's Identification (WMI) - three characters
- Vehicle Description Section (VDS) - five characters
- Check digit
- Vehicle Identification Section (VIS) - eight characters

When decoded, a VIN specifies the country and year of manufacture; make, model, and serial number; assembly plant; and even some equipment specifications.

VINs are normally located in several locations on a car, but the most common places are in the door frame of the front doors, on the engine itself, around the steering wheel, or on the dash near the window.

What do you think of the VIN as a method of vehicle identification? How do you reconcile it with the recommendation to use meaningless identifiers? If you were consulted on a new VIN system, what would advise?

The seven habits of highly effective data modelers

There is often a large gap between the performance of *average* and *expert* data modelers. An insight into the characteristics that make some data modelers more skillful than others should improve your data modeling capabilities.

Immerse

Find out what the client wants by immersing yourself in the task environment. Spend some time following the client around and participate in daily business. Firsthand experience of the problem will give you a greater understanding of the client's requirements. Observe, ask questions, reflect, and talk to a wide variety of people (e.g., managers, operations personnel, customers, and suppliers). The more you learn about the problem, the better equipped you are to create a high-fidelity data model.

Challenge

Challenge existing assumptions; dig out the exceptions. Test the boundaries of the data model. Try to think about the business problem from different perspectives (e.g., how might the industry leader tackle this problem?). Run a brainstorming session with the client to stimulate the search for breakthrough solutions.

Generalize

Reduce the number of entities whenever possible by using generalized structures (remember the Art Collection model) to simplify the data model. Simpler data models are usually easier to understand and less costly to implement. Expert data modelers can see beyond surface differences to discern the underlying similarities of seemingly different entities.

Test

Test the data model by reading it to yourself and several people intimately familiar with the problem. Test both directions of every relationship (e.g., a farmer has many cows and a cow belongs to only one farmer). Build a prototype so that the client can experiment and learn with a concrete model. Testing is very important because it costs very little to fix a data model but a great deal to repair a system based on an incorrect data model.

Limit

Set reasonable limits to the time and scope of data modeling. Don't let the data modeling phase continue forever. Discover the boundaries early in the project and stick to these unless there are compelling reasons to extend the project. Too many projects are allowed to expand because it is easier to say *yes* than *no*. In the long run, however, you do the client a disservice by promising too much and extending the life of the project. Determine the core entities and attributes that will solve most of the problems, and confine the data model to this core. Keep sight of the time and budget constraints of the project.

Integrate

Step back and reflect on how your project fits with the organization's information architecture. Integrate with existing systems where feasible and avoid duplication. How does your data model fit with the corporate data model and those of other projects? Can you use part of an existing data model? A skilled data modeler has to see both the fine-grained detail of a project data model and the big picture of the corporate data resource.

Complete

Good data modelers don't leave data models ill-defined. All entities, attributes, and relationships are carefully defined, ambiguities are resolved, and exceptions are handled. Because the full value of a data model is realized only when the system is complete, the data modeler should stay involved with the project until the system is implemented. The data modeler, who generally gets involved in the project from its earliest days, can provide continuity through the various phases and ensure the system solves the problem.

Summary

Data modeling is both a technique for modeling data and its relationships and a graphical representation of a database. It communicates a database's design. The goal is to identify the facts that must be stored in a database. Building a data model is a partnership between a client, a representative of the eventual owners of the database, and a designer. The building blocks are an entity, attribute, identifier, and relationship. A well-formed data model, which means the construction rules have been obeyed, clearly communicates information to the client. A high-fidelity data model faithfully describes the world it represents.

A data model is an evolving representation. Each change should be an incremental improvement in quality. The quality of a data model can be determined only by understanding the context in which it will be used. A data model can model historical data relationships just as readily as current ones. Cardinality specifies the precise multiplicity of a relationship. Modality indicates whether a relationship is optional or mandatory. The five different types of entities are independent, dependent, associative, aggregate, and subordinate. Expect a data model to expand and contract. A data model has no ordering. Introduce an attribute if ordering is required. An attribute must have the same meaning for every instance. An attribute is the smallest piece of data that will conceivably form part of a query. Synonyms are words that have the same meaning; homonyms are words that sound the same but have different meanings. Identifiers should generally have no meaning. Highly effective data modelers immerse, challenge, generalize, test, limit, integrate, and complete.

Key terms and concepts

Aggregate entity
Aggregation
Associative entity
Attribute
Cardinality
Composite aggregation
Data model
Data model quality
Dependent entity
Determinant
Domain
Entity
Generalization
High-fidelity image
Homonym
Identifier
Independent entity
Instance
Mandatory
Modality
Modeling
Optional
Relationship
Relationship descriptor
Relationship label
Shared aggregation
Synonym

References and additional readings

Carlis, J. V. 1991. *Logical data structures*. Minneapolis, MN: University of Minnesota.

Hammer, M. 1990. Reengineering work: Don't automate, obliterate. *Harvard Business Review* 68 (4):104–112.

Exercises

Short answers

1. What is data modeling?
2. What is a useful technique for identifying entities in a written description of a data modeling problem?
3. When do you label relationships?
4. When is a data model well formed, and when is it high-fidelity?
5. How do you handle exceptions when data modeling?
6. Describe the different types of entities.
7. Why might a data model grow?
8. Why might a data model contract?
9. How do you indicate ordering of instances in a data model?
10. What is the difference between a synonym and a homonym?

Data modeling

Create a data model from the following narratives, which are sometimes intentionally incomplete. You will have to make some assumptions. Make certain you state these alongside your data model. Define the identifier(s) and attributes of each entity.

1. The president of a book wholesaler has told you that she wants information about publishers, authors, and books.
2. A university has many subject areas (e.g., MIS, Romance languages). Professors teach in only one subject area, but the same subject area can have many professors. Professors can teach many different courses in their subject area. An offering of a course (e.g., Data Management 457, French 101) is taught by only one professor at a particular time.
3. Kids'n'Vans retails minivans for a number of manufacturers. Each manufacturer offers several models of its minivan (e.g., SE, LE, GT). Each model comes with a standard set of equipment (e.g., the Acme SE comes with wheels, seats, and an engine). Minivans can have a variety of additional equipment or accessories (radio, air-conditioning, automatic transmission, airbag, etc.), but not all accessories are available for all minivans (e.g., not all manufacturers offer a driver's side airbag). Some sets of accessories are sold as packages (e.g., the luxury package might include stereo, six speakers, cocktail bar, and twin overhead foxtails).
4. Steve operates a cinema chain and has given you the following information:
   "I have many cinemas. Each cinema can have multiple theaters. Movies are shown throughout the day starting at 11 a.m. and finishing at 1 a.m. Each movie is given a two-hour time slot. We never show a movie in more than one theater at a time, but we do shift movies among theaters because seating capacity varies. I am interested in knowing how many people, classified by adults and children, attended each showing of a movie. I vary ticket prices by movie and time slot. For instance, Lassie Get Lost is 50 cents for everyone at 11 a.m. but is 75 cents at 11 p.m."
5. A university gymnastics team can have as many as 10 gymnasts. The team competes many times during the season. A meet can have one or more opponents and consists of four events: vault, uneven

bars, beam, and floor routine. A gymnast can participate in all or some of these events though the team is limited to five participants in any event.

6. A famous Greek shipping magnate, Stell, owns many container ships. Containers are collected at one port and delivered to another port. Customers pay a negotiated fee for the delivery of each container. Each ship has a sailing schedule that lists the ports the ship will visit over the next six months. The schedule shows the expected arrival and departure dates. The daily charge for use of each port is also recorded.

7. A medical center employs several physicians. A physician can see many patients, and a patient can be seen by many physicians, though not always on the one visit. On any particular visit, a patient may be diagnosed to have one or more illnesses.

8. A telephone company offers a 10 percent discount to any customer who phones another person who is also a customer of the company. To be eligible for the discount, the pairing of the two phone numbers must be registered with the telephone company. Furthermore, for billing purposes, the company records both phone numbers, start time, end time, and date of call.

9. Global Trading (GT), Inc. is a conglomerate. It buys and sells businesses frequently and has difficulty keeping track of what strategic business units (SBUs) it owns, in what nations it operates, and what markets it serves. For example, the CEO was recently surprised to find that GT owns 25 percent of Dundee's Wild Adventures, headquartered in Zaire, that has subsidiaries operating tours of Australia, Zaire, and New York. You have been commissioned to design a database to keep track of GT's businesses. The CEO has provided you with the following information:

   SBUs are headquartered in one country, not necessarily the United States. Each SBU has subsidiaries or foreign agents, depending on local legal requirements, in a number of countries. Each subsidiary or foreign agent operates in only one country but can operate in more than one market. GT uses the standard industrial code (SIC) to identify a market (e.g., newspaper publishing). The SIC is a unique four-digit code.

   While foreign agents operate as separate legal entities, GT needs to know in what countries and markets they operate. On the other hand, subsidiaries are fully or partly owned by GT, and it is important for GT to know who are the other owners of any subsidiary and what percentage of the subsidiary they own. It is not unusual for a corporation to have shares in several of GT's subsidiary companies and for several corporations to own a portion of a subsidiary. Multiple ownership can also occur at the SBU level.

10. A real estate investment company owns many shopping malls. Each mall contains many shops. To encourage rental of its shops, the company gives a negotiated discount to retailers who have shops in more than one mall. Each shop generates an income stream that can vary from month to month because rental is based on a flat rental charge and a negotiated percentage of sales revenue. Also, each shop has monthly expenses for scheduled and unscheduled maintenance. The company uses the data to compute its monthly net income per square meter for each shop and for ad hoc querying.

11. Draw a data model for the following table taken from a magazine that evaluates consumer goods. The reports follow a standard fashion of listing a brand and model, price, overall score, and then an evaluation of a series of attributes, which can vary with the product. For example, the sample table evaluates stereo systems. A table for evaluating microwave ovens would have a similar layout, but different features would be reported (e.g., cooking quality).

Brand and model	Price	Overall score	Sound quality	Taping quality	FM tuning	CD handling	Ease of use
Phillips SC-AK103	140	62	Very good	Good	Very good	Excellent	Fair
Panasonic MC-50	215	55	Good	Good	Very good	Very good	Good
Rio G300	165	38	Good	Good	Fair	Very good	Poor

12. Draw a data model for the following freight table taken from a mail order catalog.

Merchandise subtotals	Regular delivery 7–10 days	Rush delivery 4–5 business days	Express delivery 1–2 business days
Up to $30.00	$4.95	$9.95	$12.45
$30.01–$65.00	$6.95	$11.95	$15.45
$65.01–$125.00	$8.95	$13.95	$20.45
$125.01+	$9.95	$15.95	$25.45

Reference 1: Basic Structures

Few things are harder to put up with than the annoyance of a good example.

Mark Twain, *Pudd'nhead Wilson*, 1894

Every data model is composed of the same basic structures. This is a major advantage because you can focus on a small part of a full data model without being concerned about the rest of it. As a result, translation to a relational database is very easy because you systematically translate each basic structure. This section describes each of the basic structures and shows how they are mapped to a relational database. Because the mapping is shown as a diagram and `SQL CREATE` statements, you will use this section frequently.

One entity

No relationships

The unrelated entity was introduced in Chapter 3. This is simply a flat file, and the mapping is very simple. Although it is unlikely that you will have a data model with a single entity, the single entity is covered for completeness.

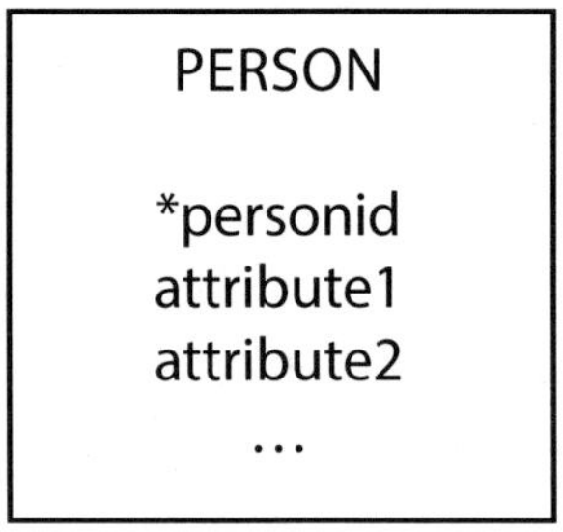

PERSON

*personid
attribute1
attribute2
...

```
CREATE TABLE person (
   personid       INTEGER,
   attribute1        … ,
   attribute2        … ,
   …
      PRIMARY KEY(personid));
```

A 1:1 recursive relationship

A recursive one-to-one (1:1) relationship is used to describe situations like current marriage. In most countries, a person can legally have zero or one current spouse. The relationship should be labeled to avoid misunderstandings. Mapping to the relational database requires that the identifier of one end of the relationship becomes a foreign key. It does not matter which one you select. Notice that when `personid` is used as a foreign key, it must be given a different column name— in this case `partner`—because two columns in the same table cannot have the same name. The foreign key constraint is not defined, because this constraint cannot refer to the table being created.

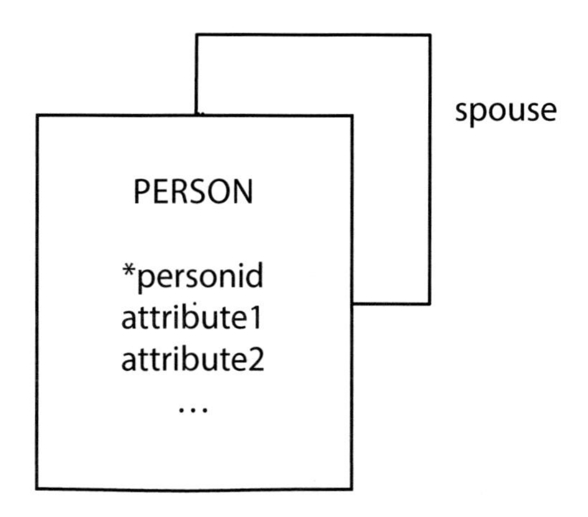


```
CREATE TABLE person (
   personid      INTEGER,
   attribute1       … ,
   attribute2       … ,
   …
   partner       INTEGER,
      PRIMARY KEY(personid));
```

A recursive 1:m relationship

A recursive one-to-many (1:m) relationship describes situations like fatherhood or motherhood. The following figure maps fatherhood. A father may have many biological children, but a child has only one biological father. The relationship is mapped like any other 1:m relationship. The identifier of the one end becomes a foreign key in the many end. Again, we must rename the identifier when it becomes a foreign key in the same table. Also, again the foreign key constraint is not defined because it cannot refer to the table being created.

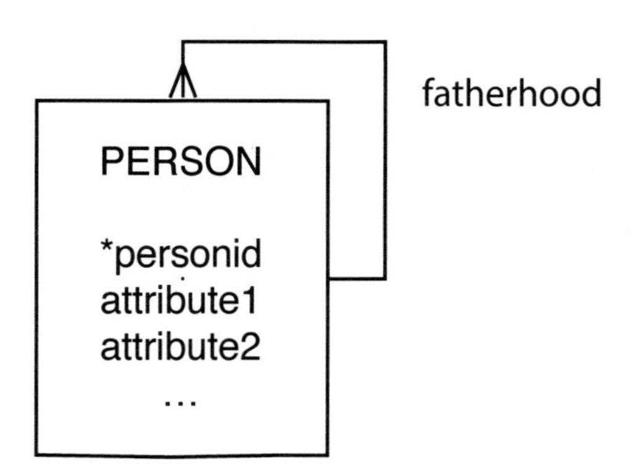


```
CREATE TABLE person (
   personid      INTEGER,
   Attribute1       … ,
   Attribute2       … ,
   …
   father    INTEGER,
      PRIMARY KEY(personid));
```

It is possible to have more than one 1:m recursive relationship. For example, details of a mother-child relationship would be represented in the same manner and result in the data model having a second 1:m recursive relationship. The mapping to the relational model would result in an additional column to contain a foreign key `mother`, the `personid` of a person's mother.

A recursive m:m relationship

A recursive many-to-many (m:m) relationship can describe a situation such as friendship. A person can have many friends and be a friend to many persons. As with m:m relationships between a pair of entities, we convert this relationship to two 1:m relationships and create an associative entity.

The resulting entity FRIENDSHIP has a composite primary key based on the identifier of PERSON, which in effect means the two components are based on *personid*. To distinguish between them, the columns are called `personid1` and `personid2`, so you can think of FRIENDSHIP as a pair of *personids*. You will see the same pattern occurring with other m:m recursive relationships. Notice both person identifiers are independent foreign keys, because they are used to map the two 1:m relationships between PERSON and FRIENDSHIP.

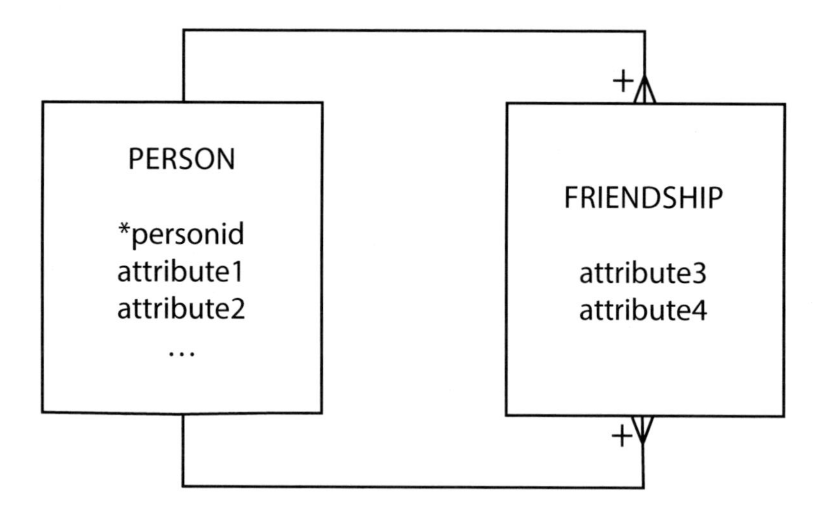


```
CREATE TABLE person (
    personid      INTEGER,
    Attribute1       … ,
    attribute2       … ,
    …
        PRIMARY KEY(personid));

CREATE TABLE friendship (
    personid1     INTEGER,
    personid2     INTEGER,
    attribute3       … ,
    attribute4       … ,
    …
        PRIMARY KEY(personid1,personid2),
        CONSTRAINT fk_friendship_person1
            FOREIGN KEY(personid1) REFERENCES person(personid),
```

```
    CONSTRAINT fk_friendship_person2
        FOREIGN KEY(personid2) REFERENCES person(personid));
```

A single entity can have multiple m:m recursive relationships. Relationships such as enmity (not enemyship) and siblinghood are m:m recursive on PERSON. The approach to recording these relationships is the same as that outlined previously.

Two entities

No relationship

When there is no relationship between two entities, the client has decided there is no need to record a relationship between the two entities. When you are reading the data model with the client, be sure that you check whether this assumption is correct both now and for the foreseeable future. When there is no relationship between two entities, map them each as you would a single entity with no relationships.

A 1:1 relationship

A 1:1 relationship sometimes occurs in parallel with a 1:m relationship between two entities. It signifies some instances of an entity that have an additional role. For example, a department has many employees (the 1:m relationship), and a department has one boss (the 1:1). The data model fragment shown in the following figure represents the 1:1 relationship.

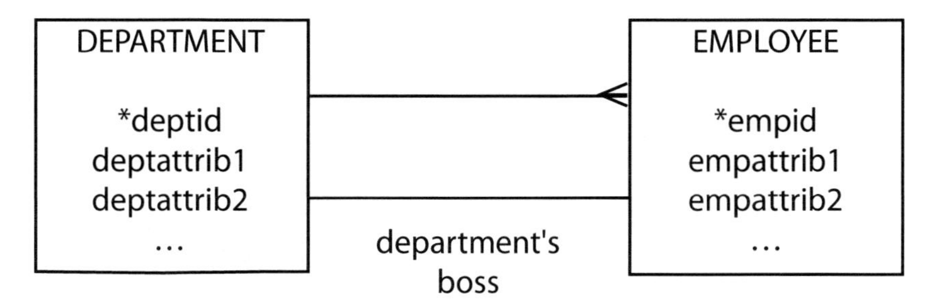

The guideline, as explained in Chapter 6, is to map the relationship to the relational model by placing the foreign key to minimize the number of instances when it will have a null value. In this case, we place the foreign key in department.

```
CREATE TABLE employee (
   empid      INTEGER,
   empattrib1       … ,
   empattrib2       … ,
   …
      PRIMARY KEY(empid));

CREATE TABLE department (
   deptid     CHAR(3),
   deptattrib1      … ,
   deptattrib2      … ,
   …
   bossid INTEGER,
      PRIMARY KEY(deptid),
```

```
    CONSTRAINT fk_department_employee
        FOREIGN KEY(bossid) REFERENCES employee(empid));
```

A 1:m relationship

The 1:m relationship is possibly the easiest to understand and map. The mapping to the relational model is very simple. The primary key of the "one" end becomes a foreign key in the "many" end.

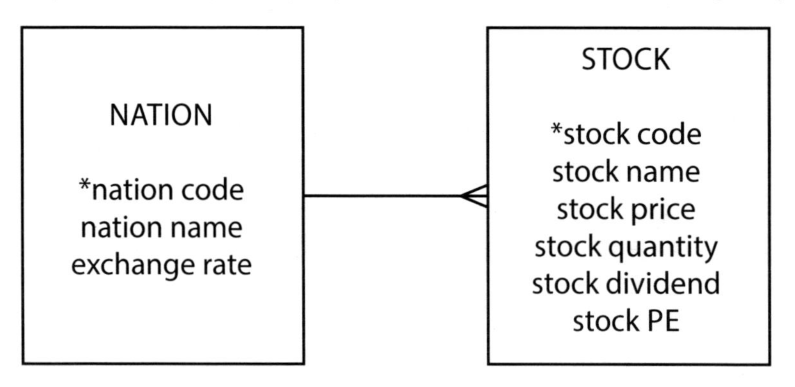


```
CREATE TABLE nation (
    natcode         CHAR(3),
    natname         VARCHAR(20),
    exchrate        DECIMAL(9,5),
        PRIMARY KEY(natcode));

CREATE TABLE stock (
    stkcode         CHAR(3),
    stkfirm         VARCHAR(20),
    stkprice        DECIMAL(6,2),
    stkqty      DECIMAL(8),
    stkdiv      DECIMAL(5,2),
    stkpe           DECIMAL(5),
    natcode         CHAR(3),
        PRIMARY KEY(stkcode),
        CONSTRAINT fk_stock_nation
            FOREIGN KEY(natcode) REFERENCES nation(natcode));
```

An m:m relationship

An m:m relationship is transformed into two 1:m relationships. The mapping is then a twofold application of the 1:m rule.

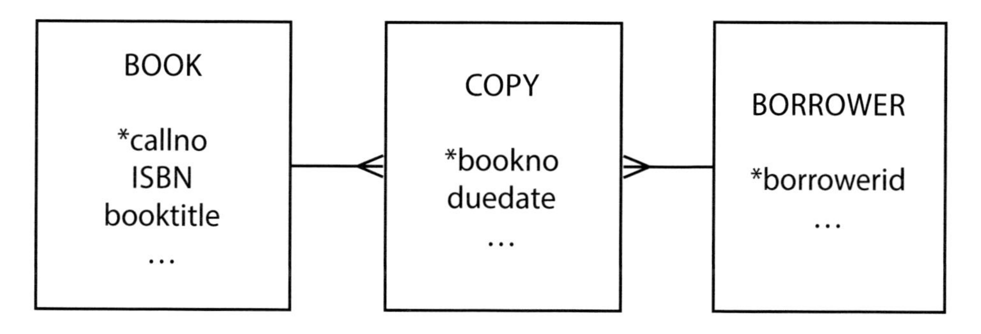

The book and borrower tables must be created first because copy contains foreign key constraints that refer to book and borrower. The column borrowerid can be null because a book need not be borrowed; if it's sitting on the shelf, there is no borrower.

```
CREATE TABLE book (
   callno     VARCHAR(12),
   isbn          … ,
   booktitle     … ,
   …
      PRIMARY KEY(callno));

CREATE TABLE borrower (
   borrowerid    INTEGER,
      …
      PRIMARY KEY(borrowerid));

CREATE TABLE copy (
   bookno        INTEGER,
   duedate       DATE,
   …
   callno        VARCHAR(12),
   borrowerid    INTEGER,
      PRIMARY KEY(bookno),
      CONSTRAINT fk_copy_book
         FOREIGN KEY(callno) REFERENCES book(callno),
      CONSTRAINT fk_copy_borrower
         FOREIGN KEY (borrowerid) REFERENCES borrower(borrowerid));
```

Another entity's identifier as part of the identifier

Using one entity's identifier as part of another entity's identifier tends to cause the most problems for novice data modelers. (One entity's identifier is part of another identifier when there is a plus sign on an arc. The plus is almost always at the crow's foot end of a 1:m relationship.) Tables are formed by applying the following rule: The primary key of the table at the other end of the relationship becomes both a foreign key and part of the primary key in the table at the plus end. The application of this rule is shown for several common data model fragments.

A weak or dependent entity

In the following figure, *regname* is part of the identifier (signified by the plus near the crow's foot) and a foreign key of `CITY` (because of the 1:m between `REGION` and `CITY`).

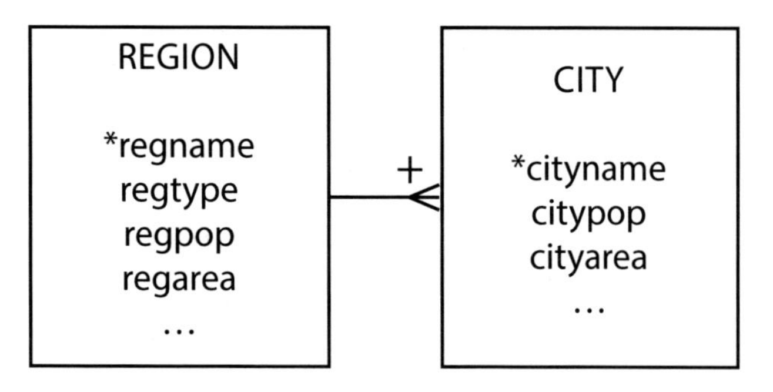


```
CREATE TABLE region (
    regname        VARCHAR(20),
    regtype        ...,
    regpop     ...,
    regarea        ...,
        ...
        PRIMARY KEY(regname));

CREATE TABLE city (
    cityname       VARCHAR(20),
    citypop        ... ,
    cityarea       ... ,
    ...
    regname        VARCHAR(20),
        PRIMARY KEY(cityname,regname),
        CONSTRAINT fk_city_region
            FOREIGN KEY(regname) REFERENCES region(regname));
```

An associative entity

In the following figure, observe that `CITYNAME` and `FIRMNAME` are both part of the primary key (signified by the plus near the crow's foot) and foreign keys (because of the two 1:m relationships) of `STORE`.

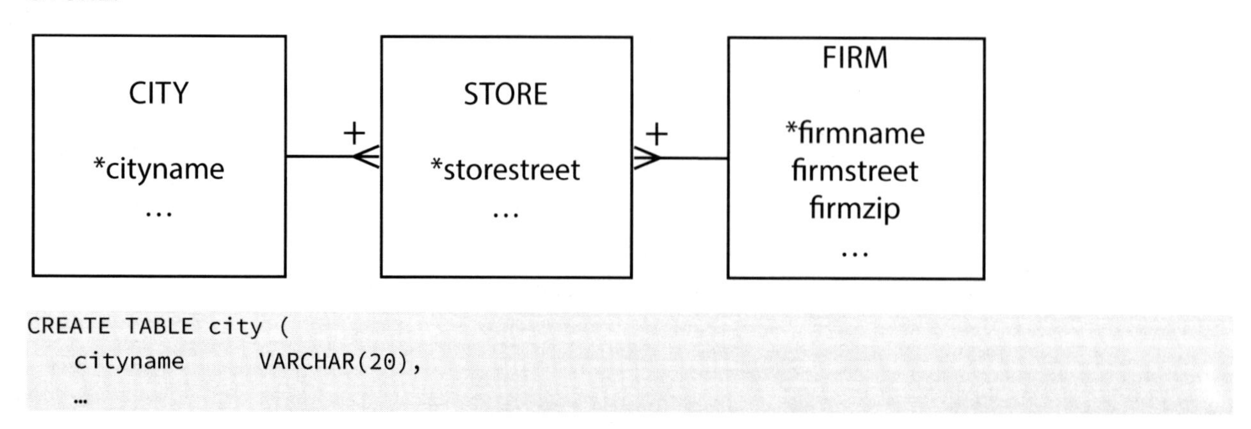


```
CREATE TABLE city (
    cityname       VARCHAR(20),
    ...
```

```
      PRIMARY KEY(cityname));
CREATE TABLE firm
   firmname        VARCHAR(15),
   firmstreet          … ,
   firmzip             … ,
   …
      PRIMARY KEY(firmname));
CREATE TABLE store (
   storestreet  VARCHAR(30),
   storezip            … ,
   …
   cityname        VARCHAR(20),
   firmname        VARCHAR(15),
      PRIMARY KEY(storestreet,cityname,firmname),
      CONSTRAINT fk_store_city
         FOREIGN KEY(cityname) REFERENCES city(cityname),
      CONSTRAINT fk_store_firm
         FOREIGN KEY(firmname) REFERENCES firm(firmname));
```

A tree structure

The interesting feature of the following figure is the primary key. Notice that the primary key of a lower level of the tree is a composite of its partial identifier and the primary key of the immediate higher level.

The primary key of DEPARTMENT is a composite of DEPTNAME, DIVNAME, and FIRMNAME. Novice modelers often forget to make this translation.

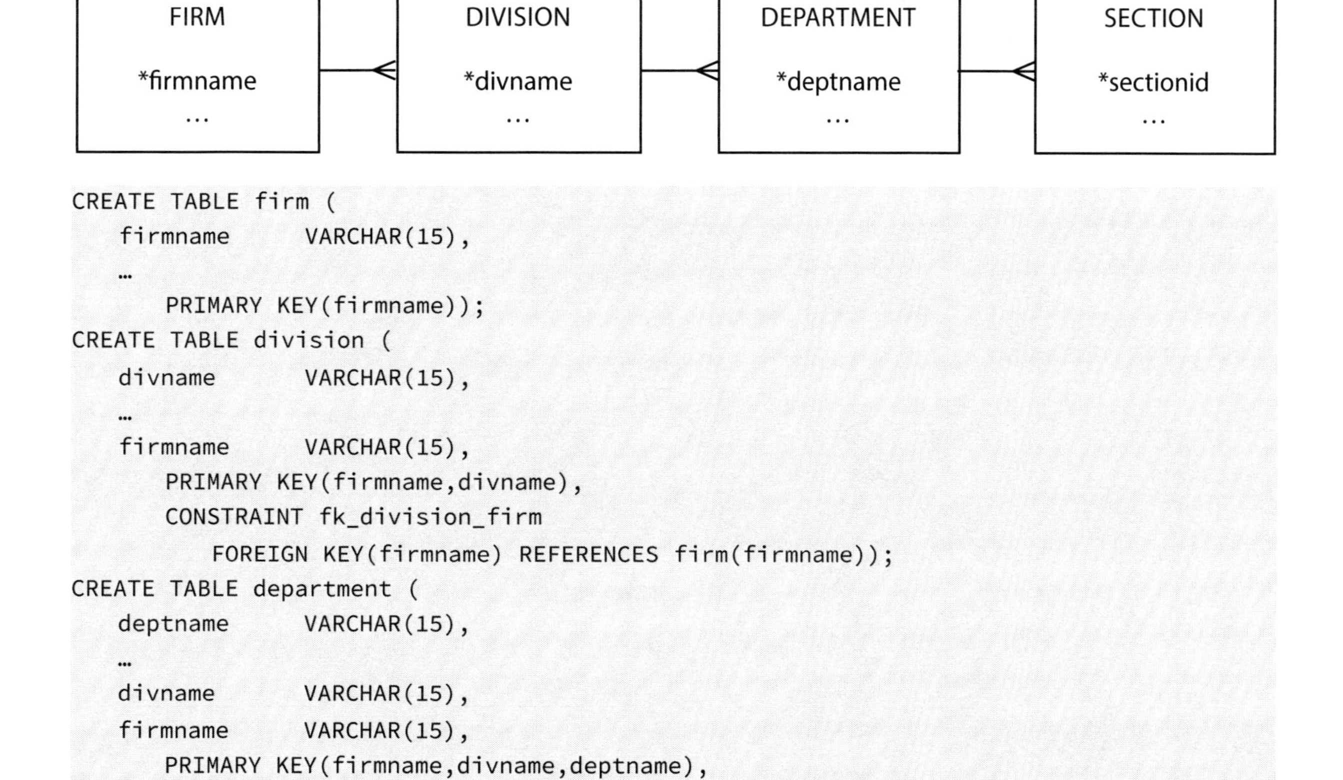


```
CREATE TABLE firm (
   firmname        VARCHAR(15),
   …
      PRIMARY KEY(firmname));
CREATE TABLE division (
   divname         VARCHAR(15),
   …
   firmname        VARCHAR(15),
      PRIMARY KEY(firmname,divname),
      CONSTRAINT fk_division_firm
         FOREIGN KEY(firmname) REFERENCES firm(firmname));
CREATE TABLE department (
   deptname        VARCHAR(15),
   …
   divname         VARCHAR(15),
   firmname        VARCHAR(15),
      PRIMARY KEY(firmname,divname,deptname),
```

```
      CONSTRAINT fk_department_division
         FOREIGN KEY (firmname,divname)
            REFERENCES division(firmname,divname));
CREATE TABLE section (
   sectionid      VARCHAR(15),
   …
   divname        VARCHAR(15),
   firmname       VARCHAR(15),
   deptname       VARCHAR(15),
      PRIMARY KEY(firmname,divname,deptname,sectionid),
      CONSTRAINT fk_department_department
         FOREIGN KEY (firmname,divname,deptname)
            REFERENCES department(firmname,divname,deptname));
```

Another approach to a tree structure

A more general approach to modeling a tree structure is to recognize that it is a series of 1:m recursive relationships. Thus, it can be modeled as follows. This model is identical in structure to that of recursive 1:m reviewed earlier in this chapter and converted to a table in the same manner. Notice that we label the relationship *superunit*, and this would be a good choice of name for the foreign key.

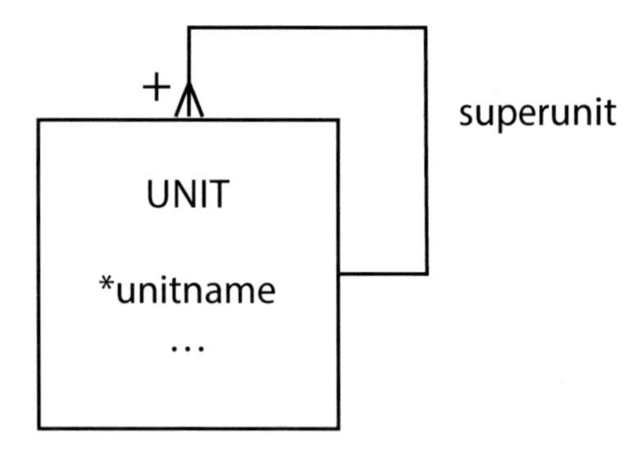

Exercises

Write the SQL CREATE statements for the following data models.

a.

b.

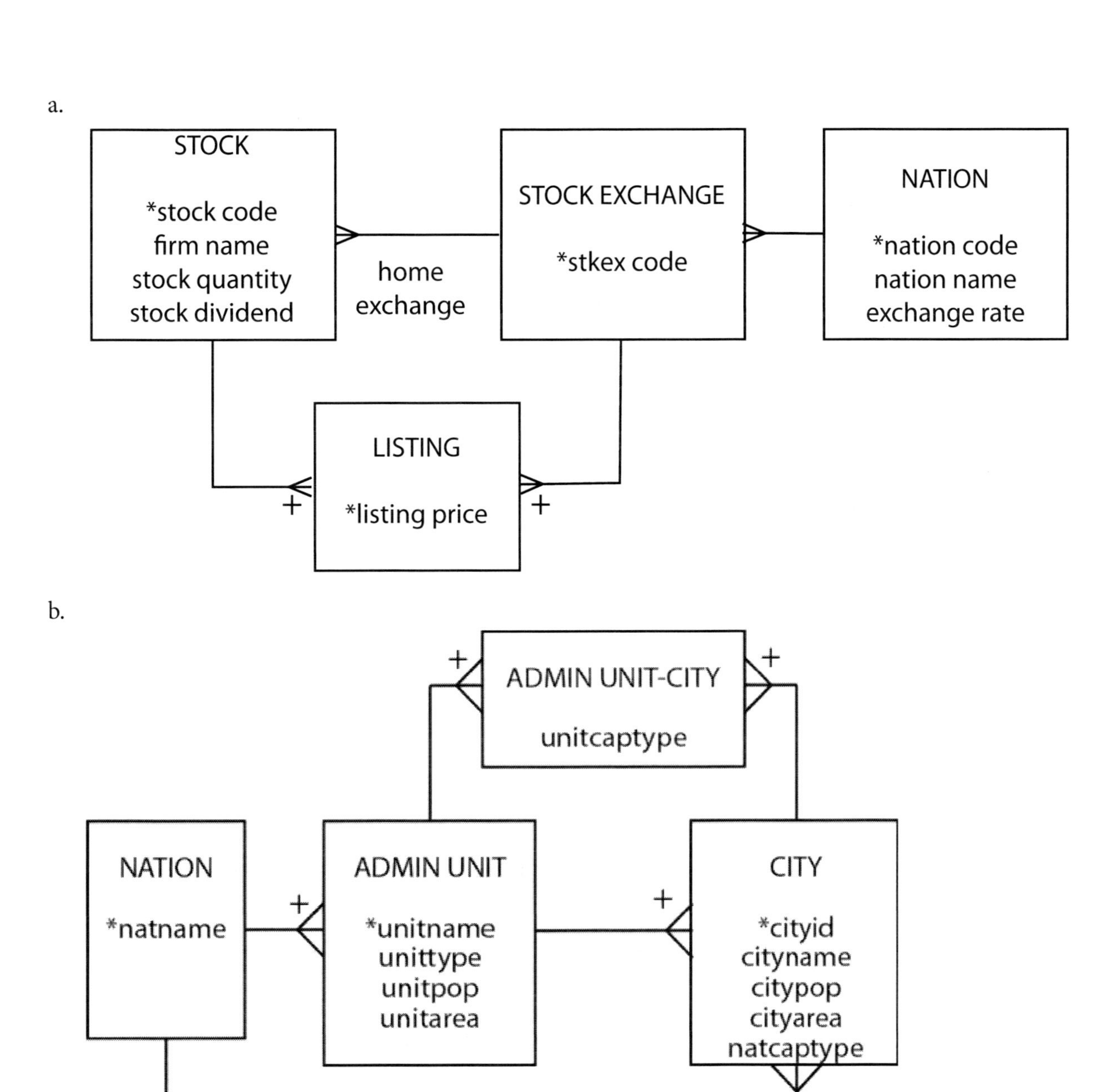
STOCK
*stock code
firm name
stock quantity
stock dividend
home
exchange
STOCK EXCHANGE
*stkex code
NATION
*nation code
nation name
exchange rate
LISTING
*listing price
ADMIN UNIT-CITY
unitcaptype
NATION
*natname
ADMIN UNIT
*unitname
unittype
unitpop
unitarea
CITY
*cityid
cityname
citypop
cityarea
natcaptype
National capital

c.

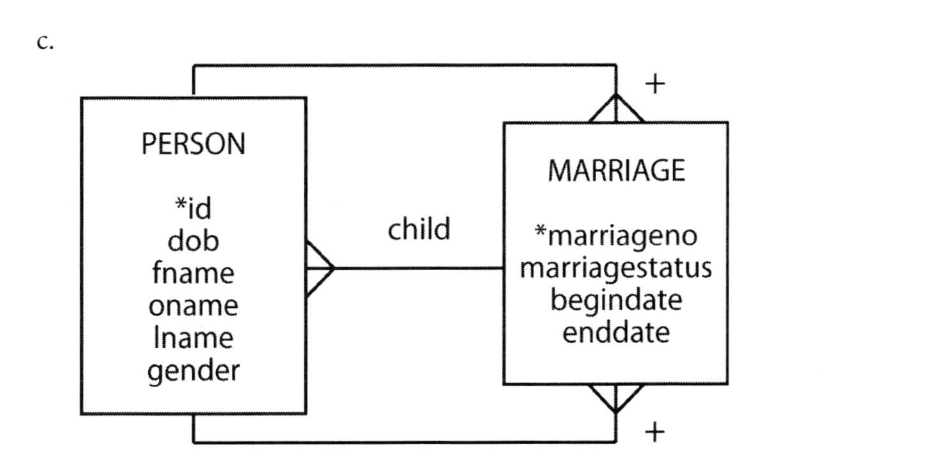
PERSON
*id
dob
fname
oname
lname
gender
child
MARRIAGE
*marriageno
marriagestatus
begindate
enddate

d.

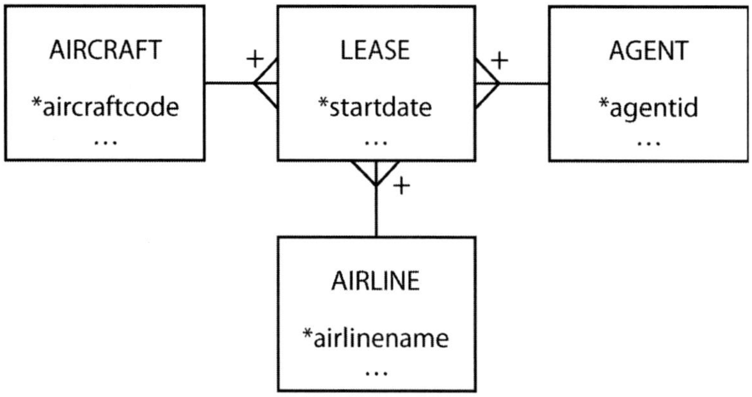

AIRCRAFT
*aircraftcode
...

LEASE
*startdate
...

AGENT
*agentid
...

AIRLINE
*airlinename
...

e. Under what circumstances might you use choose a fixed tree over a 1:m recursive data model?

8. Normalization and Other Data Modeling Methods

There are many paths to the top of the mountain, but the view is always the same.

Chinese Proverb

Learning objectives

Students completing this chapter will

- understand the process of normalization;
- be able to distinguish between different normal forms;
- recognize that different data modeling approaches, while they differ in representation methods, are essentially identical.

Introduction

There are often many ways to solve a problem, and different methods frequently produce essentially the same solution. When this is the case, the goal is to find the most efficient path. We believe that data modeling, as you have learned in the preceding chapters, is an efficient and easily learned approach to database design. There are, however, other paths to consider. One of these methods, normalization, was the initial approach to database design. It was developed as part of the theoretical foundation of the relational data model. It is useful to understand the key concepts of normalization because they advance your understanding of the relational model. Experience and research indicate, however, that normalization is more difficult to master than data modeling.

Data modeling first emerged as entity-relationship (E-R) modeling in a paper by Peter Pin-Shan Chen. He introduced two key database design concepts:

- Identify entities and the relationships between them.
- A graphical representation improves design communication.

Chen's core concepts spawned many species of data modeling. To give you an appreciation of the variation in data modeling approaches, we briefly review, later in this chapter, Chen's E-R approach and IDEF1X, an approach used by the Department of Defense.

Normalization

Normalization is a method for increasing the quality of database design. It is also a theoretical base for defining the properties of relations. The theory gradually developed to create an understanding of the desirable properties of a relation. The goal of normalization is identical to that of data modeling—a high-fidelity design. The need for normalization seems to have arisen from the conversion of prior file systems into database format. Often, analysts started with the old file design and used normalization to design the new database. Now, designers are more likely to start with a clean slate and use data modeling.

Normal forms can be arrived at in several ways. The recommended approach is data modeling, as experience strongly indicates people find it is an easier path to high quality database design. If the principles of data modeling are followed faithfully, then the outcome should be a high-fidelity model and a normalized database. In other words, if you model data correctly, you create a normalized design.

Nevertheless, modeling mistakes can occur, and normalization is a useful crosscheck for ensuring the soundness of a data model. Normalization also provides a theoretical underpinning to data modeling.

Normalization gradually converts a file design into normal form by the successive application of rules to move the design from first to fifth normal form. Before we look at these steps, it is useful to learn about functional dependency.

Functional dependency

A functional dependency is a relationship between attributes in an entity. It simply means that one or more attributes determine the value of another. For example, given a stock's code, you can determine its current PE ratio. In other words, PE ratio is functionally dependent on stock code. In addition, stock name, stock price, stock quantity, and stock dividend are functionally dependent on stock code. The notation for indicating that stock code functionally determines stock name is

```
stock code → stock name
```

An identifier functionally determines all the attributes in an entity. That is, if we know the value of stock code, then we can determine the value of stock name, stock price, and so on.

Formulae, such as yield = stock dividend/stock price*100, are a form of functional dependency. In this case, we have

```
(stock dividend, stock price) → yield
```

This is an example of **full functional dependency** because yield can be determined only from both attributes.

An attribute, or set of attributes, that fully functionally determines another attribute is called a **determinant**. Thus, stock code is a determinant because it fully functionally determines stock PE. An identifier, usually called a key when discussing normalization, is a determinant. Unlike a key, a determinant need not be unique. For example, a university could have a simple fee structure where undergraduate courses are $5000 and graduate courses are $7500. Thus, course type → fee. Since there are many undergraduate and graduate courses, course type is not unique for all records.

There are situations where a given value determines multiple values. This **multidetermination** property is denoted as A → → B and reads "A multidetermines B." For instance, a department multidetermines a course. If you know the department, you can determine the set of courses it offers. **Multivalued dependency** means that functional dependencies are multivalued.

Functional dependency is a property of a relation's data. We cannot determine functional dependency from the names of attributes or the current values. Sometimes, examination of a relation's data will indicate that a functional dependency does not exist, but it is by understanding the relationships between data elements that we determine functional dependency.

Functional dependency is a theoretical avenue for understanding relationships between attributes. If we have two attributes, say A and B, then three relations are possible, as shown in the following table.

Functional dependencies of two attributes

Relationship	Functional dependency	Relationship
They determine each other	A → B and B → A	1:1
One determines the other	A → B	1:m
They do not determine each other	A not→ B and B not→ A	m:m

One-to-one attribute relationship

Consider two attributes that determine each other (A → B and B → A), for instance a country's code and its name. Using the example of Switzerland, there is a one-to-one (1:1) relationship between CH and Switzerland: CH → Switzerland and Switzerland → CH. When two attributes have a 1:1 relationship, they must occur together in at least one table in a database so that their equivalence is a recorded fact.

One-to-many attribute relationship

Examine the situation where one attribute determines another (i.e., A→ B), but the reverse is not true (i.e., A not→ B), as is the case with country name and its currency unit. If you know a country's name, you can determine its currency, but if you know the currency unit (e.g., the euro), you cannot always determine the country (e.g., both Italy and Portugal use the euro). As a result, if A and B occur in the same table, then A must be the key. In our example, country name would be the key, and currency unit would be a nonkey column.

Many-to-many attribute relationship

The final case to investigate is when neither attribute determines the other (i.e., A not→ B and B not→ A). The relationship between country name and language is many-to-many (m:m). For example, Belgium has two languages (French and Flemish), and French is spoken in many countries. To record the m:m relationship between these attributes, a table containing both attributes as a composite key is required. This is essentially the associative entity created during data modeling when there is an m:m relationship between entities.

As you can understand from the preceding discussion, functional dependency is an explicit form of presenting some of the ideas you gained implicitly in earlier chapters on data modeling. The next step is to delve into another theoretical concept underlying the relational model: normal forms.

Normal forms

Normal forms describe a classification of relations. Initial work by Codd identified first (1NF), second (2NF), and third (3NF) normal forms. Later researchers added Boyce-Codd (BCNF), fourth (4NF), and fifth (5NF) normal forms. Normal forms are nested like a set of Russian dolls, with the innermost doll, 1NF, contained within all other normal forms. The hierarchy of normal forms is 5NF, 4NF, BCNF, 3NF, 2NF, and 1NF. Thus, 5NF is the outermost doll.

A new normal form, domain-key normal form (DK/NF), appeared in 1981. When a relation is in DK/NF, there are no modification anomalies. Conversely, any relation that is free of anomalies must be in DK/NF. The difficult part is discovering how to convert a relation to DK/NF.

First normal form

A relation is in first normal form if and only if all columns are single-valued. In other words, 1NF states that all occurrences of a row must have the same number of columns. In data modeling terms, this means that an attribute must have a single value. An attribute that can have multiple values must be represented as a one-to-many (1:m) relationship. At a minimum, a data model will be in 1NF because all attributes of an entity are required to be single-valued.

Second normal form

Second normal form is violated when a nonkey column is dependent on only a component of the primary key. This can also be stated as *a relation is in second normal form if and only if it is in first normal form, and all nonkey columns are dependent on the key.*

Consider the following table. The primary key of `order` is a composite of `itemno` and `customerid`. The problem is that `customer-credit` is a fact about `customerid` (part of the composite key) rather than the full key (itemno+customerid), or in other words, it is not fully functionally dependent on the primary key. An insert anomaly arises when you try to add a new customer to ORDER. You cannot add a customer until that person places an order, because until then you have no value for item number, and part of the primary key will be null. Clearly, this is neither an acceptable business practice nor an acceptable data management procedure.

Analyzing this problem, you realize an item can be in many orders, and a customer can order many items — an m:m relationship. By drawing the data model in the following figure, you realize that *customer-credit* is an attribute of customer, and you get the correct relational mapping.

Second normal form violation

order			
*itemno	*customerid	quantity	customer-credit
12	57	25	OK
34	679	3	POOR

Resolving second normal form violation

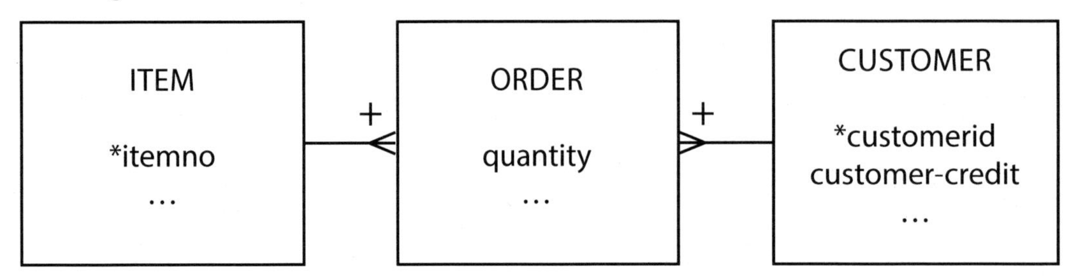

Third normal form

Third normal form is violated when a nonkey column is a fact about another nonkey column. Alternatively, *a relation is in third normal form if and only if it is in second normal form and has no transitive dependencies.*

The problem in the following table is that exchange rate is a fact about nation, a nonkey field. In the language of functional dependency, exchange rate is not fully functionally dependent on stockcode, the primary key.

The functional dependencies are stockcode → nation → exchange rate. In other words, exchange rate is transitively dependent on STOCK, since exchange rate is dependent on nation and nation is dependent on stockcode. The fundamental problem becomes very apparent when you try to add a new nation to the stock table. Until you buy at least one stock for that nation, you cannot insert the nation, because you do not have a primary key. Similarly, if you delete MG from the stock table, details of the USA exchange rate are lost. There are modification anomalies.

When you think about the data relationships, you realize that a nation has many stocks and a stock belongs to only one nation. Now the data model and relational map can be created readily .

Third normal form violation

stock		
*stockcode	nation	exchange rate
MG	USA	0.67
IR	AUS	0.46

Resolving third normal form violation

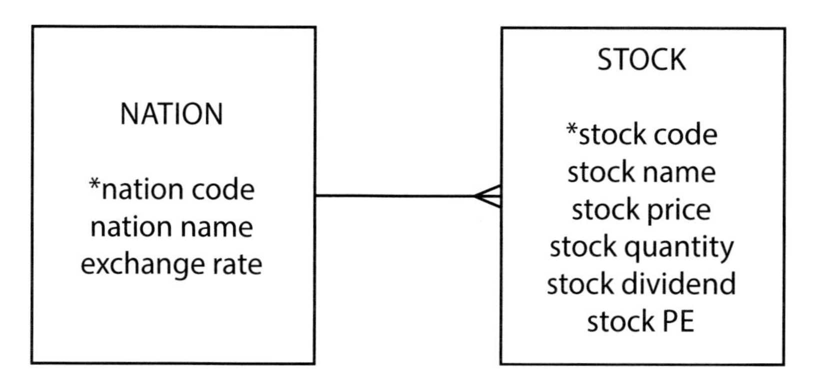

Skill builder

You have been given a spreadsheet that contains details of invoices. The column headers for the spreadsheet are *date, invoice number, invoice amount, invoice tax, invoice total, cust number, cust name, cust street, cust city, cust state, cust postal code, cust nation, product code, product price, product quantity, salesrep number, salesrep first name, salesrep last name, salesrep district, district name,* and *district size.* Normalize this spreadsheet so that it can be converted to a high-fidelity relational database.

Boyce-Codd normal form

The original definition of 3NF did not cover a situation that, although rare, can occur. So Boyce-Codd normal form, a stronger version of 3NF, was developed. BCNF is necessary because 3NF does not cover the cases when

- A relation has multiple candidate keys.
- Those candidate keys are composite.
- The candidate keys overlap because they have at least one column in common.

Before considering an example, **candidate key** needs to be defined. Earlier we introduced the idea that an entity could have more than one potential unique identifier. These identifiers become candidate keys when the data model is mapped to a relational database. One of these candidates is selected as the primary key.

Consider the following case from a management consulting firm. A client can have many types of problems (e.g., finance, personnel), and the same problem type can be an issue for many clients. Consultants specialize and advise on only one problem type, but several consultants can advise on one problem type. A consultant advises a client. Furthermore, for each problem type, the client is advised by only one consultant. If you did not use data modeling, you might be tempted to create the following table.

Boyce-Codd normal form violation

advisor		
*client	*probtype	consultant
Alpha	Marketing	Gomez
Alpha	Production	Raginiski

The column `client` cannot be the primary key because a client can have several problem types; however, a client is advised by only one consultant for a specific problem type, so the composite key `client+probtype` determines `consultant`. Also, because a consultant handles only one type of problem, the composite key `client +consultant` determines `probtype`. So, both of these composites are candidate keys. Either one can be selected as the primary key, and in this case `client+probtype` was selected. Notice that all the previously stated conditions are satisfied—there are multiple, composite candidate keys that overlap. This means the table is 3NF, but not BCNF. This can be easily verified by considering what happens if the firm adds a new consultant. A new consultant cannot be added until there is a client—an insertion anomaly. The problem is that `consultant` is a determinant, `consultant→probtype`, but is not a candidate key. In terms of the phrasing used earlier, the problem is that part of the key column is a fact about a nonkey column. The precise definition is *a relation is in Boyce-Codd normal form if and only if every determinant is a candidate key.*

This problem is avoided by creating the correct data model and then mapping to a relational model.

Resolving Boyce-Codd normal form violation

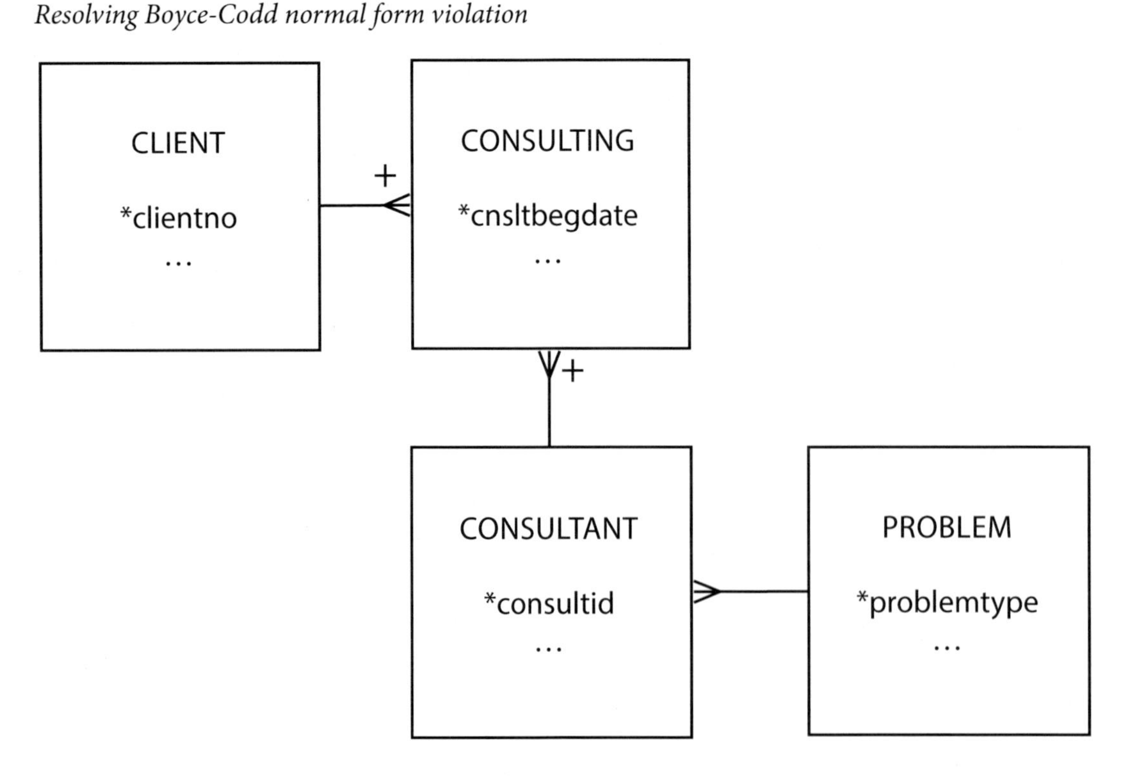

Fourth normal form

Fourth normal form requires that a row should not contain two or more independent multivalued facts about an entity. This requirement is more readily understood after investigating an example.

Consider students who play sports and study subjects. One way of representing this information is shown in the following table. Consider the consequence of trying to insert a student who did not play a sport. Sport would be null, and this is not permissible because part of the composite primary key would then be null — a violation of the entity integrity rule. You cannot add a new student until you know her sport and her subject. Modification anomalies are very apparent.

Fourth normal form violation

student			
*studentid	*sport	*subject	...
50	Football	English	...
50	Football	Music	...
50	Tennis	Botany	...
50	Karate	Botany	...

This table is not in 4NF because sport and subject are independent multivalued facts about a student. There is no relationship between sport and subject. There is an indirect connection because sport and subject are associated with a student. In other words, a student can play many sports, and the same sport can be played by many students — a many-to-many (m:m) relationship between student and sport. Similarly, there is an m:m relationship between student and subject. It makes no sense to store information about a student's sports and subjects in the same table because sport and subject are not related. The problem arises because sport and subject are multivalued dependencies of student. The solution is to convert multivalued dependencies to functional dependencies. More formally, *a relation is in fourth normal form if it is in Boyce-Codd normal form and all multivalued dependencies on the relation are functional dependencies.* A data model sorts out this problem, and note that the correct relational mapping requires five tables.

Resolving fourth normal form violation

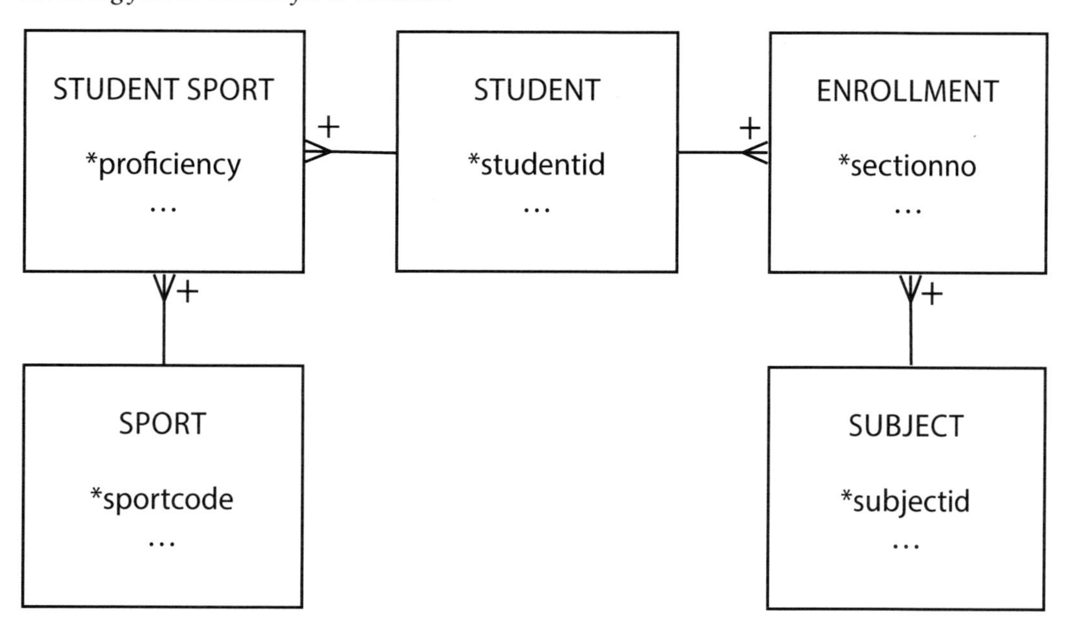

Fifth normal form

Fifth normal form deals with the case where a table can be reconstructed from other tables. The reconstruction approach is preferred, because it means less redundancy and fewer maintenance problems.

A consultants, firms, and skills problem is used to illustrate the concept of 5NF. The problem is that consultants provide skills to one or more firms and firms can use many consultants; a consultant has many skills and a skill can be used by many firms; and a firm can have a need for many skills and the same skill can be required by many firms. The data model for this problem has the ménage-à-trois structure which was introduced in Chapter 7.

The CONSULTANT-FIRM-SKILL data model without a rule

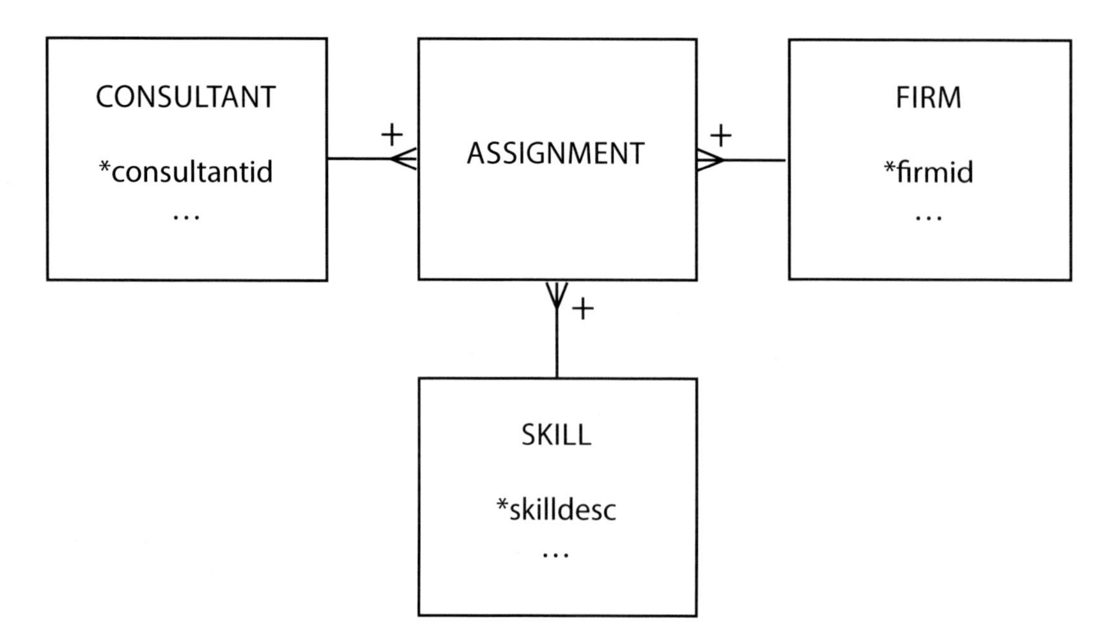

The relational mapping of the three-way relationships results in four tables, with the associative entity ASSIGNMENT mapping shown in the following table.

Relation table assignment

assignment		
*consultid	*firmid	*skilldesc
Tan	IBM	Database
Tan	Apple	Data comm

The table is in 5NF because the combination of all three columns is required to identify which consultants supply which firms with which skills. For example, we see that Tan advises IBM on database and Apple on data communications. The data in this table cannot be reconstructed from other tables.

The preceding table is not in 5NF *if there is a rule of the following form*: If a consultant has a certain skill (e.g., database) and has a contract with a firm that requires that skill (e.g., IBM), then the consultant advises that firm on that skill (i.e., he advises IBM on database). Notice that this rule means we can infer a relationship, and we no longer need the combination of the three columns. As a result, we break the single three-entity m:m relationship into three two-entity m:m relationships. Revised data model after the introduction of the rule

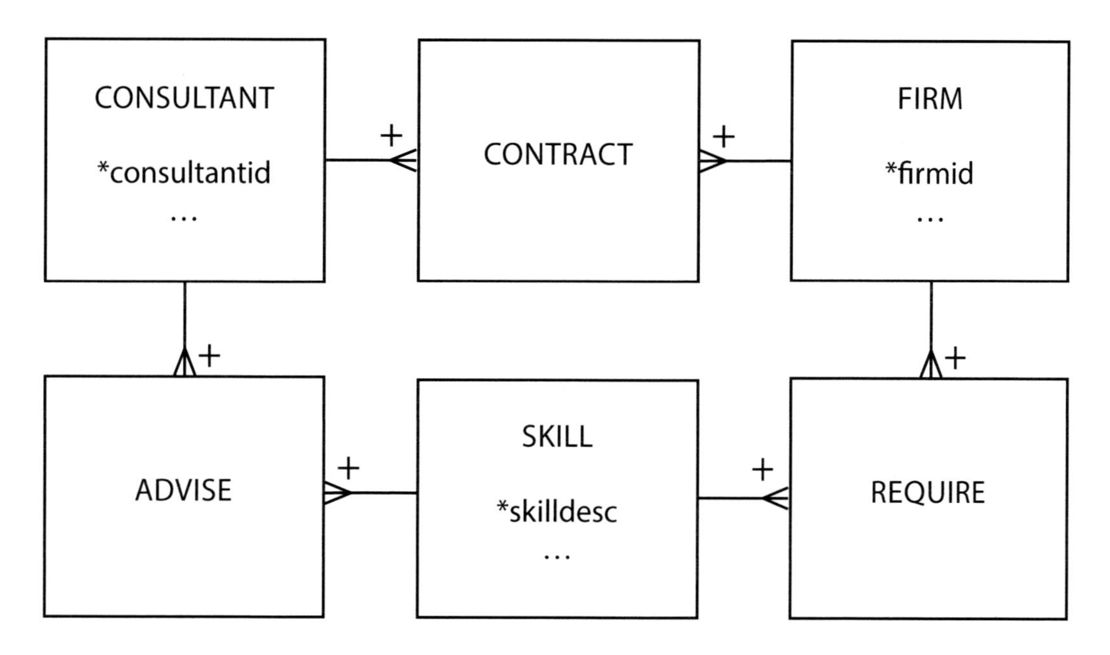

Further understanding of 5NF is gained by examining the relational tables resulting from these three associative entities. Notice the names given to each of the associative entities. The table `contract` records data about the firms a consultant advises; `advise` describes the skills a consultant has; and `require` stores data about the types of skills each firm requires. To help understand this problem, examine the three tables below.

Relational table contract

contract	
*consultid	*firmid
Gonzales	Apple
Gonzales	IBM
Gonzales	NEC
Tan	IBM
Tan	NEC
Wood	Apple

Relational table advise

advise	
*consultid	*skilldesc
Gonzales	Database
Gonzales	Data comm
Gonzales	Groupware
Tan	Database
Tan	Data comm
Wood	Data comm

Relational table require

require	
*firmid	*skilldesc
IBM	Data comm
IBM	Groupware
NEC	Data comm
NEC	Database
NEC	Groupware
Apple	Data comm

Consider joining `contract` and `advise`, the result of which we call `could advise` because it lists skills the consultant could provide if the firm required them. For example, Tan has skills in database and data communications, and Tan has a contract with IBM. If IBM required database skills, Tan could handle it. We need to look at `require` to determine whether IBM requires advice on database; it does not.

Relational table could advise

could advise		
consultid	firmid	skilldesc
Gonzales	Apple	Database
Gonzales	Apple	Data comm
Gonzales	Apple	Groupware
Gonzales	IBM	Database
Gonzales	IBM	Data comm
Gonzales	IBM	Groupware
Gonzales	NEC	Database
Gonzales	NEC	Data comm
Gonzales	NEC	Groupware
Tan	IBM	Database
Tan	IBM	Data comm
Tan	NEC	Database
Tan	NEC	Data comm
Wood	Apple	Data comm

Relational table can advise

The join of `could advise` with `require` gives details of a firm's skill needs that a consultant can provide. The table `can advise` is constructed by directly joining `contract, advise`, and `require`. Because we can construct `can advise` from three other tables, the data are in 5NF.

can advise		
consultid	firmid	skilldesc
Gonzales	IBM	Data comm
Gonzales	IBM	Groupware

can advise		
consultid	firmid	skilldesc
Gonzales	NEC	Database
Gonzales	NEC	Data comm
Gonzales	NEC	Groupware
Tan	IBM	Data comm
Tan	NEC	Database
Tan	NEC	Data comm
Wood	Apple	Data comm

Since data are stored in three separate tables, updating is easier. Consider the case where IBM requires database skills. We only need to add one row to `require` to record this fact. In the case of `can advise`, we would have to add two rows, one for Gonzales and one for Tan.

Now we can give 5NF a more precise definition: *A relation is in fifth normal form if and only if every join dependency of the relation is a consequence of the candidate keys of the relation.*

Up to this point, data modeling has enabled you to easily avoid normalization problems. Fifth normal form introduces a complication, however. How can you tell when to use a single three-way associative entity or three two-way associative entities? If there is a constraint or rule that is applied to the relationships between entities, consider the possibility of three two-way associative entities. Question the client carefully whenever you find a ménage-à-trois. Check to see that there are no rules or special conditions.

Domain-key normal form

The definition of DK/NF builds on three terms: key, constraint, and domain. You already know a key is a unique identifier. A **constraint** is a rule governing attribute values. It must be sufficiently well-defined so that its truth can be readily evaluated. Referential integrity constraints, functional dependencies, and data validation rules are examples of constraints. A **domain** is a set of all values of the same data type. With the help of these terms, the concept of DK/NF is easily stated. *A relation is in domain-key normal form if and only if every constraint on the relation is a logical consequence of the domain constraints and the key constraints that apply to the relation.*

Note that DK/NF does not involve ideas of dependency; it just relies on the concepts of key, constraint, and domain. The problem with DK/NF is that while it is conceptually simple, no algorithm has been developed for converting relations to DK/NF. Hence, database designers must rely on their skills to create relations that are in DK/NF.

Conclusion

Normalization provides designers with a theoretical basis for understanding what they are doing when modeling data. It also alerts them to be aware of problems that they might not ordinarily detect when modeling. For example, 5NF cautions designers to investigate situations carefully (i.e., look for special rules), when their data model contains a ménage-à-trois.

Other data modeling methods

As mentioned previously, there are many species of data modeling. Here we consider two methods.

The E-R model

One of the most widely known data modeling methods is the E-R model developed by Chen. There is no standard for the E-R model, and it has been extended and modified in a number of ways.

As the following figure shows, an E-R model looks like the data models with which you are now familiar. Entities are shown by rectangles, and relationships are depicted by diamonds within which the cardinality of the relationship is indicated (e.g., there is a 1:m relationship between NATION and STOCK EXCHANGE). Cardinality can be 1:1, 1:m (conventionally shown as 1:N in E-R diagrams), and m:m (shown as M:N). Recursive relationships are also readily modeled. One important difference is that an m:m relationship is not shown as an associative entity; thus the database designer must convert this relationship to a table.

An E-R diagram

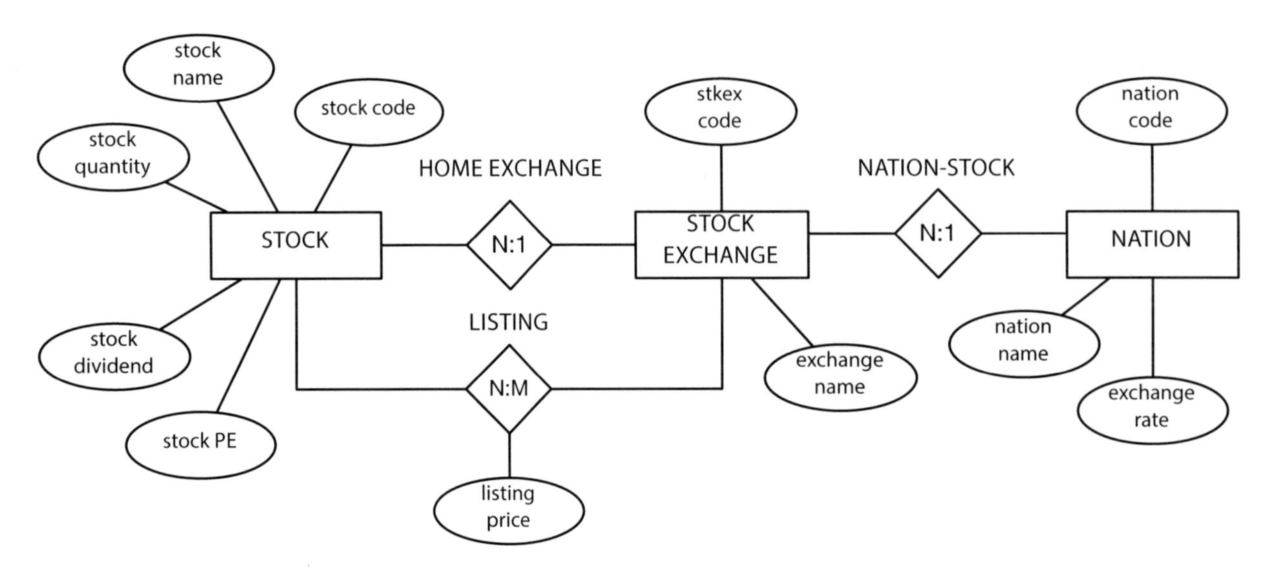

Attributes are shown as ellipses connected to the entity or relationship to which they belong (e.g., *stock name* is an attribute of STOCK). Relationships are named, and the name is shown above or below the relationship symbol. For example, LISTING is the name of the m:m relationship between STOCK and STOCK EXCHANGE. The method of recording attributes consumes space, particularly for large data models. It is more efficient to record attributes within an entity rectangle, as in the data modeling method you have learned.

IDEF1X

IDEF1X (Integrated DEFinition 1X) was conceived in the late 1970s and refined in the early 1980s. It is based on the work of Codd and Chen. A quick look at some of the fundamental ideas of IDEF1X will show you how similar it is to the method you have learned. Entities are represented in a similar manner, as the following figure shows. The representation is a little different (e.g., the name of the entity is in separate box within a rectangle), but the information is the same.

IDEF1X adds additional notation to an attribute to indicate whether it is an alternate key or inversion entry. An **alternate key** is a possible primary key (identifier) that was not selected as the primary key. *Stock name*, assuming it is unique, is an alternate key. An **inversion entry** indicates a likely frequent way of accessing the entity, so it should become an index. For instance, the entity may be accessed frequently via *stock name*. Alternate keys are shown as AK1, AK2,... and inversion entries as IE1, IE2, ... Thus, *stock name* is both an alternate key and an inversion entry.

The representation of relationships is also similar. A dot is used rather than a crow's foot. Also, notice that relationships are described, whereas you learned to describe only relationships when they are not readily inferred.

An IDEF1X relationship

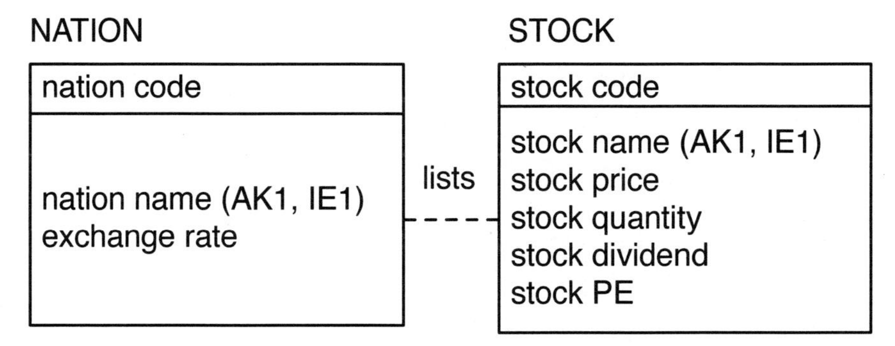

Skill builder

Ken is a tour guide for trips that can last several weeks. He has requested a database that will enable him to keep track of the people who have been on the various tours he has led. He wants to be able to record his comments about some of his clients to remind him of extra attention he might need to pay to them on a future trip (e.g., Bill tends to get lost, Daisy has problems getting along with others). Some times people travel with a partner, and Ken wants to record this in his database.

Design a database using MySQL Workbench. Experiment with Object Notation and Relationship Notation (options of the Model tab) to get a feel for different forms of representing data models.

Conclusion

The underlying concepts of different data modeling methods are very similar. They all have the same goal: improving the quality of database design. They share common concepts, such as entity and relationship mapping. We have adopted an approach that is simple and effective. The method can be learned rapidly, novice users can create high-quality models, and it is similar to the representation style of MySQL Workbench. Also, if you have learned one data modeling method, you can quickly adapt to the nuances of other methods.

Summary

Normalization gradually converts a file design into normal form by successive application of rules to move the design from first to fifth normal form. Functional dependency means that one or more attributes determine the value of another. An attribute, or set of attributes, that fully functionally determines another attribute is called a *determinant*. First normal form states that all occurrences of a row must have the same number of columns. Second normal form is violated when a nonkey column is a fact about a component of the prime key. Third normal form is violated when a nonkey column is a fact about another nonkey column. Boyce-Codd normal form is a stronger version of third normal form. Fourth normal form requires that a row should not contain two or more independent multivalued facts about an entity. Fifth normal form deals with the case where a table can be reconstructed from data in other tables. Fifth normal form arises when there are special rules or conditions. A high-fidelity data model will be of high normal form.

One of the most widely known methods of data modeling is the E-R model. The basic concepts of most data modeling methods are very similar. All aim to improve database design.

Key terms and concepts

Alternate key
Boyce-Codd normal form (BCNF)
Constraint
Data model
Determinant
Domain
Domain-key normal form (DK/NF)
Entity
Entity-relationship (E-R) model
Fifth normal form (5NF)
First normal form (1NF)
Fourth normal form (1NF)
Full functional dependency
Functional dependency
Generalization hierarchy
IDEF1X
Inversion entry
Many-to-many attribute relationship
Multidetermination
Multivalued dependency
Normalization
One-to-many attribute relationship
One-to-one attribute relationship
Relationship
Second normal form (2NF)
Third normal form (3NF)

References and additional readings

Chen, P. 1976. The entity-relationship model—toward a unified view of data. *ACM Transactions on Database Systems* 1 (1):9–36.

Kent, W. 1983. A simple guide to five normal forms in relational database theory. *Communications of the ACM 26 (2): 120-125.*

Exercises

Short answers

1. What is normalization, and what is its goal?
2. How does DK/NF differ from earlier normal forms?
3. How do E-R and IDEF1X differ from the data modeling method of this text?

Normalization and data modeling

Using normalization, E-R, or IDEF1X, create data models from the following narratives, which are sometimes intentionally incomplete. You will have to make some assumptions. Make certain you state these assumptions alongside your data model. Define the identifier(s) and attributes of each entity.

1. The president of a book wholesaler has told you that she wants information about publishers, authors, and books.
2. A university has many subject areas (e.g., MIS, Romance languages). Professors teach in only one subject area, but the same subject area can have many professors. Professors can teach many different courses in their subject area. An offering of a course (e.g., Data Management 457, French 101) is taught by only one professor at a particular time.
3. Kids'n'Vans retails minivans for a number of manufacturers. Each manufacturer offers several models of its minivan (e.g., SE, LE, GT). Each model comes with a standard set of equipment (e.g.,

the Acme SE comes with wheels, seats, and an engine). Minivans can have a variety of additional equipment or accessories (radio, air conditioning, automatic transmission, airbag, etc.), but not all accessories are available for all minivans (e.g., not all manufacturers offer a driver's side airbag). Some sets of accessories are sold as packages (e.g., the luxury package might include stereo, six speakers, cocktail bar, and twin overhead foxtails).

4. Steve operates a cinema chain and has given you the following information:
   "I have many cinemas. Each cinema can have multiple theaters. Movies are shown throughout the day starting at 11 a.m. and finishing at 1 a.m. Each movie is given a two-hour time slot. We never show a movie in more than one theater at a time, but we do shift movies among theaters because seating capacity varies. I am interested in knowing how many people, classified by adults and children, attended each showing of a movie. I vary ticket prices by movie and time slot. For instance, *Lassie Get Lost* is 50 cents for everyone at 11 a.m. but is 75 cents at 11 p.m."

5. A telephone company offers a 10 percent discount to any customer who phones another person who is also a customer of the company. To be eligible for the discount, the pairing of the two phone numbers must be registered with the telephone company. Furthermore, for billing purposes, the company records both phone numbers, start time, end time, and date of call.

9. The Relational Model and Relational Algebra

Nothing is so practical as a good theory.

K. Lewin, 1945

Learning objectives

Students completing this chapter will

- know the structures of the relational model;
- understand relational algebra commands;
- be able to determine whether a DBMS is completely relational.

Background

The relational model, developed as a result of recognized shortcomings of hierarchical and network DBMSs, was introduced by Codd in 1970. As the major developer, Codd believed that a sound theoretical model would solve most practical problems that could potentially arise.

In another article, Codd expounds the case for adopting the relational over earlier database models. There is a threefold thrust to his argument. *First,* earlier models force the programmer to code at a low level of structural detail. As a result, application programs are more complex and take longer to write and debug. *Second,* no commands are provided for processing multiple records at one time. Earlier models do not provide the set-processing capability of the relational model. The set-processing feature means that queries can be more concisely expressed. *Third,* the relational model, through a query language such as structured query language (SQL), recognizes the clients' need to make ad hoc queries. The relational model and SQL permits an IS department to respond rapidly to unanticipated requests. It can also mean that analysts can write their own queries. Thus, Codd's assertion that the relational model is a practical tool for increasing the productivity of IS departments is well founded.

The productivity increase arises from three of the objectives that drove Codd's research. The *first* was to provide a clearly delineated boundary between the logical and physical aspects of database management. Programming should be divorced from considerations of the physical representation of data. Codd labels this the **data independence objective**. The *second* objective was to create a simple model that is readily understood by a wide range of analysts and programmers. This **communicability objective** promotes effective and efficient communication between clients and IS personnel. The *third* objective was to increase processing capabilities from record-at-a-time to multiple-records-at-a-time—the **set-processing objective**. Achievement of these objectives means fewer lines of code are required to write an application program, and there is less ambiguity in client-analyst communication.

The relational model has three major components:

- Data structures
- Integrity rules
- Operators used to retrieve, derive, or modify data

Data structures

Like most theories, the relational model is based on some key structures or concepts. We need to understand these in order to understand the theory.

Domain

A domain is a set of values all of the same data type. For example, the domain of nation name is the set of all possible nation names. The domain of all stock prices is the set of all currency values in, say, the range $0 to $10,000,000. You can think of a domain as all the legal values of an attribute.

In specifying a domain, you need to think about the smallest unit of data for an attribute defined on that domain. In Chapter 8, we discussed how a candidate attribute should be examined to see whether it should be segmented (e.g., we divide name into first name, other name, and last name, and maybe more). While it is unlikely that name will be a domain, it is likely that there will be a domain for first name, last name, and so on. Thus, a domain contains values that are in their *atomic* state; they cannot be decomposed further.

The practical value of a domain is to define what comparisons are permissible. Only attributes drawn from the same domain should be compared; otherwise it is a bit like comparing bananas and strawberries. For example, it makes no sense to compare a stock's PE ratio to its price. They do not measure the same thing; they belong to different domains. Although the domain concept is useful, it is rarely supported by relational model implementations.

Relations

A relation is a table of *n* columns (or *attributes*) and *m* rows (or *tuples*). Each column has a unique name, and all the values in a column are drawn from the same domain. Each row of the relation is uniquely identified. The order of columns and rows is immaterial.

The **cardinality** of a relation is its number of rows. The **degree** of a relation is the number of columns.

For example, the relation `NATION` (see the following figure) is of degree 3 and has a cardinality of 4. Because the cardinality of a relation changes every time a row is added or deleted, you can expect cardinality to change frequently. The degree changes if a column is added to a relation, but in terms of relational theory, it is considered to be a new relation. So, only a relation's cardinality changes.

A relational database with tables nation and stock

nation		
*natcode	natname	exchrate
AUS	Australia	0.46
IND	India	0.0228
UK	United Kingdom	1
USA	United States	0.67

stock						
*stkcode	stkfirm	stkprice	stkqty	stkdiv	stkpe	*natcode*
FC	Freedonia Copper	27.5	10,529	1.84	16	UK
PT	Patagonian Tea	55.25	12,635	2.5	10	UK
AR	Abyssinian Ruby	31.82	22,010	1.32	13	UK
SLG	Sri Lankan Gold	50.37	32,868	2.68	16	UK
ILZ	Indian Lead & Zinc	37.75	6,390	3	12	UK
BE	Burmese Elephant	0.07	154,713	0.01	3	UK

stock						
*stkcode	stkfirm	stkprice	stkqty	stkdiv	stkpe	*natcode*
BS	Bolivian Sheep	12.75	231,678	1.78	11	UK
NG	Nigerian Geese	35	12,323	1.68	10	UK
CS	Canadian Sugar	52.78	4,716	2.5	15	UK
ROF	Royal Ostrich Farms	33.75	1,234,923	3	6	UK
MG	Minnesota Gold	53.87	816,122	1	25	USA ←
GP	Georgia Peach	2.35	387,333	0.2	5	USA ←
NE	Narembeen Emu	12.34	45,619	1	8	AUS
QD	Queensland Diamond	6.73	89,251	0.5	7	AUS
IR	Indooroopilly Ruby	15.92	56,147	0.5	20	AUS
BD	Bombay Duck	25.55	167,382	1	12	IN

Relational database

A relational database is a collection of relations or tables. The distinguishing feature of the relational model, when compared to the hierarchical and network models, is that there are no explicit linkages between tables. Tables are linked by common columns drawn on the same domain; thus, the portfolio database (see the preceding tables) consists of tables `stock` and `nation`. The 1:m relationship between the two tables is represented by the column `natcode` that is common to both tables. Note that the two columns need not have the same name, but they must be drawn on the same domain so that comparison is possible. In the case of an m:m relationship, while a new table must be created to represent the relationship, the principle of linking tables through common columns on the same domain remains in force.

Primary key

A relation's primary key is its unique identifier; for example, the primary key of `nation` is `natcode`. As you already know, a primary key can be a composite of several columns. The primary key guarantees that each row of a relation can be uniquely addressed.

Candidate key

In some situations, there may be several attributes, known as candidate keys, that are potential primary keys. Column `natcode` is unique in the `nation` relation, for example. We also can be fairly certain that two nations will not have the same name. Therefore, `nation` has multiple candidate keys: `natcode` and `natname`.

Alternate key

When there are multiple candidate keys, one is chosen as the primary key, and the remainder are known as alternate keys. In this case, we selected `natcode` as the primary key, and `natname` is an alternate key.

Foreign key

The foreign key is an important concept of the relational model. It is the way relationships are represented and can be thought of as the glue that holds a set of tables together to form a relational database. A foreign key is an attribute (possibly composite) of a relation that is also a primary key of a relation. The foreign key and primary key may be in different relations or the same relation, but both keys must be drawn from the same domain.

Integrity rules

The integrity section of the relational model consists of two rules. The **entity integrity rule** ensures that each instance of an entity described by a relation is identifiable in some way. Its implementation means that each row in a relation can be uniquely distinguished. The rule is

No component of the primary key of a relation may be null.

Null in this case means that the component cannot be undefined or unknown; it must have a value. Notice that the rule says "component of a primary key." This means every part of the primary key must be known. If a part cannot be defined, it implies that the particular entity it describes cannot be defined. In practical terms, it means you cannot add a nation to `nation` unless you also define a value for `natcode`.

The definition of a foreign key implies that there is a corresponding primary key. The **referential integrity rule** ensures that this is the case. It states

A database must not contain any unmatched foreign key values.

Simply, this rule means that you cannot define a foreign key without first defining its matching primary key. In practical terms, it would mean you could not add a Canadian stock to the `stock` relation without first creating a row in `nation` for Canada.

Notice that the concepts of foreign key and referential integrity are intertwined. There is not much sense in having foreign keys without having the referential integrity rule. Permitting a foreign key without a corresponding primary key means the relationship cannot be determined. Note that the referential integrity rule does not imply a foreign key cannot be null. There can be circumstances where a relationship does not exist for a particular instance, in which case the foreign key is null.

Manipulation languages

There are four approaches to manipulating relational tables, and you are already familiar with SQL. Query-by-example (QBE), which describes a general class of graphical interfaces to a relational database to make it easier to write queries, is also commonly used. Less widely used manipulation languages are relational algebra and relational calculus. These languages are briefly discussed, with some attention given to relational algebra in the next section and SQL in the following chapter.

Query-by-example (QBE) is not a standard, and each vendor implements the idea differently. The general principle is to enable those with limited SQL skills to interrogate a relational database. Thus, the analyst might select tables and columns using drag and drop or by clicking a button. The following screen shot shows the QBE interface to LibreOffice, which is very similar to that of MS Access. It shows selection of the `stock` table and two of its columns, as well as a criterion for `stkqty`. QBE commands are translated to SQL prior to execution, and many systems allow you to view the generated SQL. This can be handy. You can use QBE to generate a portion of the SQL for a complex query and then edit the generated code to fine-tune the query.

Query-by-example with LibreOffice

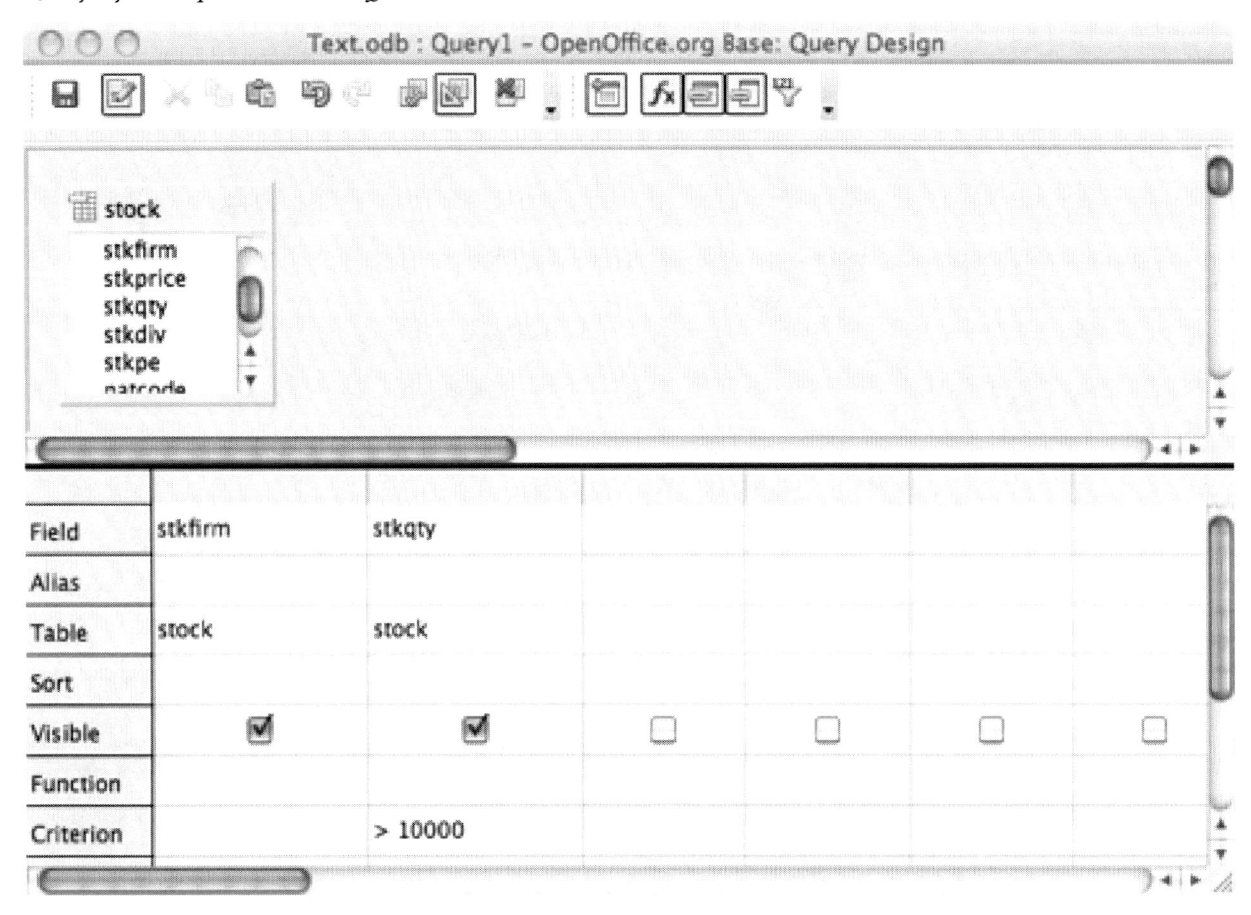

Relational algebra has a set of operations similar to traditional algebra (e.g., add and multiply) for manipulating tables. Although relational algebra can be used to resolve queries, it is seldom employed, because it requires you to specify both what you want and how to get it. That is, you have to specify the operations on each table. This makes relational algebra more difficult to use than SQL, where the focus is on specifying what is wanted.

Relational calculus overcomes some of the shortcomings of relational algebra by concentrating on what is required. In other words, there is less need to specify how the query will operate. Relational calculus is classified as a nonprocedural language, because you do not have to be overly concerned with the procedures by which a result is determined.

Unfortunately, relational calculus can be difficult to learn, and as a result, language designers developed **SQL** and **QBE**, which are nonprocedural and more readily mastered. You have already gained some skills in SQL. Although QBE is generally easier to use than SQL, IS professionals need to master SQL for two reasons. *First,* SQL is frequently embedded in other programming languages, a feature not available with QBE. *Second,* it is very difficult, or impossible, to express some queries in QBE (e.g., divide). Because SQL is important, it is the focus of the next chapter.

Relational algebra

The relational model includes a set of operations, known as relational algebra, for manipulating relations. Relational algebra is a standard for judging data retrieval languages. If a retrieval language, such as SQL, can be used to express every relational algebra operator, it is said to be **relationally complete**.

Relational algebra operators

Operator	Description
Restrict	Creates a new table from specified rows of an existing table
Project	Creates a new table from specified columns of an existing table
Product	Creates a new table from all the possible combinations of rows of two existing tables
Union	Creates a new table containing rows appearing in one or both tables of two existing tables
Intersect	Creates a new table containing rows appearing in both tables of two existing tables
Difference	Creates a new table containing rows appearing in one table but not in the other of two existing tables
Join	Creates a new table containing all possible combinations of rows of two existing tables satisfying the join condition
Divide	Creates a new table containing x_i such that the pair (x_i, y_i) exists in the first table for every y_i in the second table

There are eight relational algebra operations that can be used with either one or two relations to create a new relation. The assignment operator (:=) indicates the name of the new relation. The relational algebra statement to create relation A, the union of relations B and C, is expressed as

```
A := B UNION C
```

Before we begin our discussion of each of the eight operators, note that the first two, restrict and project, operate on a single relation and the rest require two relations.

Restrict

Restrict extracts specified rows from a single relation. As the shaded rows in the following figure depict, restrict takes a horizontal slice through a relation.

Relational operation restrict—a horizontal slice

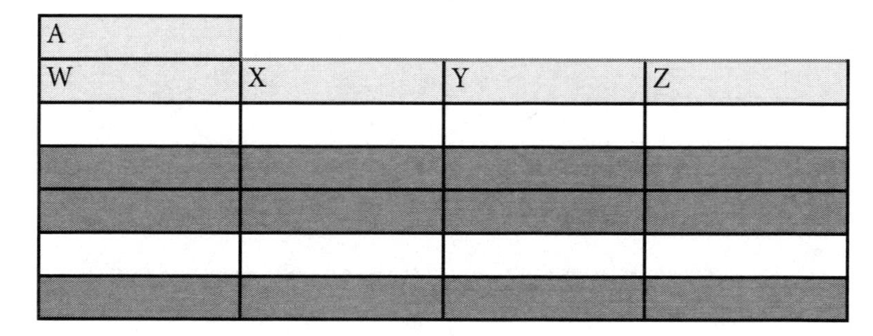

The relational algebra command to create the new relation using restrict is

```
tablename WHERE column1 theta column2
```

or

```
tablename WHERE column1 theta literal
```

where THETA can be =, <>, >, >=, <, or <=.

For example, the following relational algebra command creates a new table from `stock` containing all rows with a nation code of `'US'`:

```
usstock := stock WHERE natcode = 'US'
```

Project

Project extracts specified columns from a table. As the following figure shows, project takes a vertical slice through a table.

Relational operator project—a vertical slice

A			
W	X	Y	Z

The relational algebra command to create the new relation using project is

```
tablename [columnname, …]
```

So, in order to create a new table from nation that contains the nation's name and its exchange rate, for example, you would use this relational algebra command:

```
rates := nation[natname, exchrate]
```

Product

Product creates a new relation from all possible combinations of rows in two other relations. It is sometimes called `TIMES` or `MULTIPLY`. The relational command to create the product of two tables is

```
tablename1 TIMES tablename2
```

The operation of product is illustrated in the following figure, which shows the result of A TIMES B.

Relational operator product

A	
V	W
v_1	w_1
v_2	w_2
v_3	w_3

B		
X	Y	Z
x_1	y_1	z_1
x_2	y_2	z_2

A TIMES B				
V	W	X	Y	Z
v_1	w_1	x_1	y_1	z_1
v_1	w_1	x_2	y_2	z_2
v_2	w_2	x_1	y_1	z_1
v_2	w_2	x_2	y_2	z_2
v_3	w_3	x_1	y_1	z_1
v_3	w_3	x_2	y_2	z_2

Union

The union of two relations is a new relation containing all rows appearing in one or both relations. The two relations must be **union compatible,** which means they have the same column names, in the same order, and drawn on the same domains. Duplicate rows are automatically eliminated—they must be, or the relational model is no longer satisfied. The relational command to create the union of two tables is

```
tablename1 UNION tablename2
```

Union is illustrated in the following figure. Notice that corresponding columns in tables A and B have the same names. While the sum of the number of rows in relations A and B is five, the union contains four rows, because one row (x_2, y_2) is common to both relations.

Relational operator union

A	
X	Y
x_1	y_1
x_2	y_2
x_3	y_3

B	
X	Y
x_2	y_2
x_4	y_4

A UNION B	
X	Y
x_1	y_1
x_2	y_2
x_3	y_3
x_4	y_4

Intersect

The intersection of two relations is a new relation containing all rows appearing in both relations. The two relations must be union compatible. The relational command to create the intersection of two tables is

```
tablename1 INTERSECT tablename2
```

The result of `A INTERSECT B` is one row, because only one row (x_2, y_2) is common to both relations A and B.

Relational operator intersect

A	
X	Y
x_1	y_1
x_2	y_2
x_3	y_3

B	
X	Y
x_2	y_2
x_4	y_4

A INTERSECT B	
X	Y
x_2	y_2

Difference

The difference between two relations is a new relation containing all rows appearing in the first relation but not in the second. The two relations must be **union compatible**. The relational command to create the difference between two tables is

```
tablename1 MINUS tablename2
```

The result of A MINUS B is two rows (see the following figure). Both of these rows are in relation A but not in relation B. The row containing (x1, y2) appears in both A and B and thus is not in A MINUS B.

Relational operator difference

A	
X	Y
x_1	y_1
x_2	y_2
x_3	y_3

B	
X	Y
x_2	y_2
x_4	y_4

A MINUS B	
X	Y
x_1	y_1
x_3	y_3

Join

Join creates a new relation from two relations for all combinations of rows satisfying the join condition. The general format of join is

```
tablename1 JOIN tablename2 WHERE tablename1.columnname1 theta
   tablename2.columnname2
```

where THETA can be =, <>, >, >=, <, or <=.

The following figure illustrates `A JOIN B WHERE W = Z`, which is an equijoin because theta is an equals sign. Tables A and B are matched when values in columns W and Z in each relation are equal. The matching columns should be drawn from the same domain. You can also think of join as a product followed by restrict on the resulting relation. So the join can be written

```
(A TIMES B) WHERE W theta Z
```

Relational operator join

A	
V	W
v_1	wz_1
v_2	wz_2
v_3	wz_3

B		
X	Y	Z
x_1	y_1	wz_1
x_2	y_2	wz_2

A EQUIJOIN B				
V	W	X	Y	Z
v_1	wz_1	x_1	y_1	wz_1
v_3	wz_3	x_2	y_2	wz_3

Divide

Divide is the hardest relational operator to understand. Divide requires that A and B have a set of attributes, in this case Y, that are common to both relations. Conceptually, A divided by B asks the question, "Is there a value in the X column of A (e.g., x_1) that has a value in the Y column of A for every value of y in the Y column of B?" Look first at B, where the Y column has values y_1 and y_2. When you examine the X column of A, you find there are rows (x_1, y_1) and (x_1, y_2). That is, for x_1, there is a value in the Y column of A for every value of y in the Y column of B. Thus, the result of the division is a new relation with a single row and column containing the value x_1.

Relational operator divide

A	
X	Y
x_1	y_1
x_1	y_3
x_1	y_2
x_2	y_1
x_3	y_3

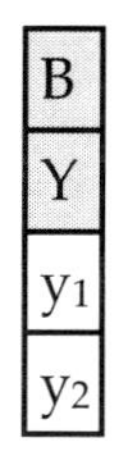

B
Y
y_1
y_2

A DIVIDE B
X
x_1

Querying with relational algebra

A few queries will give you a taste of how you might use relational algebra. To assist in understanding these queries, the answer is expressed, side-by-side, in both relational algebra (on the left) and SQL (on the right).

- *List all data in share.*

A very simple report of all columns in the table SHARE.

share	SELECT * FROM share

- *Report a firm's name and price-earnings ratio.*

A projection of two columns from share.

share [shrfirm, shrpe]	SELECT shrfirm, shrpe FROM share

- *Get all shares with a price-earnings ratio less than 12.*

A restriction of `share` on column `shrpe`.

`share WHERE shrpe < 12`	`SELECT * FROM share` ` WHERE shrpe < 12`

- *List details of firms where the share holding is at least 100,000.*

A restriction and projection combined. Notice that the restriction is expressed within parentheses.

`(share WHERE shrqty >= 100000)`
`[shrfirm, shrprice, shrqty,`
`shrdiv]`
	`SELECT shrfirm, shrprice,`
	`      shrqty, shrdiv FROM share`
	`            WHERE shrqty >= 100000`

- *Find all shares where the PE is 12 or higher and the share holding is less than 10,000.*

Intersect is used with two restrictions to identify shares satisfying both conditions.

`(share WHERE shrpe >= 12)`
`INTERSECT`
`(share WHERE shrqty < 10000)`
	`SELECT * FROM share`
	` WHERE shrpe >= 12`
	` AND shrqty < 10000`

- *Report all shares other than those with the code CS or PT.*

Difference is used to subtract those shares with the specified codes. Notice the use of union to identify shares satisfying either condition.

`share MINUS`
` ((share WHERE shrcode = 'CS')`
`UNION`
` (share WHERE shrcode = 'PT'))`
	`SELECT * FROM share WHERE shrcode`
	`NOT IN ('CS', 'PT')`

- *Report the value of each stock holding in UK pounds.*

A join of `stock` and `nation` and projection of specified attributes.

`(stock JOIN nation`
` WHERE stock.natcode =`
`  nation.natcode )`
`[natname, stkfirm,`
` stkprice, stkqty, exchrate,`
` stkprice*stkqty*exchrate]`
	`SELECT natname, stkfirm,`
	` stkprice, stkqty, exchrate,`
	` stkprice*stkqty*exchrate`
	`  FROM stock JOIN nation`
	`   ON stock.natcode =`
	`    nation.natcode`

- *Find the items that have appeared in all sales.*

Divide reports the item numbers of items appearing in all sales, and this result is then joined with `item` to report these items.

((lineitem[itemno, saleno]
 DIVIDEBY sale[saleno])
  JOIN item) [itemno, itemname]
	SELECT itemno, itemname FROM item
	 WHERE NOT EXISTS
	  (SELECT * FROM sale
	   WHERE NOT EXISTS
	    (SELECT * FROM lineitem
	     WHERE lineitem.itemno =
	      item.itemno
	     AND lineitem.saleno =
	      sale.saleno))

Skill builder

Write relational algebra statements for the following queries:

Report a firm's name and dividend.

Find the names of firms where the holding is greater than 10,000.

A primitive set of relational operators

The full set of eight relational operators is not required. As you have already seen, JOIN can be defined in terms of product and restrict. Intersection and divide can also be defined in terms of other commands. Indeed, only five operators are required: restrict, project, product, union, and difference. These five are known as *primitives* because these are the minimal set of relational operators. None of the primitive operators can be defined in terms of the other operators. The following table illustrates how each primitive can be expressed as an SQL command, which implies that SQL is relationally complete.

Comparison of relational algebra primitive operators and SQL

Operator	Relational algebra	SQL
Restrict	A WHERE CONDITION	SELECT * FROM A WHERE CONDITION
Project	A [X]	SELECT X FROM A
Product	A TIMES B	SELECT * FROM A, B
Union	A UNION B	SELECT * FROM A UNION SELECT * FROM B
Difference	A MINUS B	SELECT * FROM A WHERE NOT EXISTS (SELECT * FROM B WHERE A.X = B.X AND A.Y = B.Y AND …*

* Essentially where all columns of A are equal to all columns of B

A fully relational database

The three components of a relational database system are structures (domains and relations), integrity rules (primary and foreign keys), and a manipulation language (relational algebra). A **fully relational database** system provides complete support for each of these components. Many commercial systems support SQL but do not provide support for domains. Such systems are not fully relational but are relationally complete.

In 1985, Codd established the 12 commandments of relational database systems (see the following table). In addition to providing some insights into Codd's thinking about the management of data, these rules can also be used to judge how well a database fits the relational model. The major impetus for these rules was uncertainty in the marketplace about the meaning of "relational DBMS." Codd's rules are a checklist for establishing the completeness of a relational DBMS. They also provide a short summary of the major features of the relational DBMS.

Codd's rules for a relational DBMS

Rules
The information rule
The guaranteed access rule
Systematic treatment of null values
Active online catalog of the relational model
The comprehensive data sublanguage rule
The view updating rule
High-level insert, update, and delete
Physical data independence
Logical data independence
Integrity independence
Distribution independence
The nonsubversion rule

The information rule

There is only one logical representation of data in a database. All data must appear to be stored as values in a table.

The guaranteed access rule

Every value in a database must be addressable by specifying its table name, column name, and the primary key of the row in which it is stored.

Systematic treatment of null values

There must be a distinct representation for unknown or inappropriate data. This must be unique and independent of data type. The DBMS should handle null data in a systematic way. For example, a zero or a blank cannot be used to represent a null. This is one of the more troublesome areas because null can have several meanings (e.g., missing or inappropriate).

Active online catalog of the relational model

There should be an online catalog that describes the relational model. Authorized users should be able to access this catalog using the DBMS's query language (e.g., SQL).

The comprehensive data sublanguage rule

There must be a relational language that supports data definition, data manipulation, security and integrity constraints, and transaction processing operations. Furthermore, this language must support both interactive querying and application programming and be expressible in text format. SQL fits these requirements.

The view updating rule

The DBMS must be able to update any view that is theoretically updatable.

High-level insert, update, and delete

The system must support set-at-a-time operations. For example, multiple rows must be updatable with a single command.

Physical data independence

Changes to storage representation or access methods will not affect application programs. For example, application programs should remain unimpaired even if a database is moved to a different storage device or an index is created for a table.

Logical data independence

Information-preserving changes to base tables will not affect application programs. For instance, no applications should be affected when a new table is added to a database.

Integrity independence

Integrity constraints should be part of a database's definition rather than embedded within application programs. It must be possible to change these constraints without affecting any existing application programs.

Distribution independence

Introduction of a distributed DBMS or redistribution of existing distributed data should have no impact on existing applications.

The nonsubversion rule

It must not be possible to use a record-at-a-time interface to subvert security or integrity constraints. You should not be able, for example, to write a Java program with embedded SQL commands to bypass security features.

Codd issued an additional higher-level rule, rule 0, which states that a relational DBMS must be able to manage databases entirely through its relational capacities. In other words, a DBMS is either totally relational or it is not relational.

Summary

The relational model developed as a result of recognized shortcomings of hierarchical and network DBMSs. Codd created a strong theoretical base for the relational model. Three objectives drove relational database research: data independence, communicability, and set processing. The relational model has domain structures, integrity rules, and operators used to retrieve, derive, or modify data. A domain is a set of values all of the same data type. The practical value of a domain is to define what comparisons are permissible. A relation is a table of *n* columns and *m* rows. The cardinality of a relation is its number of rows. The degree of a relation is the number of columns.

A relational database is a collection of relations. The distinguishing feature of the relational model is that there are no explicit linkages between tables. A relation's primary key is its unique identifier. When there are multiple candidates for the primary key, one is chosen as the primary key and the remainder are known as alternate keys. The foreign key is the way relationships are represented and can be thought of as the glue that binds a set of tables together to form a relational database. The purpose of the entity integrity rule is to ensure that each entity described by a relation is identifiable in some way. The referential integrity rule ensures that you cannot define a foreign key without first defining its matching primary key.

The relational model includes a set of operations, known as relational algebra, for manipulating relations. There are eight operations that can be used with either one or two relations to create a new relation. These operations are restrict, project, product, union, intersect, difference, join, and divide. Only five operators, known as primitives, are required to define all eight relational operations. An SQL statement can be translated to relational algebra and vice versa. If a retrieval language can be used to express every relational algebra operator, it is said to be relationally complete. A fully relational database system provides complete support for domains, integrity rules, and a manipulation language. Codd set forth 12 rules that can be used to judge how well a database fits the relational model. He also added rule 0—a DBMS is either totally relational or it is not relational.

Key terms and concepts

Alternate key

Candidate key

Cardinality

Catalog

Communicability objective

Data independence

Data structures

Data sublanguage rule

Degree

Difference

Distribution independence

Divide

Domain

Entity integrity

Foreign key

Fully relational database

Guaranteed access rule

Information rule

Integrity independence

Integrity rules

Intersect

Join

Logical data independence

Nonsubversion rule

Null

Operators

Physical data independence

Primary key

Product

Project

Query-by-example (QBE)

Referential integrity

Relational algebra

Relational calculus

Relational database

Relational model

Relationally complete

Relations

Restrict

Set processing objective

Union

View updating rule

References and additional readings

Codd, E. F. 1982. Relational database: A practical foundation for productivity. *Communications of the ACM* 25 (2):109–117.

Exercises

1. What reasons does Codd give for adopting the relational model?
2. What are the major components of the relational model?
3. What is a domain? Why is it useful?
4. What are the meanings of cardinality and degree?

5. What is a simple relational database?
6. What is the difference between a primary key and an alternate key?
7. Why do we need an entity integrity rule and a referential integrity rule?
8. What is the meaning of "union compatible"? Which relational operations require union compatibility?
9. What is meant by the term "a primitive set of operations"?
10. What is a fully relational database?
11. How well does OpenOffice's database satisfy Codd's rules.
12. Use relational algebra with the given data model to solve the following queries:

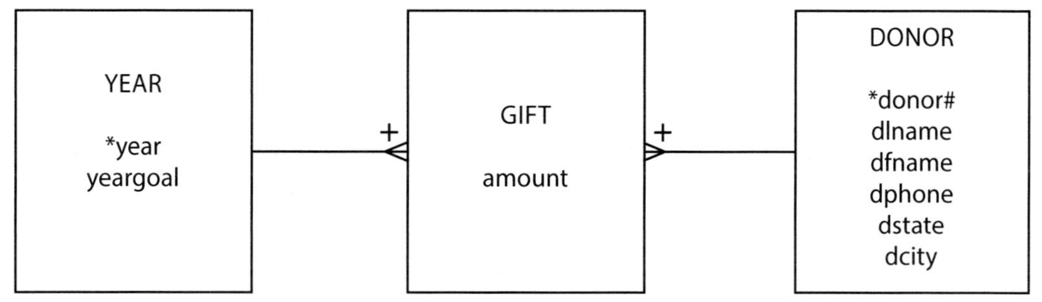

a. List all donors.
b. List the first and last names of all donors.
c. List the phone numbers of donors Hays and Jefts.
d. List the amount given by each donor for each year.
e. List the donors who have made a donation every year.
f. List the names of donors who live in Georgia or North Carolina.
g. List the names of donors whose last name is Watson and who live in Athens, GA.

10. SQL

The questing beast.

Sir Thomas Malory, *Le Morte D'Arthur,* 1470

Learning objectives

Students completing this chapter will have a detailed knowledge of SQL.

Introduction

Structured query language (SQL) is widely used as a relational database language, and SQL skills are essential for data management in a world that is increasingly reliant on relational technology. SQL originated in the IBM Research Laboratory in San Jose, California. Versions have since been implemented by commercial database vendors and open source teams for a wide range of operating systems. Both the American National Standards Institute (ANSI) and the International Organization for Standardization (ISO) have designated SQL as a standard language for relational database systems.

SQL is a **complete database language**. It is used for defining a relational database, creating views, and specifying queries. In addition, it allows for rows to be inserted, updated, and deleted. In database terminology, it is both a **data definition language** (DDL), a **data manipulation language** (DML), and a **data control language** (DCL). SQL, however, is not a complete programming language like PHP and Java. Because SQL statements can be embedded into general-purpose programming languages, SQL is often used in conjunction with such languages to create application programs. The **embedded SQL** statements handle the database processing, and the statements in the general-purpose language perform the necessary tasks to complete the application.

SQL is a declarative language, because you declare the desired results. Languages such as Java are procedural languages, because the programmer specifies each step the computer must execute. The SQL programmer can focus on defining what is required rather than detailing the process to achieve what is required. Thus, SQL programs are much shorter than their procedural equivalents.

You were introduced to SQL in Chapters 3 through 6. This chapter provides an integrated coverage of the language, pulling together the various pieces presented previously.

Data definition

The DDL part of SQL encompasses statements to operate on tables, views, and indexes. Before we proceed, however, the term "base table" must be defined. A **base table** is an autonomous, named table. It is autonomous because it exists in its own right; it is physically stored within the database. In contrast, a view is not autonomous because it is derived from one or more base tables and does not exist independently. A view is a virtual table. A base table has a name by which it can be referenced. This name is chosen when the base table is generated using the `CREATE` statement. Short-lived temporary tables, such as those formed as the result of a query, are not named.

Keys

The concept of a **key** occurs several times within SQL. In general, a key is one or more columns identified as such in the description of a table, an index, or a referential constraint. The same column can be part of more than one key. For example, it can be part of a primary key and a foreign key. A **composite key** is an ordered set of columns of the same table. In other words, the primary key of `lineitem` is always the composite of (`lineno, saleno`) in that order, which cannot be changed.

Comparing composite keys actually means that corresponding components of the keys are compared. Thus, application of the referential integrity rule—*the value of a foreign key must be equal to a value of the primary key*—means that each component of the foreign key must be equal to the corresponding component of the composite primary key.

So far, you have met primary and foreign keys. A **unique key** is another type of key. Its purpose is to ensure that no two values of a given column are equal. This constraint is enforced by the DBMS during the execution of `INSERT` and `UPDATE` statements. A unique key is part of the index mechanism.

Indexes

Indexes are used to accelerate data access and ensure uniqueness. An **index** is an ordered set of pointers to rows of a base table. Think of an index as a table that contains two columns. The first column contains values for the index key, and the second contains a list of addresses of rows in the table. Since the values in the first column are ordered (i.e., in ascending or descending sequence), the index table can be searched quickly. Once the required key has been found in the table, the row's address in the second column can be used to retrieve the data quickly. An index can be specified as being unique, in which case the DBMS ensures that the corresponding table does not have rows with identical index keys.

An example of an index

itemtype index

itemtype	
C	
C	
C	
C	
E	
E	
F	
N	
N	
N	

item

itemno	itemname	itemtype	itemcolor
1	Pocket knife–Nile	E	Brown
2	Pocket knife–Thames	E	Brown
3	Compass	N	–
4	Geo positioning system	N	–
5	Map measure	N	–
6	Hat–polar explorer	C	Red
7	Hat–polar explorer	C	White
8	Boots–snakeproof	C	Green
9	Boots–snakeproof	C	Black
10	Safari hat	F	Khaki

Notation

A short primer on notation is required before we examine SQL commands.

1. Text in uppercase is required as is.
2. Text in lowercase denotes values to be selected by the query writer.
3. Statements enclosed within square brackets are optional.
4. | indicates a choice.
5. An ellipsis (...) indicates that the immediate syntactic unit may be repeated optionally more than once.

Creating a table

CREATE TABLE is used to define a new base table, either interactively or by embedding the statement in a host language. The statement specifies a table's name, provides details of its columns, and provides integrity checks. The syntax of the command is

```
CREATE TABLE base-table
   column-definition-block
   [primary-key-block]
   [referential-constraint-block]
   [unique-block];
```

Column definition

The column definition block defines the columns in a table. Each column definition consists of a column name, data type, and optionally the specification that the column cannot contain null values. The general form is

```
(column-definition [, …])
```

where column-definition is of the form

```
column-name data-type [NOT NULL]
```

The NOT NULL clause specifies that the particular column must have a value whenever a new row is inserted.

Constraints

A constraint is a rule defined as part of CREATE TABLE that defines valid sets of values for a base table by placing limits on INSERT, UPDATE, and DELETE operations. Constraints can be named (e.g., fk_stock_nation) so that they can be turned on or off and modified. The four constraint variations apply to primary key, foreign key, unique values, and range checks.

Primary key constraint

The primary key constraint block specifies a set of columns that constitute the primary key. Once a primary key is defined, the system enforces its uniqueness by checking that the primary key of any new row does not already exist in the table. A table can have only one primary key. While it is not mandatory to define a primary key, it is good practice always to define a table's primary key, though it is not that common to name the constraint. The general form of the constraint is

```
[primary-key-name] PRIMARY KEY(column-name [asc|desc] [, …])
```

The optional ASC or DESC clause specifies whether the values from this key are arranged in ascending or descending order, respectively.

For example:

```
pk_stock PRIMARY KEY(stkcode)
```

Foreign key constraint

The referential constraint block defines a foreign key, which consists of one or more columns in the table that together must match a primary key of the specified table (or else be null). A foreign key value is null when any one of the columns in the row constituting the foreign key is null. Once the foreign key constraint is defined, the DBMS will check every insert and update to ensure that the constraint is observed. The general form is

```
CONSTRAINT constraint-name FOREIGN KEY(column-name [,…])
   REFERENCES table-name(column-name [,…])
      [ON DELETE (RESTRICT | CASCADE | SET NULL)]
```

The constraint-name defines a referential constraint. You cannot use the same constraint-name more than once in the same table. Column-name identifies the column or columns that comprise the foreign key. The data type and length of foreign key columns must match exactly the data type and length of the primary key columns. The clause `REFERENCES table-name` specifies the name of an existing table, and its primary key, that contains the primary key, which cannot be the name of the table being created.

The `ON DELETE` clause defines the action taken when a row is deleted from the table containing the primary key. There are three options:

1. `RESTRICT` prevents deletion of the primary key row until all corresponding rows in the related table, the one containing the foreign key, have been deleted. `RESTRICT` is the default and the cautious approach for preserving data integrity.
2. `CASCADE` causes all the corresponding rows in the related table also to be deleted.
3. `SET NULLS` sets the foreign key to null for all corresponding rows in the related table.

For example:

```
CONSTRAINT fk_stock_nation FOREIGN KEY(natcode)
   REFERENCES nation(natcode)
```

Unique constraint

A unique constraint creates a unique index for the specified column or columns. A unique key is constrained so that no two of its values are equal. Columns appearing in a unique constraint must be defined as `NOT NULL`. Also, these columns should not be the same as those of the table's primary key, which is guaranteed uniqueness by its primary key definition. The constraint is enforced by the DBMS during execution of `INSERT` and `UPDATE` statements. The general format is

```
UNIQUE constraint-name (column-name [ASC|DESC] [, …])
```

An example follows:

```
CONSTRAINT unq_stock_stkname UNIQUE(stkname)
```

Check constraint

A check constraint defines a set of valid values and comes in three forms: table, column, or domain constraint.

Table constraints are defined in `CREATE TABLE` and `ALTER TABLE` statements. They can be set for one or more columns. A table constraint, for example, might limit `itemcode` to values less than 500.

```
CREATE TABLE item (
   itemcode integer,
      CONSTRAINT chk_item_itemcode check(itemcode <500));
```

A **column constraint** is defined in a `CREATE TABLE` statement and must be for a single column. Once created, a column constraint is, in effect, a table constraint. The following example shows that there is little difference between defining a table and column constraint. In the following case, `itemcode` is again limited to values less than 500.

```
CREATE TABLE item (
   itemcode INTEGER CONSTRAINT chk_item_itemcode CHECK(itemcode <500),
   itemcolor VARCHAR(10));
```

A **domain constraint** defines the legal values for a type of object (e.g., an item's color) that can be defined in one or more tables. The following example sets legal values of color.

```
CREATE domain valid_color AS char(10)
CONSTRAINT chk_qitem_color check(
    IN ('Bamboo','Black','Brown','Green','Khaki','White'));
```

In a `CREATE TABLE` statement, instead of defining a column's data type, you specify the domain on which the column is defined. The following example shows that `itemcolor` is defined on the domain `valid_color`.

```
CREATE table item (
   itemcode INTEGER,
   itemcolor valid_color);
```

Data types

Some of the variety of data types that can be used are depicted in the following figure and described in more detail in the following pages.

Data types

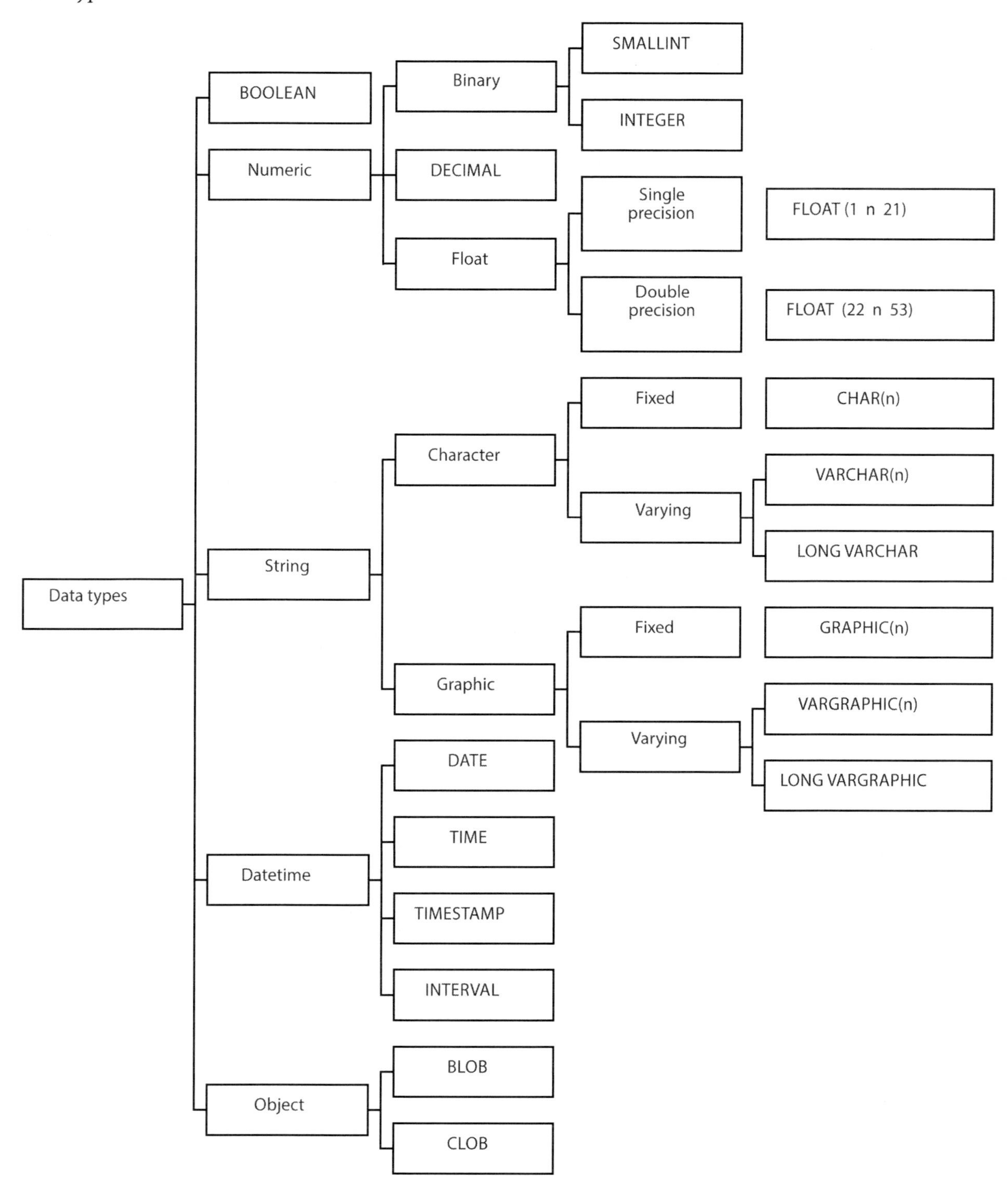

BOOLEAN

`BOOLEAN` data types can have the values true, false, or unknown.

SMALLINT and INTEGER

Most commercial computers have a 32-bit word, where a word is a unit of storage. An integer can be stored in a full word or half a word. If it is stored in a full word (`INTEGER`), then it can be 31 binary digits in length. If half-word storage is used (`SMALLINT`), then it can be 15 binary digits long. In each case, one bit is used for the sign of the number. A column defined as `INTEGER` can store a number in the range -2^{31} to 2^{31}-1 or –2,147,483,648 to 2,147,483,647. A column defined as `SMALLINT` can store a number in the range -2^{15} to 2^{15}-1 or –32,768 to 32,767. Just remember that `INTEGER` is good for ±2 billion and `SMALLINT` for ±32,000.

FLOAT

Scientists often deal with numbers that are very large (e.g., Avogadro's number is 6.02252×10^{23}) or very small (e.g., Planck's constant is 6.6262×10^{-34} joule sec). The `FLOAT` data type is used for storing such numbers, often referred to as *floating-point* numbers. A single-precision floating-point number requires 32 bits of storage and can represent numbers in the range -7.2×10^{75} to -5.4×10^{-79}, 0, 5.4×10^{-79} to 7.2×10^{75} with a precision of about 7 decimal digits. A double-precision floating-point number requires 64 bits. The range is the same as for a single-precision floating-point number. The extra 32 bits are used to increase precision to about 15 decimal digits.

In the specification `FLOAT(N)`, if *n* is between 1 and 21 inclusive, single-precision floating-point is selected. If *n* is between 22 and 53 inclusive, the storage format is double-precision floating-point. If *n* is not specified, double-precision floating-point is assumed.

DECIMAL

Binary is the most convenient form of storing data from a computer's perspective. People, however, work with a decimal number system. The `DECIMAL` data type is convenient for business applications because data storage requirements are defined in terms of the maximum number of places to the left and right of the decimal point. To store the current value of an ounce of gold, you would possibly use `DECIMAL(6,2)` because this would permit a maximum value of $9,999.99. Notice that the general form is `DECIMAL(P,Q)`, where `P` is the total number of digits in the column, and `Q` is the number of digits to the right of the decimal point.

CHAR and VARCHAR

Nonnumeric columns are stored as character strings. A person's family name is an example of a column that is stored as a character string. `CHAR(N)` defines a column that has a fixed length of n characters, where n can be a maximum of 255.

When a column's length can vary greatly, it makes sense to define the field as `VARCHAR`. A column defined as `VARCHAR` consists of two parts: a header indicating the length of the character string and the string. If a table contains a column that occasionally stores a long string of text (e.g., a message field), then defining it as `VARCHAR` makes sense. `VARCHAR` can store strings up to 65,535 characters long.

Why not store all character columns as `VARCHAR` and save space? There is a price for using `VARCHAR` with some relational systems. *First,* additional space is required for the header to indicate the length of the string. *Second,* additional processing time is required to handle a variable-length string compared to a fixed-length string. Depending on the DBMS and processor speed, these might be important considerations, and some systems will automatically make an appropriate choice. For example, if you use both data types in the same table, MySQL will automatically change `CHAR` into `VARCHAR` for compatibility reasons.

There are some columns where there is no trade-off decision because all possible entries are always the same length. Canadian postal codes, for instance, are always six characters (e.g., the postal code for Ottawa is K1A0A1).

Data compression is another approach to the *space wars* problem. A database can be defined with generous allowances for fixed-length character columns so that few values are truncated. Data compression can be used to compress the file to remove *wasted* space. Data compression, however, is slow and will increase the time to process queries. You save space at the cost of time, and save time at the cost of space. When dealing with character fields, the database designer has to decide whether time or space is more important.

Times and dates

Columns that have a data type of `DATE` are stored as *yyyymmdd* (e.g., 2012-11-04 for November 4, 2012). There are two reasons for this format. *First,* it is convenient for sorting in chronological order. The common American way of writing dates (*mmddyy*) requires processing before chronological sorting. *Second*, the full form of the year should be recorded for exactness.

For similar reasons, it makes sense to store times in the form *hhmmss* with the understanding that this is 24-hour time (also known as European time and military time). This is the format used for data type `TIME`.

Some applications require precise recording of events. For example, transaction processing systems typically record the time a transaction was processed by the system. Because computers operate at high speed, the `TIMESTAMP` data type records date and time with microsecond accuracy. A timestamp has seven parts—year, month, day, hour, minute, second, and microsecond. Date and time are defined as previously described (i.e., *yyyymmdd* and *hhmmss*, respectively). The range of the microsecond part is 000000 to 999999.

Although times and dates are stored in a particular format, the formatting facilities that generally come with a DBMS usually allow tailoring of time and date output to suit local standards. Thus for a U.S. firm, date might appear on a report in the form *mm/dd/yy*; for a European firm following the ISO standard, date would appear as *yyyy-mm-dd.*

SQL-99 introduced the `INTERVAL` data type, which is a single value expressed in some unit or units of time (e.g., 6 years, 5 days, 7 hours).

BLOB (binary large object)

`BLOB` is a large-object data type that stores any kind of binary data. Binary data typically consists of a saved spreadsheet, graph, audio file, satellite image, voice pattern, or any digitized data. The `BLOB` data type has no maximum size.

CLOB (character large object)

`CLOB` is a large-object data type that stores any kind of character data. Text data typically consists of reports, correspondence, chapters of a manual, or contracts. The `CLOB` data type has no maximum size.

Skill builder

What data types would you recommend for the following?

1. A book's ISBN
2. A photo of a product
3. The speed of light (2.9979×10^8 meters per second)

4. A short description of an animal's habitat
5. The title of a Japanese book
6. A legal contract
7. The status of an electrical switch
8. The date and time a reservation was made
9. An item's value in euros
10. The number of children in a family

Collation sequence

A DBMS will typically support many character sets. so it can handle text in different languages. While many European languages are based on an alphabet, they do not all use the same alphabet. For example, Norwegian has some additional characters (e.g., æ ,ø, å) compared to English, and French accents some letters (e.g., é, ü, and â), which does not occur in English. Alphabet based languages have a collating sequence, which defines how to sort individual characters in a particular language. For English, it is the familiar A B C … X Y Z. Norwegian's collating sequence includes three additional symbols, and the sequence is A B C … X Y Z Æ Ø Å. When you define a database you need to define its collating sequence. Thus, a database being set up for exclusive use in Chile would opt for a Spanish collating sequence. You can specify a collation sequence at the database, table, and, column level. The usual practice is to specify at the database level.

```
CREATE DATABASE ClassicModels COLLATE latin1_general_cs;
```

The `LATIN1_GENERAL` character set is suitable for Western European languages. The `CS` suffix indicates that comparisons are **case sensitive**. In other words, a query will see the two strings 'abc' and 'Abc' as different, whereas if case sensitivity is turned off, the strings are considered identical. Case sensitivity is usually the right choice to ensure precision of querying

Scalar functions

Most implementations of SQL include functions that can be used in arithmetic expressions, and for data conversion or data extraction. The following sampling of these functions will give you an idea of what is available. You will need to consult the documentation for your version of SQL to determine the functions it supports. For example, Microsoft SQL Server has more than 100 additional functions.

Some examples of SQL's built-in scalar functions

Function	Description
`CURRENT_DATE()`	Retrieves the current date
`EXTRACT(DATE_TIME_PART FROM EXPRESSION)`	Retrieves part of a time or date (e.g., YEAR, MONTH, DAY, HOUR, MINUTE, or SECOND)
`SUBSTRING(STR, POS, LEN)`	Retrieves a string of length *len* starting at position pos from string *str*

Some examples:

```
SELECT extract(day FROM CURRENT_DATE());
SELECT SUBSTRING(`person first`, 1,1), `person last` FROM person;
```

A vendor's additional functions can be very useful. Remember, though, that use of a vendor's extensions might limit portability.

- *How many days' sales are stored in the sale table?*

This sounds like a simple query, but you have to do a self-join and also know that there is a function, DATEDIFF, to determine the number of days between any two dates. Consult your DBMS manual to learn about other functions for dealing with dates and times.

```
SELECT DISTINCT DATEDIFF(late.saledate,early.saledate) AS difference
   FROM sale late JOIN sale early
      ON late.saledate =
         (SELECT MAX(saledate) FROM sale)
      AND early.saledate =
         (SELECT MIN(saledate) FROM sale);
```

DIFFERENCE
1

The preceding query is based on the idea of joining `sale` with a copy of itself. The matching column from `late` is the latest sale's date (or MAX), and the matching column from `early` is the earliest sale's date (or MIN). As a result of the join, each row of the new table has both the earliest and latest dates.

Formatting

You will likely have noticed that some queries report numeric values with a varying number of decimal places. The FORMAT function gives you control over the number of decimal places reported, as illustrated in the following example where yield is reported with two decimal places.

```
SELECT shrfirm, shrprice, shrqty, FORMAT(shrdiv/shrprice*100,2) AS yield
   FROM share;
```

When you use FORMAT you create a string, but you often want to sort on the numeric value of the formatted field. The following example illustrates how to do this.

```
SELECT shrfirm, shrprice, shrqty, FORMAT(shrdiv/shrprice*100,2) AS yield FROM SHARE
   ORDER BY shrdiv/shrprice*100 DESC
```

To see the difference, run the following code

```
SELECT shrfirm, shrprice, shrqty, FORMAT(shrdiv/shrprice*100,2) AS yield FROM SHARE
   ORDER BY yield DESC
```

Altering a table

The ALTER TABLE statement has two purposes. *First,* it can add a single column to an existing table. *Second,* it can add, drop, activate, or deactivate primary and foreign key constraints. A base table can be altered by adding one new column, which appears to the right of existing columns. The format of the command is

```
ALTER TABLE base-table ADD column data-type;
```

Notice that there is no optional NOT NULL clause for column-definition with ALTER TABLE. It is not allowed because the ALTER TABLE statement automatically fills the additional column with null in every case. If you want to add multiple columns, you repeat the command. ALTER TABLE does not permit changing the width of a column or amending a column's data type. It can be used for deleting an unwanted column.

```
ALTER TABLE stock ADD stkrating CHAR(3);
```

`ALTER TABLE` is also used to change the status of referential constraints. You can deactivate constraints on a table's primary key or any of its foreign keys. Deactivation also makes the relevant tables unavailable to all users except the table's owner or someone possessing database management authority. After the data are loaded, referential constraints must be reactivated before they can be automatically enforced again. Activating the constraints enables the DBMS to validate the references in the data.

Dropping a table

A base table can be deleted at any time by using the `DROP` statement. The format is

```
DROP TABLE base-table;
```

The table is deleted, and any views or indexes defined on the table are also deleted.

Creating a view

A view is a virtual table. It has no physical counterpart but appears to the client as if it really exists. A view is defined in terms of other tables that exist in the database. The syntax is

```
CREATE VIEW view [column [,column] …)]
   AS subquery;
```

There are several reasons for creating a view. *First,* a view can be used to restrict access to certain rows or columns. This is particularly important for sensitive data. An organization's `PERSON` table can contain both private data (e.g., annual salary) and public data (e.g., office phone number). A view consisting of public data (e.g., person's name, department, and office telephone number) might be provided to many people. Access to all columns in the table, however, might be confined to a small number of people. Here is a sample view that restricts access to a table.

```
CREATE VIEW stklist
   AS SELECT stkfirm, stkprice FROM stock;
```

Handling derived data is a `SECOND` reason for creating a view. A column that can be computed from one or more other columns should always be defined by a view. Thus, a stock's yield would be computed by a view rather than defined as a column in a base table.

```
CREATE VIEW stk
   (stkfirm, stkprice, stkqty, stkyield)
   AS SELECT stkfirm, stkprice, stkqty, stkdiv/stkprice*100
      FROM stock;
```

A *third* reason for defining a view is to avoid writing common SQL queries. For example, there may be some joins that are frequently part of an SQL query. Productivity can be increased by defining these joins as views. Here is an example:

```
CREATE VIEW stkvalue
   (nation, firm, price, qty, value)
   AS SELECT natname, stkfirm, stkprice*exchrate, stkqty,
      stkprice*exchrate*stkqty FROM stock JOIN nation
         ON stock.natcode = nation.natcode;
```

The preceding example demonstrates how CREATE VIEW can be used to rename columns, create new columns, and involve more than one table. The column nation corresponds to natname, firm to stkfirm, and so forth. A new column, price, is created that converts all share prices from the local currency to British pounds.

Data conversion is a *fourth* useful reason for a view. The United States is one of the few countries that does not use the metric system, and reports for American managers often display weights and measures in pounds and feet, respectively. The database of an international company could record all measurements in metric format (e.g., weight in kilograms) and use a view to convert these measures for American reports.

When a CREATE VIEW statement is executed, the definition of the view is entered in the systems catalog. The subquery following AS, the view definition, is executed only when the view is referenced in an SQL command. For example, the following command would enable the subquery to be executed and the view created:

```
SELECT * FROM stkvalue WHERE price > 10;
```

In effect, the following query is executed:

```
SELECT natname, stkfirm, stkprice*exchrate, stkqty, stkprice*exchrate*stkqty
   FROM stock JOIN nation
      ON stock.natcode = nation.natcode
      WHERE stkprice*exchrate > 10;
```

Any table that can be defined with a SELECT statement is a potential view. Thus, it is possible to have a view that is defined by another view.

Dropping a view

DROP VIEW is used to delete a view from the system catalog. A view might be dropped because it needs to be redefined or is no longer used. It must be dropped first before a revised version of the view is created. The syntax is

```
DROP VIEW view;
```

Remember, if a base table is dropped, all views based on that table are also dropped.

Creating an index

An index helps speed up retrieval (a more detailed discussion of indexing is covered later in this book). A column that is frequently referred to in a WHERE clause is a possible candidate for indexing. For example, if data on stocks were frequently retrieved using stkfirm, then this column should be considered for an index. The format for CREATE INDEX is

```
CREATE [UNIQUE] INDEX indexname
   ON base-table (column [order] [,column, [order]] …)
   [CLUSTER];
```

This next example illustrates use of CREATE INDEX.

```
CREATE UNIQUE INDEX stkfirmindx ON stock(stkfirm);
```

In the preceding example, an index called `stkfirmindx` is created for the table `stock`. Index entries are ordered by ascending (the default order) values of `stkfirm`. The optional clause `UNIQUE` specifies that no two rows in the base table can have the same value for `stkfirm`, the indexed column. Specifying `UNIQUE` means that the DBMS will reject any insert or update operation that would create a duplicate value for `stkfirm`.

A composite index can be created from several columns, which is often necessary for an associative entity. The following example illustrates the creation of a composite index.

```
CREATE INDEX lineitemindx ON lineitem (lineno, saleno);
```

Dropping an index

Indexes can be dropped at any time by using the `DROP INDEX` statement. The general form of this statement is

```
DROP INDEX index;
```

Data manipulation

SQL supports four DML statements—`SELECT`, `INSERT`, `UPDATE`, and `DELETE`. Each of these will be discussed in turn, with most attention focusing on `SELECT` because of the variety of ways in which it can be used. First, we need to understand why we must qualify column names and temporary names.

Qualifying column names

Ambiguous references to column names are avoided by qualifying a column name with its table name, especially when the same column name is used in several tables. Clarity is maintained by prefixing the column name with the table name. The following example demonstrates qualification of the `natcode`, which appears in both `stock` and `nation`.

```
SELECT stkfirm, stkprice FROM stock JOIN nation
   ON stock.natcode = nation.natcode;
```

Temporary names

A table or view can be given a temporary name, or alias, that remains current for a query. Temporary names are used in a self-join to distinguish the copies of the table.

```
SELECT wrk.empfname
   FROM emp wrk JOIN emp boss
      ON wrk.bossno = boss.empno;
```

A temporary name also can be used as a shortened form of a long table name. For example, `L` might be used merely to avoid having to enter `lineitem` more than once. If a temporary name is specified for a table or view, any qualified reference to a column of the table or view must also use that temporary name.

SELECT

The `SELECT` statement is by far the most interesting and challenging of the four DML statements. It reveals a major benefit of the relational model—powerful interrogation capabilities. It is challenging because mastering the power of `SELECT` requires considerable practice with a wide range of queries. The major varieties of `SELECT` are presented in this section. The SQL Playbook, which follows this chapter, reveals the full power of the command.

The general format of SELECT is

```
SELECT [DISTINCT] item(s) FROM table(s)
   [WHERE condition]
   [GROUP BY column(s)] [HAVING condition]
   [ORDER BY column(s)];
```

Alternatively, we can diagram the structure of SELECT.

Structure of SELECT

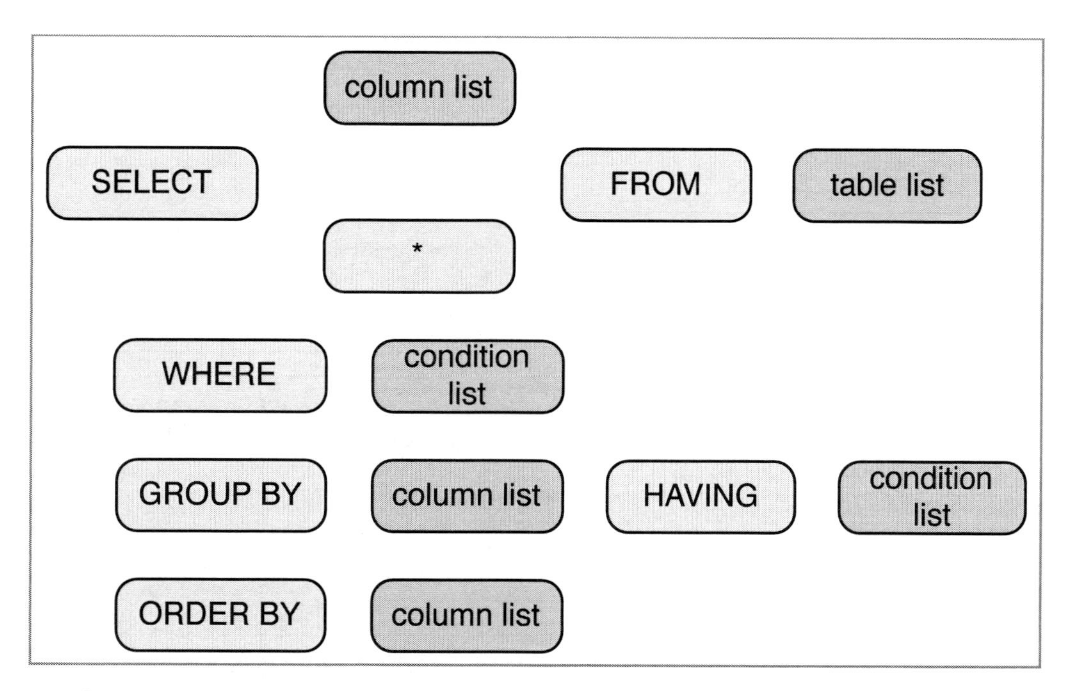

Product

Product, or more strictly Cartesian product, is a fundamental operation of relational algebra. It is rarely used by itself in a query; however, understanding its effect helps in comprehending join. The product of two tables is a new table consisting of all rows of the first table concatenated with all possible rows of the second table. For example:

- *Form the product of stock and nation.*

```
SELECT * FROM stock, nation;
```

The new table contains 64 rows (16*4), where `stock` has 16 rows and `nation` has 4 rows. It has 10 columns (7 + 3), where `stock` has 7 columns and `nation` has 3 columns. The result of the operation is shown in the following table. Note that each row in `stock` is concatenated with each row in `nation`.

Product of stock and nation

stkcode	stkfirm	stkprice	stkqty	stkdiv	stkpe	natcode	natcode	natname	exchrate
FC	Freedonia Copper	27.5	10529	1.84	16	UK	UK	United Kingdom	1.0000
FC	Freedonia Copper	27.5	10529	1.84	16	UK	US	United States	0.6700
FC	Freedonia Copper	27.5	10529	1.84	16	UK	AUS	Australia	0.4600
FC	Freedonia Copper	27.5	10529	1.84	16	UK	IND	India	0.0228
PT	Patagonian Tea	55.25	12635	2.5	10	UK	UK	United Kingdom	1.0000

stkcode	stkfirm	stkprice	stkqty	stkdiv	stkpe	natcode	natcode	natname	exchrate
PT	Patagonian Tea	55.25	12635	2.5	10	UK	US	United States	0.6700
PT	Patagonian Tea	55.25	12635	2.5	10	UK	AUS	Australia	0.4600
PT	Patagonian Tea	55.25	12635	2.5	10	UK	IND	India	0.0228
AR	Abyssinian Ruby	31.82	22010	1.32	13	UK	UK	United Kingdom	1.0000
AR	Abyssinian Ruby	31.82	22010	1.32	13	UK	US	United States	0.6700
AR	Abyssinian Ruby	31.82	22010	1.32	13	UK	AUS	Australia	0.4600
AR	Abyssinian Ruby	31.82	22010	1.32	13	UK	IND	India	0.0228
SLG	Sri Lankan Gold	50.37	32868	2.68	16	UK	UK	United Kingdom	1.0000
SLG	Sri Lankan Gold	50.37	32868	2.68	16	UK	US	United States	0.6700
SLG	Sri Lankan Gold	50.37	32868	2.68	16	UK	AUS	Australia	0.4600
SLG	Sri Lankan Gold	50.37	32868	2.68	16	UK	IND	India	0.0228
ILZ	Indian Lead & Zinc	37.75	6390	3	12	UK	UK	United Kingdom	1.0000
ILZ	Indian Lead & Zinc	37.75	6390	3	12	UK	US	United States	0.6700
ILZ	Indian Lead & Zinc	37.75	6390	3	12	UK	AUS	Australia	0.4600
ILZ	Indian Lead & Zinc	37.75	6390	3	12	UK	IND	India	0.0228
BE	Burmese Elephant	0.07	154713	0.01	3	UK	UK	United Kingdom	1.0000
BE	Burmese Elephant	0.07	154713	0.01	3	UK	US	United States	0.6700
BE	Burmese Elephant	0.07	154713	0.01	3	UK	AUS	Australia	0.4600
BE	Burmese Elephant	0.07	154713	0.01	3	UK	IND	India	0.0228
BS	Bolivian Sheep	12.75	231678	1.78	11	UK	UK	United Kingdom	1.0000
BS	Bolivian Sheep	12.75	231678	1.78	11	UK	US	United States	0.6700
BS	Bolivian Sheep	12.75	231678	1.78	11	UK	AUS	Australia	0.4600
BS	Bolivian Sheep	12.75	231678	1.78	11	UK	IND	India	0.0228
NG	Nigerian Geese	35	12323	1.68	10	UK	UK	United Kingdom	1.0000
NG	Nigerian Geese	35	12323	1.68	10	UK	US	United States	0.6700
NG	Nigerian Geese	35	12323	1.68	10	UK	AUS	Australia	0.4600
NG	Nigerian Geese	35	12323	1.68	10	UK	IND	India	0.0228
CS	Canadian Sugar	52.78	4716	2.5	15	UK	UK	United Kingdom	1.0000
CS	Canadian Sugar	52.78	4716	2.5	15	UK	US	United States	0.6700
CS	Canadian Sugar	52.78	4716	2.5	15	UK	AUS	Australia	0.4600
CS	Canadian Sugar	52.78	4716	2.5	15	UK	IND	India	0.0228
ROF	Royal Ostrich Farms	33.75	1234923	3	6	UK	UK	United Kingdom	1.0000
ROF	Royal Ostrich Farms	33.75	1234923	3	6	UK	US	United States	0.6700
ROF	Royal Ostrich Farms	33.75	1234923	3	6	UK	AUS	Australia	0.4600
ROF	Royal Ostrich Farms	33.75	1234923	3	6	UK	IND	India	0.0228
MG	Minnesota Gold	53.87	816122	1	25	US	UK	United Kingdom	1.0000
MG	Minnesota Gold	53.87	816122	1	25	US	US	United States	0.6700
MG	Minnesota Gold	53.87	816122	1	25	US	AUS	Australia	0.4600
MG	Minnesota Gold	53.87	816122	1	25	US	IND	India	0.0228
GP	Georgia Peach	2.35	387333	0.2	5	US	UK	United Kingdom	1.0000
GP	Georgia Peach	2.35	387333	0.2	5	US	US	United States	0.6700
GP	Georgia Peach	2.35	387333	0.2	5	US	AUS	Australia	0.4600
GP	Georgia Peach	2.35	387333	0.2	5	US	IND	India	0.0228
NE	Narembeen Emu	12.34	45619	1	8	AUS	UK	United Kingdom	1.0000
NE	Narembeen Emu	12.34	45619	1	8	AUS	US	United States	0.6700
NE	Narembeen Emu	12.34	45619	1	8	AUS	AUS	Australia	0.4600
NE	Narembeen Emu	12.34	45619	1	8	AUS	IND	India	0.0228

stkcode	stkfirm	stkprice	stkqty	stkdiv	stkpe	natcode	natcode	natname	exchrate
QD	Queensland Diamond	6.73	89251	0.5	7	AUS	UK	United Kingdom	1.0000
QD	Queensland Diamond	6.73	89251	0.5	7	AUS	US	United States	0.6700
QD	Queensland Diamond	6.73	89251	0.5	7	AUS	AUS	Australia	0.4600
QD	Queensland Diamond	6.73	89251	0.5	7	AUS	IND	India	0.0228
IR	Indooroopilly Ruby	15.92	56147	0.5	20	AUS	UK	United Kingdom	1.0000
IR	Indooroopilly Ruby	15.92	56147	0.5	20	AUS	US	United States	0.6700
IR	Indooroopilly Ruby	15.92	56147	0.5	20	AUS	AUS	Australia	0.4600
IR	Indooroopilly Ruby	15.92	56147	0.5	20	AUS	IND	India	0.0228
BD	Bombay Duck	25.55	167382	1	12	IND	UK	United Kingdom	1.0000
BD	Bombay Duck	25.55	167382	1	12	IND	US	United States	0.6700
BD	Bombay Duck	25.55	167382	1	12	IND	AUS	Australia	0.4600
BD	Bombay Duck	25.55	167382	1	12	IND	IND	India	0.0228

- *Find the percentage of Australian stocks in the portfolio.*

To answer this query, you need to count the number of Australian stocks, count the total number of stocks in the portfolio, and then compute the percentage. Computing each of the totals is a straightforward application of `COUNT`. If we save the results of the two counts as views, then we have the necessary data to compute the percentage. The two views each consist of a single-cell table (i.e., one row and one column). We create the product of these two views to get the data needed for computing the percentage in one row. The SQL is

```
CREATE VIEW austotal (auscount) AS
  SELECT COUNT(*) FROM nation JOIN stock
    ON natname = 'Australia'
    WHERE nation.natcode = stock.natcode;

CREATE VIEW total (totalcount) AS
  SELECT COUNT(*) FROM stock;

SELECT auscount/totalcount*100
  AS percentage FROM austotal, total;

CREATE VIEW total (totalcount) AS
   SELECT COUNT(*) FROM stock;
SELECT auscount*100/totalcount as Percentage
   FROM austotal, total;
```

The result of a COUNT is always an integer, and SQL will typically create an integer data type in which to store the results. When two variables have a data type of integer, SQL will likely use integer arithmetic for all computations, and all results will be integer. To get around the issue of integer arithmetic, we first multiply the number of Australian stocks by 100 before dividing by the total number of stocks. Because of integer arithmetic, you might get a different answer if you used the following SQL.

```
SELECT auscount/totalcount*100 as Percentage
   FROM austotal, total;
```

The preceding example was used to show when you might find product useful. You can also write the query as

```
SELECT (SELECT COUNT(*) FROM stock WHERE natcode = 'AUS')*100/
   (SELECT COUNT(*) FROM stock) as Percentage;
```

Inner join

Inner join, often referred to as join, is a powerful and frequently used operation. It creates a new table from two existing tables by matching on a column common to both tables.An **equijoin** is the simplest form of join; in this case, columns are matched on equality. The shaded rows in the preceding table indicate the result of the following join example.

```
SELECT * FROM stock JOIN nation
   ON stock.natcode = nation.natcode;
```

There are other ways of expressing join that are more concise. For example, we can write

```
SELECT * FROM stock INNER JOIN nation USING (natcode);
```

The preceding syntax implicitly recognizes the frequent use of the same column name for matching primary and foreign keys.

A further simplification is to rely on the primary and foreign key definitions to determine the join condition, so we can write

```
SELECT * FROM stock NATURAL JOIN nation;
```

An equijoin creates a new table that contains two identical columns. If one of these is dropped, then the remaining table is called a natural join.

As you now realize, join can be thought of as a product with a condition clause. There is no reason why this condition needs to be restricted to equality. There could easily be another comparison operator between the two columns. This general version is called a theta-join because theta is a variable that can take any value from the set [=, <>, >, >=, <, <=].

As you discovered earlier, there are occasions when you need to join a table to itself. To do this, make two copies of the table first and give each of these copies a unique name.

- *Find the names of employees who earn more than their boss.*

```
SELECT wrk.empfname
   FROM emp wrk JOIN emp boss
      ON wrk.bossno = boss.empno
      WHERE wrk.empsalary > boss.empsalary;
```

Outer join

An inner join reports those rows where the primary and foreign keys match. There are also situations where you might want an **outer join**, which comes in three flavors as shown in the following figure.

Types of joins

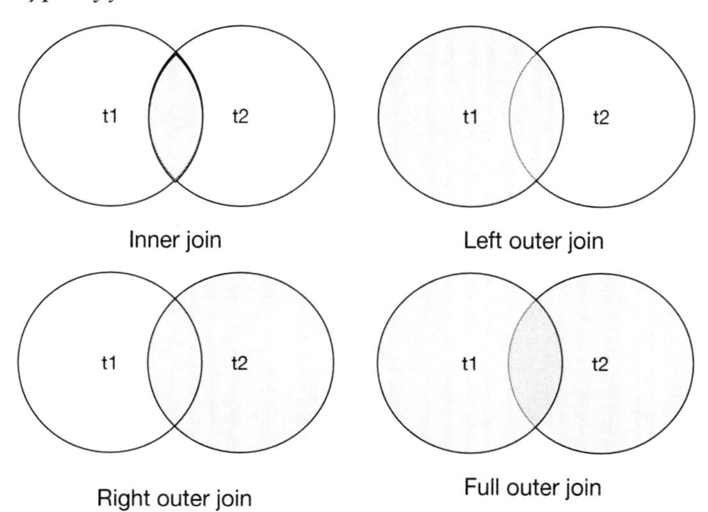

An outer join reports these matching rows and others depending on which form is used, as the following examples illustrate for a sample pair of tables.

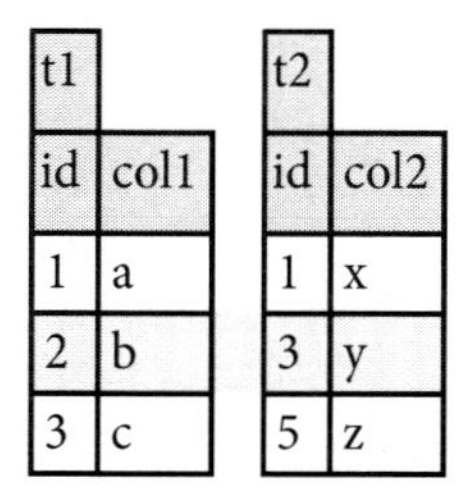

t1

id	col1
1	a
2	b
3	c

t2

id	col2
1	x
3	y
5	z

A **left outer join** is an inner join plus those rows from t1 not included in the inner join.

```
SELECT id, col1, col2 FROM t1 LEFT JOIN t2 USING (id)
```

id	col1	col2
1	a	x
2	b	null
3	c	y

Here is an example to illustrate the use of a left join.

- *For all brown items, report each sale. Include in the report those brown items that have appeared in no sales.*

```
SELECT itemname, saleno, lineqty FROM item
    LEFT JOIN lineitem USING (itemno)
        WHERE itemcolor = 'Brown'
        ORDER BY itemname;
```

itemname	saleno	lineqty
Map case	null	null
Pocket knife - Avon	1	1
Pocket knife - Avon	3	1
Pocket knife - Avon	2	1
Pocket knife - Avon	5	1
Pocket knife - Avon	4	1
Pocket knife - Nile	null	null
Stetson	null	null

A **right outer join** is an inner join plus those rows from t2 not included in the inner join.

```
SELECT id, col1, col2 FROM t1 RIGHT JOIN t2 USING (id);
```

Id	col1	col2
1	a	x
3	c	y
5	null	z

A **full outer join** is an inner join plus those rows from t1 and t2 not included in the inner join.

```
SELECT id, col1, col2 FROM t1 FULL JOIN t2 USING (id);
```

id	col1	col2
1	a	x
2	b	null
3	c	y
5	null	z

MySQL does not support a full outer join, rather you must use a union of a left and right outer joins.

```
SELECT id, col1, col2 FROM t1 LEFT JOIN t2 USING (id)
UNION
SELECT id, col1, col2 FROM t1 RIGHT JOIN t2 USING (id);
```

Simple subquery

A subquery is a query within a query. There is a SELECT statement nested inside another SELECT statement. Simple subqueries were used extensively in earlier chapters. For reference, here is a simple subquery used earlier:

```
SELECT stkfirm FROM stock
   WHERE natcode IN
      (SELECT natcode FROM nation
         WHERE natname = 'Australia');
```

Correlated subquery

A correlated subquery differs from a simple subquery in that the inner query must be evaluated more than once. Consider the following example described previously:

- *Find those stocks where the quantity is greater than the average for that country.*

```
SELECT natname, stkfirm, stkqty FROM stock JOIN nation
ON stock.natcode = nation.natcode
WHERE stkqty >
   (SELECT AVG(stkqty) FROM stock
      WHERE stock.natcode = nation.natcode);
```

The requirement to compare a column against a function (e.g., average or count) of some column of specified rows of is usually a clue that you need to write a correlated subquery. In the preceding example, the stock quantity for each row is compared with the average stock quantity for that row's country.

Aggregate functions

SQL's aggregate functions increase its retrieval power. These functions were covered earlier and are only mentioned briefly here for completeness. The five aggregate functions are shown in the following table. Nulls in the column are ignored in the case of `SUM`, `AVG`, `MAX`, and `MIN`. `COUNT(*)` does not distinguish between null and non-null values in a column. Use `COUNT(columnname)` to exclude a null value in `columnname`.

Aggregate functions

Function	Description
COUNT	Counts the number of values in a column
SUM	Sums the values in a column
AVG	Determines the average of the values in a column
MAX	Determines the largest value in a column
MIN	Determines the smallest value in a column

GROUP BY and HAVING

The `GROUP BY` clause is an elementary form of control break reporting and supports grouping of rows that have the same value for a specified column and produces one row for each different value of the grouping column. For example,

- *Report by nation the total value of stockholdings.*

```
SELECT natname, SUM(stkprice*stkqty*exchrate) AS total
   FROM stock JOIN nation ON stock.natcode = nation.natcode
      GROUP BY natname;
```

gives the following results:

natname	total
Australia	946430.65
India	97506.71

natname	total
United Kingdom	48908364.25
United States	30066065.54

The HAVING clause is often associated with GROUP BY. It can be thought of as the WHERE clause of GROUP BY because it is used to eliminate rows for a GROUP BY condition. Both GROUP BY and HAVING are dealt with in-depth in Chapter 4.

REGEXP

The REGEXP clause supports pattern matching to find a defined set of strings in a character column (CHAR or VARCHAR). Refer to Chapters 3 and 4 for more details.

INSERT

There are two formats for INSERT. The first format is used to insert one row into a table.

Inserting a single record

The general form is

```
INSERT INTO table [(column [,column] …)]
   VALUES (literal [,literal] …);
```

For example,

```
INSERT INTO stock
   (stkcode,stkfirm,stkprice,stkqty,stkdiv,stkpe)
   VALUES ('FC','Freedonia Copper',27.5,10529,1.84,16);
```

In this example, stkcode is given the value "FC," stkfirm is "Freedonia Copper," and so on. In general, the *n*th column in the table is the *n*th value in the list.

When the value list refers to all field names in the left-to-right order in which they appear in the table, then the columns list can be omitted. So, it is possible to write the following:

```
INSERT INTO stock
   VALUES ('FC','Freedonia Copper',27.5,10529,1.84,16);
```

If some values are unknown, then the INSERT can omit these from the list. Undefined columns will have nulls. For example, if a new stock is to be added for which the price, dividend, and PE ratio are unknown, the following INSERT statement would be used:

```
INSERT INTO stock
   (stkcode, stkfirm, stkqty)
   VALUES ('EE','Elysian Emeralds',0);
```

Inserting multiple records using a query

The second form of INSERT operates in conjunction with a subquery. The resulting rows then are inserted into a table. Imagine the situation where stock price information is downloaded from an information service into a table. This table could contain information about all stocks and may contain additional columns that are not required for the stock table. The following INSERT statement could be used:

```
INSERT INTO stock
   (stkcode, stkfirm, stkprice, stkdiv, stkpe)
   SELECT code, firm, price, div, pe
   FROM download WHERE code IN
   ('FC','PT','AR','SLG','ILZ','BE','BS','NG','CS','ROF');
```

Think of INSERT with a subquery as a way of copying a table. You can select the rows and columns of a particular table that you want to copy into an existing or new table.

UPDATE

The UPDATE command is used to modify values in a table. The general format is

```
UPDATE table
   SET column = scalar expression
      [, column = scalar expression] …
   [WHERE condition];
```

Permissible scalar expressions involve columns, scalar functions (see the section on scalar functions in this chapter), or constants. No aggregate functions are allowable.

Updating a single row

UPDATE can be used to modify a single row in a table. Suppose you need to revise your data after 200,000 shares of Minnesota Gold are sold. You would code the following:

```
UPDATE stock
   SET stkqty = stkqty - 200000
   WHERE stkcode = 'MG';
```

Updating multiple rows

Multiple rows in a table can be updated as well. Imagine the situation where several stocks change their dividend to £2.50. Then the following statement could be used:

```
UPDATE stock
   SET stkdiv = 2.50
   WHERE stkcode IN ('FC','BS','NG');
```

Updating all rows

All rows in a table can be updated by simply omitting the WHERE clause. To give everyone at The Expeditioner a 5 percent raise, use

```
UPDATE emp
   SET empsalary = empsalary*1.05;
```

Updating with a subquery

A subquery can also be used to specify which rows should be changed. Consider the following example. The employees in the departments on the fourth floor of The Expeditioner have won a productivity improvement bonus of 10 percent. The following SQL statement would update their salaries:

```
UPDATE emp
   SET empsalary = empsalary*1.10
   WHERE deptname IN
      (SELECT deptname FROM dept WHERE deptfloor = 4);
```

DELETE

The `DELETE` statement erases one or more rows in a table. The general format is

```
DELETE FROM table
   [WHERE condition];
```

Delete a single record

If all stocks with `stkcode` equal to "BE" were sold, then this row can be deleted using

```
DELETE FROM stock WHERE stkcode = 'BE';
```

Delete multiple records

If all Australian stocks were liquidated, then the following command would delete all the relevant rows:

```
DELETE FROM stock
   WHERE natcode in
   (SELECT natcode FROM nation WHERE natname = 'Australia');
```

Delete all records

All records in a table can be deleted by omitting the `WHERE` clause. The following statement would delete all rows if the entire portfolio were sold:

```
DELETE FROM stock;
```

This command is not the same as `DROP TABLE` because, although the table is empty, it still exists.

Delete with a subquery

Despite their sterling efforts in the recent productivity drive, all the employees on the fourth floor of The Expeditioner have been fired (the rumor is that they were fiddling the tea money). Their records can be deleted using

```
DELETE FROM emp
   WHERE deptname IN
      (SELECT deptname FROM dept WHERE deptfloor = 4);
```

SQL routines

SQL provides two types of routines—functions and procedures—that are created, altered, and dropped using standard SQL. Routines add flexibility, improve programmer productivity, and facilitate the enforcement of business rules and standard operating procedures across applications.

SQL function

A function is SQL code that returns a value when invoked within an SQL statement. It is used in a similar fashion to SQL's built-in functions. Consider the case of an Austrian firm with a database in which all measurements are in SI units (e.g., meters). Because its U.S. staff is not familiar with SI,[21] it decides to implement a series of user-defined functions to handle the conversion. Here is the function for converting from kilometers to miles.

```
CREATE FUNCTION km_to_miles(km REAL)
   RETURNS REAL
   RETURN 0.6213712*km;
```

The preceding function can be used within any SQL statement to make the conversion. For example:

```
SELECT km_to_miles(100);
```

Skill builder

Create a table containing the average daily temperature in Tromsø, Norway, then write a function to convert Celsius to Fahrenheit (F = C*1.8 + 32), and test the function by reporting temperatures in C and F.

Month	Jan	Feb	Mar	Apr	May	Jun	Jul	Aug	Sep	Oct	Nov	Dec
ºC	-4.7	-4.1	-1.9	1.1	5.6	10.1	12.7	11.8	7.7	2.9	-1.5	-3.7

SQL procedure

A procedure is SQL code that is dynamically loaded and executed by a CALL statement, usually within a database application. We use an accounting system to demonstrate the features of a stored procedure, in which a single accounting transaction results in two entries (one debit and one credit). In other words, a transaction has multiple entries, but an entry is related to only one transaction. An account (e.g., your bank account) has multiple entries, but an entry is for only one account. Considering this situation results in the following data model.

A simple accounting system

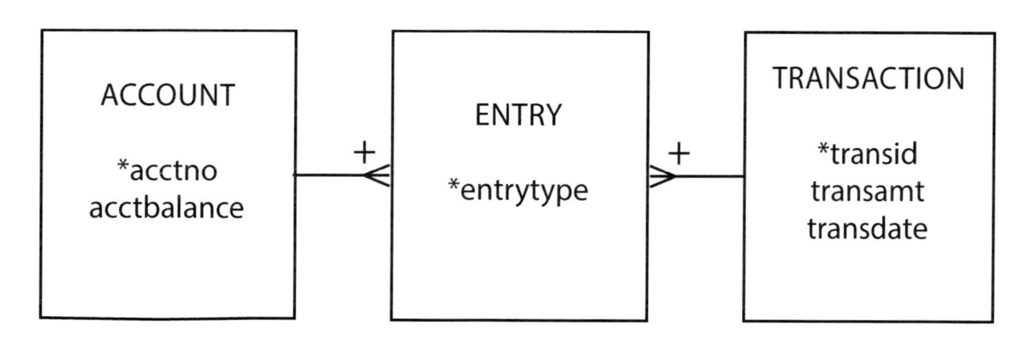

21 The international system of units of measurement. SI is the from French *Système International.*

The following are a set of steps for processing a transaction (e.g., transferring money from a checking account to a money market account):

1. Write the transaction to the `transaction` table so you have a record of the transaction.
2. Update the account to be credited by incrementing its balance in the `account` table.
3. Insert a row in the `entry` table to record the credit.
4. Update the account to be debited by decrementing its balance in the `account` table.
5. Insert a row in the `entry` table to record the debit.

Here is the code for a stored procedure to execute these steps. Note that the first line sets the delimiter to // because the default delimiter for SQL is a semicolon (;), which we need to use to delimit the multiple SQL commands in the procedure. The last statement in the procedure is thus `END //` to indicate the end of the procedure.

```
DELIMITER //
CREATE PROCEDURE transfer (
IN `Credit account` INTEGER,
IN `Debit account` INTEGER,
IN Amount DECIMAL(9,2),
IN `Transaction ID` INTEGER)
LANGUAGE SQL
DETERMINISTIC
BEGIN
INSERT INTO transaction VALUES (`Transaction ID`, Amount, CURRENT_DATE);
UPDATE account
SET acctbalance = acctbalance + Amount
WHERE acctno = `Credit account`;
INSERT INTO entry VALUES (`Transaction ID`, `Credit account`, 'cr');
UPDATE account
SET acctbalance = acctbalance - Amount
WHERE acctno = `Debit account`;
```

INSERT INTO entry VALUES (`Transaction ID`, `Debit account`, 'db');

```
END//
```

A `CALL` statement executes a stored procedure. The generic `CALL` statement for the preceding procedure is

```
CALL transfer(cracct, dbacct, amt, transno);
```

Thus, imagine that transaction 1005 transfers $100 to account 1 (the credit account) from account 2 (the debit account). The specific call is

```
CALL transfer(1,2,100,1005);
```

Alternatively, you can use an automatically generated pop-up window to run the procedure by clicking on the rightmost icon for the procedure under the Stored Procedures header.

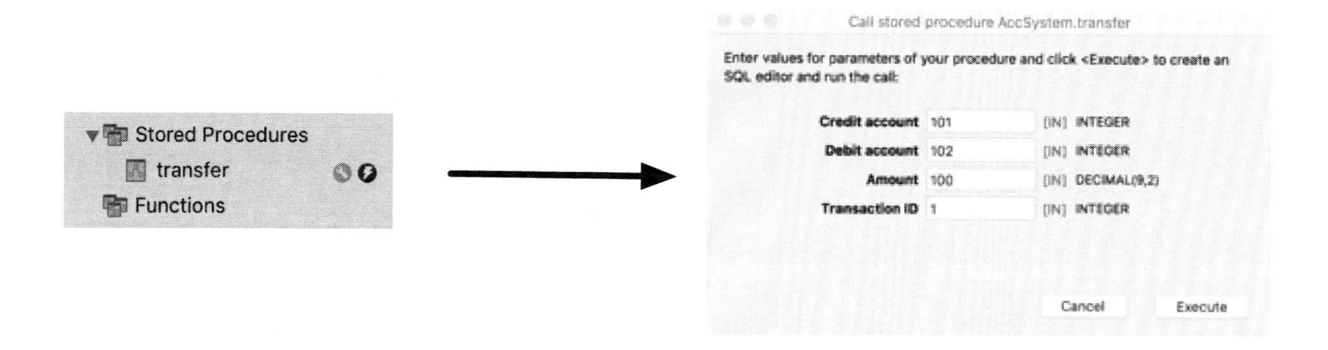

Skill builder

1. After verifying that the system you use for learning SQL supports stored procedures, create the tables for the preceding data model and enter the code for the stored procedure. Now, test the stored procedure and query the tables to verify that the procedure has worked.
2. Write a stored procedure to add details of a gift to the donation database (see exercises in Chapter 5).

Triggers

Triggers are a form of stored procedure that execute automatically when a table's rows are modified. Triggers can be defined to execute either before or after rows are inserted into a table, when rows are deleted from a table, and when columns are updated in the rows of a table. Triggers can include virtual tables that reflect the row image before and after the operation, as appropriate. Triggers can be used to enforce business rules or requirements, integrity checking, and automatic transaction logging.

Consider the case of recording all updates to the `stock` table (see Chapter 4). First, you must define a table in which to record details of the change.

```
CREATE TABLE stock_log (
stkcode        CHAR(3),
old_stkprice DECIMAL(6,2),
new_stkprice DECIMAL(6,2),
old_stkqty         DECIMAL(8),
new_stkqty         DECIMAL(8),
update_stktime   TIMESTAMP NOT NULL,
   PRIMARY KEY(update_stktime));
```

The trigger writes a record to `stock_log` every time an update is made to `stock`. Use is made of two virtual tables (`old` and `new`) to access the prior and current values of stock price (`old.stkprice` and `new.stkprice` and stock quantity (`old.stkprice` and `new.stkprice`). The `INSERT` statement also writes the stock's identifying code and the time of the transaction.

```
CREATE TRIGGER stock_update
AFTER UPDATE ON stock
FOR EACH ROW BEGIN
INSERT INTO stock_log VALUES
  (old.stkcode, old.stkprice, new.stkprice, old.stkqty, new.stkqty,
CURRENT_TIMESTAMP);
END
```

Skill builder

Why is the primary key of `stock_log` not the same as that of `stock`?

Nulls—much ado about missing information

Nulls are overworked in SQL because they can represent several situations. Null can represent unknown information. For example, you might add a new stock to the database, but lacking details of its latest dividend, you leave the field null. Null can be used to represent a value that is inapplicable. For instance, the `EMPLOYEE` table contains a null value in `BOSSNO` for Alice because she has no boss. The value is not unknown; it is not applicable for that field. In other cases, null might mean "no value supplied" or "value undefined." Because null can have multiple meanings, the user must infer which meaning is appropriate to the circumstances.

Do not confuse null with blank or zero, which are values. In fact, null is a marker that specifies that the value for the particular column is null. Thus, null represents no value.

The well-known database expert Chris Date has been outspoken in his concern about the confusion caused by nulls. His advice is that nulls should be explicitly avoided by specifying `NOT NULL` for all columns and by using codes to make the meaning of a value clear (e.g., "U" means "unknown," "I" means "inapplicable," "N" means "not supplied").

Security

Data are a valuable resource for nearly every organization. Just as an organization takes measures to protect its physical assets, it also needs to safeguard its electronic assets—its organizational memory, including databases. Furthermore, it often wants to limit the access of authorized users to particular parts of a database and restrict their actions to particular operations.

Two SQL features are used to administer security procedures. A view, discussed earlier in this chapter, can restrict a client's access to specified columns or rows in a table, and authorization commands can establish a user's privileges.

The authorization subsystem is based on the concept of a privilege—the authority to perform an operation. For example, a person cannot update a table unless she has been granted the appropriate update privilege. The database administrator (DBA) is a master of the universe and has the highest privilege. The DBA can perform any legal operation. The creator of an object, say a base table, has full privileges for that object. Those with privileges can then use `GRANT` and `REVOKE`, commands included in SQL's data control language (DCL) to extend privileges to or rescind them from other users.

GRANT

The `GRANT` command defines a user's privileges. The general format of the statement is

```
GRANT privileges ON object TO users [WITH GRANT OPTION];
```

where "privileges" can be a list of privileges or the keyword `ALL PRIVILEGES`, and "users" is a list of user identifiers or the keyword `PUBLIC`. An "object" can be a base table or a view.

The following privileges can be granted for tables and views: `SELECT`, `UPDATE`, `DELETE`, and `INSERT`.

The `UPDATE` privilege specifies the particular columns in a base table or view that may be updated. Some privileges apply *only* to base tables. These are `ALTER` and `INDEX`.

The following examples illustrate the use of GRANT:

- *Give Alice all rights to the stock table.*

```
GRANT ALL PRIVILEGES ON stock TO alice;
```

- *Permit the accounting staff, Todd and Nancy, to update the price of a stock.*

```
GRANT UPDATE (stkprice) ON stock TO todd, nancy;
```

- *Give all staff the privilege to select rows from item.*

```
GRANT SELECT ON item TO PUBLIC;
```

- *Give Alice all rights to view stk.*

```
GRANT SELECT, UPDATE, DELETE, INSERT ON stk TO alice;
```

The WITH GRANT OPTION clause

The WITH GRANT OPTION command allows a client to transfer his privileges to another client, as this next example illustrates:

- *Give Ned all privileges for the item table and permit him to grant any of these to other staff members who may need to work with item.*

```
GRANT ALL PRIVILEGES ON item TO ned WITH GRANT OPTION;
```

This means that Ned can now use the GRANT command to give other staff privileges. To give Andrew permission for select and insert on ITEM, for example, Ned would enter

```
GRANT SELECT, INSERT ON item TO andrew;
```

REVOKE

What GRANT granteth, REVOKE revoketh. Privileges are removed using the REVOKE statement. The general format of this statement is

```
REVOKE privileges ON object FROM users;
```

These examples illustrate the use of REVOKE.

- *Remove Sophie's ability to select from item.*

```
REVOKE SELECT ON item FROM sophie;
```

- *Nancy is no longer permitted to update stock prices.*

```
REVOKE UPDATE ON stock FROM nancy;
```

Cascading revoke

When a REVOKE statement removes a privilege, it can result in more than one revocation. An earlier example illustrated how Ned used his WITH GRANT OPTION right to authorize Andrew to select and insert rows on item. The following REVOKE command

```
REVOKE INSERT ON item FROM ned;
```

automatically revokes Andrew's insert privilege.

The system catalog

The system catalog describes a relational database. It contains the definitions of base tables, views, indexes, and so on. The catalog itself is a relational database and can be interrogated using SQL. Tables in the catalog are called *system tables* to distinguish them from base tables, though conceptually these tables are the same. In MySQL, the system catalog is called INFORMATION_SCHEMA. Some important system tables in this schema are tables, and columns, and these are used in the following examples. Note that the names of the system catalog tables vary with DBMS implementations, so while the following examples illustrate use of system catalog tables, it is likely that you will have to change the table names for other DBMSs.

The table TABLES contains details of all tables in the database. There is one row for each table in the database.

- *Find the table(s) with the most columns.*

```
SELECT table_name, table_rows
   FROM information_schema.tables
      WHERE table_rows = (SELECT MAX(table_rows)
         FROM information_schema.tables);
```

The COLUMN table stores details about each column in the database.

- *What columns in what tables store dates?*

```
SELECT table_name, column_name
   FROM information_schema.columns
      WHERE DATA_TYPE = 'date'
         ORDER BY table_name, column_name;
```

As you can see, querying the catalog is the same as querying a database. This is a useful feature because you can use SQL queries on the catalog to find out more about a database.

Natural language processing

Infrequent inquirers of a relational database may be reluctant to use SQL because they don't use it often enough to remain familiar with the language. While the QBE approach can make querying easier, a more natural approach is to use standard English. In this case, natural language processing (NLP) is used to convert ordinary English into SQL so the query can be passed to the relational database. The example in the table below shows the successful translation of a query to SQL. A natural language processor must translate a request to SQL and request clarification where necessary.

An example of natural language processing

English	SQL generated for MS Access
Which movies have won best foreign film sorted by year?	SELECT DISTINCT [Year], [Title] FROM [Awards] INNER JOIN [Movies] ON [Movies].[Movie ID] = [Awards].[Movie ID] WHERE [Category]='Best Foreign Film' and [Status]='Winner' ORDER BY [Year] ASC;

Connectivity and ODBC

Over time and because of differing needs, an organization is likely to purchase DBMS software from a variety of vendors. Also, in some situations, mergers and acquisitions can create a multivendor DBMS environment. Consequently, the SQL Access Group developed SQL Call-Level Interface (CLI), a unified standard for remote database access. The intention of CLI is to provide programmers with a generic approach for writing software that accesses a database. With the appropriate CLI database driver, any DBMS server can provide access to client programs that use the CLI. On the server side, the DBMS CLI driver is responsible for translating the CLI call into the server's access language. On the client side, there must be a CLI driver for each database to which it connects. CLI is not a query language but a way of wrapping SQL so it can be understood by a DBMS. In 1996, CLI was adopted as an international standard and renamed X/Open CLI.

Open database connectivity (ODBC)

The de facto standard for database connectivity is **Open Database Connectivity** (ODBC), an extended implementation of CLI developed by Microsoft. This application programming interface (API) is cross-platform and can be used to access any DBMS or DBMS server that has an ODBC driver.This enables a software developer to build and distribute an application without targeting a specific DBMS. Database drivers are then added to link the application to the client's choice of DBMS. For example, a microcomputer running under Windows can use ODBC to access an Oracle DBMS running on a Unix box.

There is considerable support for ODBC. Application vendors like it because they do not have to write and maintain code for each DBMS; they can write one API. DBMS vendors support ODBC because they do not have to convince application vendors to support their product. For database systems managers, ODBC provides vendor and platform independence. Although the ODBC API was originally developed to provide database access from MS Windows products, many ODBC driver vendors support Linux and Macintosh clients.

Most vendors also have their own SQL APIs. The problem is that most vendors, as a means of differentiating their DBMS, have a more extensive native API protocol and also add extensions to standard ODBC. The developer who is tempted to use these extensions threatens the portability of the database.

ODBC introduces greater complexity and a processing overhead because it adds two layers of software. As the following figure illustrates, an ODBC-compliant application has additional layers for the ODBC API and ODBC driver. As a result, ODBC APIs can never be as fast as native APIs.

ODBC layers

Application
ODBC API
ODBC driver manager
Service provider API
Driver for DBMS server
DBMS server

Embedded SQL

SQL can be used in two modes. *First,* SQL is an interactive query language and database programming language. `SELECT` defines queries; `INSERT`, `UPDATE`, and `DELETE` maintain a database. *Second*, any interactive SQL statement can be embedded in an application program.

This dual-mode principle is a very useful feature. It means that programmers need to learn only one database query language, because the same SQL statements apply for both interactive queries and application statements. Programmers can also interactively examine SQL commands before embedding them in a program, a feature that can substantially reduce the time to write an application program.

Because SQL is not a complete programming language, however, it must be used with a traditional programming language to create applications. Common complete programming languages, such as PHP and Java, support embedded SQL. If you are need to write application programs using embedded SQL, you will need training in both the application language and the details of how it communicates with SQL.

User-defined types

Versions of SQL prior to the SQL-99 specification had predefined data types, and programmers were limited to selecting the data type and defining the length of character strings. One of the basic ideas behind the object extensions of the SQL standard is that, in addition to the normal built-in data types defined by SQL, **user-defined data types** (UDTs) are available. A UDT is used like a predefined type, but it must be set up before it can be used.

The future of SQL

Since 1986, developers of database applications have benefited from an SQL standard, one of the most successful standardization stories in the software industry. Although most database vendors have implemented proprietary extensions of SQL, standardization has kept the language consistent, and SQL code is highly portable. Standardization was relatively easy when focused on the storage and retrieval of numbers and characters. Objects have made standardization more difficult.

Summary

Structured Query Language (SQL), a widely used relational database language, has been adopted as a standard by ANSI and ISO. It is a data definition language (DDL), data manipulation language (DML), and data control language (DCL). A base table is an autonomous, named table. A view is a virtual table. A key is one or more columns identified as such in the description of a table, an index, or a referential constraint. SQL supports primary, foreign, and unique keys. Indexes accelerate data access and ensure uniqueness. `CREATE TABLE` defines a new base table and specifies primary, foreign, and unique key constraints. Numeric, string, date, or graphic data can be stored in a column. `BLOB` and `CLOB` are data types for large fields. `ALTER TABLE` adds one new column to a table or changes the status of a constraint. `DROP TABLE` removes a table from a database. `CREATE VIEW` defines a view, which can be used to restrict access to data, report derived data, store commonly executed queries, and convert data. A view is created dynamically. `DROP VIEW` deletes a view. `CREATE INDEX` defines an index, and `DROP INDEX` deletes one.

Ambiguous references to column names are avoided by qualifying a column name with its table name. A table or view can be given a temporary name that remains current for a query. SQL has four data manipulation statements — `SELECT`, `INSERT`, `UPDATE`, and `DELETE`. `INSERT` adds one or more rows to a table. `UPDATE` modifies a table by changing one or more rows. `DELETE` removes one or more rows from a table. `SELECT` provides powerful interrogation facilities. The product of two tables is a new table consisting of all rows of the first table concatenated with all possible rows of the second table. Join creates a new table from two existing tables by matching on a column common to both tables. A subquery is a query within a query. A correlated subquery differs from a simple subquery in that the inner query is evaluated multiple times rather than once.

SQL's aggregate functions increase its retrieval power. `GROUP BY` supports grouping of rows that have the same value for a specified column. The `REXEXP` clause supports pattern matching. SQL includes scalar functions that can be used in arithmetic expressions, data conversion, or data extraction. Nulls cause problems because they can represent several situations—unknown information, inapplicable information, no value supplied, or value undefined. Remember, a null is not a blank or zero. The SQL commands,

GRANT and REVOKE, support data security. GRANT authorizes a user to perform certain SQL operations, and REVOKE removes a user's authority. The system catalog, which describes a relational database, can be queried using SELECT. SQL can be used as an interactive query language and as embedded commands within an application programming language. Natural language processing (NLP) and open database connectivity (ODBC) are extensions to relational technology that enhance its usefulness.

Key terms and concepts

Aggregate functions
ALTER TABLE
ANSI
Base table
Complete database language
Complete programming language
Composite key
Connectivity
Correlated subquery
CREATE FUNCTION
CREATE INDEX
CREATE PROCEDURE
CREATE TABLE
CREATE TRIGGER
CREATE VIEW
Cursor
Data control language (DCL)
Data definition language (DDL)
Data manipulation language (DML)
Data types
DELETE
DROP INDEX
DROP TABLE
DROP VIEW
Embedded SQL
Foreign key
GRANT
GROUP BY
Index
INSERT
ISO
Join
Key
Natural language processing (NLP)
Null
Open database connectivity (ODBC)
Primary key
Product
Qualified name
Referential integrity rule
REVOKE
Routine
Scalar functions
Security
SELECT
Special registers
SQL
Subquery
Synonym
System catalog
Temporary names
Unique key
UPDATE
View

References and additional readings

Date, C. J. 2003. *An introduction to database systems*. 8th ed. Reading, MA: Addison-Wesley.

Exercises

1. Why is it important that SQL was adopted as a standard by ANSI and ISO?
2. What does it mean to say "SQL is a complete database language"?
3. Is SQL a complete programming language? What are the implications of your answer?
4. List some operational advantages of a DBMS.
5. What is the difference between a base table and a view?
6. What is the difference between a primary key and a unique key?
7. What is the purpose of an index?
8. Consider the three choices for the `ON DELETE` clause associated with the foreign key constraint. What are the pros and cons of each option?
9. Specify the data type (e.g., `DECIMAL(6,2)`) you would use for the following columns:
    a. The selling price of a house
    b. A telephone number with area code
    c. Hourly temperatures in Antarctica
    d. A numeric customer code
    e. A credit card number
    f. The distance between two cities
    g. A sentence using Chinese characters
    h. The number of kilometers from the Earth to a given star
    i. The text of an advertisement in the classified section of a newspaper
    j. A basketball score
    k. The title of a CD
    l. The X-ray of a patient
    m. A U.S. zip code
    n. A British or Canadian postal code
    o. The photo of a customer
    p. The date a person purchased a car
    q. The time of arrival of an e-mail message
    r. The number of full-time employees in a small business
    s. The text of a speech
    t. The thickness of a layer on a silicon chip

10. What is the difference between `DROP TABLE` and deleting all the rows in a table?
11. Give some reasons for creating a view.
12. When is a view created?
13. Write SQL codes to create a unique index on firm name for the `SHARE` table defined in Chapter 3. Would it make sense to create a unique index for PE ratio in the same table?
14. What is the difference between product and join?
15. What is the difference between an equijoin and a natural join?
16. You have a choice between executing two queries that will both give the same result. One is written as a simple subquery and the other as a correlated subquery. Which one would you use and why?
17. What function would you use for the following situations?
    a. Computing the total value of a column
    b. Finding the minimum value of a column
    c. Counting the number of customers in the `customer` table
    d. Displaying a number with specified precision
    e. Reporting the month part of a date
    f. Displaying the second part of a time
    g. Retrieving the first five characters of a city's name
    h. Reporting the distance to the sun in feet
18. Write SQL statements for the following:
    a. Let Hui-Tze query and add to the nation table.
    b. Give Lana permission to update the phone number column in the customer table.
    c. Remove all of William's privileges.
    d. Give Chris permission to grant other users authority to select from the address table.
    e. Find the name of all tables that include the word sale.
    f. List all the tables created last year.
    g. What is the maximum length of the column city in the ClassicModels database? Why do you get two rows in the response?
    h. Find all columns that have a data type of `SMALLINT`.
19. What are the two modes in which you can use SQL?
20. Using the Classic Models database, write an SQL procedure to change the credit limit of all customers in a specified country by a specified amount. Provide before and after queries to show your procedure works.
21. How do procedural programming languages and SQL differ in the way they process data? How is this difference handled in an application program? What is embedded SQL?

22. Using the Classic Models database, write an SQL procedure to change the MSRP of all products in a product line by a specified percentage. Provide before and after queries to show your procedure works.

Reference 2: SQL Playbook

This section has been moved to the book's web site.

See http://richardtwatson.com/dm6e/Reader/sql/playbook.pdf

Section 3: Advanced Data Management

Advancement only comes with habitually doing more than you are asked.

Gary Ryan Blair.[22]

The wiring of the world has given us ubiquitous networks and broadened the scope of issues that data management must now embrace. In everyday life, you might use a variety of networks (e.g., the Internet, 4G, WiFi, and bluetooth) to gain access to information wherever you might be and whatever time it is. As a result, data managers need to be concerned with both the **spatial and temporal** dimensions of data. In a highly connected world, massive amounts of data are exchanged every minute between computers to enable a high level of global integration in economic and social activity. **XML** has emerged as the foundation for data exchange across many industries. It is a core technology for global economic development. On the social side, every day people generate millions of messages, photos, and videos that fuel services such as Twitter, Flickr, and YouTube. Many organizations are interested in analyzing these data streams to learn about social trends, customers' opinions, and entrepreneurial opportunities. Organizations need skills in collecting, processing, and interpreting the myriad data flows that intersect with their everyday business. **Organizational or business intelligence** is the general term for describing an enterprise's efforts to collect, store, process, and interpret data from internal and external sources. It is the first stage of data-driven decision making. Once data have been captured and stored in an organizational repository, there are several techniques that can be applied.

In a world awash with data, **visualization** has become increasingly important for enabling executives to make sense of the business environment, to identify problems, and highlight potential new directions. **Text mining** is a popular tool for trying to make sense of data streams emanating from tweets and blogs. The many new sources of data and their high growth rate have made it more difficult to support real time analysis of the torrents of data that might contain valuable insights for an organization's managers. Fortunately, **Hadoop distributed file system (HDFS)** and **MapReduce** are a breakthrough in storing and processing data that enable faster processing at lower cost. **Dashboards** are widely used for presenting key information. Furthermore, the open source statistics and graphical package, **R**, provides a common foundation for handling text mining, data visualization, HDFS, and MapReduce. It has become another component of the data manager's toolkit.

The section covers the following topics.

- Spatial and temporal data management
- XML
- Organizational intelligence
- Introduction to R
- Data visualization
- Text mining
- HDFS and MapReduce
- Dashboards

22 http://www.garyryanblair.com

11. Spatial and Temporal Data Management

Nothing puzzles me more than time and space; and yet nothing troubles me less, as I never think about them.

Charles Lamb, 1810.

Learning objectives

Students completing this chapter will

- be able to define and use a spatial database;
- be familiar with the issues surrounding the management of temporal data.

Introduction

The introduction of ubiquitous networks and smartphones has led to the advent of location-based services. Customers expect information delivered based on, among other things, where they are. For example, a person touring the historic German city of Regensburg could receive information about its buildings and parks via her mobile phone in the language of her choice. Her smartphone will determine her location and then select from a database details of her immediate environment. Data managers need to know how to manage the **spatial** data necessary to support location-based services.

Some aspect of time is an important fact to remember for many applications. Banks, for example, want to remember what dates customers made payments on their loans. Airlines need to recall for the current and future days who will occupy seats on each flight. Thus, the management of time-varying, or **temporal,** data would be assisted if a database management system had built-in temporal support. As a result, there has been extensive research on temporal data models and DBMSs for more than a decade. The management of temporal data is another skill required of today's data management specialist.

The Open Geospatial Consortium, Inc. (OGC) is a nonprofit international organization developing standards for geospatial and location-based services. Its goal is to create open and extensible software application programming interfaces for geographic information systems (GIS) and other geospatial technologies. DBMS vendors (e.g., MySQL) have implemented some of OGC's recommendations for adding spatial features to SQL. MySQL is gradually adding further GIS features as it develops its DBMS.

Managing spatial data

A spatial database is a data management system for the collection, storage, manipulation, and output of spatially referenced information. Also known as a geographic information system (GIS), it is an extended form of DBMS. Geospatial modeling is based on three key concepts: theme, geographic object, and map.

A **theme** refers to data describing a particular topic (e.g., scenic lookouts, rivers, cities) and is the spatial counterpart of an entity. When a theme is presented on a screen or paper, it is commonly seen in conjunction with a **map**. Color may be used to indicate different themes (e.g., blue for rivers and black for roads). A map will usually have a scale, legend, and possibly some explanatory text.

A **geographic object** is an instance of a theme (e.g., a river). Like an instance of an entity, it has a set of attributes. In addition, it has spatial components that can describe both geometry and topology. Geometry refers to the location-based data, such as shape and length, and topology refers to spatial relationships among objects, such as adjacency. Management of spatial data requires some additional data types to represent a point, line, and region.

Generic spatial data types

Data type	Dimensions	Example
Point	0	Scenic lookout
Line	1	River
Region	2	County

Consider the case where we want to create a database to store some details of political units. A political unit can have many boundaries. The United States, for example, has a boundary for the continental portion, one for Alaska, one for Hawaii, and many more to include places such as American Samoa. In its computer form, a boundary is represented by an ordered set of line segments (a path).

Data model for political units

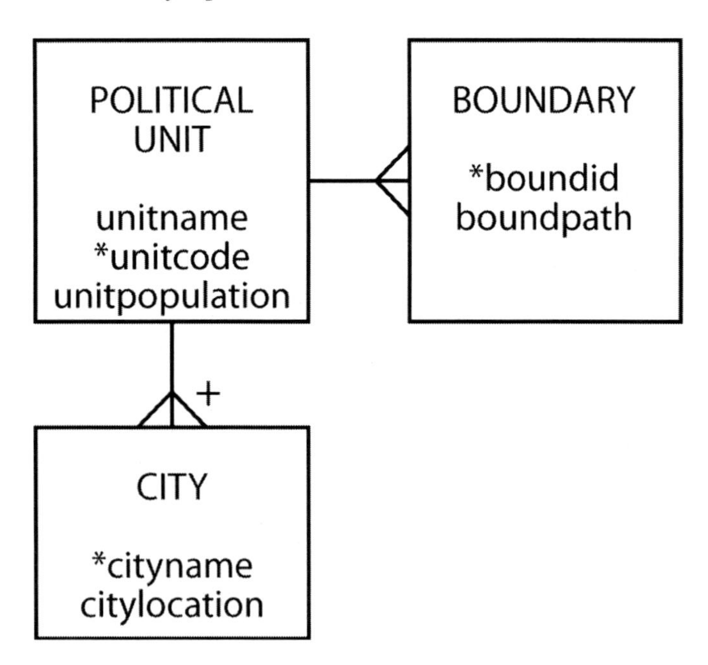

SQL/MM Spatial

SQL/MM, also known as ISO 13249, is an extension of SQL to handle spatial data. It uses the prefix ST_ for tables, views, data types, and function names. Originally, this prefix meant *Spatial* and *Temporal*, because the intention was to develop a standard that combined spatial and temporal extensions to SQL. However, it was realized that temporal required a broader perspective and should be separate standard. Thus, think of ST_ as meaning *Spatial Type*.

MySQL has data types for storing geometric data, which are:

Type	Representation	Description
Point	`POINT(x y)`	A point in space (e.g., a city's center)
LineString	`LINESTRING(x1 y1,x2 y2, …)`	A sequence of points with linear interpolation between points (e.g., a road)

Type	Representation	Description
Polygon	`POLYGON((x1 y1,x2 y2,…), (x1 y1,x2 y2,…))`	A polygon (e.g., a boundary) which has a single exterior boundary and zero or more interior boundaries (i.e., holes)

The data model in the figure is mapped to MySQL using the statements listed in the political unit database definition table. In the preceding definition of the database's tables, note two things. The `boundpath` column of boundary is defined with a data type of `POLYGON`. The `cityloc` column of `city` is defined as a `POINT`. Otherwise, there is little new in the set of statements to create the tables.

Political unit database definition

```
CREATE TABLE political_unit (
  Unitname VARCHAR(30) NOT NULL,
  Unitcode CHAR(2),
  Unitpop  DECIMAL(6,2),
  PRIMARY KEY(unitcode));
CREATE TABLE boundary (
  Boundid  INTEGER,
  Boundpath   POLYGON NOT NULL,
  Unitcode CHAR(2),
  PRIMARY KEY(boundid),
  CONSTRAINT fk_boundary_polunit FOREIGN KEY(unitcode)
    REFERENCES political_unit(unitcode));
CREATE TABLE city (
  Cityname VARCHAR(30),
  Cityloc  POINT NOT NULL,
  Unitcode CHAR(2),
  PRIMARY KEY(unitcode,cityname),
  CONSTRAINT fk_city_polunit FOREIGN KEY(unitcode)
    REFERENCES political_unit(unitcode));
```

We now use the geographic entity of Ireland to demonstrate the application of spatial concepts. The island has two political units. The Republic of Ireland (Eire) governs the south, while Northern Ireland, a part of the United Kingdom, is in the north.

Map of Ireland

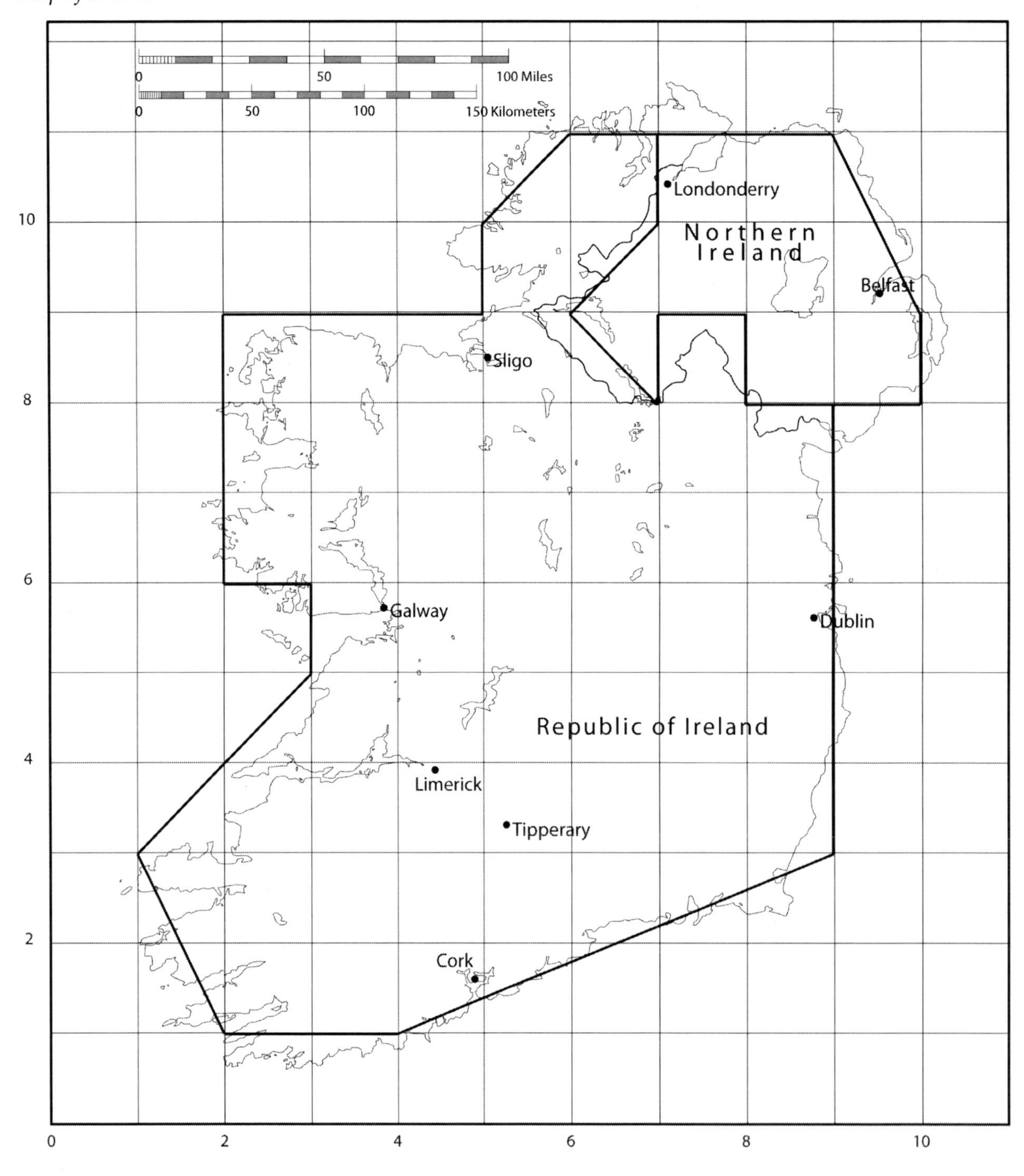

To represent these two political units within a spatial database, we need to define their boundaries. Typically, this is done by approximating the boundary by a single exterior polygon. In the preceding figure, you see a very coarse representation of the island based on connecting intersection points of the overlay grid.

Insert statements for populating database

```
INSERT INTO political_unit VALUES ('Republic of Ireland','ie', 3.9);
```

```
INSERT INTO political_unit VALUES ('Northern Ireland','ni', 1.7);
INSERT INTO boundary VALUES
   (1,ST_GeomFromText('polygon((9 8, 9 3, 4 1, 2 2, 1 3, 3 5, 3 6, 2 6,
   2 9, 5 9, 5 10, 6 11, 7 11, 7 10, 6 9, 7 8, 7 9, 8 9, 8 8, 9 8))
   '),'ie');
INSERT INTO boundary VALUES
   (2,ST_GeomFromText('polygon((7 11, 9 11, 10 9, 10 8, 8 8, 8 9, 7 9,
    7 8, 6 9, 7 10, 7 11))'),'ni');
INSERT INTO city VALUES ('Dublin',ST_GeomFromText('POINT(9 6)'),'ie');
INSERT INTO city VALUES ('Cork',ST_GeomFromText('POINT(5 2)'),'ie');
INSERT INTO city VALUES ('Limerick',ST_GeomFromText('POINT(4 4)'),'ie');
INSERT INTO city VALUES ('Galway',ST_GeomFromText('POINT(4 6)'),'ie');
INSERT INTO city VALUES ('Sligo',ST_GeomFromText('POINT(5 8)'),'ie');
INSERT INTO city VALUES ('Tipperary',ST_GeomFromText('POINT(5 3)'),'ie');
INSERT INTO city VALUES ('Belfast',ST_GeomFromText('POINT(9 9)'),'ni');
INSERT INTO city VALUES ('Londonderry',ST_GeomFromText('POINT(7 10)'),'ni');
```

The two sets of values for the column `boundary` define the boundaries of the Republic of Ireland and Northern Ireland. Because of the coarseness of this sample mapping, the Republic of Ireland has only one boundary. A finer-grained mapping would have multiple boundaries, such as one to include the Arran Islands off the west coast near Galway. Each city's location is defined by a point or pair of coordinates. `ST_GeomFromText` is an MySQL function to convert text into a geometry data form.

Workbench can show you the boundaries for you spatial database. See the following screenshot.

Boundary path as displayed by Workbench

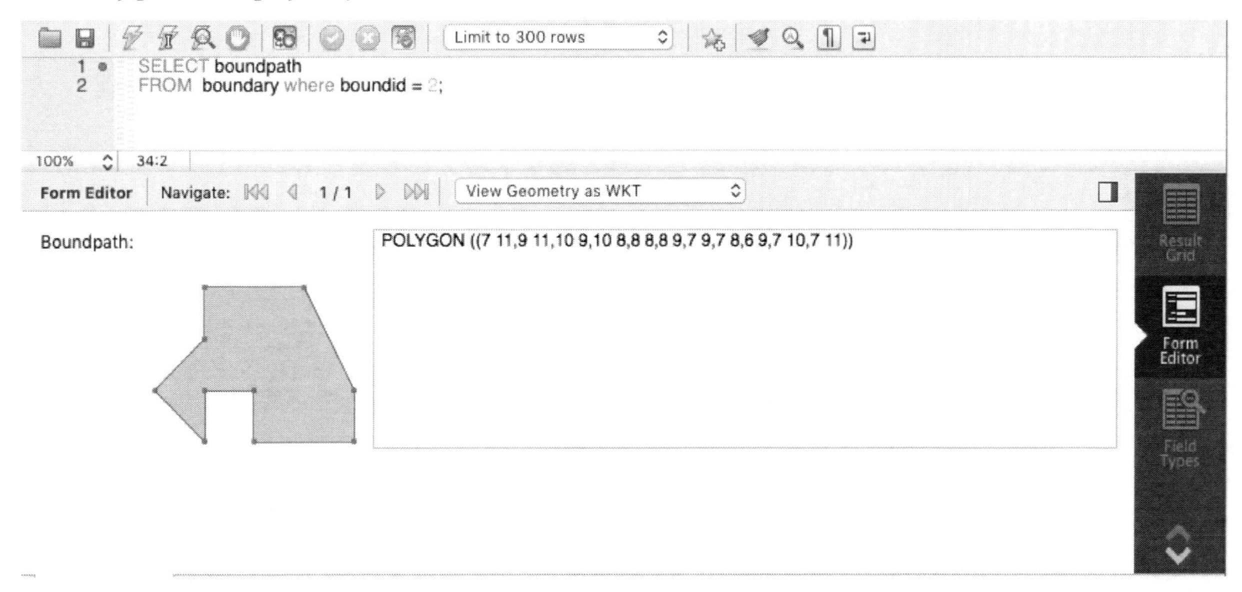

Skill builder

Create the three tables for the example and insert the rows listed in the preceding SQL code.

MySQL includes a number of geometry functions and operators for processing spatial data that simplify the writing of queries. All calculations are done assuming a flat or planar surface. In other words, MySQL uses Euclidean geometry. For illustrative purposes, just a few of the geometric functions are described.

Some MySQL geometric functions

Function	Description
`ST_X(Point)`	The x-coordinate of a point
`ST_Y(Point)`	The y-coordinate of a point
`ST_Length(LineString)`	The length of a linestring
`ST_NumPoints(LineString)`	The number of points in a linestring
`ST_Area(Polygon)`	The area of a polygon

Once the database is established, we can do some queries to gain an understanding of the additional capability provided by the spatial additions. Before starting, examine the scale on the map and note that one grid unit is about 37.5 kilometers (23 miles). This also means that the area of one grid unit is 1406 km^2 (526 square miles).

- *What is the area of the Republic of Ireland?*

Because we approximate the border by a polygon, we use the area function and multiply the result by 1406 to convert to square kilometers.

```
SELECT ST_AREA(boundpath)*1406
   AS 'Area (km^2)' FROM political_unit JOIN boundary
   ON political_unit.unitcode = boundary.unitcode
   WHERE unitname = 'Republic of Ireland';
```

Area(km^2)
71706

- *How far, as the crow flies, is it from Sligo to Dublin?*

MySQL has not yet implemented a distance function to measure how far it is between two points. However, we can get around this limitation by using the function `GLENGTH` to compute the length of a linestring. We will not get into the complications of how the linestring is created, but you should understand that a one segment linestring is created, with its end points being the locations of the two cities. Notice also that the query is based on a self-join.

```
SELECT ST_Distance(orig.cityloc,dest.cityloc)*37.5
AS 'Distance (kms)'
   FROM city orig, city dest
      WHERE orig.cityname = 'Sligo'
      AND dest.cityname = 'Dublin';
```

Distance (kms)
167.71

- *What is the closest city to Limerick?*

This query has a familiar structure. The inner query determines the minimum distance between Limerick and other cities. Notice that there is a need to exclude comparing the distance from Limerick to itself, which is zero.

```
SELECT dest.cityname FROM city orig, city dest
WHERE orig.cityname = 'Limerick'
AND ST_Distance(orig.cityloc,dest.cityloc)=
   (SELECT MIN(ST_Distance(orig.cityloc,dest.cityloc))
   FROM city orig, city dest
   WHERE orig.cityname = 'Limerick' AND dest.cityname <> 'Limerick');
```

cityname
Tipperary

- *What is the westernmost city in Ireland?*

The first thing to recognize is that by convention the west is shown on the left side of the map, which means the westernmost city will have the smallest x-coordinate.

```
SELECT west.cityname FROM city west
WHERE NOT EXISTS
   (SELECT * FROM city other WHERE ST_X(other.cityloc) < ST_X(west.cityloc));
```

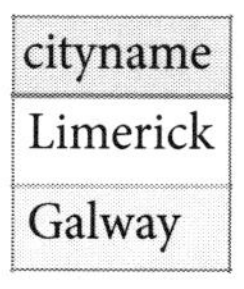

cityname
Limerick
Galway

Skill builder

1. What is the area of Northern Ireland? Because Northern Ireland is part of the United Kingdom and miles are still often used to measure distances, report the area in square miles.
2. What is the direct distance from Belfast to Londonderry in miles?
3. What is the northernmost city of the Republic of Ireland?

Geometry collections

A geometry collection is a data type for describing one or more geometries. It covers multiple points, strings, polygons, as well as their possible combinations.

MULTIPOINT

The `MULTIPOINT` data type records information about a set of points, such as the bus stops on campus. For example:

```
MULTIPOINT(9.0 6.1, 8.9 6.0)
```

MULTILINESTRING

The MULTILINESTRING data type records information about a set of line strings, such as the bus routes on campus. For example:

```
MULTILINESTRING((9 6, 4 6), (9 6, 5 2))
```

MULTIPOLYGON

The MULTIPOLYGON data type records information about a set of polygons, such as the shapes of the buildings on campus. For example:

```
MULTIPOLYGON(((0 0,10 0,10 10,0 10,0 0)),((5 5,7 5,7 7,5 7, 5 5)))
```

GEOMETRYCOLLECTION

The GEOMETRYCOLLECTION data type records information about a collection of geometries, such as the bus routes and stops on campus. For example:

```
GEOMETRYCOLLECTION(LINESTRING(15 15, 20 20), POINT(10 10), POINT(30 30))
```

You can insert data using ST_GeomCollFromText, as the following example illustrates:

```
INSERT INTO table VALUES ST_GeomCollFromText('GEOMETRYCOLLECTION(POINT(1
1),LINESTRING(0 0,1 1,2 2,3 3,4 4))');
```

Skill builder

Modify the example database design to include:

1. Historic buildings in a city
2. Walking paths in a city
3. Use of the MULTIPOLYGON data type to indicate a political region's boundary

Geocoding using Google Maps

To get the latitude and longitude of a location, you can use Google Maps by following this procedure.

1. Go to maps.google.com.
2. Enter your address, zip code, airport code, or whatever you wish to geocode.
3. Click on the link that says 'link to this page.' It is on the right side, just above the upper right corner of the map.
4. The address bar (URL) will change. Copy the full link. For example: http://maps.google.com/maps?f=q&source=s_q&hl=en&geocode=&q=ahn&aq=&sll=37.0625,-95.677068&sspn=48.822589,67.763672&ie=UTF8&hq=&hnear=Athens+Ben+Epps+Airport-Ahn,+1010+Ben+Epps+Dr,+Athens,+Georgia+30605&ll=33.953791,-83.323746&spn=0.025168,0.033088&z=15&iwloc=A.

5. The latitude and longitude are contained in the URL following &ll. In this case, latitude is: 33.953791 and longitude: -83.323746.

R-tree

Conventional DBMSs were developed to handle one-dimensional data (numbers and text strings). In a spatial database, points, lines, and rectangles may be used to represent the location of retail outlets, roads, utilities, and land parcels. Such data objects are represented by sets of *x, y* or *x, y, z* coordinates. Other applications requiring the storage of spatial data include computer-aided design (CAD), robotics, and computer vision.

The B-tree, often used to store data in one-dimensional databases, can be extended to *n* dimensions, where $n \geq 2$. This extension of the B-tree is called an **R-tree**. As well as storing pointers to records in the sequence set, an R-tree also stores boundary data for each object. For a two-dimensional application, the boundary data are the **x** and **y** coordinates of the lower left and upper-right corners of the *minimum bounding* rectangle, the smallest possible rectangle enclosing the object. The index set, which contains pointers to lower-level nodes as in a B-tree, also contains data for the minimum bounding rectangle enclosing the objects referenced in the node. The data in an R-tree permit answers to such problems as *Find all pizza stores within 5 miles of the dorm.*

How an R-tree stores data is illustrated in the following figure, which depicts five two-dimensional objects labeled A, B, C, D, and E. Each object is represented by its minimum bounding rectangle (the objects could be some other form, such as a circle). Data about these objects are stored in the sequence set. The index set contains details of two intermediate rectangles: X and Y. X fully encloses A, B, and C. Y fully encloses D and E.

An R-tree with sample spatial data

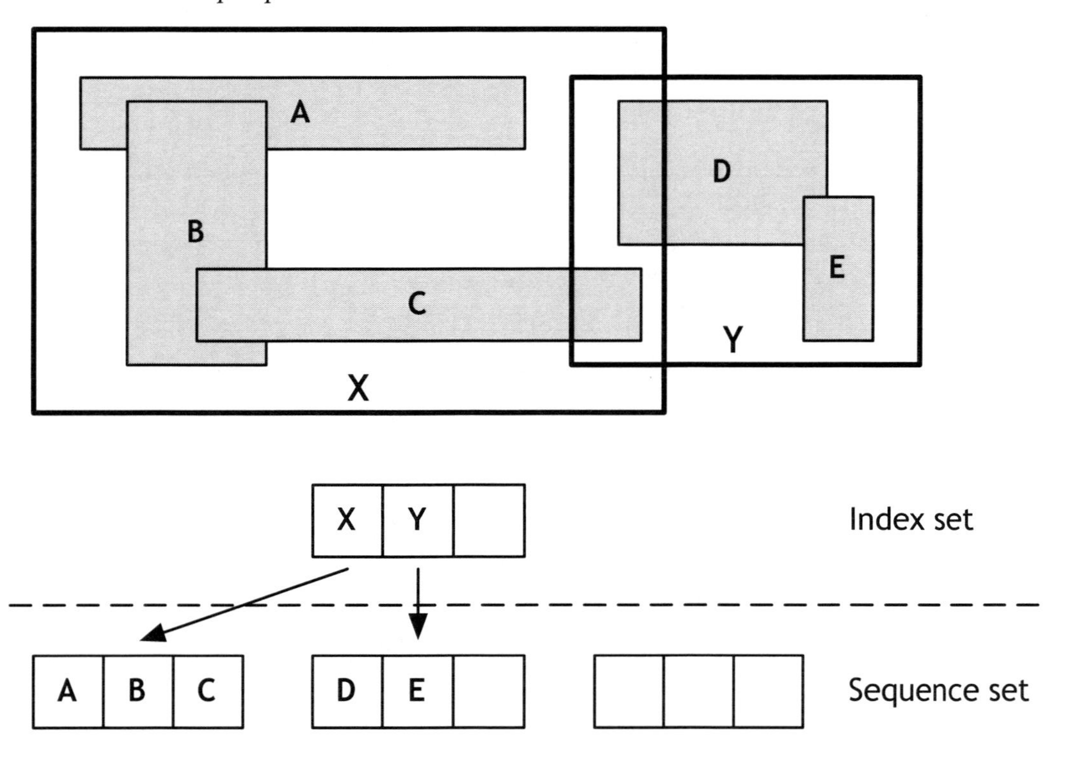

An example demonstrates how these data are used to accelerate searching. Using a mouse, a person could outline a region on a map displayed on a screen. The minimum bounding rectangle for this region would then be calculated and the coordinates used to locate geographic objects falling within the minimum boundary. Because an R-tree is an index, geographic objects falling within a region can be found rapidly. In the following figure, the drawn region (it has a bold border) completely covers object E. The R-tree software would determine that the required object falls within intermediate region Y, and thus takes the middle node at the next level of the R-tree. Then, by examining coordinates in this node, it would determine that E is the required object.

Searching an R-tree

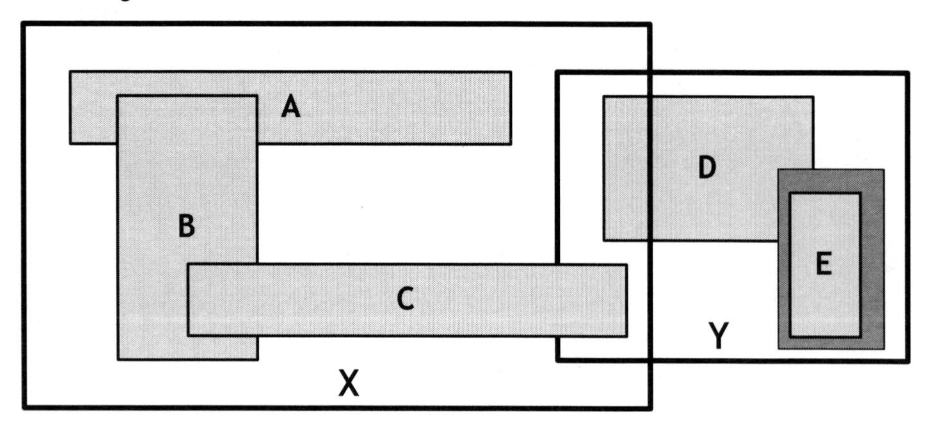

As the preceding example illustrates, an R-tree has the same index structure as a B-tree. An R-tree stores data about n-dimensional objects in each node, whereas a B-tree stores data about a one-dimensional data type in each node. Both also store pointers to the next node in the tree (the index set) or the record (the sequence set).

This short introduction to spatial data has given you some idea of how the relational model can be extended to support geometric information. Most of the major DBMS vendors support management of spatial data.

Managing temporal data

With a temporal database, stored data have an associated time period indicating when the item was valid or stored in the database. By attaching a timestamp to data, it becomes possible to store and identify different database states and support queries comparing these states. Thus, you might be able to determine the number of seats booked on a flight by 3 p.m. on January 21, 2011, and compare that to the number booked by 3 p.m. on January 22, 2011.

To appreciate the value of a temporal database, you need to know the difference between transaction and valid time and that bitemporal data combines both valid and transaction time.

- **Transaction time** is the timestamp applied by the system when data are entered and cannot be changed by an application. It can be applied to a particular item or row. For example, when changing the price of a product, the usual approach would be to update the existing product row with the new price. The old price would be lost unless it was stored explicitly. In contrast, with a temporal database, the old and new prices would automatically have separate timestamps. In effect, an additional row is inserted to store the new price and the time when the insert occurred.
- **Valid time** is the actual time at which an item was or will be a valid or true value. Consider the case where a firm plans to increase its prices on a specified date. It might post new prices some time before their effective date. Valid time can be changed by an application.

- **Bitemporal data** records both the valid time and transaction time for a fact. It usually requires four extra columns to record the upper and lower bounds for valid time and transaction time.

Valid time records when the change takes effect, and transaction time records when the change was entered. Storing transaction time is essential for database recovery because the DMBS can roll back the database to a previous state. Valid time provides a historical record of the state of the database. Both forms of time are necessary for a temporal database.

As you might expect, a temporal database will be somewhat larger than a traditional database because data are never discarded and new timestamped values are inserted so that there is a complete history of the values of an instance (e.g., the price of a product since it was first entered in the database). Thus, you can think of most of the databases we have dealt with previously as snapshots of a particular state of the database, whereas a temporal database is a record of all states of the database. As disk storage becomes increasingly cheaper and firms recognize the value of business intelligence, we are likely to see more attention paid to temporal database technology.

Times remembered

SQL supports several different data types for storing numeric values (e.g., integer and float), and a temporal database also needs a variety of data types for storing time values. The first level of distinction is to determine whether the time value is anchored or unanchored. **Anchored time** has a defined starting point (e.g., October 15, 1582), and **unanchored time** is a block of time with no specified start (e.g., 45 minutes).

Types of temporal data[23]

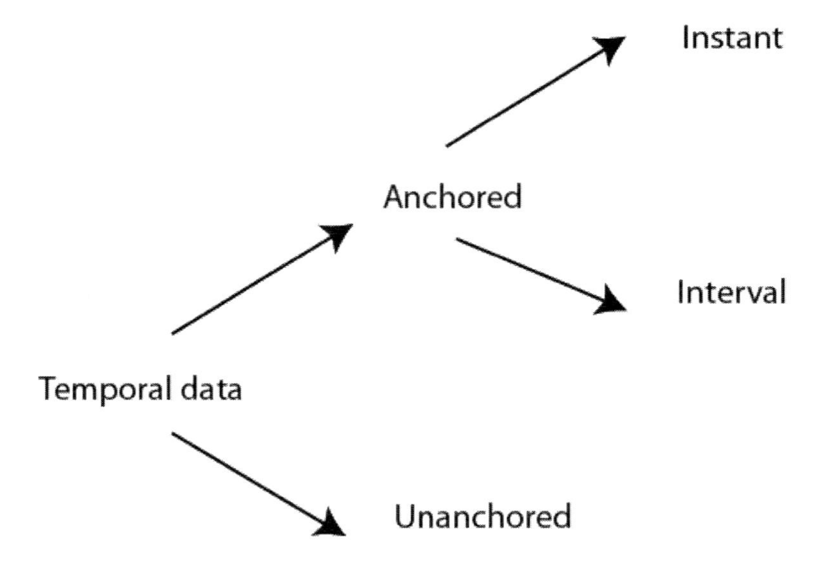

Anchored time is further split into an instant or interval. An **instant** is a moment in time (e.g., a date and time). In SQL, an instant can be represented by a date, time, or timestamp data type. An **interval** is the time between two specified instants, and can be defined as a value or a range with an upper and lower

23 Adapted from: Goralwalla, I. A., M. T. Özsu, and D. Szafron. 1998. *An object-oriented framework for temporal data models. In Temporal databases: research and practice*, edited by O. Etzion, S. Jajoda, and S. Sripada. Berlin: Springer-Verlag

bound instant. For example, [2011-01-01, 2011-01-23] defines an interval in 2011 beginning January 1 and ending January 23.

Interval

SQL-99 introduced the `INTERVAL` data type, which has not yet been implemented in MySQL. `INTERVAL` is a single value expressed in some unit or units of time (e.g., 6 years, 5 days, 7 hours). A small example illustrates the use of `INTERVAL` for time values. Consider the rotational and orbital periods of the planets . The `CREATE` statement for this table is

```
CREATE TABLE planet (
   pltname      VARCHAR(7),
   pltday       INTERVAL,
   pltyear      INTERVAL,
   PRIMARY KEY(pltname));
```

Planetary data

Planet	Rotational period (hours)	Orbital period (years)
Mercury	1407.51	0.24
Venus	–5832.44[a]	0.62
Earth	23.93	1
Mars	24.62	1.88
Jupiter	9.92	11.86
Saturn	10.66	29.45
Uranus	17.24	84.02
Neptune	16.11	164.79
Pluto	153.28	247.92

a. Rotates in the opposite direction to the other planets

To insert the values for Mercury, you would use

```
INSERT INTO planet VALUES ('Mercury','1407.51 hours','0.24 years');
```

Modeling temporal data

You already have the tools for modeling temporal values. For example, the project management data model discussed in Chapter 7 and reproduced in the following figure contains temporal data.

A project management data model

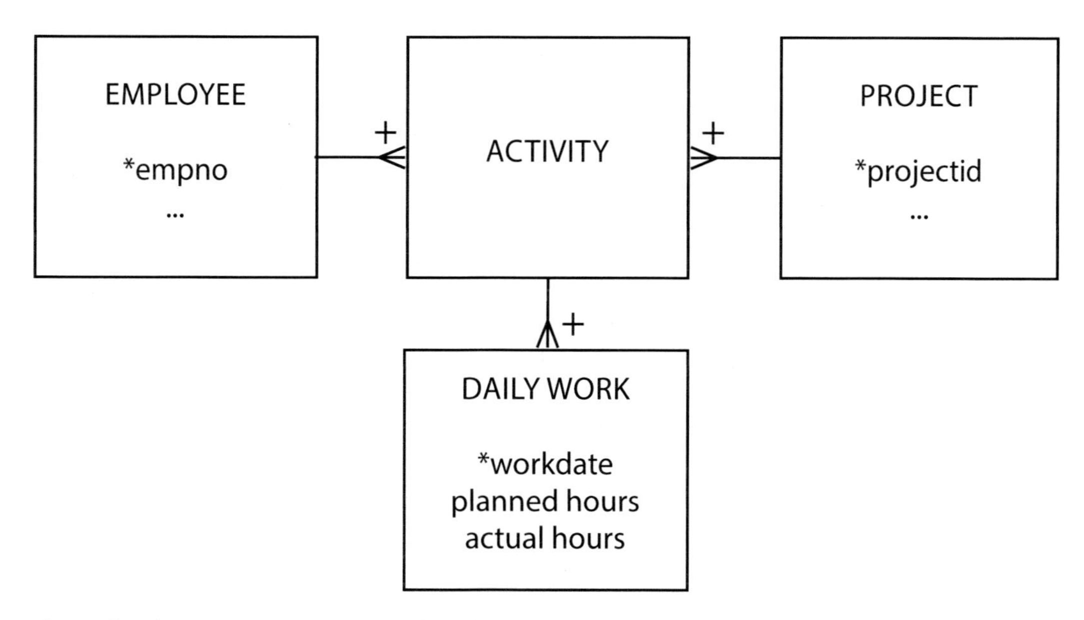

If we take the SHARE entity introduced very early in your data modeling experience, we can add temporal information to record the history of all values that are time-varying (i.e., price, quantity, dividend, and earnings). The data model to record temporal data is displayed. Firms pay dividends and report earnings only a few times per year, so we can associate a date with each value of dividend and earnings. Recording the history of trading transactions requires a timestamp, because a person can make multiple trades in a day. Every time a share is bought or sold, a new row is inserted containing both the transaction quantity and price. The number owned can be derived by using `SUM`.

A temporal model of SHARE

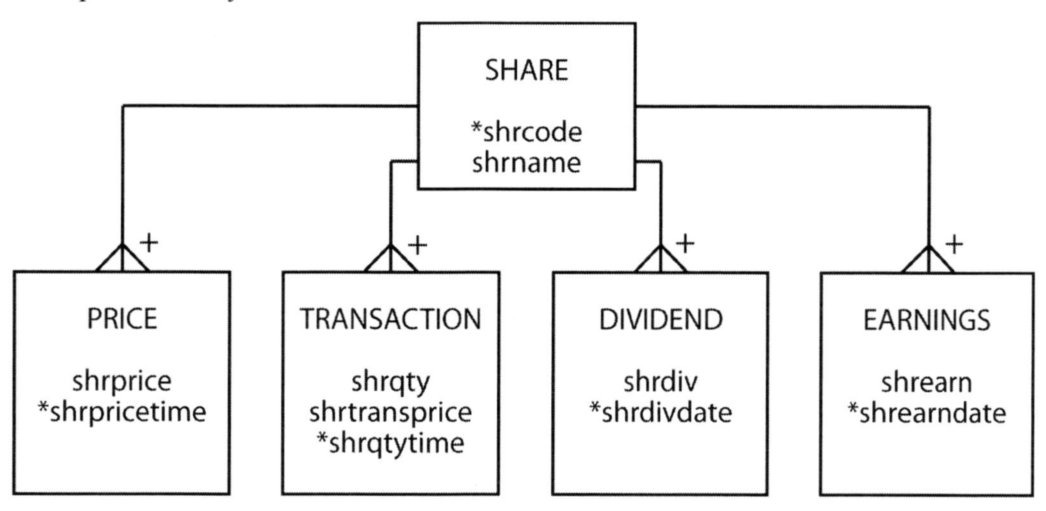

Recording the share's price requires further consideration. If the intention is to record every change in price, then a time stamp is required as there will be multiple price changes in a day, and even in an hour, in busy trading. If there is less interest in the volatility of the stock and only the closing price for each day is of interest, then a date would be recorded.

You can add additional attributes to tables in a relational database to handle temporal data, but doing so does not make it a temporal database. The problem is that current relational implementations do not have built-in functions for querying time-varying data. Such queries can also be difficult to specify in SQL.

A temporal database has additional features for temporal data definition, constraint specification, data manipulation, and querying. A step in this direction is the development of TSQL (Temporal Structured Query Language). Based on SQL, TSQL supports querying of temporal databases without specifying time-varying criteria. SQL:2011, the seventh revision of the SQL standard, has improved support for temporal data.

Summary

Spatial database technology stores details about items that have geometric features. It supports additional data types to describe these features, and has functions and operators to support querying. The new data types support point, line, and region values. Spatial technology is likely to develop over the next few years to support organizations offering localized information services.

Temporal database technology provides data types and functions for managing time-varying data. Transaction time and valid time are two characteristics of temporal data. Times can be anchored or unanchored and measured as an instant or as an interval.

Key terms and concepts

Anchored time

Geographic object

Geographic information system (GIS)

Interval

Map

R-tree

Spatial data

Temporal data

Theme

Transaction time

Valid time

References and additional readings

Gibson, Rich, and Schuyler Erle. 2006. *Google maps hacks*. Sebastopol, CA: O'Reilly.

Gregersen, H., and C. S. Jensen. 1999. Temporal entity-relationship models—a survey. *IEEE Transactions on Knowledge and Engineering* 11 (3):464–497.

Rigaux, P., M. O. Scholl, and A. Voisard. 2002. *Spatial databases: with application to GIS*, The Morgan Kaufmann series in data management systems. San Francisco: Morgan Kaufmann Publishers.

Exercises

1. What circumstances will lead to increased use of spatial data?

2. A national tourist bureau has asked you to design a database to record details of items of interest along a scenic road. What are some of the entities you might include? How would you model a road? Draw the data model.

3. Using the map of the Iberian peninsula in the following figure, populate the spatial database with details of Andorra, Portugal, and Spain. Answer the following questions.

   a. What is the direct distance, or bee line, from Lisbon to Madrid?

b. What is the farthest Spanish city from Barcelona?

c. Imagine you get lost in Portugal and your geographic positioning system (GPS) indicates that your coordinates are (3,9). What is the nearest city?

d. Are there any Spanish cities west of Braga?

e. What is the area of Portugal?

f. What is the southernmost city of Portugal?

4. Redesign the data model for political units assuming that your relational database does not support point and polygon data types.

5. For more precision and to meet universal standards, it would be better to use latitude and longitude to specify points and paths. You should also recognize that Earth is a globe and not flat. How would you enter latitude and longitude in MySQL?

6. When might you use transaction time and when might you use valid time?

7. Design a database to report basketball scores. How would you record time?

8. A supermarket chain has asked you to record what goods customers buy during each visit. In other words, you want details of each shopping basket. It also wants to know when each purchase was made. Design the database.

9. An online auction site wants to keep track of the bids for each item that a supplier sells. Design the database.

10. Complete the Google maps lab exercise listed on the book's Web site.

Iberian Peninsula

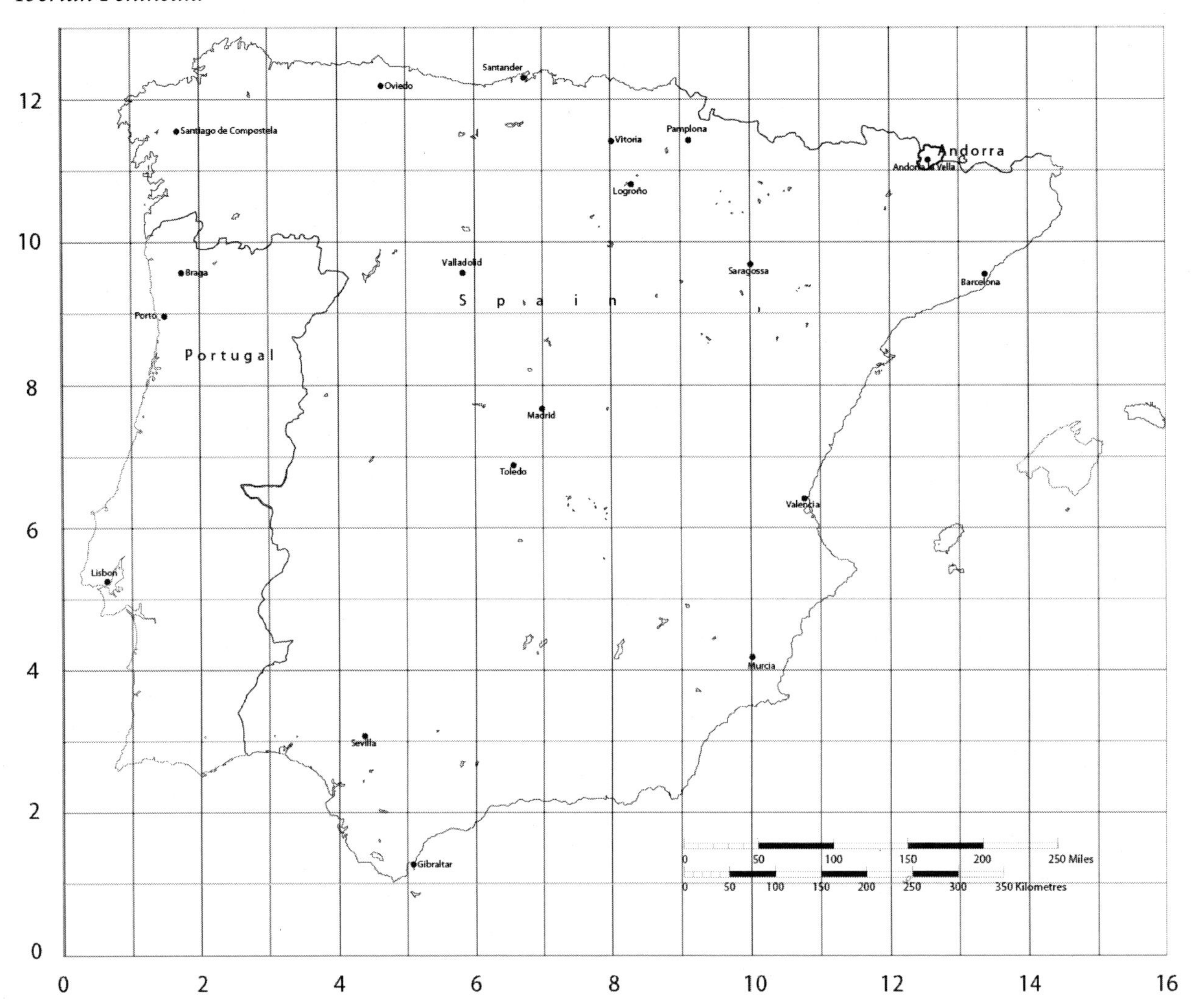

12. XML: Managing Data Exchange

Words can have no single fixed meaning. Like wayward electrons, they can spin away from their initial orbit and enter a wider magnetic field. No one owns them or has a proprietary right to dictate how they will be used.

David Lehman, End of the Word, 1991

Learning objectives

Students completing this chapter will be able to

- define the purpose of XML;
- create an XML schema;
- code data in XML format;
- create an XML stylesheet;
- discuss data management options for XML documents.

Introduction

There are four central problems in data management: capture, storage, retrieval, and exchange. The focus for most of this book has been on storage (i.e., data modeling) and retrieval (i.e., SQL). Now it is time to consider capture and exchange. Capture has always been an important issue, and the guiding principle is to capture data once in the cheapest possible manner.

SGML

The Standard Generalized Markup Language (SGML) was designed to reduce the cost and increase the efficiency of document management. Its child, XML, has essentially replaced SGML. For example, the second edition of the Oxford English Dictionary was specified in SGML, and the third edition is stored in XML format.[24]

A markup language embeds information about a document in the text. In the following table, the markup tags indicate that the text contains CD liner notes. Note also that the titles and identifiers of the mentioned CDs are explicitly identified.

Markup language

```
<cdliner>This uniquely creative collaboration between Miles Davis and Gil
Evans has already resulted in two extraordinary albums—<cdtitle>Miles Ahead</
cdtitle><cdid>CL 1041</cdid> and <cdtitle>Porgy and Bess</cdtitle><cdid>CL
1274</cdid>.</cdliner>
```

SGML is an International Standard (ISO 8879) that defines the structure of documents. It is a vendor-independent language that supports cross-system portability and publication for all media. Developed in 1986 to manage software documentation, SGML was widely accepted as the markup language for a

24 Cowlishaw, M. F. (1987). Lexx—a programmable structured editor. *IBM Journal of Research and Development*, 31(1), 73-80.

number of information-intensive industries. As a metalanguage, SGML is the mother of both HTML and XML..

SGML illustrates four major advantages a markup language provides for data management:

- **Reuse**: Information can be created once and reused over and over. By storing critical documents in markup format, firms do not need to duplicate efforts when there are changes to documents. For example, a firm might store all its legal contracts in SGML.
- **Flexibility**: SGML documents can be published in any medium for a wide variety of audiences. Because SGML is content-oriented, presentation decisions are delayed until the output format is known. Thus, the same content could be printed, presented on the Web in HTML, or written to a DVD as a PDF.
- **Revision**: SGML enhances control over revision and enables version control. When stored in an SGML database, original data are archived alongside any changes. That means you know exactly what the original document contained and what changes were made.
- **Format independence**: SGML files are stored as text and can be read by many programs on all operating systems. Thus, it preserves textual information independent of how and when it is presented. SGML protects a firm's investment in documentation for the long term. Because it is now possible to display documentation using multiple media (e.g., Web and iPad), firms have become sensitized to the need to store documents in a single, independent manner that can then be converted for display by a particular medium.

SGML's power is derived from its recording of both text and the meaning of that text. A short section of SGML demonstrates clearly the features and strength of SGML. The tags surrounding a chunk of text describe its meaning and thus support presentation and retrieval. For example, the pair of tags `<TITLE>` and `</TITLE>` surrounding "XML: Managing Data Exchange" indicates that it is the chapter title.

SGML code

```
<chapter>
<no>18</no>
<title>XML: Managing Data Exchange</title>
<section>
<quote><emph type = '2'>Words can have no single fixed meaning. Like wayward
electrons, they can spin away from their initial orbit and enter a wider
magnetic field. No one owns them or has a proprietary right to dictate how
they will be used.</emph>
</quote>
</section>
</chapter>
```

Taking this piece of SGML, it is possible, using an appropriate stylesheet, to create a print version where the title of the chapter is displayed in Times, 16 point, bold, or a HTML version where the title is displayed in red, Georgia, 14 point, italics. Furthermore, the database in which this text is stored can be searched for any chapters that contain "Exchange" in their title.

Now, consider the case where the text is stored as HTML. How do you, with complete certainty, identify the chapter title? Do you extract all text contained by <H1> and </H1> tags? You will then retrieve "18" as a possible chapter title. What happens if there is other text displayed using <H1> and </H1> tags? The problem with HTML is that it defines presentation and has very little meaning. A similar problem exists for documents prepared with a word processor.

HTML code

```
<html>
<body>
<h1><b>18 </b></h1>
<h1><b>XML: Managing Data Exchange</b></h1>
<p><i>Words can have no single fixed meaning. Like wayward electrons, they can
spin away from their initial orbit and enter a wider magnetic field. No one
owns them or has a proprietary right to dictate how they will be used.</i>
</body>
</html>
```

By using embedded tags to record meaning, SGML makes a document platform-independent and greatly improves the effectiveness of searching. Despite its many advantages, there are some features of SGML that make implementation difficult and also limit the ability to create tools for information management and exchange. As a result, XML, a derivative of SGML, was developed.

XML

Extensible Markup Language (XML), a language designed to make information self-describing, retains the core ideas of SGML. You can think of XML as SGML for electronic and mobile commerce. Since the definition of XML was completed in early 1998 by the World Wide Web Consortium (W3C), the standard has spread rapidly because it solves a critical data management problem. XML is a metalanguage—a language to generate languages.

Despite having the same parent, there are major differences between XML and HTML.

XML vs. HTML

XML	HTML
Structured text	Formatted text
User-definable structure (extensible)	Predefined formats (not extensible)
Context-sensitive retrieval	Limited retrieval
Greater hypertext linking	Limited hypertext linking

HTML, an electronic-publishing language, describes how a Web browser should display text and images on a computer screen. It tells the browser nothing about the meaning of the data. For example, the browser does not know whether a piece of text represents a price, a product code, or a delivery date. Humans infer meaning from the context (e.g., August 8, 2012, is recognized as a date). Given the explosive growth of the Web, HTML clearly works well enough for exchanging data between computers and humans. It does not, however, work for exchanging data between computers, because computers are not smart enough to deduce meaning from context.

Successful data exchange requires that the meaning of the exchanged data be readily determined by a computer. The XML solution is to embed tags in a file to describe the data (e.g., insert tags into an order to indicate attributes such as price, size, quantity, and color). A browser, or program for that matter, can then recognize this document as a customer order. Consequently, it can do far more than just display the price. For example, it can convert all prices to another currency. More importantly, the data can be exchanged between computers and understood by the receiving system.

XML consists of rules (that anyone can follow to create a markup language (e.g., a markup language for financial data such as XBRL). Hence, the "eXtensible" in the XML name, indicating that the language can be easily extended to include new tags. In contrast, HTML is not extensible and its set of tags is fixed, which is one of the major reasons why HTML is easy to learn. The XML rules ensure that a type of computer program known as a parser can process any extension or addition of new tags.

XML rules

- Elements must have both an opening and a closing tag.
- Elements must follow a strict hierarchy with only one root element.
- Elements must not overlap other elements.
- Element names must obey XML naming conventions.
- XML is case sensitive.

Consider the credit card company that wants to send you your latest statement via the Internet so that you can load it into your financial management program. Since this is a common problem for credit card companies and financial software authors, these industry groups have combined to create Open Financial Exchange (OFX),[25] a language for the exchange of financial data across the Internet.

XML has a small number of rules. Tags always come in pairs, as in HTML. A pair of tags surrounds each piece of data (e.g., <PRICE>89.12</PRICE>) to indicate its meaning, whereas in HTML ,they indicate how the data are presented. Tag pairs can be nested inside one another to multiple levels, which effectively creates a tree or hierarchical structure. Because XML uses Unicode (see the discussion in Chapter 11), it enables the exchange of information not only between different computer systems, but also across language boundaries.

The differences between HTML and XML are captured in the following examples for each markup language. Note that in the following table, HTML incorporates formatting instructions (i.e., the course code is bold), whereas XML describes the meaning of the data.

Comparison of HTML and XML coding

HTML	XML
<P><B>MIST7600</B>	<COURSE>
DATA MANAGEMENT 	<CODE>MIST7600</CODE>
3 CREDIT HOURS</P>	<TITLE>DATA MANAGEMENT</TITLE>
</COURSE>	<CREDIT>3</CREDIT>

25 www.ofx.net

XML enables a shift of processing from the server to the browser. At present, most processing has to be done by the server because that is where knowledge about the data is stored. The browser knows nothing about the data and therefore can only present but not process. However, when XML is implemented, the browser can take on processing that previously had to be handled by the server.

Imagine that you are selecting a shirt from a mail-order catalog. The merchant's Web server sends you data on 20 shirts (100 Kbytes of text and images) with prices in U.S. dollars. If you want to see the prices in euros, the calculation will be done by the server, and the full details for the 20 shirts retransmitted (i.e., another 100 Kbytes are sent from the server to the browser). However, once XML is in place, all that needs to be sent from the server to the browser is the conversion rate of U.S. dollars to euros and a program to compute the conversion at the browser end. In most cases, less data will be transmitted between a server and browser when XML is in place. Consequently, widespread adoption of XML will reduce network traffic.

Execution of HTML and XML code

HTML	XML
Retrieve shirt data with prices in USD.	Retrieve shirt data with prices in USD.
Retrieve shirt data with prices in EUR.	Retrieve conversion rate of USD to EUR.
	Retrieve Java program to convert currencies.
	Compute prices in EUR.

XML can also make searching more efficient and effective. At present, search engines look for matching text strings, and consequently return many links that are completely irrelevant. For instance, if you are searching for details on the Nomad speaker system, and specify "nomad" as the sought text string, you will get links to many items that are of no interest (e.g., The Fabulous Nomads Surf Band). Searching will be more precise when you can specify that you are looking for a product name that includes the text "nomad." The search engine can then confine its attention to text contained with the tags `<PRODUCTNAME>` and `</PRODUCTNAME>`, assuming these tags are the XML standard for representing product names.

The major expected gains from the introduction of XML are

- Store once and format many ways—Data stored in XML format can be extracted and reformatted for multiple presentation styles (e.g., printed report, DVD).
- Hardware and software independence—One format is valid for all systems. Capture once and exchange many times—Data are captured as close to the source as possible and never again (i.e., no rekeying).
- Accelerated targeted searching—Searches are more precise and faster because they use XML tags.
- Less network congestion—The processing load shifts from the server to the browser.

XML language design

XML lets developers design application-specific vocabularies. To create a new language, designers must agree on three things:

- The allowable tags
- The rules for nesting tagged elements

- Which tagged elements can be processed

The first two, the language's vocabulary and structure, are typically defined in an XML schema. Developers use the XML schema to understand the meaning of tags so they can write software to process an XML file.

XML tags describe meaning, independent of the display medium. An XML stylesheet, another set of rules, defines how an XML file is automatically formatted for various devices. This set of rules is called an Extensible Stylesheet Language (XSL). Stylesheets allow data to be rendered in a variety of ways, such as Braille or audio for visually impaired people.

XML schema

An XML schema (or just schema for brevity) is an XML file associated with an XML document that informs an application how to interpret markup tags and valid formats for tags. The advantage of a schema is that it leads to standardization. Consistently named and defined tags create conformity and support organizational efficiency. They avoid the confusion and loss of time when the meaning of data is not clear. Also, when validation information is built into a schema, some errors are detected before data are exchanged.

XML does not require the creation of a schema. If a document is well formed, XML will interpret it correctly. A well-formed document follows XML syntax and has tags that are correctly nested.

A schema is a very strict specification, and any errors will be detected when parsing. A schema defines:

- The names and contents of all elements that are permissible in a certain document
- The structure of the document
- How often an element may appear
- The order in which the elements must appear
- The type of data the element can contain

DOM

The **Document Object Model** (DOM) is the model underlying XML. It is based on a tree (i.e., it directly supports one-to-one and one-to-many, but not many-to-many relationships). A document is modeled as a hierarchical collection of nodes that have parent/child relationships. The node is the primary object and can be of different types (such as document, element, attribute, text). Each document has a single document node, which has no parent, and zero or more children that are element nodes. It is a good practice to create a visual model of the XML document and then convert this to a schema, which is XML's formal representation of the DOM.

At this point, an example is the best way to demonstrate XML, schema, and DOM concepts. We will use the familiar CD problem that was introduced in Chapter 3. In keeping with the style of this text, we define a minimal amount of XML to get you started, and then more features are added once you have mastered the basics.

CD library case

The CD library case gradually develops, over several chapters, a data model for recording details of a CD collection, culminating in the model at the end of Chapter 6. Unfortunately, we cannot quickly convert this final model to an XML document model, because a DOM is based on a tree model. Thus, we must start afresh.

The model , in this case, is based on the observation that a CD library has many CDs, and a CD has many tracks.

CD library tree data model

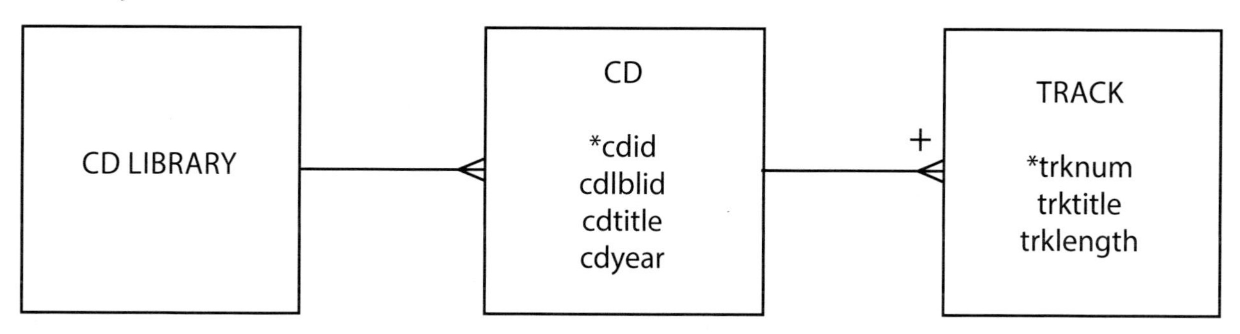

A model is then mapped into a schema using the following procedure.

- Each entity becomes a complex element type.
- Each data model attribute becomes a simple element type.
- The one-to-many (1:m) relationship is recorded as a sequence.

The schema for the CD library follows. For convenience of exposition, the source code lines have been numbered, but these numbers are not part of a schema.[26]

26 You should use an XML editor for creating XML files. I have not found a good open source product. Among the commercial products, my preference is for Oxygen, which you can get for a 30 days trial.

Schema for CD library (cdlib.xsd)

```
1   <?xml version="1.0" encoding="UTF-8"?>
2   <xsd:schema xmlns:xsd='http://www.w3.org/2001/XMLSchema'>
3   <!--CD library-->
4      <xsd:element name="cdlibrary">
5         <xsd:complexType>
6            <xsd:sequence>
7               <xsd:element name="cd" type="cdType" minOccurs="1"
8                  maxOccurs="unbounded"/>
9            </xsd:sequence>
10        </xsd:complexType>
11     </xsd:element>
12  <!--CD-->
13     <xsd:complexType name="cdType">
14        <xsd:sequence>
15           <xsd:element name="cdid" type="xsd:string"/>
16           <xsd:element name="cdlabel" type="xsd:string"/>
17           <xsd:element name="cdtitle" type="xsd:string"/>
18           <xsd:element name="cdyear" type="xsd:integer"/>
19           <xsd:element name="track" type="trackType" minOccurs="1"
20              maxOccurs="unbounded"/>
21        </xsd:sequence>
22     </xsd:complexType>
23  <!--Track-->
24     <xsd:complexType name="trackType">
25        <xsd:sequence>
26           <xsd:element name="trknum" type="xsd:integer"/>
27           <xsd:element name="trktitle" type="xsd:string"/>
28           <xsd:element name="trklen" type="xsd:time"/>
29        </xsd:sequence>
30     </xsd:complexType>
31  </xsd:schema>
```

There are several things to observe about the schema.

- All XML documents begin with an XML declaration {1}.[27] The encoding attribute (i.e., ENCODING="UTF-8") specifies what form of Unicode is used (in this case the 8-bit form).
- The XSD Schema namespace[28] is declared {2}.
- Comments are placed inside the tag pair <!-- AND --> {3}.
- The CD library is defined {4–10} as a complex element type, which essentially means that it can have embedded elements, which are a sequence of CDs in this case.
- A sequence is a series of child elements embedded in a parent, as illustrated by a CD library containing a sequence of CDs {7}, and a CD containing elements of CD identifier, label, and so forth {15–20}. The order of a sequence must be maintained by any XML document based on the schema.
- A sequence can have a specified range of elements. In this case, there must be at least one CD (MINOCCURS="1") but there is no upper limit (MAXOCCURS= "UNBOUNDED") on how many CDs there can be in the library {7}.
- An element that has a child (e.g., CDLIBRARY, which is at the 1 end of a 1:m) or possesses attributes (e.g., TRACK) is termed a complex element type.
- A CD is represented by a complex element type {13–20}, and has the name CDTYPE {13}.
- The element CD is defined by specifying the name of the complex type (i.e., CDTYPE) containing its specification {7}.
- A track is represented by a complex type because it contains elements of track number, title, and length {24–30}. The name of this complex type is TRACKTYPE {24}.
- Notice the reference within the definition of CD to the complex type TRACKTYPE, used to specify the element TRACK {19}.
- Simple types (e.g., CDID and CDYEAR) do not contain any elements, and thus the type of data they store must be defined. Thus, CDID is a text string and CDYEAR is an integer.

The purpose of a schema is to define the contents and structure of an XML file. It is also used to verify that an XML file has a valid structure and that all elements in the XML file are defined in the schema.

If you use an editor, you can possibly create a visual view of the schema.

A visual depiction of a schema as created by Oxygen

Some common data types are shown in the following table. The meaning is obvious in most cases for those familiar with SQL, except for URIREFERENCE. A Uniform Resource Identifier (URI) is a generalization of the URL concept.

27 In this chapter, numbers in {} refer to line numbers in the corresponding XML code.

28 A namespace is a collection of valid names of attributes, types, and elements for a schema. Think of a namespace as a dictionary.

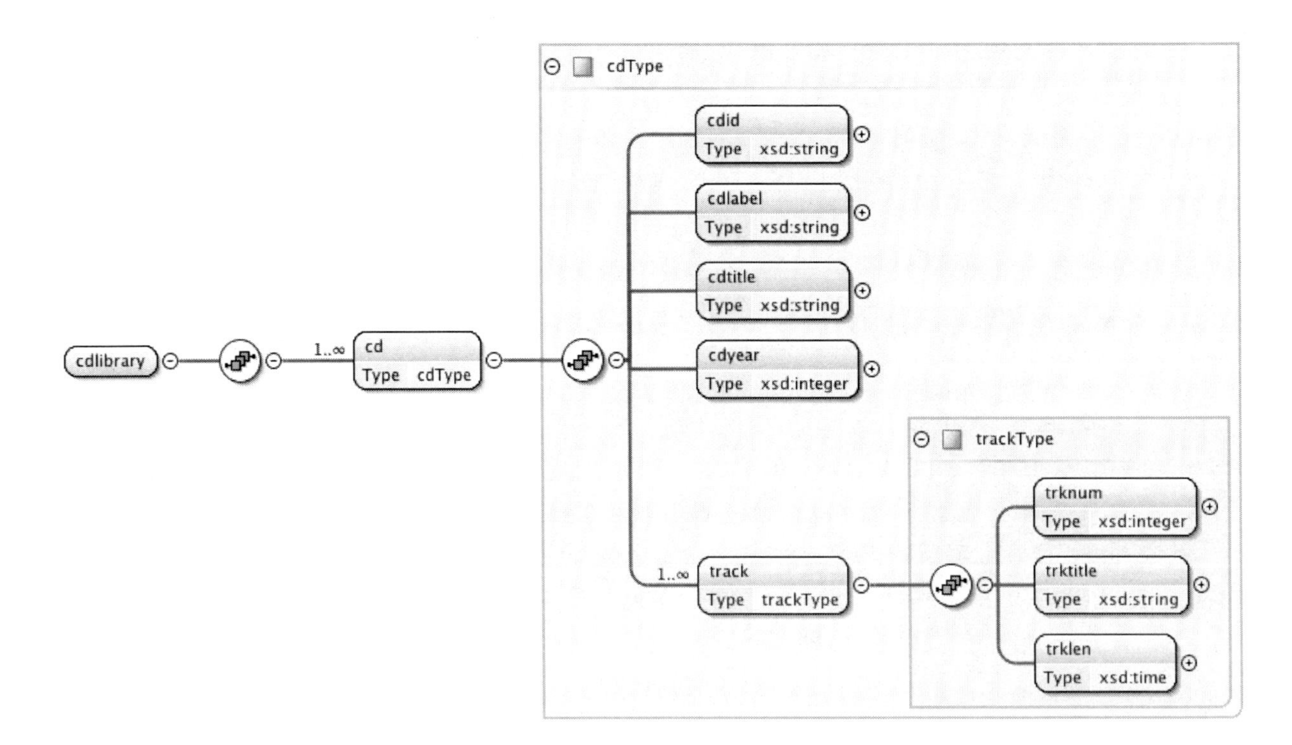

Some common data types

Data type
string
boolean
anyURI
decimal
float
integer
time
date

We can now use the recently defined CDlibrary schema to describe a small CD library containing the CD information given in the following table.

Data for a small CD library

Id	A2 1325	D136705
Label	Atlantic	Verve
Title	Pyramid	Ella Fitzgerald
Year	1960	2000

Track	Title	Length	Title	Length
1	Vendome	2:30	A tisket, a tasket	2:37
2	Pyramid	10:46	Vote for Mr. Rhythm	2:25
3			Betcha nickel	2:52

The XML for describing the CD library follows. There are several things to observe:

- All XML documents begin with an XML declaration.
- The declaration immediately following the XML declaration identifies the root element of the document (i.e., CDLIBRARY) and the schema (i.e., CDLIB.XSD).
- Details of a CD are enclosed by the tags <cd> and </cd>.
- Details of a track are enclosed by the tags <track> and </track>.

XML for describing a CD (cdlib.xml)

```
<?xml version="1.0" encoding="UTF-8"?>
<cdlibrary xmlns:xsi="http://www.w3.org/2001/XMLSchema-instance"
   xsi:noNamespaceSchemaLocation="cdlib.xsd">
    <cd>
        <cdid>A2 1325</cdid>
        <cdlabel>Atlantic</cdlabel>
        <cdtitle>Pyramid</cdtitle>
        <cdyear>1960</cdyear>
        <track>
            <trknum>1</trknum>
            <trktitle>Vendome</trktitle>
            <trklen>00:02:30</trklen>
        </track>
        <track>
            <trknum>2</trknum>
            <trktitle>Pyramid</trktitle>
            <trklen>00:10:46</trklen>
        </track>
    </cd>
    <cd>
        <cdid>D136705</cdid>
        <cdlabel>Verve</cdlabel>
        <cdtitle>Ella Fitzgerald</cdtitle>
        <cdyear>2000</cdyear>
        <track>
            <trknum>1</trknum>
            <trktitle>A tisket, a tasket</trktitle>
            <trklen>00:02:37</trklen>
        </track>
        <track>
            <trknum>2</trknum>
            <trktitle>Vote for Mr. Rhythm</trktitle>
```

```
            <trklen>00:02:25</trklen>
        </track>
        <track>
            <trknum>3</trknum>
            <trktitle>Betcha nickel</trktitle>
            <trklen>00:02:52</trklen>
        </track>
    </cd>
</cdlibrary>
```

As you now realize, the definition of an XML document is relatively straightforward. It is a bit tedious with all the typing of tags to surround each data element. Fortunately, there are XML editors that relieve this tedium.

Skill builder

1. Use the Firefox browser[29] to access this book's Web site, link to the Support > XML section, and click on customerpayments.xml. You will see how this browser displays XML. Investigate what happens when you click on the '-' and '+' signs next to some entries.
2. Again, using Firefox, save the displayed XML code (Save Page As ...) as customerpayments.xml, and open it in a text editor.
3. Now, add details of the customer and payment data displayed in the following table to the beginning of the XML file. Open the saved file with Firefox, and verify your work.

Customer and payment data

AA Souvenirs Yallingup Australia		
Check	Amount	Date
QP45901	9387.45	2005-03-16
AG9984	3718.67	2005-07-24

XSL

As you now know from the prior exercise, the browser display of XML is not particularly useful. What is missing is a stylesheet that tells the browser how to display an XML file. The eXtensible Stylesheet Language (XSL) is a language for defining the rendering of an XML file. An XSL document defines the rules for presenting an XML document's data. XSL is an application of XML, and an XSL file is also an XML file.

The power of XSL is demonstrated by applying the stylesheet that follows to the preceding XML.

29 Overall, Firefox seems to do the best job of displaying xml files.

Result of applying a stylesheet to CD library data

Complete List of Songs

Pyramid, Atlantic, 1960.5 [A2 1325]

1	Vendome	00:02:30
2	Pyramid	00:10:46

Ella Fitzgerald, Verve, 2000 [D136705]

1	A tisket, a tasket	00:02:37
2	Vote for Mr. Rhythm	00:02:25
3	Betcha nickel	00:02:52

Stylesheet for displaying an XML file of CD data (cdlib.xsl)

```
<?xml version="1.0" encoding="UTF-8"?>
<xsl:stylesheet version="1.0"
      xmlns:xsl="http://www.w3.org/1999/XSL/Transform">
   <xsl:output encoding="UTF-8" indent="yes" method="html" />
   <xsl:template match="/">
   <html>
      <head>
         <title> Complete List of Songs </title>
      </head>
   <body>
      <h1> Complete List of Songs </h1>
      <xsl:apply-templates select="cdlibrary" />
      </body>
      </html>
   </xsl:template>
   <xsl:template match="cdlibrary">
      <xsl:for-each select="cd">
         <br/>
         <font color="maroon">
            <xsl:value-of select="cdtitle" />
            ,
            <xsl:value-of select="cdlabel" />
            ,
            <xsl:value-of select="cdyear" />
            [
            <xsl:value-of select="cdid" />
            ] </font>
         <br/>
         <table>
            <xsl:for-each select="track">
               <tr>
                  <td align="left">
                     <xsl:value-of select="trknum" />
                  </td>
                  <td>
                     <xsl:value-of select="trktitle" />
                  </td>
                  <td align="center">
                     <xsl:value-of select="trklen" />
                  </td>
               </tr>
            </xsl:for-each>
         </table>
         <br/>
      </xsl:for-each>
   </xsl:template>
</xsl:stylesheet>
```

To use a stylesheet with an XML file, you must add a line of code to point to the stylesheet file. In this case, you add the following:

```
<?xml-stylesheet type="text/xsl" href="cdlib.xsl" media="screen"?>
```

as the second line of cdlib.xml (i.e., it appears before `<CDLIBRARY … >`). The added line of code points to cdlib.xsl as the stylesheet. This means that when the browser loads cdlib.xml, it uses the contents of cdlib.xsl to determine how to render the contents of cdlib.xml.

We now need to examine the contents of cdlib.xsl so that you can learn some basics of creating XSL commands. You will soon notice that all XSL commands are preceded by xsl:.

- Tell the browser it is processing an XML file {1}
- Specify that the file is a stylesheet {2}
- Specify a template, which identifies which elements should be processed and how they are processed. The match attribute {4} indicates the template applies to the source node. Process the template {11} defined in the file {15–45}. A stylesheet can specify multiple templates to produce different reports from the same XML input.
- Specify a template to be applied when the XSL processor encounters the `<CDLIBRARY>` node {15}.
- Create an outer loop for processing each CD {16–44}.
- Define the values to be reported for each CD (i.e., title, label, year, and id) {19, 21, 23, 25}. The respective XSL commands select the values. For example, `<XSL:VALUE-OF SELECT="CDTITLE" />` specifies selection of `CDTITLE`.
- Create an inner loop for processing the tracks on a particular CD {29–41}.
- Present the track data in tabular form using HTML table commands interspersed with XSL {28–42}.

Skill builder

1. Use the Firefox browser to access this book's Web site, navigate to the XML page, and download cdlib.xml and cdlib.xsl to a directory or folder on your machine. Use Save Page As … for downloading.
2. Using a text editor, change the saved copy of cdlib.xml by inserting the following as the second line:
```
<?XML-STYLESHEET TYPE="TEXT/XSL" HREF="CDLIB.XSL" MEDIA="SCREEN"?>
```
3. Save the edited file in the same directory or folder as cdlib.xsl. Open the saved XML file with Firefox.

Converting XML

There are occasions when there is a need to convert an XML file:

- **Transformation**: conversion from one XML vocabulary to another (e.g., between financial languages FPML and finML)
- **Manipulation**: reordering, filtering, or sorting parts of a document
- **Rendering in another language**: rendering the XML file using another format

You have already seen how XSL can be used to transform XML for rendering as HTML. The original XSL has been split into three languages:

- XSLT for transformation and manipulation
- XSLT for rendition
- XPath for accessing the structure of an XML file

For a data management course, this is as far as you need to go with learning about XSL. Just remember that you have only touched the surface. To become proficient in XML, you will need an entire course on the topic.

XPath for navigating an XML document

XPath is a navigation language for an XML document. It defines how to select nodes or sets of nodes in a document. The first step to understanding XPath is to know about the different types of nodes. In the following XML document, the document node is `<CDLIBRARY>` {1}, `<TRKTITLE>VENDOME</TRKTITLE>` {9} is an example of an element node, and `<TRACK LENGTH="00:02:30">` {7} is an instance of an attribute node.

An XML document

```
1   <cdlibrary>
2       <cd>
3           <cdid>A2 1325</cdid>
4           <cdlabel>Atlantic</cdlabel>
5           <cdtitle>Pyramid</cdtitle>
6           <cdyear>1960</cdyear>
7           <track length="00:02:30">
8               <trknum>1</trknum>
9               <trktitle>Vendome</trktitle>
10          </track>
11          <track length="00:10:46">
12              <trknum>2</trknum>
13              <trktitle>Pyramid</trktitle>
14          </track>
15      </cd>
16  </cdlibrary>
```

A family of nodes

Each element and attribute has one **parent node**. In the preceding XML document, CD is the parent of CDID, CDLABEL, CDYEAR, and TRACK. Element nodes may have zero or more **children nodes**. Thus CDID, CDLABEL, CDYEAR, and TRACK are the children of CD. **Ancestor nodes** are the parent, parent's parent, and so forth of a node. For example, CD and CDLIBRARY are ancestors of CDTITLE. Similarly, we have **descendant nodes**, which are the children, children's children, and so on of a node. The descendants of CD include CDID, CDTITLE, TRACK, TRKNUM, and TRKTITLE. Finally, we have **sibling nodes**, which are nodes that share a parent. In the sample document, CDID, CDLABEL, CDYEAR, and TRACK are siblings.

Navigation examples

The examples in the following table give you an idea of how you can use XPath to extract data from an XML document. Our preference is to answer such queries using SQL, but if your data are in XML format, then XPath provides a means of interrogating the file.

XPath examples

Example	Result
`/cdlibrary/cd[1]`	Selects the first CD
`//trktitle`	Selects all the titles of all tracks
`/cdlibrary/cd[last() -1]`	Selects the second last CD
`/cdlibrary/cd[last()]/track[last()]/trklen`	The length of the last track on the last CD
`/cdlibrary/cd[cdyear=1960]`	Selects all CDs released in 1960
`/cdlibrary/cd[cdyear>1950]/cdtitle`	Titles of CDs released after 1950

XQuery for querying an XML document

XQuery is a query language for XML, and thus it plays a similar role that SQL plays for a relational database. It is used for finding and extracting elements and attributes of an XML document. It builds on XPath's method for specifying navigation through the elements of an XML document. As well as being used for querying, XQuery can also be used for transforming XML to XHTML.

The first step is to specify the location of the XML file. In this case, we will use the cdlib.xml file, which is stored on the web site for this book. Using XPath notation it is relatively straightforward to list some values in an XML file

- *List the titles of CDs.*

```
doc("http://richardtwatson.com/xml/cdlib.xml")/cdlibrary/cd/cdtitle
```

```
<?xml version="1.0" encoding="UTF-8"?>
<cdtitle>Pyramid</cdtitle>
<cdtitle>Ella Fitzgerald</cdtitle>
```

You can also use an XPath expression to select particular information.

- *List the titles of CDs released under the Verve label*

```
doc("http://richardtwatson.com/xml/cdlib.xml")/cdlibrary/cd[cdlabel='Verve']/
cdtitle
```

```
<?xml version="1.0" encoding="UTF-8"?>
<cdtitle>Ella Fitzgerald</cdtitle>
```

XQuery commands can also be written using an SQL like structure, which is called FLWOR, an easy way to remember 'For, Let, Where, Order by, Return.' Here is an example.

- *List the titles of tracks longer than 5 minutes*

```
for $x in doc("http://richardtwatson.com/xml/cdlib.xml")/cdlibrary/cd/track
where $x/'track length' > "00:05:00"
order by $x/'trktitle'
return $x
```

```
<?xml version="1.0" encoding="UTF-8"?>
<track>
   <trknum>2</trknum>
   <trktitle>Pyramid</trktitle>
   <trklen>00:10:46</trklen>
</track>
```

If you want to just report the data without the tags, use `RETURN DATA($X)`.

XML and databases

XML is more than a document-processing technology. It is also a powerful tool for data management. For database developers, XML can be used to facilitate middle-tier data integration and schemas. Most of the major DBMS producers have XML-centric extensions for their product lines.

Many XML documents are stored for the long term, because they are an important repository of organizational memory. A data exchange language, XML is a means of moving data between databases, which means a need for tools for exporting and importing XML.

XML documents can be stored in the same format as you would store a word processing or HTML file: You just place them in an appropriately named folder. File systems, however, have limitations that become particularly apparent when a large number of files need to be stored, as in the corporate setting.

What is needed is a DBMS for storing, retrieving, and manipulating XML documents. Such a DBMS should:

- Be able to store many documents
- Be able to store large documents
- Support access to portions of a document (e.g., the data for a single CD in a library of 20,000 CDs)
- Enable concurrent access to a document but provide locking mechanisms to prevent the lost update problem, which is discussed later in this book
- Keep track of different versions of a document
- Integrate data from other sources (e.g., insert the results of an SQL query formatted as an XML fragment into an XML document)

Two possible solutions for XML document management are a relational database management system (RDBMS) or an XML database.

RDBMS

An XML document can stored within an RDBMS. Storing an intact XML document as a CLOB is a sensible strategy if the XML document contains static content that will only be updated by replacing the entire document. Examples include written text such as articles, advertisements, books, or legal contracts. These document-centric files (e.g., articles and legal contracts) are retrieved and updated in their entirety.

For more dynamic data-centric XML files (e.g., orders, price lists, airline schedules), the RDBMS must be extended to support the structure of the data so that portions of the document (e.g., an element such as the price for a product) can be retrieved and updated.

XML database

A second approach is to build a special-purpose XML database. Tamino[30] is an example of such an approach.

MySQL and XML

MySQL has functions for storing, searching, and maintaining XML documents. Any XML file can be stored as a document, and each XML fragment is stored as a character string. Creation of a table is straightforward:

```
CREATE TABLE cdlib (
   docid     INT AUTO_INCREMENT,
   doc       VARCHAR(10000),
   PRIMARY KEY(docid));
```

Insertion of an XML fragment follows the familiar pattern:

```
INSERT INTO cdlib (doc) VALUES
('<cd>
   <cdid>A2 1325</cdid>
   <cdlabel>Atlantic</cdlabel>
   <cdtitle>Pyramid</cdtitle>
   <cdyear>1960</cdyear>
   <track length="00:02:30">
      <trknum>1</trknum>
      <trktitle>Vendome</trktitle>
   </track>
   <track length="00:10:46">
      <trknum>2</trknum>
      <trktitle>Pyramid</trktitle>
   </track>
</cd>');
```

Querying XML

`EXTRACTVALUE` is the MySQL function for retrieving data from a node. It requires specification of the XML fragment to be retrieved and the XPath expression to locate the required node.

- *Report the title of the first CD in the library*

```
SELECT ExtractValue(doc,'/cd/cdtitle[1]') FROM cdlib;
```

30 www.softwareag.com/Corporate/products/wm/tamino/default.asp

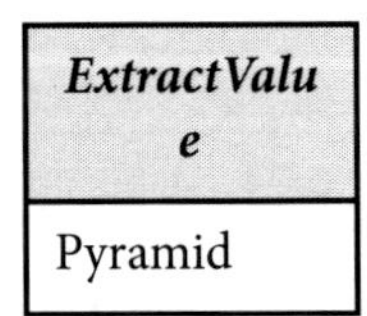

ExtractValue
Pyramid

- *Report the title of the first track on the first CD.*

```
SELECT ExtractValue(doc,'//cd[1]/track[1]/trktitle') FROM cdlib;
```

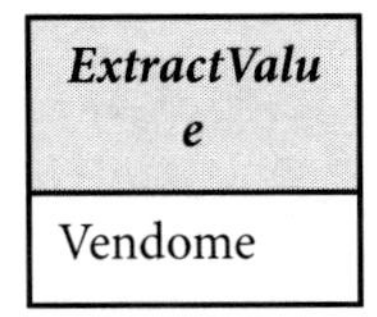

ExtractValue
Vendome

Updating XML

`UPDATEXML` replaces a single fragment of XML with another fragment. It requires specification of the XML fragment to be replaced and the XPath expression to locate the required node.

- *Change the title for the CD with identifier A2 1325 to Stonehenge*

```
UPDATE cdlib SET doc = UpdateXML(doc,'//cd[cdid="A2 1325"]/cdtitle',
'<cdtitle>Stonehenge</cdtitle>');
```

You can repeat the previous query to find the title of the first CD to verify the update.

Generating XML

A set of user defined functions (UDF)[31] has been developed to convert the results of an SQL query into XML. The xql_element function is used to define the name and value of the XML element to be reported. Here is an example.

- *List in XML format the name of all stocks with a PE ratio greater than 14.*

```
SELECT xql_element ('firm',shrfirm) FROM share WHERE shrpe > 14
```

Query output

```
<firm>Canadian Sugar</firm>
<firm>Freedonia Copper</firm>
<firm>Sri Lankan Gold</firm>
```

The preceding is just a brief example of what can be done. Read the documentation on UDF to learn how to handle more elaborate transformations.

31 The lib_mysqludf_xql library is housed at www.mysqludf.org. It might not be installed for the version of MySQL you are using.

Conclusion

XML has two main roles. The first is to facilitate the exchange of data between organizations and within those organizations that do not have integrated systems. Its second purpose is to support exchange between servers.

Mastery of XML is well beyond the scope of a single chapter. Indeed, it is a book-length topic, and hundreds of books have been written on XML. It is important to remember that the prime goal of XML is to support data interchange. If you would like to continue learning about XML, then consider the open content textbook (en.wikibooks.org/wiki/XML), which was created by students and is under continual revision. You might want to contribute to this book.

Summary

Electronic data exchange became more important with the introduction of the Internet. SGML, a precursor of XML, defines the structure of documents. SGML's value derives from its reusability, flexibility, support for revision, and format independence. XML, a derivative of SGML, is designed to support electronic commerce and overcome some of the shortcomings of SGML. XML supports data exchange by making information self-describing. It is a metalanguage because it is a language for generating other languages (e.g., finML). It provides substantial gains for the management and distribution of data. The XML language consists of an XML schema, document object model (DOM), and XSL. A schema defines the structure of a document and how an application should interpret XML markup tags. The DOM is a tree-based data model of an XML document. XSL is used to specify a stylesheet for displaying an XML document. XML documents can be stored in either a RDBMS or XML database.

Key terms and concepts

Document object model (DOM)
Document type definition (DTD)
Electronic data interchange (EDI)
Extensible markup language (XML)
Extensible stylesheet language (XSL)
Hypertext markup language (HTML)
Markup language
Occurrence indicators
Standard generalized markup language (SGML)
XML database
XML schema

References and additional readings

Anderson, R. 2000. *Professional XML*. Birmingham, UK; Chicago: Wrox Press.

Watson, R. T., and others. 2004. *XML: managing data exchange*: http://en.wikibooks.org/wiki/XML_-_Managing_Data_Exchange.

Exercises

1. A business has a telephone directory that records the first and last name, telephone number, and e-mail address of everyone working in the firm. Departments are the main organizing unit of the firm, so the telephone directory is typically displayed in department order, and shows for each department the contact phone and fax numbers and e-mail address.
    a. Create a hierarchical data model for this problem.
    b. Define the schema.

c. Create an XML file containing some directory data.
d. Create an XSL file for a stylesheet and apply the transformation to the XML file.

2. Create a schema for your university or college's course bulletin.
3. Create a schema for a credit card statement.
4. Create a schema for a bus timetable.
5. Using the portion of ClassicModels that has been converted to XML,[32] answer the following questions using XPath.
   a. List all customers.
   b. Who is the last customer in the file?
   c. Select all customers in Sweden.
   d. List the payments of more than USD 100,000.
   e. Select the first payments by Toys4GrownUps.com.
   f. What was the payment date for check DP677013?
   g. Who paid with check DP677013?
   h. What payments were received on 2003-12-04?
   i. Who made payments on 2003-12-04?
   j. List the numbers of all checks from customers in Denmark.
6. Using the portion of ClassicModels that has been converted to XML, answer the following questions using XQuery.
   a. List all customers.
   b. Who is the last customer in the file?
   c. Select all customers in Sweden sorted by customer name.
   d. List the payments of more than USD 100,000.
   e. Select the first payments by Toys4GrownUps.com.
   f. What was the payment date for check DP677013?
   g. Who paid with check DP677013?
   h. What payments were received on 2003-12-04?
   i. Who made payments on 2003-12-04?
   j. List the numbers of all checks from customers in Denmark.

32 http://richardtwatson.com/xml/customerpayments.xml

13. Organizational Intelligence

There are three kinds of intelligence: One kind understands things for itself, the other appreciates what others can understand, the third understands neither for itself nor through others. This first kind is excellent, the second good, and the third kind useless.

Machiavelli, *The Prince*, 1513

Learning objectives

Students completing this chapter will be able to

- understand the principles of organizational intelligence;
- decide whether to use verification or discovery for a given problem;
- select the appropriate data analysis technique(s) for a given situation;

Introduction

Too many companies are *data rich* but *information poor*. They collect vast amounts of data with their transaction processing systems, but they fail to turn these data into the necessary information to support managerial decision making. Many organizations make limited use of their data because they are scattered across many systems rather than centralized in one readily accessible, integrated data store. Technologies exist to enable organizations to create vast repositories of data that can be then analyzed to inform decision making and enhance operational performance.

Organizational intelligence is the outcome of an organization's efforts to collect, store, process, and interpret data from internal and external sources. The conclusions or clues gleaned from an organization's data stores enable it to identify problems or opportunities, which is the first stage of decision making.

Organizational intelligence technology is in transition. In this chapter, we deal with the the older version, which is still in place in many organizations. In the latter chapter on HDFS and MapReduce, we cover the newer approach that some firms have already adopted. It is likely that a mix of the two sets of technologies will exist in parallel for some time.

An organizational intelligence system

Transaction processing systems (TPSs) are a core component of organizational memory and thus an important source of data. Along with relevant external information, the various TPSs are the bedrock of an organizational intelligence system. They provide the raw facts that an organization can use to learn about itself, its competitors, and the environment. A TPS can generate huge volumes of data. In the United States, a telephone company may generate millions of records per day detailing the telephone calls it has handled. The hundreds of million credit cards on issue in the world generate billions of transactions per year. A popular Web site can have a hundred million hits per day. TPSs are creating a massive torrent of data that potentially reveals to an organization a great deal about its business and its customers.

Unfortunately, many organizations are unable to exploit, either effectively or efficiently, the massive amount of data generated by TPSs. Data are typically scattered across a variety of systems, in different database technologies, in different operating systems, and in different locations. The fundamental problem is that organizational memory is highly fragmented. Consequently, organizations need a technology that can accumulate a considerable proportion of organizational memory into one readily accessible system. Making these data available to decision makers is crucial to improving organizational performance, providing first-class customer service, increasing revenues, cutting costs, and preparing for the future. For many organizations, their memory is a major untapped resource—*an underused intelligence system containing undetected key facts about customers.* To take advantage of the mass of available raw data, an organization first needs to organize these data into one logical collection and then use software to sift through this collection to extract meaning.

The **data warehouse**, a subject-oriented, integrated, time-variant, and nonvolatile set of data that supports decision making, has emerged as the key device for harnessing organizational memory. *Subject* databases are designed around the essential entities of a business (e.g., customer) rather than applications (e.g., auto insurance). *Integrated* implies consistency in naming conventions, keys, relationships, encoding, and translation (e.g., gender is always coded as `M` or `F` in all relevant fields). *Time-variant* means that data are organized by various time periods (e.g., by months). Because a data warehouse is updated with a bulk upload, rather than as transactions occur, it contains *nonvolatile* data.

Data warehouses are enormous collections of data, often measured in terabytes, compiled by mass marketers, retailers, and service companies from the transactions of their millions of customers. Associated with a data warehouse are data management aids (e.g., data extraction), analysis tools (e.g., OLAP), and applications (e.g., executive information system).

The data warehouse environment

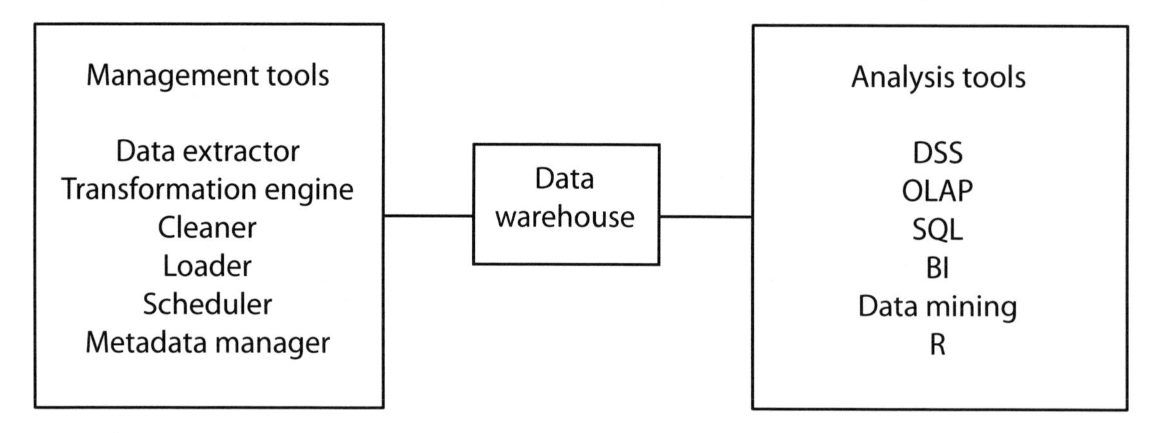

The data warehouse

Creating and maintaining the data warehouse

A data warehouse is a snapshot of an organization at a particular time. In order to create this snapshot, data must be extracted from existing systems, transformed, cleaned, and loaded into the data warehouse. In addition, regular snapshots must be taken to maintain the usefulness of the warehouse.

Extraction

Data from the operational systems, stored in operational data stores (ODS), are the raw material of a data warehouse. Unfortunately, it is not simply a case of pulling data out of ODSs and loading them into the warehouse. Operational systems were often written many years ago at different times. There was no plan to merge these data into a single system. Each application is independent or shares little data with others. The same data may exist in different systems with different names and in different formats. The extraction of data from many different systems is time-consuming and complex. Furthermore, extraction is not a one-time process. Data must be extracted from operational systems on an ongoing basis so that analysts can work with current data.

Transformation

Transformation is part of the data extraction process. In the warehouse, data must be standardized and follow consistent coding systems. There are several types of transformation:

- Encoding: Non-numeric attributes must be converted to a common coding system. Gender may be coded, for instance, in a variety of ways (e.g., m/f, 1/0, or M/F) in different systems. The extraction program must transform data from each application to a single coding system (e.g., m/f).
- Unit of measure: Distance, volume, and weight can be recorded in varying units in different systems (e.g., centimeters or inches) and must be converted to a common system.
- Field: The same attribute may have different names in different applications (e.g., sales-date, sdate, or saledate), and a standard name must be defined.
- Date: Dates are stored in a variety of ways. In Europe the standard for date is *dd/mm/yy*, in the U.S. it is *mm/dd/yy*, whereas the ISO standard is *yyyy-mm-dd.*

Cleaning

Unfortunately, some of the data collected from applications may be *dirty*—they contain errors, inconsistencies, or redundancies. There are a variety of reasons why data may need cleaning:

- The same record is stored by several departments. For instance, both Human Resources and Production have an employee record. Duplicate records must be deleted.
- Multiple records for a company exist because of an acquisition. For example, the record for Sun Microsystems should be removed because it was acquired by Oracle.
- Multiple entries for the same entity exist because there are no corporate data entry standards. For example, FedEx and Federal Express both appear in different records for the same company.
- Data entry fields are misused. For example, an address line field is used to record a second phone number.

Data cleaning starts with determining the dirtiness of the data. An analysis of a sample should indicate the extent of the problem and whether commercial data-cleaning tools are required. Data cleaning is unlikely to be a one-time process. All data added to the data warehouse should be validated in order to maintain the integrity of the warehouse. Cleaning can be performed using specialized software or custom-written code.

Loading

Data that have been extracted, transformed, and cleaned can be loaded into the warehouse. There are three types of data loads:

- **Archival:** Historical data (e.g., sales for the period 2005–2012) that is loaded once. Many organizations may elect not to load these data because of their low value relative to the cost of loading.
- **Current:** Data from current operational systems.
- **Ongoing:** Continual revision of the warehouse as operational data are generated. Managing the ongoing loading of data is the largest challenge for warehouse

management. This loading is done either by completely reloading the data warehouse or by just updating it with the changes.

Scheduling

Refreshing the warehouse, which can take many hours, must be scheduled as part of a data center's regular operations. Because a data warehouse supports medium- to long-term decision making, it is unlikely that it would need to be refreshed more frequently than daily. For shorter decisions, operational systems are available. Some firms may decide to schedule less frequently after comparing the cost of each load with the cost of using data that are a few days old.

Metadata

A data dictionary is a reference repository containing *metadata* (i.e., *data about data*). It includes a description of each data type, its format, coding standards (e.g., volume in liters), and the meaning of the field. For the data warehouse setting, a data dictionary is likely to include details of which operational system created the data, transformations of the data, and the frequency of extracts. Analysts need access to metadata so that they can plan their analyses and learn about the contents of the data warehouse. If a data dictionary does not exist, it should be established and maintained as part of ensuring the integrity of the data warehouse.

Data warehouse technology

Selecting an appropriate data warehouse system is critical to support significant data mining or online analytical processing. Data analysis often requires intensive processing of large volumes of data, and large main memories are necessary for good performance. In addition, the system should be scalable so that as the demand for data analysis grows, the system can be readily upgraded.

In recent years, there has been a shift to Hadoop, which is covered in the next chapter, as the foundation for a data warehouse. It offers speed and cost advantages over the technology that had predominated for some years.

Exploiting data stores

Two approaches to analyzing a data store (i.e., a database or data warehouse) are data mining and online analytical processing (OLAP). Before discussing each of these approaches, it is helpful to recognize the fundamentally different approaches that can be taken to exploiting a data store.

Verification and discovery

The **verification** approach to data analysis is driven by a hypothesis or conjecture about some relationship (e.g., customers with incomes in the range of $50,000–75,000 are more likely to buy minivans). The analyst then formulates a query to process the data to test the hypothesis. The resulting report will either support or disconfirm the theory. If the theory is disconfirmed, the analyst may continue to propose and test hypotheses until a target customer group of likely prospects for minivans is identified. Then, the minivan firm may market directly to this group because the likelihood of converting them to customers is higher than mass marketing to everyone. The verification approach is highly dependent on a persistent analyst eventually finding a useful relationship (i.e., who buys minivans?) by testing many hypotheses. OLAP, DSS, EIS, and SQL-based querying systems support the verification approach.

Data mining uses the **discovery** approach. It sifts through the data in search of frequently occurring patterns and trends to report generalizations about the data. Data mining tools operate with minimal guidance from the client. Data mining tools are designed to yield useful facts about business relationships efficiently from a large data store. The advantage of discovery is that it may uncover important relationships that no amount of conjecturing would have revealed and tested.

A useful analogy for thinking about the difference between verification and discovery is the difference between conventional and open-pit gold mining. A conventional mine is worked by digging shafts and tunnels with the intention of intersecting the richest gold vein. Verification is like conventional mining—some parts of the gold deposit may never be examined. The company drills where it believes there will be gold. In open-pit mining, everything is excavated and processed. Discovery is similar to open-pit mining—everything is examined. Both verification and discovery are useful; it is not a case of selecting one or the other. Indeed, analysts should use both methods to gain as many insights as possible from the data.

Comparison of verification and discovery

Verification	Discovery
What is the average sale for in-store and catalog customers?	What is the best predictor of sales?
What is the average high school GPA of students who graduate from college compared to those who do not?	What are the best predictors of college graduation?

OLAP

Edgar F. Codd, the father of the relational model, and colleagues (including, notably, Sharon B. Codd, his wife) proclaimed in 1993 that RDBMSs were never intended to provide powerful functions for data synthesis, analysis, and consolidation. This was the role of spreadsheets and special-purpose applications. They argued that analysts need data analysis tools that complement RDBMS technology, and they put forward the concept of **online analytical processing (OLAP):** the analysis of business operations with the intention of making timely and accurate analysis-based decisions.

Instead of rows and columns, OLAP tools provide multidimensional views of data, as well as some other differences. OLAP means fast and flexible access to large volumes of derived data whose underlying inputs may be changing continuously.

Comparison of TPS and OLAP applications

TPS	OLAP
Optimized for transaction volume	Optimized for data analysis
Process a few records at a time	Process summarized data
Real-time update as transactions occur	Batch update (e.g., daily)
Based on tables	Based on hypercubes
Raw data	Aggregated data
SQL is widely used	MDX becoming a standard

For instance, an OLAP tool enables an analyst to view how many widgets were shipped to each region by each quarter in 2012. If shipments to a particular region are below budget, the analyst can find out which customers in that region are ordering less than expected. The analyst may even go as far as examining the data for a particular quarter or shipment. As this example demonstrates, the idea of OLAP is to give analysts the power to view data in a variety of ways at different levels. In the process of investigating data anomalies, the analyst may discover new relationships. The operations supported by the typical OLAP tool include

- Calculations and modeling across dimensions, through hierarchies, or across members
- Trend analysis over sequential time periods

- Slicing subsets for on-screen viewing
- Drill-down to deeper levels of consolidation
- Drill-through to underlying detail data
- Rotation to new dimensional comparisons in the viewing area

An OLAP system should give fast, flexible, shared access to analytical information. Rapid access and calculation are required if analysts are to make ad hoc queries and follow a trail of analysis. Such quick-fire analysis requires computational speed and fast access to data. It also requires powerful analytic capabilities to aggregate and order data (e.g., summarizing sales by region, ordered from most to least profitable). Flexibility is another desired feature. Data should be viewable from a variety of dimensions, and a range of analyses should be supported.

MDDB

OLAP is typically used with an MDDB, a data management system in which data are represented by a multidimensional structure. The MDDB approach is to mirror and extend some of the features found in spreadsheets by moving beyond two dimensions. These tools are built directly into the MDDB to increase the speed with which data can be retrieved and manipulated. These additional processing abilities, however, come at a cost. The dimensions of analysis must be identified prior to building the database. In addition, MDDBs have size limitations that RDBMSs do not have and, in general, are an order of magnitude smaller than a RDBMS.

MDDB technology is optimized for analysis, whereas relational technology is optimized for the high transaction volumes of a TPS. For example, SQL queries to create summaries of product sales by region, region sales by product, and so on, could involve retrieving many of the records in a marketing database and could take hours of processing. A MDDB could handle these queries in a few seconds. TPS applications tend to process a few records at a time (e.g., processing a customer order may entail one update to the customer record, two or three updates to inventory, and the creation of an order record). In contrast, OLAP applications usually deal with summarized data.

Fortunately, RDBMS vendors have standardized on SQL, and this provides a commonality that allows analysts to transfer considerable expertise from one relational system to another. Similarly, MDX, originally developed by Microsoft to support multidimensional querying of an SQL server, has been implemented by a number of vendors for interrogating an MDDB. More details are provided later in this chapter.

The current limit of MDDB technology is approximately 10 dimensions, which can be millions to trillions of data points.

ROLAP

An alternative to a physical MDDB is a *relational OLAP* (or ROLAP), in which case a multidimensional model is imposed on a relational model. As we discussed earlier, this is also known as a logical MDDB. Not surprisingly, a system designed to support OLAP should be superior to trying to retrofit relational technology to a task for which it was not specifically designed.

The **star schema** is used by some MDDBs to represent multidimensional data within a relational structure. The center of the star is a table storing multidimensional ***facts*** derived from other tables. Linked to this central table are the ***dimensions*** (e.g., region) using the familiar primary-key/foreign-key approach of the relational model. The following figure depicts a star schema for an international automotive company. The advantage of the star model is that it makes use of a RDBMS, a mature technology capable of handling massive data stores and having extensive data management features (e.g., backup and recovery). However, if the fact table is very large, which is often the case, performance may be slow. A typical query is a join between the fact table and some of the dimensional tables.

A star schema

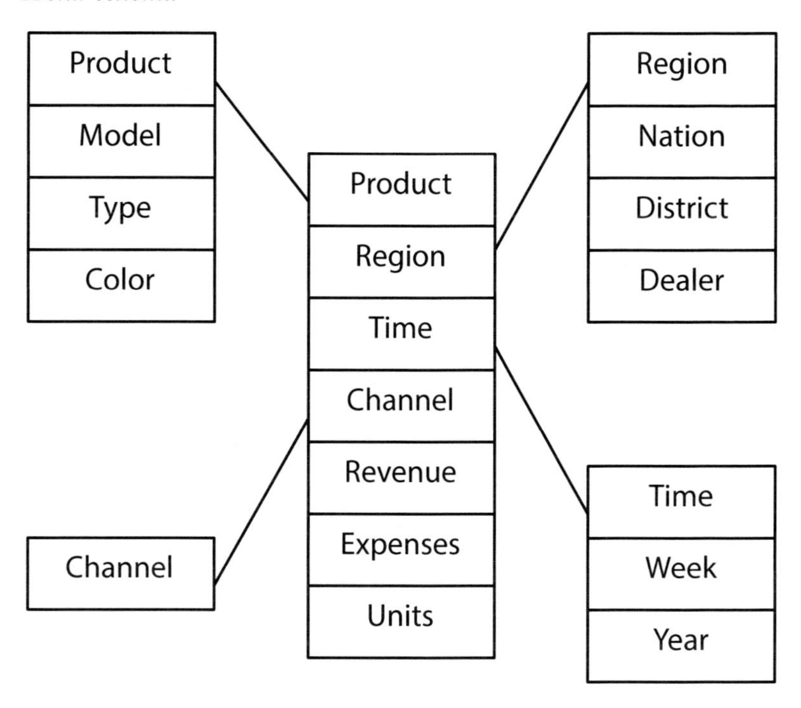

A **snowflake schema**, more complex than a star schema, resembles a snowflake. Dimensional data are grouped into multiple tables instead of one large table. Space is saved at the expense of query performance because more joins must be executed. Unless you have good reasons, you should opt for a star over a snowflake schema.

A snowflake schema

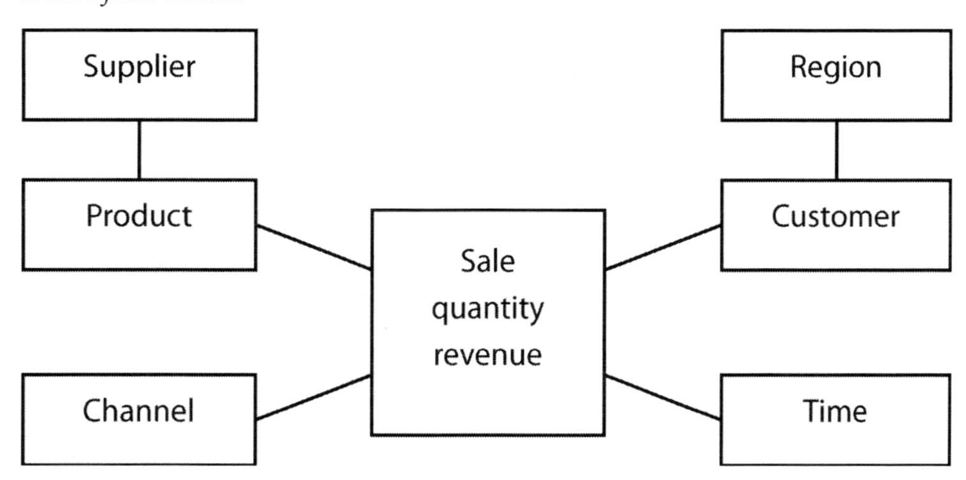

Rotation, drill-down, and drill-through

MDDB technology supports **rotation** of data objects (e.g., changing the view of the data from "by year" to "by region" as shown in the following figure) and **drill-down** (e.g., reporting the details for each nation in a selected region as shown in the Drill Down figure), which is also possible with a relational system. Drill-down can slice through several layers of summary data to get to finer levels of detail. The Japanese data, for instance, could be dissected by region (e.g., Tokyo), and if the analyst wants to go further, Tokyo could be analyzed by store. In some systems, an analyst can **drill through** the summarized data to examine the source data within the organizational data store from which the MDDB summary data were extracted.

Rotation

		Region			
Year	Data	Asia	Europe	North America	Grand total
2010	Sum of hardware	97	23	198	318
	Sum of software	83	41	425	549
2011	Sum of hardware	115	28	224	367
	Sum of software	78	65	410	553
2012	Sum of hardware	102	25	259	386
	Sum of software	55	73	497	625
Total sum of hardware		314	76	681	1,071
Total sum of software		216	179	1,322	1717

		Year			
Region	Data	2010	2011	2012	Grand total
Asia	Sum of hardware	97	115	102	314
	Sum of software	83	78	55	216
Europe	Sum of hardware	23	28	25	76
	Sum of software	41	65	73	179
North America	Sum of hardware	198	224	259	681
	Sum of software	425	410	497	1,332
Total sum of hardware		318	367	386	1,071
Total sum of software		549	553	625	1,727

Drill-down

Region	***Sales variance***
Africa	105%
Asia	57%
Europe	122%
North America	97%
Pacific	85%
South America	163%

Nation	Sales variance
China	123%
Japan	52%
India	87%
Singapore	95%

The hypercube

From the analyst's perspective, a fundamental difference between MDDB and RDBMS is the representation of data. As you know from data modeling, the relational model is based on tables, and analysts must think in terms of tables when they manipulate and view data. The relational world is two-dimensional. In contrast, the **hypercube** is the fundamental representational unit of a MDDB. Analysts can move beyond two dimensions. To envisage this change, consider the difference between the two-dimensional blueprints of a house and a three-dimensional model. The additional dimension provides greater insight into the final form of the building.

A hypercube

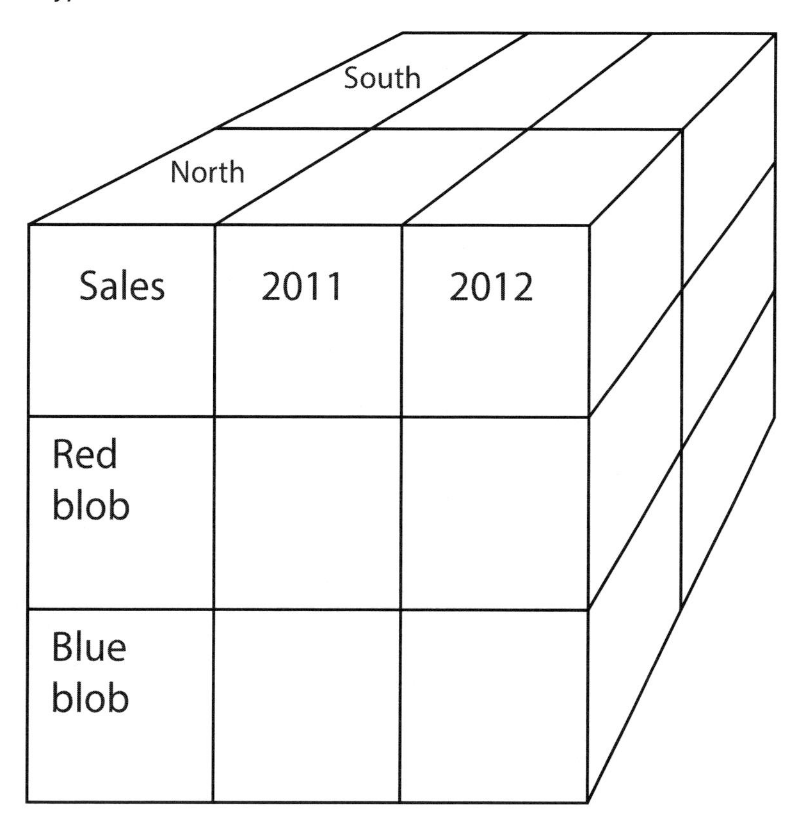

Of course, on a screen or paper only two dimensions can be shown. This problem is typically overcome by selecting an attribute of one dimension (e.g., North region) and showing the other two dimensions (i.e., product sales by year). You can think of the third dimension (i.e., region in this case) as the page dimension—each page of the screen shows one region or slice of the cube.

A three-dimensional hypercube display

Page				Columns
Region: North				Sales
		Red blob	Blue blob	Total
	2011			
Rows	2012			
Year	Total			

A hypercube can have many dimensions. Consider the case of a furniture retailer who wants to capture six dimensions of data. Although it is extremely difficult to visualize a six-dimensional hypercube, it helps to think of each cell of the cube as representing a fact (e.g., the Atlanta store sold five Mt. Airy desks to a business in January).

A six-dimensional hypercube

Dimension	Example
Brand	Mt. Airy
Store	Atlanta
Customer segment	Business
Product group	Desks
Period	January
Variable	Units sold

Similarly, a six-dimensional hypercube can be represented by combining dimensions (e.g., brand and store can be combined in the row dimension by showing stores within a brand).

A quick inspection of the table of the comparison of TPS and OLAP applications reveals that relational and multidimensional database technologies are designed for very different circumstances. Thus, the two technologies should be considered as complementary, not competing, technologies. Appropriate data can be periodically extracted from an RDBMS, aggregated, and loaded into an MDDB. Ideally, this process is automated so that the MDDB is continuously updated. Because analysts sometimes want to drill down to low-level aggregations and even drill through to raw data, there must be a connection from the MDDB to the RDBMS to facilitate access to data stored in the relational system.

The relationship between RDBMS and MDDB

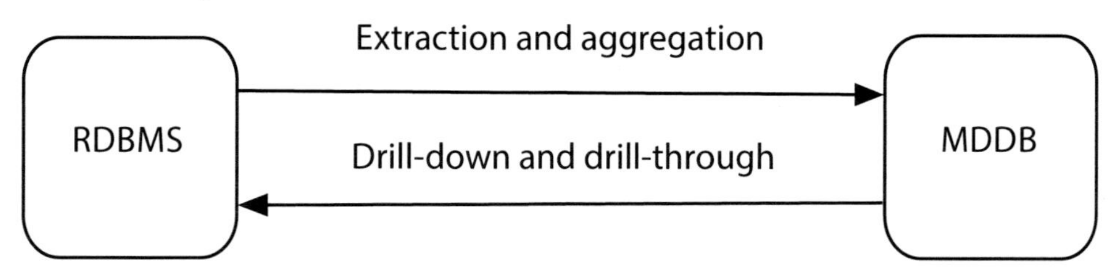

Designing a multidimensional database

The multidimensional model, based on the hypercube, requires a different design methodology from the relational model. At this stage, there is no commonly used approach, such as the entity-relationship principle of the relational model. However, the method proposed by Thomsen (1997) deserves consideration.

The starting point is to identify what must be tracked (e.g., sales for a retailer or revenue per passenger mile for a transportation firm). A collection of tracking variables is called a **variable dimension**.

The next step is to consider what types of analyses will be performed on the variable dimension. In a sales system, these may include sales by store by month, comparison of this month's sales with last month's for each product, and sales by class of customer. These types of analyses cannot be conducted unless the instances of each variable have an identifying tag. In this case, each sale must be tagged with time, store, product, and customer type. Each set of identifying factors is an **identifier dimension**. As a cross-check for identifying either type of dimension, use these six basic prompts.

Basic prompts for determining dimensions

Prompt	Example	Source
When?	June 5, 2013 10:27am	Transaction data
Where?	Paris	
What?	Tent	
How?	Catalog	
Who?	Young adult woman	Face recognition or credit card issuer
Why?	Camping trip to Bolivia	Social media
Outcome?	Revenue of €624.00	Transaction data

Most of the data related to the prompts can be extracted from transactional data. Face recognition software could be used to estimate the age and gender of the buyer in a traditional retail establishment. If the buyer uses a credit card, then such data, with greater precision, could be obtained from the bank issuing the credit card. In the case of why, the motivation for the purchase, the retailer can mine social exchanges made by the customer. Of course, this requires the retailer to be able to uniquely identify the buyer, through a store account or credit card, and use this identification to mine social media.

Variables and identifiers are the key concepts of MDDB design. The difference between the two is illustrated in the following table. Observe that time, an identifier, follows a regular pattern, whereas sales do not. Identifiers are typically known in advance and remain constant (e.g., store name and customer type), while variables change. It is this difference that readily distinguishes between variables and identifiers. Unfortunately, when this is not the case, there is no objective method of discriminating between the two. As a result, some dimensions can be used as both identifiers and variables.

A sales table

Identifier	Variable
time (hour)	sales (dollars)
10:00	523
11:00	789
12:00	1,256
13:00	4,128
14:00	2,634

There can be a situation when your intuitive notion of an identifier and variable is not initially correct. Consider a Web site that is counting the number of hits on a particular page. In this case, the identifier is hit and time is the variable because the time of each hit is recorded.

A hit table

Identifier	Variable
hit	time (hh:mm:ss)
1	9:34:45
2	9:34:57
3	9:36:12
4	9:41:56

The next design step is to consider the form of the dimensions. You will recall from statistics that there are three types of variables (dimensions in MDDB language): nominal, ordinal, and continuous. A nominal variable is an unordered category (e.g., region), an ordinal variable is an ordered category (e.g., age group), and a continuous variable has a numeric value (e.g., passenger miles). A hypercube is typically a combination of several types of dimensions. For instance, the identifier dimensions could be product and store (both nominal), and the variable dimensions could be sales and customers. A dimension's type comes into play when analyzing relationships between identifiers and variables, which are known as independent and dependent variables in statistics. The most powerful forms of analysis are available when both dimensions are continuous. Furthermore, it is always possible to recode a continuous variable into ordinal categories. As a result, wherever feasible, data should be collected as a continuous dimension.

Relationship of dimension type to possible analyses

		Identifier dimension	
		Continuous	*Nominal or ordinal*
Variable dimension	*Continuous*	Regression and curve fitting *Sales over time*	Analysis of variance *Sales by store*
	Nominal or ordinal	Logistic regression *Customer response (yes or no) to the level of advertising*	Contingency table analysis *Number of sales by region*

This brief introduction to multidimensionality modeling has demonstrated the importance of distinguishing between types of dimensions and considering how the form of a dimension (e.g., nominal or continuous) will affect the choice of analysis tools. Because multidimensional modeling is a relatively new concept, you can expect design concepts to evolve. If you become involved in designing an MDDB, then be sure to review carefully current design concepts. In addition, it would be wise to build some prototype systems, preferably with different vendor implementations of the multidimensional concept, to enable analysts to test the usefulness of your design.

Skill builder

A national cinema chain has commissioned you to design a multidimensional database for its marketing department. What identifier and variable dimensions would you select?

Multidimensional expressions (MDX)

MDX is a language for reporting data stored in a multidimensional database. On the surface, it looks like SQL because the language includes SELECT, FROM, and WHERE. The reality is that MDX is quite different. MDX works with cubes of data, and the result of an MDX query is a cube, just as SQL produces a table.The basic syntax of MDX statement is

```
SELECT {member selection} ON columns
   FROM [cube name]
```

Thus, a simple query would be

```
SELECT {[measures].[unit sales] } ON columns
   FROM [sales]
```

Measures
Unit Sales
266,773

The first step in using MDX is to define a cube, which has dimensions and measures. Dimensions are the categories for reporting measures. The dimensions of product for a supermarket might be food, drink, and non-consumable. Measures are the variables that typically report outcomes. For the supermarket example, the measures could be unit sales, cost, and revenue. Once a cube is defined, you can write MDX queries for it. If we want to see sales broken down by the product dimensions, we would write

```
SELECT {[measures].[unit sales] ON columns,
   {[product].[all products].[food]} ON rows
   FROM [sales]
```

In this section, we use the foodmart dataset to illustrate MDX. To start simple, we just focus on a few dimensions and measures. We have a product dimension containing three members: food, drink, and non-consumable, and we will use the measures of unit sales and store sales. Here is the code to report the data for all products.

```
SELECT {[measures].[unit sales],[measures].[store sales]} ON columns,
   {[product].[all products]} ON rows
   FROM [sales]
```

	Measures	
Product	Unit Sales	Store Sales
All Products	266,773	565,238.13

To examine the unit sales and store sales by product, use the following code:

```
SELECT {[measures].[unit sales], [measures].[store sales]} ON columns,
   {[product].[all products],
   [product].[all products].[drink],
   [product].[all products].[food],
   [product].[all products].[non-consumable]} ON rows
   FROM [sales]
```

	Measures	
Product	Unit Sales	Store Sales
All Products	266,773	565,238.13
Drink	24,597	48,836.21
Food	191,940	409,035.59
Non-consumable	50,236	107,366.33

These few examples give you some idea of how to write an MDX query and the type of output it can produce. In practice, most people will use a GUI for defining queries. The structure of a MDDB makes it well suited to the select-and-click generation of an MDX query.

Data mining

Data mining is the search for relationships and global patterns that exist in large databases but are hidden in the vast amounts of data. In data mining, an analyst combines knowledge of the data with advanced machine learning technologies to discover nuggets of knowledge hidden in the data. Data mining software can find meaningful relationships that might take years to find with conventional techniques. The software is designed to sift through large collections of data and, by using statistical and artificial intelligence techniques, identify hidden relationships. The mined data typically include electronic point-of-sale records, inventory, customer transactions, and customer records with matching demographics, usually obtained from an external source. Data mining does not require the presence of a data warehouse. An organization can mine data from its operational files or independent databases. However, data mining in independent files will not uncover relationships that exist between data in different files. Data mining will usually be easier and more effective when the organization accumulates as much data as possible in a single data store, such as a data warehouse. Recent advances in processing speeds and lower storage costs have made large-scale mining of corporate data a reality.

Database marketing, a common application of data mining, is also one of the best examples of the effective use of the technology. Database marketers use data mining to develop, test, implement, measure, and modify tailored marketing programs. The intention is to use data to maintain a lifelong relationship with a customer. The database marketer wants to anticipate and fulfill the customer's needs as they emerge. For example, recognizing that a customer buys a new car every three or four years and with each purchase gets an increasingly more luxurious car, the car dealer contacts the customer during the third year of the life of the current car with a special offer on its latest luxury model.

Data mining uses

There are many applications of data mining:

- Predicting the probability of default for consumer loan applications. Data mining can help lenders reduce loan losses substantially by improving their ability to predict bad loans.
- Reducing fabrication flaws in VLSI chips. Data mining systems can sift through vast quantities of data collected during the semiconductor fabrication process to identify conditions that cause yield problems.
- Predicting audience share for television programs. A market-share prediction system allows television programming executives to arrange show schedules to maximize market share and increase advertising revenues.
- Predicting the probability that a cancer patient will respond to radiation therapy. By more accurately predicting the effectiveness of expensive medical procedures, health care costs can be reduced without affecting quality of care.
- Predicting the probability that an offshore oil well is going to produce oil. An offshore oil well may cost $30 million. Data mining technology can increase the probability that this investment will be profitable.
- Identifying quasars from trillions of bytes of satellite data. This was one of the earliest applications of data mining systems, because the technology was first applied in the scientific community.

Data mining functions

Based on the functions they perform, five types of data mining functions exist:

Associations

An association function identifies affinities existing among the collection of items in a given set of records. These relationships can be expressed by rules such as "72 percent of all the records that contain items A, B, and C also contain items D and E." Knowing that 85 percent of customers who buy a certain brand of wine also buy a certain type of pasta can help supermarkets improve use of shelf space and promotional offers. Discovering that fathers, on the way home on Friday, often grab a six-pack of beer after buying some diapers, enabled a supermarket to improve sales by placing beer specials next to diapers.

Sequential patterns

Sequential pattern mining functions identify frequently occurring sequences from given records. For example, these functions can be used to detect the set of customers associated with certain frequent buying patterns. Data mining might discover, for example, that 32 percent of female customers within six months of ordering a red jacket also buy a gray skirt. A retailer with knowledge of this sequential pattern can then offer the red-jacket buyer a coupon or other enticement to attract the prospective gray-skirt buyer.

Classifying

Classifying divides predefined classes (e.g., types of customers) into mutually exclusive groups, such that the members of each group are as *close* as possible to one another, and different groups are as *far* as possible from one another, where distance is measured with respect to specific predefined variables. The classification of groups is done before data analysis. Thus, based on sales, customers may be first categorized as *infrequent, occasional,* and *frequent.* A classifier could be used to identify those attributes, from a given set, that discriminate among the three types of customers. For example, a classifier might identify frequent customers as those with incomes above $50,000 and having two or more children. Classification functions have been used extensively in applications such as credit risk analysis, portfolio selection, health risk analysis, and image and speech recognition. Thus, when a new customer is recruited, the firm can use the classifying function to determine the customer's sales potential and accordingly tailor its market to that person.

Clustering

Whereas classifying starts with predefined categories, clustering starts with just the data and discovers the *hidden* categories. These categories are derived from the data. Clustering divides a dataset into mutually exclusive groups such that the members of each group are as *close* as possible to one another, and different groups are as *far* as possible from one another, where distance is measured with respect to all available variables. The goal of clustering is to identify categories. Clustering could be used, for instance, to identify natural groupings of customers by processing all the available data on them. Examples of applications that can use clustering functions are market segmentation, discovering affinity groups, and defect analysis.

Prediction

Prediction calculates the future value of a variable. For example, it might be used to predict the revenue value of a new customer based on that person's demographic variables.

These various data mining techniques can be used together. For example, a sequence pattern analysis could identify potential customers (e.g., red jacket leads to gray skirt), and then classifying could be used to distinguish between those prospects who are converted to customers and those who are not (i.e., did not follow the sequential pattern of buying a gray skirt). This additional analysis should enable the retailer to refine its marketing strategy further to increase the conversion rate of red-jacket customers to gray-skirt purchasers.

Data mining technologies

Data miners use technologies that are based on statistical analysis and data visualization.

Decision trees

Tree-shaped structures can be used to represent decisions and rules for the classification of a dataset. As well as being easy to understand, tree-based models are suited to selecting important variables and are best when many of the predictors are irrelevant. A decision tree, for example, can be used to assess the risk of a prospective renter of an apartment.

A decision tree

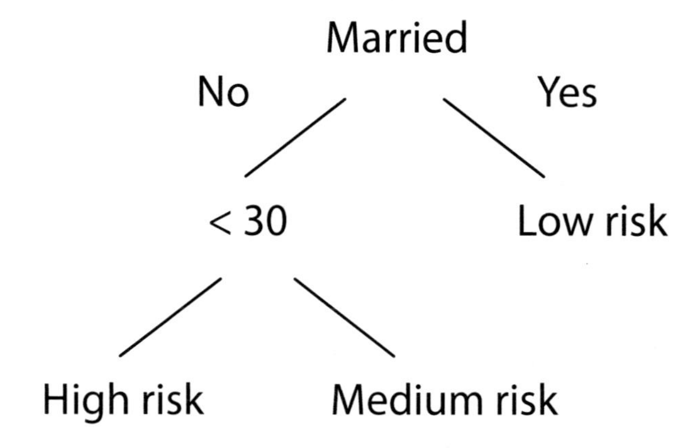

Genetic algorithms

Genetic algorithms are optimization techniques based on the concepts of biological evolution, and use processes such as genetic combination, mutation, and natural selection. Possible solutions for a problem compete with each other. In an evolutionary struggle of the survival of the fittest, the best solution survives the battle. Genetic algorithms are suited for optimization problems with many candidate variables (e.g., candidates for a loan).

K-nearest-neighbor method

The nearest-neighbor method is used for clustering and classification. In the case of clustering, the method first plots each record in *n*-dimensional space, where *n* attributes are used in the analysis. Then, it adjusts the weights for each dimension to cluster together data points with similar goal features. For instance, if the goal is to identify customers who frequently switch phone companies, the *k*-nearest-neighbor method would adjust weights for relevant variables (such as monthly phone bill and percentage of non-U.S. calls) to cluster switching customers in the same neighborhood. Customers who did not switch would be clustered some distance apart.

Neural networks

A neural network, mimicking the neurophysiology of the human brain, can learn from examples to find patterns in data and classify data. Although neural networks can be used for classification, they must first be trained to recognize patterns in a sample dataset. Once trained, a neural network can make predictions from new data. Neural networks are suited to combining information from many predictor variables; they work well when many of the predictors are partially redundant. One shortcoming of a neural network is that it can be viewed as a black box with no explanation of the results provided. Often managers are reluctant to apply models they do not understand, and this can limit the applicability of neural networks.

Data visualization

Data visualization can make it possible for the analyst to gain a deeper, intuitive understanding of data. Because they present data in a visual format, visualization tools take advantage of our capability to discern visual patterns rapidly. Data mining can enable the analyst to focus attention on important patterns and trends and explore these in depth using visualization techniques. Data mining and data visualization work especially well together.

SQL-99 and OLAP

SQL-99 includes extensions to the `GROUP BY` clause to support some of the data aggregation capabilities typically required for OLAP. Prior to SQL-99, the following questions required separate queries:

1. Find the total revenue.
2. Report revenue by location.
3. Report revenue by channel.
4. Report revenue by location and channel.

Skill builder

Write SQL to answer each of the four preceding queries using the exped table, which is a sample of 1,000 sales transactions from The Expeditioner.

Writing separate queries is time-consuming for the analyst and is inefficient because it requires multiple passes of the table. SQL-99 introduced `GROUPING SETS`, `ROLLUP`, and `CUBE` as a means of getting multiple answers from a single query and addressing some of the aggregation requirements necessary for OLAP.

Grouping sets

The `GROUPING SETS` clause is used to specify multiple aggregations in a single query and can be used with all the SQL aggregate functions. In the following SQL statement, aggregations by location and channel are computed. In effect, it combines questions 2 and 3 of the preceding list.

```
SELECT location, channel, SUM(revenue)
  FROM exped
    GROUP BY GROUPING SETS (location, channel);
```

location	channel	revenue
null	Catalog	108762
null	Store	347537
null	Web	27166
London	null	214334
New York	null	39123
Paris	null	143303
Sydney	null	29989
Tokyo	null	56716

The query sums revenue by channel or location. The null in a cell implies that there is no associated location or channel value. Thus, the total revenue for catalog sales is 108,762, and that for Tokyo sales is 56,716.

Although `GROUPING SETS` enables multiple aggregations to be written as a single query, the resulting output is hardly elegant. It is not a relational table, and thus a view based on `GROUPING SETS` should not be used as a basis for further SQL queries.

Rollup

The `ROLLUP` option supports aggregation across multiple columns. It can be used, for example, to cross-tabulate revenue by channel and location.

```
SELECT location, channel, SUM(revenue)
```

```
    FROM exped
        GROUP BY ROLLUP (location, channel);
```

location	channel	revenue
null	null	483465
London	null	214334
New York	null	39123
Paris	null	143303
Sydney	null	29989
Tokyo	null	56716
London	Catalog	50310
London	Store	151015
London	Web	13009
New York	Catalog	8712
New York	Store	28060
New York	Web	2351
Paris	Catalog	32166
Paris	Store	104083
Paris	Web	7054
Sydney	Catalog	5471
Sydney	Store	21769
Sydney	Web	2749
Tokyo	Catalog	12103
Tokyo	Store	42610
Tokyo	Web	2003

In the columns with null for location and channel, the preceding query reports a total revenue of 483,465. It also reports the total revenue for each location and revenue for each combination of location and channel. For example, Tokyo Web revenue totaled 2,003.

Cube

CUBE reports all possible values for a set of reporting variables. If SUM is used as the aggregating function, it will report a grand total, a total for each variable, and totals for all combinations of the reporting variables.

```
SELECT location, channel, SUM(revenue)
    FROM exped
        GROUP BY cube (location, channel);
```

location	channel	revenue
null	Catalog	108762
null	Store	347537
null	Web	27166
null	null	483465
London	null	214334

location	channel	revenue
New York	null	39123
Paris	null	143303
Sydney	null	29989
Tokyo	null	56716
London	Catalog	50310
London	Store	151015
London	Web	13009
New York	Catalog	8712
New York	Store	28060
New York	Web	2351
Paris	Catalog	32166
Paris	Store	104083
Paris	Web	7054
Sydney	Catalog	5471
Sydney	Store	21769
Sydney	Web	2749
Tokyo	Catalog	12103
Tokyo	Store	42610
Tokyo	Web	2003

MySQL

MySQL supports a variant of CUBE. The MySQL format for the preceding query is

```
SELECT location, channel, FORMAT(SUM(revenue),0) FROM exped
   GROUP BY location, channel WITH ROLLUP;
```

Skill builder

Using MySQL's ROLLUP capability, report the sales by location and channel for each item.

The SQL-99 extensions to GROUP BY are useful, but they certainly do not give SQL the power of a multidimensional database. It would seem that CUBE could be used as the default without worrying about the differences among the three options.

Conclusion

Data management is a rapidly evolving discipline. Where once the spotlight was clearly on TPSs and the relational model, there are now multiple centers of attention. In an information economy, the knowledge to be gleaned from data collected by routine transactions can be an important source of competitive advantage. The more an organization can learn about its customers by studying their behavior, the more likely it can provide superior products and services to retain existing customers and lure prospective buyers. As a result, data managers now have the dual responsibility of administering databases that keep the organization in business today and tomorrow. They must now master the organizational intelligence technologies described in this chapter.

Summary

Organizations recognize that data are a key resource necessary for the daily operations of the business and its future success. Recent developments in hardware and software have given organizations the capability to store and process vast collections of data. Data warehouse software supports the creation and management of huge data stores. The choice of architecture, hardware, and software is critical to establishing a data warehouse. The two approaches to exploiting data are verification and discovery. DSS and OLAP are mainly data verification methods. Data mining, a data discovery approach, uses statistical analysis techniques to discover *hidden* relationships. The relational model was not designed for OLAP, and MDDB is the appropriate data store to support OLAP. MDDB design is based on recognizing variable and identifier dimensions. MDX is a language for interrogating an MDDB. SQL-99 includes extensions to `GROUP BY` to improve aggregation reporting.

Key terms and concepts

Association
Classifying
Cleaning
Clustering
`CUBE`
Continuous variable
Database marketing
Data mining
Data visualization
Data warehouse
Decision support system (DSS)
Decision tree
Discovery
Drill-down
Drill-through
Extraction
Genetic algorithm
`GROUPING SETS`
Hypercube
Identifier dimension
Information systems cycle
K-nearest-neighbor method
Loading
Management information system (MIS)
Metadata
Multidimensional database (MDDB)
Multidimensional expressions (MDX)
Neural network
Nominal variable
Object-relational
Online analytical processing (OLAP)
Operational data store (ODS)
Ordinal variable
Organizational intelligence
Prediction
Relational OLAP (ROLAP)
`ROLLUP`
Rotation
Scheduling
Sequential pattern
Star model
Transaction processing system (TPS)
Transformation
Variable dimension
Verification

References and additional readings

Codd, E. F., S. B. Codd, and C. T. Salley. 1993. Beyond decision support. *Computerworld*, 87–89.

Thomsen, E. 1997. *OLAP solutions: Building multidimensional information systems*. New York, NY: Wiley.

Spofford, George. 2001. *MDX solutions*. New York, NY: Wiley,

Exercises

1. Identify data captured by a TPS at your university. Estimate how much data are generated in a year.
2. What data does your university need to support decision making? Does the data come from internal or external sources?
3. If your university were to implement a data warehouse, what examples of dirty data might you expect to find?
4. How frequently do you think a university should refresh its data warehouse?
5. Write five data verification questions for a university data warehouse.
6. Write five data discovery questions for a university data warehouse.
7. Imagine you work as an analyst for a major global auto manufacturer. What techniques would you use for the following questions?
    a. How do sports car buyers differ from other customers?
    b. How should the market for trucks be segmented?
    c. Where does our major competitor have its dealers?
    d. How much money is a dealer likely to make from servicing a customer who buys a luxury car?
    e. What do people who buy midsize sedans have in common?
    f. What products do customers buy within six months of buying a new car?
    g. Who are the most likely prospects to buy a luxury car?
    h. What were last year's sales of compacts in Europe by country and quarter?
    i. We know a great deal about the sort of car a customer will buy based on demographic data (e.g., age, number of children, and type of job). What is a simple visual aid we can provide to sales personnel to help them show customers the car they are most likely to buy?
8. An international airline has commissioned you to design an MDDB for its marketing department. Choose identifier and variable dimensions. List some of the analyses that could be performed against this database and the statistical techniques that might be appropriate for them.
9. A telephone company needs your advice on the data it should include in its MDDB. It has an extensive relational database that captures details (e.g., calling and called phone numbers, time of day, cost, length of call) of every call. As well, it has access to extensive demographic data so that it can allocate customers to one of 50 lifestyle categories. What data would you load into the MDDB? What aggregations would you use? It might help to identify initially the identifier and variable dimensions.
10. What are the possible dangers of data mining? How might you avoid these?
11. Download the file exped.xls from the book's web site and open it as a spreadsheet in LibreOffice. This file is a sample of 1,000 sales transactions for The Expeditioner. For each sale, there is a row

recording when it was sold, where it was sold, what was sold, how it was sold, the quantity sold, and the sales revenue. Use the PivotTable Wizard (Data>PivotTable) to produce the following report:

Sum of REVENUE	HOW			
WHERE	Catalog	Store	Web	Grand total
London	50,310	151,015	13,009	214,334
New York	8,712	28,060	2,351	39,123
Paris	32,166	104,083	7,054	143,303
Sydney	5,471	21,769	2,749	29,989
Tokyo	12,103	42,610	2,003	56,716
Grand Total	108,762	347,537	27,166	483,465

Continue to use the PivotTable Wizard to answer the following questions:

a. What was the value of catalog sales for London in the first quarter?

b. What percent of the total were Tokyo web sales in the fourth quarter?

c. What percent of Sydney's annual sales were catalog sales?

d. What was the value of catalog sales for London in January? Give details of the transactions.

e. What was the value of camel saddle sales for Paris in 2002 by quarter?

f. How many elephant polo sticks were sold in New York in each month of 2002?

14. Introduction to R

Statistics are no substitute for judgment

Henry Clay, U.S. congressman and senator

Learning objectives

Students completing this chapter will:

- Be able to use R for file handling and basic statistics;
- Be competent in the use of RStudio.

The R project

The R project supports ongoing development of R, a free software environment for statistical computing, data visualization, and data analytics. It is a highly-extensible platform, the R programming language is object-oriented, and R runs on the common operating systems. There is evidence to indicate that adoption of R has grown in recent years, and is now the most popular analytics platform.[33]

RStudio is a commonly used integrated development environment (IDE) for R. It contains four windows. The upper-left window contains scripts, one or more lines of R code that constitute a task. Scripts can be saved and reused. It is good practice to save scripts as you will find you can often edit an existing script to meet the needs of a new task. The upper-right window provides details of all datasets created. It also useful for importing datasets and reviewing the history of R commands executed. The lower-left window displays the results of executed scripts. If you want to clear this window, then press control-L. The lower-right window can be used to show files on your system, plots you have created by executing a script, packages installed and loaded, and help information.

33 http://r4stats.com/articles/popularity/

RStudio interface

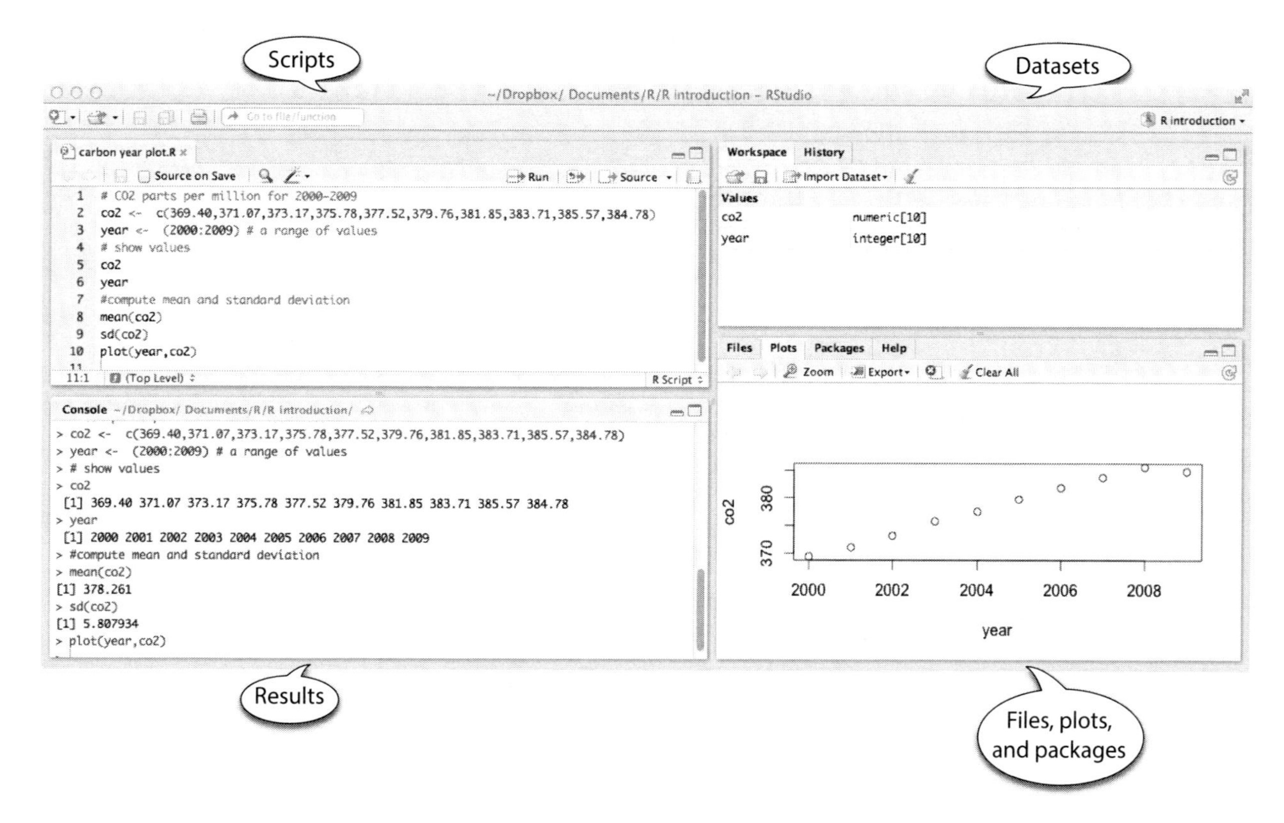

Creating a project

It usually makes sense to store all R scripts and data in the same folder or directory. Thus, when you first start RStudio, create a new project.

Project > Create Project...

RStudio remembers the state of each window, so when you quit and reopen, it will restore the windows to their prior state. You can also open an existing project, which sets the path to the folder for that project. As a result, all saved scripts and files during a session are stored in that folder.

Scripting

A script is a set of R commands. You can also think of it as a short program.

```
# CO2 parts per million (ppm) for 2000-2009
co2 <-
c(369.40,371.07,373.17,375.78,377.52,379.76,381.85,383.71,385.57,384.78)
year <-  (2000:2009) # a range of values
# show values
co2
year
# compute mean and standard deviation
mean(co2)
sd(co2)
```

```
plot(year,co2)
```

The previous script

- Creates an object co2 with the values 369.40, 371.07, … , 348.78.
- Creates an object year with values 2000 through 2009.
- Displays in the lower-left window the values stored in these two objects.
- Computes the mean for each variable.
- Creates an x-y plot of year and co2, which is shown in the lower-right window.

Note the use of <- for assigning values to an object and that c is short for combine in the expression:[34]

```
co2 <- c(369.40,371.07,373.17,375.78,377.52,379.76,381.85,383.71,385.57,384.78)
```

Skill builder

Plot kWh per square foot by year for the following University of Georgia data.

year	square feet	kWh
2007	14,214,216	2,141,705
2008	14,359,041	2,108,088
2009	14,752,886	2,150,841
2010	15,341,886	2,211,414
2011	15,573,100	2,187,164
2012	15,740,742	2,057,364

```
# Data in R format
year <- (2007:2012)
sqft <- c(14214216, 14359041, 14752886, 15341886, 15573100, 15740742)
kwh <- c(2141705, 2108088, 2150841, 2211414, 2187164, 2057364)
```

[34] You might want to set up a shortcut for '<-' using a text expansion utility for your OS. Mine is ';;'

Smart editing

It is not uncommon to find that a dataset you would like to use is in electronic format, but not in a format that matches your need. In most cases, you can use a word processor, spreadsheet, or text editor to reformat the data. In the case of the data in the previous table, here is a recipe for reformatting the data.

1. Copy each column to a word processor
2. Use the convert table to text command
3. Search and replace commas with nulls (i.e, “”)
4. Search and replace returns with commas
5. Edit to put R code around numbers

In some cases, you might find it quicker to copy a table to a spreadsheet, select each column within the spreadsheet, and then proceed as described above. This technique works well when the original table is in a pdf document.

Datasets

An R dataset is the familiar table of the relational model. There is one row for each observation, and the columns contain the observed values or facts about each observation. R supports multiple data structures and data types.

Vector

A vector is a single row table where data are all of the same type (e.g., character, logical, numeric). In the following sample code, two numeric vectors are created.

```
co2 <-
c(369.40,371.07,373.17,375.78,377.52,379.76,381.85,383.71,385.57,384.78)
year <-  (2000:2009)
co2[2] # show the second value
```

Matrix

A matrix is a table where all data are of the same type. Because it is a table, a matrix has two dimensions, which need to be specified when defining the matrix. The sample code creates a matrix with 4 rows and 3 columns, as the results of the executed code illustrate.

```
m <- matrix(1:12, nrow=4,ncol=3)
m[4,3] # show the value in row 4, column 3
```

```
     [,1] [,2] [,3]
[1,]    1    5    9
[2,]    2    6   10
```

```
[3,]    3    7   11
[4,]    4    8   12
```

Skill builder

Create a matrix with 6 rows and 3 columns containing the numbers 1 through 18.

Array

An array is a multidimensional table. It extends a matrix beyond two dimensions. Review the results of running the following code by displaying the array created.

```
a <-  array(letters[seq(1:24)], c(4,3,2))
a[1,1,1] # show the first value in the array
```

Data frame

While vectors, matrices, and arrays are all forms of a table, they are restricted to data of the same type (e.g., numeric). A data frame, like a relational table, can have columns of different data types. The sample code creates a data frame with character and numeric data types.

```
gender <- c("m","f","f")
age <- c(5,8,3)
df <-  data.frame(gender,age)
# show some data frame values
df[1,2] # a cell
df[1,] # a row
df[,2] # a column
```

List

The most general form of data storage is a list, which is an ordered collection of objects. It permits you to store a variety of objects together under a single name. In other words, a list is an object that contains other objects. Retrieve a list member with a *single square bracket* []. To reference a list member directly, use a *double square bracket* [[]].

```
l <-  list(co2,m,df)
# show a list member
l[3] # retrieves list member
l[[3]] # reference a list member
l[[1]][2] # second element of list 1
```

Logical operators

R supports the common logical operators, as shown in the following table.

Logical operator	Symbol
EQUAL	==
AND	&
OR	|

Logical operator	Symbol
NOT	!

Object

In R, an object is anything that can be assigned to a variable. It can be a constant, a function, a data structure, a graph, a times series, and so on. You find that the various packages in R support the creation of a wide range of objects and provide functions to work with these and other objects. A variable is a way of referring to an object. Thus, we might use the variable named *l* to refer to the list object defined in the preceding subsection.

Types of data

R can handle the four types of data: nominal, ordinal, interval, and ratio. Nominal data, typically character strings, are used for classification (e.g., high, medium, or low). Ordinal data represent an ordering and thus can be sorted (e.g., the seeding or ranking of players for a tennis tournament). The intervals between ordinal data are not necessarily equal. Thus, the top seeded tennis play (ranked 1) might be far superior to the second seeded (ranked 2), who might be only marginally better than the third seed (ranked 3). Interval and ratio are forms of measurement data. The interval between the units of measure for interval data are equal. In other words, the distance between 50cm and 51cm is the same as the distance between 105cm and 106cm. Ratio data have equal intervals and a natural zero point. Time, distance, and mass are examples of ratio scales. Celsius and Fahrenheit are interval data types, but not ratio, because the zero point is arbitrary. As a result, 10° C is not twice as hot as 5° C. Whereas, Kelvin is a ratio data type because nothing can be colder than 0° K, a natural zero point.

In R, nominal and ordinal data types are also known as *factors*. Defining a column as a factor determines how its data are analyzed and presented. By default, factor levels for character vectors are created in alphabetical order, which is often not what is desired. To be precise, specify using the levels option.

```
rating <-   c('high','medium','low')
rating <-  factor(rating, order=T, levels = c('high','medium','low'))
```

Thus, the preceding code will result in changing the default reporting of factors from alphabetical order (i.e., high, low, and medium) to listing them in the specified order (i.e., high, medium, and low).

Missing values

Missing values in R are represented as NA, meaning not available. Infeasible values, such as the result of dividing by zero, are indicated by NaN, meaning not a number. Any arithmetic expression or function operating on data containing missing values will return NA. Thus `sum(c(1,NA,2))` will return NA.

To exclude missing values from calculations, use the option `NA.RM = T`, which specifies the removal of missing values prior to calculations. Thus, `sum(c(1,NA,2),NA.RM=T)` will return 3.

You can remove rows with missing data by using `na.omit()`, which will delete those rows containing missing values.

```
gender <- c("m","f","f","f")
age <- c(5,8,3,NA)
df <-  data.frame(gender,age)
```

```
df2 <-  na.omit(df)
```

Packages

A major advantage of R is that the basic software can be easily extended by installing additional packages, of which more than 6,000 exist. You can consult the R package directory to help find a package that has the functions you need.[35] RStudio has an interface for finding and installing packages. See the Packages tab on RStudio's lower-right window.

Before running a script, you need to indicate which packages it needs, beyond the default packages that are automatically loaded. The require or library statement specifies that a package is required for execution. The following example uses the measurements package to handle the conversion of Fahrenheit to Celsius. The package's documentation provides details of how to use its various conversion options.

```
library(measurements) # previously installed
# convert F to C
conv_unit(100,'F','C')
```

Skill builder

Install the measurements package and run the preceding code.

Reading a file

Files are the usual form of input for R scripts. Fortunately, R can handle a wide variety of input formats, including text (e.g., CSV), statistical package (e.g., SAS), and XML. A common approach is to use a spreadsheet to prepare a data file, export it as CSV, and then read it into R.

Files can be read from the local computer on which R is installed or the Internet, as the following sample code illustrates. We will use the readr library for handling files, so you will need to install it before running the following code.

```
library(readr)
# read a local file (this will not work on your computer)
t <- read_delim("Documents/R/Data/centralparktemps.txt", delim=“,")
```

You can also read a remote file using a URL.

```
library(readr)
# read using a URL
url <-  'http://people.terry.uga.edu/rwatson/data/centralparktemps.txt'
t <- read_delim(url, delim=',')
```

You must define the separator for data fields with the delim keyword (e.g., for a tab use `delim='\t'`).

35 http://cran.r-project.org/web/packages/

Learning about a file

After reading a file, you might want to learn about its contents. First, you can click on the file's name in the top-right window. This will open the file in the top-left window. If it is a long file, only a portion, such as the first 1000 rows, will be displayed. Second, you can execute some R commands, as shown in the following code, to show the first few and last few rows. You can also report the dimensions of the file, its structure, and the type of object.

```
library(readr)
url <-  'http://people.terry.uga.edu/rwatson/data/centralparktemps.txt'
t <- read_delim(url, delim=',')
head(t) #  first few rows
tail(t) #  last few rows
dim(t) # dimension
str(t) # structure of a dataset
class(t) #type of object
```

Referencing columns

Columns within a table are referenced by using the format tablename$columnname. This is similar to the qualification method used in SQL. The following code shows a few examples. It also illustrates how to add a new column to an existing table.

```
library(measurements)
library(readr)
url <-  'http://people.terry.uga.edu/rwatson/data/centralparktemps.txt'
t <- read_delim(url, delim=',')
# qualify with table name to reference a column
mean(t$temperature)
max(t$year)
range(t$month)
# create a new column with the temperature in Celsius
t$Ctemp = round(conv_unit(t$temperature,'F','C'),1) # round to one decimal
```

Recoding

Some analyses might be facilitated by the recoding of data. For instance, you might want to split a continuous measure into two categories. Imagine you decide to classify all temperatures greater than or equal to 25°C as 'hot' and the rest as 'other.' Here are the R command to create a new column in table t called Category.

```
t$Category <- ifelse(t$Ctemp >= 30, 'Hot','Other')
```

Deleting a column

You can delete a column by setting each of its values to NULL.

```
t$Category <-  NULL
```

Reshaping data

Data are not always in the shape that you want. For example, you might have a spreadsheet that, for each year, lists the quarterly observations (i.e., year and four quarterly values in one row). For analysis, you typically need to have a row for each distinct observation (e.g., year, quarter, value). *Gathering* converts a document from what is commonly called *wide* to *narrow* format. It is akin to normalization in that the new table has year and quarter as the identifier of each observation.

Once the file has been read, you create appropriate column names for the converted file using the colnames() function applied to the file t. Thus, the input file has column names of Q1, Q2, etc, but it makes sense for these to be integers in the new file in a new column called quarter.

The gather command specifies the file to be gathered, the column names of the new file, and the columns of the input file to be gathered (i.e., the four quarters in columns 2 through 5). Note that you also need convert the column quarter from character to integer.

Reshaping data with gathering and spreading

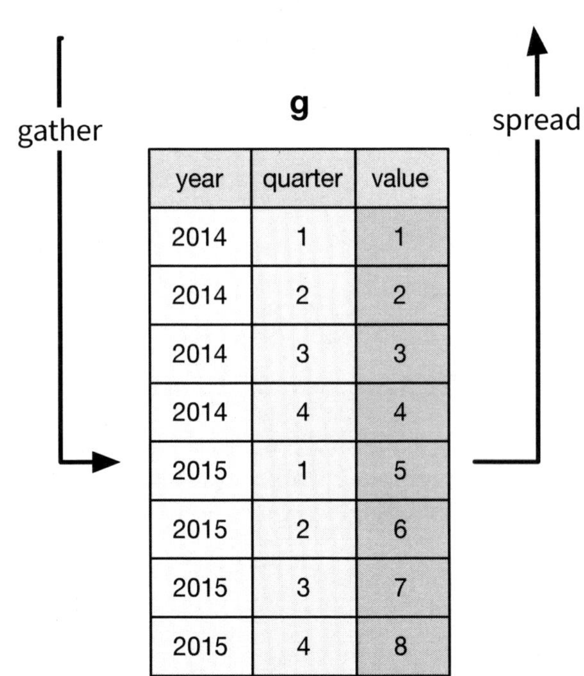


```
library(readr)
library(tidyr)
url <-  'http://people.terry.uga.edu/rwatson/data/gatherExample.csv'
t <- read_csv(url)
```

```
t
colnames(t) <- c('year',1:4)
t
# gather with data in columns 2 through 5
g <- gather(t,'quarter','value',2:5)
g$quarter <- as.integer(g$quarter)
g

# spread
s <- spread(g,quarter,value)
s
colnames(s) <- c('year', 'Q1','Q2','Q3','Q4')
s
```

Spreading takes a narrow file and converts it to wide or spreadsheet format. This is the reverse of gathering.

```
s <- spread(g,quarter,value)
s
colnames(s) <- c('year', 'Q1','Q2','Q3','Q4')
s
```

Writing files

R can write data in a few file formats. We just focus on text format in this brief introduction. The following code illustrates how to create a new column containing the temperature in C and renaming an existing column. The revised table is then written as a csv text file to the project's folder.

```
library(measurements)
library(readr)
url <-  'http://people.terry.uga.edu/rwatson/data/centralparktemps.txt'
t <- read_delim(url, delim=',')
# compute Celsius and round to one decimal place
t$Ctemp = round(conv_unit(t$temperature,'F','C'),1)
colnames(t)[3] <-  'Ftemp' # rename third column to indicate Fahrenheit
write_csv(t,"centralparktempsCF.txt")
```

Data manipulation with dplyr

The dplyr package[36] provides functions for efficiently manipulating data sets. By providing a series of basic data handling functions and use of the pipe function (%>%),[37] dplyr implements a grammar for data manipulation. The pipe function is used to pass data from one operation to the next.

---

[36] http://cran.rstudio.com/web/packages/dplyr/vignettes/introduction.html

[37] You might want to set up a shortcut for '%>%' using a text expansion utility for your OS. Mine is '..'

Some dplyr functions

Function	Purpose
filter()	Select rows
select()	Select columns
arrange()	Sort rows
summarize()	Compute a single summary statistic
group_by()	Pair with summarize() to analyze groups within a dataset
inner_join()	Join two tables
mutate()	Create a new column

Here are some examples using dplyr with data frame t.

```
library(dplyr)
library(readr)
url <-  'http://people.terry.uga.edu/rwatson/data/centralparktemps.txt'
t <- read_delim(url, delim=',')
# a row subset
trow <-  filter(t, year == 1999)
# a column subset
tcol <-  select(t, year)
```

The following example illustrates application of the pipe function. The data frame t is piped to select(), and the results of select method passed onto filter(). The final output is stored in trowcol, which contains year, month, and Celsius temperature for the years 1990 through 1999.

```
# a combo subset and use of the pipe function
trowcol <-  t %>% select(year, month, temperature) %>% filter(year > 1989 & year <
2000)
```

Sorting

You can also use dplyr for sorting.

```
t <-  arrange(t, desc(year),month)
```

Skill builder

- View the web page of yearly CO_2 emissions (million metric tons) since the beginning of the industrial revolution.
- Create a new text file using R
- Clean up the file for use with R and save it as CO2.txt
- Import (Import Dataset) the file into R
- Plot year versus CO_2 emissions

Summarizing data

The dplyr function can be used for summarizing data in a specified way (e.g., mean, minimum, standard deviation). In the sample code, a file containing the mean temperature for each year is created. Notice the use of the pipe function to first group the data by year and then compute the mean for each group.

```
library(dplyr)
url <- 'http://people.terry.uga.edu/rwatson/data/centralparktemps.txt'
t <- read_delim(url, delim=',')
w <-  t  %>% group_by(year) %>% summarize(averageF = mean(temperature))
```

Adding a column

The following example shows how to add a column and compute its average.

```
# add column
t <-  mutate(t, CTemp = (temperature-32)*5/9)
# summarize
summarize(t, mean(CTemp))
```

Skill builder

Create a file with the maximum temperature for each year.

Merging files

If there is a common column in two files, then they can be merged using dplyr::inner_join().[38] This is the same as joining two tables using a primary key and foreign key match. In the following code, a file is created with the mean temperature for each year, and it is merged with CO2 readings for the same set of years.

```
library(dplyr)
Library(readr)
url <-  'http://people.terry.uga.edu/rwatson/data/centralparktemps.txt'
t <- read_delim(url, delim=',')
# average monthly temp for each year
a <-  t %>% group_by(year) %>% summarize(mean = mean(temperature))
# read yearly carbon data (source: http://co2now.org/Current-CO2/CO2-Now/noaa-mauna-
loa-co2-data.html)
url <-  'http://people.terry.uga.edu/rwatson/data/carbon1959-2011.txt'
carbon <- read_delim(url,  delim=',')
m <-  inner_join(a,carbon)
head(m)
```

Data manipulation with sqldf

The sqldf package enables use of the broad power of SQL to extract rows or columns from a data frame to meet the needs of a particular analysis. It provides essentially the same data manipulation capabilities as dplyr. The following example illustrates use of sqldf.

---

[38] This is the R notation for indicating the package (dplyr) to which a method (inner_join) belongs.

```
library(sqldf)
library(readr)
options(sqldf.driver = "SQLite") # to avoid a conflict with RMySQL
url <-  'http://people.terry.uga.edu/rwatson/data/centralparktemps.txt'
t <- read_delim(url, delim=',')
# a row subset
trowcol <-  sqldf("select year, month, temperature from t where year > 1989 and year <
2000")
```

However, sqldf does not enable you to embed R functions within an SQL command. For this reason, dplyr is the recommended approach for data manipulation.

Correlation coefficient

A correlation coefficient is a measure of the covariation between two sets of observations. In this case, we are interested in whether there is a statistically significant covariation between temperature and the level of atmospheric CO_2.[39]

```
cor.test(m$mean, m$CO2)
```

The following results indicate that a correlation of .40 is a statistically significant as the p-value is less than 0.05, the common threshold for significance testing. Thus, we conclude that, because there is a small chance ($p = .002997$) of observing by chance such a value for the correlation coefficient, there is a relationship between mean temperature and the level of atmospheric CO_2. Given that global warming theory predicts an increase in temperature with an increase in atmospheric CO_2, we can also state that the observations support this theory. In other words, an increase in CO_2 increases temperatures in Central Park.

```
        Pearson's product-moment correlation


data:  m$mean and m$CO2
t = 3.1173, df = 51, p-value = 0.002997
95 percent confidence interval:
 0.1454994 0.6049393
sample estimates:
      cor
0.4000598
```

When reporting correlation coefficients, you can you use the terms small, moderate, and large in accordance with the values specified in following table.

Correlation coefficient	Effect size
.10 - .30	Small
.30 - .50	Moderate

[39] As I continually revise data files to include the latest observations, your answer might differ for this and other analyses.

Correlation coefficient	Effect size
> .50	Large

If we want to understand the nature of the relationship, we could fit a linear model.

```
lm(m$mean ~ m$CO2)
summary(mod)
```

The following results indicate a linear model is significant ($p < .05$), and it explains 14.36% (adjusted multiple R-squared) of the variation between temperature and atmospheric CO_2. The linear equation is

temperature = 48.29 + 0.019208* CO_2

As CO_2 emissions are measured in parts per millions (ppm), an increase of 1.0 ppm predicts an annual mean temperature increase in Central Park of .01920 F°. Currently CO_2 emissions are increasing at about 2.0 ppm per year.

As a linear model explains about 14% of the variation, this suggests that there might other variables that should be considered (e.g., level of volcanic activity) and that the relationship might not be linear.

```
Coefficients:
            Estimate Std. Error t value Pr(>|t|)
(Intercept) 48.291319   2.149937  22.462   <2e-16 ***
m$CO2        0.019208   0.006162   3.117    0.003 **
---
Signif. codes:  0 '***' 0.001 '**' 0.01 '*' 0.05 '.' 0.1 ' ' 1

Residual standard error: 1.016 on 51 degrees of freedom
Multiple R-squared:   0.16,   Adjusted R-squared:  0.1436
F-statistic: 9.718 on 1 and 51 DF,  p-value: 0.002997
```

Database access

The DBI package provides a convenient way for a direct connection between R and a relational database, such as MySQL or PostgreSQL. Once the connection has been made, you can run an SQL query and load the results into a R data frame.

The dbConnect() function makes the connection. You specify the type of relational database, url, database name, userid, and password, as shown in the following code.[40]

```
library(DBI)
conn <- dbConnect(RMySQL::MySQL(), "richardtwatson.com", dbname="Weather",
user="db2", password="student")
# Query the database and create file t for use with R
t <- dbGetQuery(conn,"select * from record;")
```

[40] You will likely get a warning message of the form "unrecognized MySQL field type 7 in column 0 imported as character," because R does does recognize the first column as a timestamp. You can ignore it, but later you might need to convert the column to R's timestamp format.

```
head(t)
            timestamp airTemp humidity precipitation
1 2010-01-01 03:00:00      44       93             0
2 2010-01-01 04:00:00      44       89             0
3 2010-01-01 05:00:00      43       89             0
4 2010-01-01 06:00:00      42       83             0
5 2010-01-01 07:00:00      41       86             0
6 2010-01-01 08:00:00      40       79             0
```

For security reasons, it is not a good idea to put database access details in your R code. They should be hidden in a file. I recommend that you create a csv file within your R code folder to containing database access parameters. First, create a new directory or folder (File > New Project > New Directory > Empty Project), called dbaccess for containing you database access files. Then, create a csvfile (Use File > New File > Text File) with the name weather_richardtwatson.csv in the newly created folder containing the following data:

```
url,dbname,user,password
richardtwatson.com,Weather,db2,student
```

The R code will now be:

```
# Database access
library(readr)
library(DBI)
url <-  'dbaccess/weather_richardtwatson.csv'
d <-  read_csv(url)
conn <- dbConnect(RMySQL::MySQL(), d$url, dbname=d$dbname, user=d$user,
password=d$password)
t <- dbGetQuery(conn,"SELECT timestamp, airTemp from record;")
head(t)
```

Despite the prior example, I will continue to show database access parameters because you need them for to run the sample code. However, in practice you should follow the security advice given.

Timestamps

Many data sets include a timestamp to indicate when an observation was recorded. A timestamp will show the data and time to the second or microsecond. The format of a standard timestamp is yyyy-mm-dd hh:mm:ss (e.g., 2010-01-31 03:05:46).

Some R functions, including read those in the lubridate package, can detect a standard format timestamp, and thus support operations for extracting parts of it, such as the year, month, day, hour, and so forth. The following example shows how to use lubridate to extract the month and year from a character string in standard timestamp format.

```
library(lubridate)
library(DBI)
conn <- dbConnect(RMySQL::MySQL(), "richardtwatson.com", dbname="Weather", user="db2",
password="student")
# Query the database and create file t for use with R
t <- dbGetQuery(conn,"select * from record;")
```

```
t$year <- year(t$timestamp)
t$month <- month(t$timestamp)
head(t)
```

Excel files

There are a number of packages with methods for reading a file. Of these, readxl seems to be the simplest. However, it can handle only files stored locally, which is the case with most of the packages examined. If the required Excel spreadsheet is stored remotely, then first download it and then store locally.

```
library(readxl)
library(httr)
# read remote file and store on disk
url <-  'http://people.terry.uga.edu/rwatson/data/GDP.xlsx'
GET(url,write_disk('temp.xlsx',overwrite = TRUE))
e <-  read_excel('temp.xlsx',sheet = 1,col_names = TRUE)
```

R resources

The vast number of packages makes the learning of R a major challenge. The basics can be mastered quite quickly, but you will soon find that many problems require special features or the data need some manipulation. Once you learn how to solve a particular problem, make certain you save the script, with some embedded comments, so you can reuse the code for a future problem. There are books that you might find useful to have in electronic format on your computer or tablet, and two of these are listed at the end of the chapter. There are, however, many more books,[41] both of a general and specialized nature. The *R Reference Card* is handy to keep nearby when you are writing scripts. I printed and laminated a copy, and it's in the top drawer of my desk. A useful website is Quick-R, which is an online reference for common tasks, such as those covered in this chapter. For a quick tutorial, you can Try R.

R and data analytics

R is a platform for a wide variety of data analytics, including

- Statistical analysis
- Data visualization
- HDFS and MapReduce
- Text mining
- Energy Informatics
- Dashboards

You have probably already completed an introductory statistical analysis course, and you can now use R for all your statistical needs. In subsequent chapters, we will discuss data visualization, HDFS and MapReduce, and text mining. Energy Informatics is concerned with optimizing energy flows, and R is an appropriate tool for analysis and optimization of energy systems. R is being used extensively to analyze climate change data.[42]

R is also a programming language. You might find that in some situations, R provides a quick method for reshaping files and performing calculations.

---

41 See http://www.r-project.org/doc/bib/R-books.html

42 http://chartsgraphs.wordpress.com/category/r-climate-data-analysis-tool/

Summary

R is a free software environment for statistical computing, data visualization, and data analytics. RStudio is a commonly used integrated development environment (IDE) for R. A script is a set of R commands. Store all R scripts and data for a project in the same folder or directory. An R dataset is a table that can be stored as a vector, matrix, array, data frame, and list. In R, an object is anything that can be assigned to a variable. R can handle the four types of data: nominal, ordinal, interval, and ratio. Nominal and ordinal data types are also known as factors. Defining a column as a factor determines how its data are analyzed and presented. Missing values are indicated by NA. R can handle a wide variety of input formats, including text (e.g., CSV), statistical package (e.g., SAS), and XML. Data can be reshaped. Gathering converts a document from what is commonly called wide to narrow format. Spreading takes a narrow file and converts it to wide format. Columns within a table are referenced using the format tablename$columname. R can write data to a file. A major advantage of R is that the basic software can be easily extended by installing additional packages. Subsetting enables selection of individual rows or columns from a table specifying row and column selection criteria. Sorting specifies the columns containing the sort keys. The aggregate function is used for summarizing data in a specified way. Files can be merged if they have a common column. RMySQL provides a direct connection between R and MySQL. Learning R is a major challenge because of the many packages available.

Key terms and concepts

Aggregate
Array
Data frame
Data type
Factor
Gather
List
Matrix
Package
R
Reshape
Script
Spread
SQL
Tibble
Subset
Vector

References

Adler, J. (2010). R in a Nutshell. Sebastopol, CA: O'Reilly Media.

Kabacoff, R. I. (2009). R in action: data analysis and graphics with R. Greenwich, CT: Manning Publications.

Exercises

1. Access people.terry.uga.edu/rwatson/data/manheim.txt which contains details of the sales of three car models: X, Y, and Z.

a. Use the table function to compute the number of sales for each type of model.

b. Use the table function to compute the number of sales for each type of sale.

c. Report the mean price for each model.

d. Report the mean price for each type of sale.

2. Use the 'Import Dataset' feature of RStudio to read http://people.terry.uga.edu/rwatson/data/electricityprices.csv, which contains details of electricity prices for a major city.[43]

 a. What is the maximum cost?

 b. What is the minimum cost?

 c. What is the mean cost?

 d. What is the median cost?

3. Read the table people.terry.uga.edu/rwatson/data/wealth.csv containing details of the wealth of various countries and complete the following exercises.

 a. Sort the table by GDP per capita.

 b. What is the average GDP per capita?

 c. Compute the ratio of US GDP per capita to the average GDP per capita.

 d. What's the correlation between GDP per capita and wealth per capita?

4. Merge the data for weather (database weather at richardtwatson.com discussed in the chapter) and electricity prices (Use RStudio's 'Import Dataset' to read http://people.terry.uga.edu/rwatson/data/electricityprices.csv) and compute the correlation between temperature and electricity price. **Hint**: MySQL might return a timestamp with decimal seconds (e.g., 2010-01-01 01:00:00.0), and you can remove the rightmost two characters using substr(),[44] so that the two timestamp columns are of the same format and length. Also, you need to ensure that the timestamps from the two data frames are of the same data type (e.g., both character).

5. Get the list of failed US banks from https://explore.data.gov/Banking-Finance-and-Insurance/FDIC-Failed-Bank-List/pwaj-zn2n.

 a. Determine how many banks failed in each state.

 b. How many banks were not acquired (hint: nrow() will count rows in a table)?

 c. How many banks were closed each year (hint: use strptime() and the lubridate package)?

6. Use Table01 of U.S. broccoli data on farms and area harvested from http://usda.mannlib.cornell.edu/MannUsda/viewDocumentInfo.do?documentID=1816. Get rid of unwanted rows to create a spreadsheet for the area harvested with one header row and the 50 states. Change cells without integer values to 0 and save the file in CSV format for reading with R.

 a. Reshape the data so that each observation contains state name, year, and area harvested.

 b. Add hectares as a column in the table. Round the calculation to two decimal places.

 c. Compute total hectares harvested each year for which data are available.

---

43 Note these prices have been transformed from the original values, but are still representative of the changes over time.

44 http://www.statmethods.net/management/functions.html

d. Save the reshaped file.

15. Data visualization

The commonality between science and art is in trying to see profoundly - to develop strategies of seeing and showing.

Edward Tufte, *The Visual Display of Quantitative Information*

Learning objectives

- Students completing this chapter will:
- Understand the principles of the grammar of graphics;
- Be able to use ggvis to create common business graphics;
- Be able to select data from a database for graphic creation;
- Be able to depict locations on a Google map.

Visual processing

Humans are particularly skilled at processing visual information because it is an innate capability, compared to reading which is a learned skill. When we evolved on the Savannah of Africa, we had to be adept at processing visual information (e.g., recognizing danger) and deciding on the right course of action (fight or flight). Our ancestors were those who were efficient visual processors and quickly detected threats and used this information to make effective decisions. They selected actions that led to survival. Those who were inefficient visual information processors did not survive to reproduce. Even those with good visual skills failed to propagate if they made poor decisions relative to detected dangers. Consequently, we should seek opportunities to present information visually and play to our evolved strength. As people vary in their preference for visual and textual information, it often makes sense to support both types of reporting.

In order to learn how to visualize data, you need to become familiar with the grammar of graphics and ggvis (an R extension for graphics). In line with the learning of data modeling and SQL, we will take an intertwined spiral approach. First we will tackle the grammar of graphics (the abstract) and then move to ggvis (the concrete). You will also learn how to take the output of an SQL query and feed it directly into ggvis. The end result will be a comprehensive set of practical skills for data visualization.

The grammar of graphics

A grammar is a system of rules for generating valid statements in a language. A grammar makes it possible for people to communicate accurately and concisely. English has a rather complex grammar, as do most languages. In contrast, computer-related languages, such as SQL, have a relatively simple grammar because they are carefully designed for expressive power, consistency, and ease of use. Each of the dialects of data modeling also has a grammar, and each of these grammars is quite similar in that they all have the same foundational elements: entities, attributes, identifiers, and relationships.

A grammar has been designed by Wilkinson[45] for creating graphics to enhance their expressiveness and comprehensiveness. From a mathematical perspective, a graph is a set of points. A graphic is a physical representation of a graph. Thus, a graph can have many physical representations, and one of the skills you need to gain is to be able to judge what is a meaningful graphic for your clients. A grammar for graphics provides you with many ways of creating a graphic, just as the grammar of English gives you many ways of

45 Wilkinson, L. (2005). *The grammar of graphics* (2nd ed.). New York: Springer.

writing a sentence. Of course, we differ in our ability to convey information in English. Similarly, we also differ in our skills in representing a graph in a visual format. The aesthetic attributes of a graph determine its ability to convey information. For a graphic, aesthetics are specified by elements such as size and color. One of the most applauded graphics is Minard's drawing in 1861 of Napoleon's advance on and retreat from Russia during 1812. The dominating aspect of the graphic is the dwindling size of the French army as it battled winter, disease, and the Russian army. The graph shows the size of the army, its location, and the direction of its movement. The temperature during the retreat is drawn at the bottom of graphic.

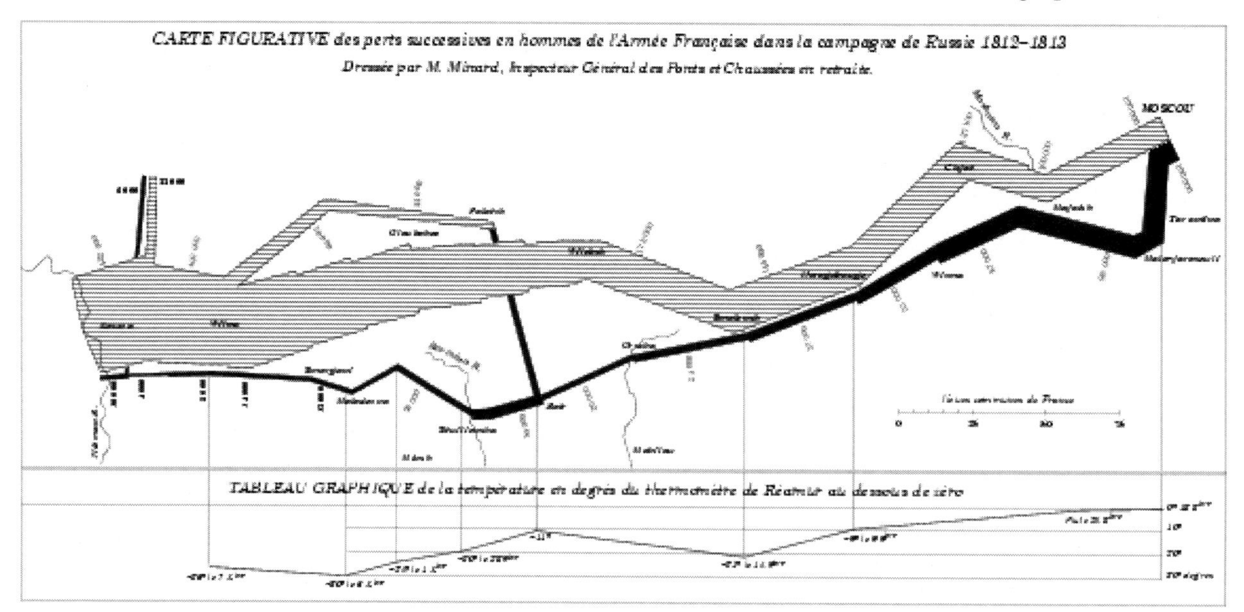

Charles Joseph Minard's graphic of Napoleon's Russian expedition in 1812

Wilkinson's grammar is based on six elements:

1. *Data*: a set of data operations that creates variables from datasets;
2. *Trans*: variable transformations;
3. *Scale*: scale transformations;
4. *Coord*: a coordinate system;
5. *Element*: a graph and its aesthetic attributes;
6. *Guide*: one or more guides.

Interest in the grammar of data visualization has increased in recent years because of the growth in data. There is ongoing search to find ways to reveal clearly the information in large data sets. Vega[46] is a recent formulation building on Wilkinson's work. In Vega, a visualization consists of basic properties of a graph (such as the width and height) and definitions of the data to visualize, scales for mapping to data to a visual form, axes for representing scales, and graphic marks (such as rectangles, circles, and arcs) for displaying the data.

[46] http://trifacta.github.io/vega/

ggvis

ggvis is an implementation of Vega in R Because it is based on a grammar, ggvis is a very powerful graphics tool for creating both simple and complex graphs. It is well-suited for generating multi-layered graphs because of its grammatical foundation. As a result, using ggvis you specify a series of steps (think of them as sentences) to create a graphic (think of it as a paragraph) to visually convey your message. ggvis is a descendant of ggplot2 and adds new features to support interactive visualization using shiny, another R package, which we will cover later.

Data

Most structured data, which is what we require for graphing, are typically stored in spreadsheets or databases. In the prior chapter introducing R, you learned how to read a spreadsheet exported as a CSV file and access a database. These are also the skills you need for reading a file containing data to be visualized.

Transformation

A transformation converts data into a format suitable for the intended visualization. In this case, we want to depict the relative change in carbon levels since pre-industrial periods, when the value was 280 ppm. Here are sample R commands.

```
# compute a new column in carbon containing the relative change in CO2
carbon$relCO2 = (carbon$CO2-280)/280
```

There are many ways that you might want to transform data. The preceding example just illustrates the general nature of a transformation. You can also think of SQL as a transformation process as it can compute new values from existing columns.

Coord

A coordinate system describes where things are located. A geopositioning system (GPS) reading of latitude and longitude describes where you are on the surface of the globe. It is typically layered onto a map so you can see where you are relative to your current location. Most graphs are plotted on a two-dimensional (2D) grid with x (horizontal) and y (vertical) coordinates. ggvis currently supports six 2D coordinate systems, as shown in the following table. The default coordinate system for most packages is Cartesian.

Coordinate systems

Name	Description
cartesian	Cartesian coordinates
equal	Equal scale Cartesian coordinates
flip	Flipped Cartesian coordinates
trans	Transformed Cartesian coordinates
map	Map projections
polar	Polar coordinates

Element

An element is a graph and its attributes. Let's start with a simple scatterplot of year against CO_2 emissions. We do this in two steps applying the ggvis approach of building a graphic by adding layers. ggvis uses the pipe function, %>%, to specify a linear sequence of data processing. The output of each process is the input to the next.

- The foundation is the ggvis function, which identifies the source of the data and what is to be plotted. In the following example, the file carbon is fed into the ggvis function, which selects year and CO2 as the x and y, respectively, dimensions of the graph
- A graph consists of a number of layers, and in the following example the points layer receives the x and y values from ggvis and plots each pair of points with a red dot.

```
library(ggvis)
library(readr)
library(dplyr)
url <- 'http://people.terry.uga.edu/rwatson/data/carbon.txt'
carbon <- read_delim(url, delim=',')
# Select year(x) and CO2(y) to create a x-y point plot
# Specify red points, as you find that aesthetically pleasing
carbon %>%
  ggvis(~year,~CO2) %>%
  layer_points(fill:='red')
# Notice how '%>%' is used for creating a pipeline of commands
```

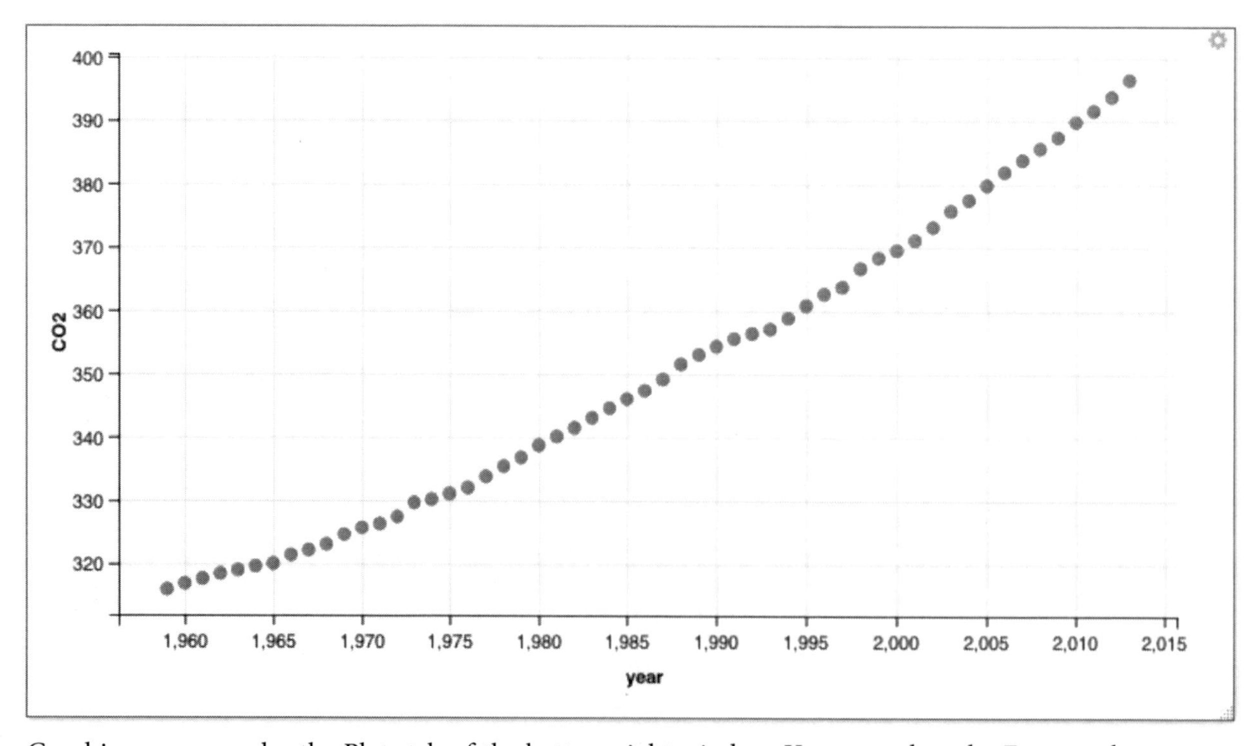

Graphics appear under the Plots tab of the bottom right window. You can select the Export tab to save a graphic to a file or the clipboard. Also, notice that you can use left and right arrows of the Plots tab to cycle through graphics created in a session.

Scale

Scales control the visualization of data. It is usually a good idea to have a zero point for the y axis so you don't distort the slope of the line. In the following example, the scale_numeric function is specifies a zero point for the y axis with the command zero=T.

```
carbon %>%
  ggvis(~year,~CO2) %>%
  layer_points(fill:='red') %>%
  scale_numeric('y',zero=T)
```

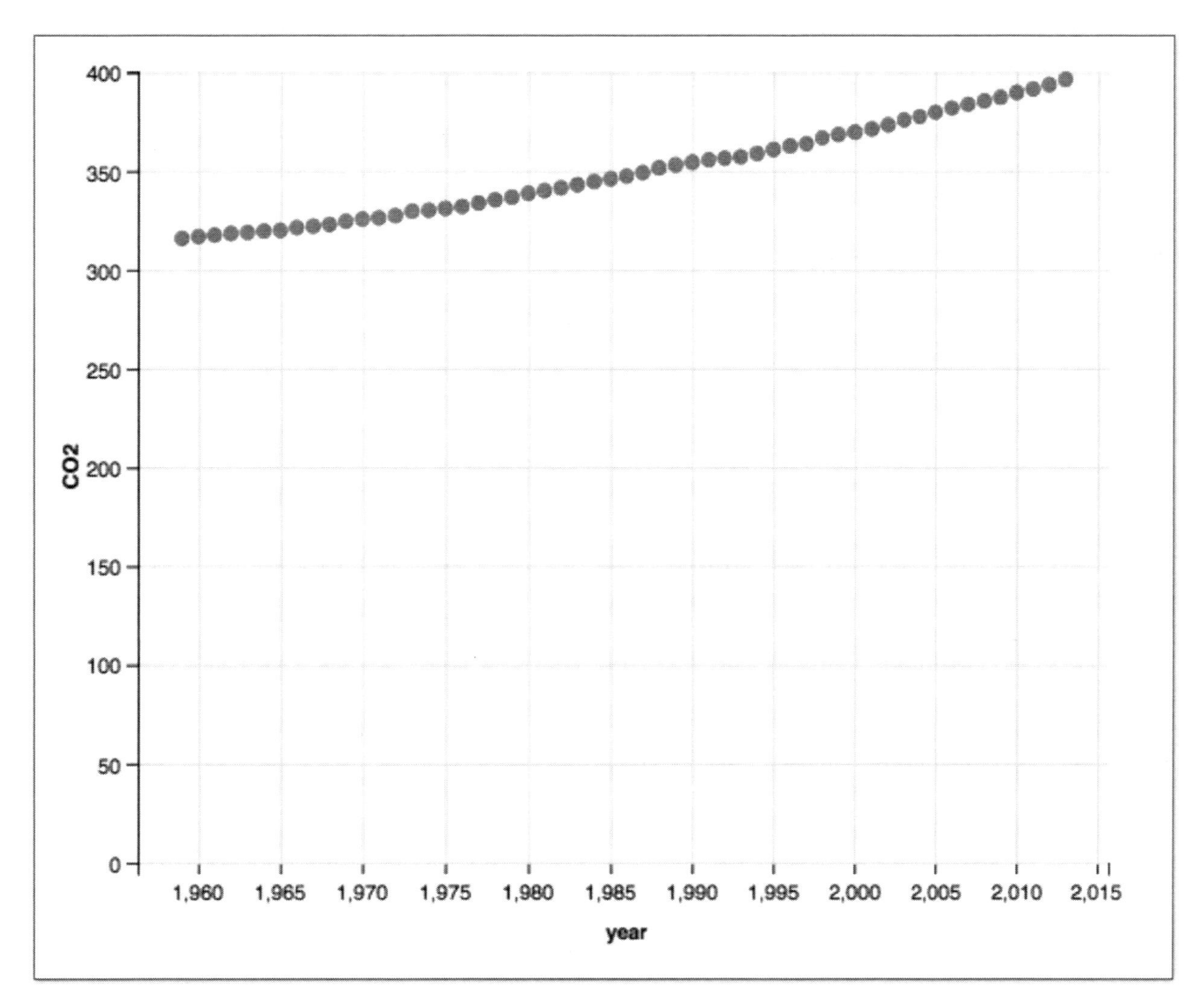

Axes and legends

Axes and legends are both forms of guides, which help the reader to understand a graphic. Readability enhancers such as axes, title, and tick marks are generated automatically based on the parameters specified in the ggvis command. You can override the defaults.

Let's redo the graphic with the relative change since the beginning of industrialization in the mid 18th century, when the level of CO2 was around 280 ppm. This time, we will create a line plot. We also add some titles for the axes and specify a format of four consecutive digits for displaying a year on the x-axis. We also move or offset the title for the y-axis a bit to the left to improve readability.

```
# compute a new column containing the relative change in CO2
carbon %>%
  mutate(relCO2 = (CO2-280)/280) %>% # transformation
  ggvis(~year,~relCO2) %>%
  layer_lines(stroke:="blue") %>%
  scale_numeric('y',zero=T) %>%
  add_axis('y', title = "CO2 ppm of the atmosphere", title_offset=50) %>%
  add_axis('x', title ='Year', format='####')
```

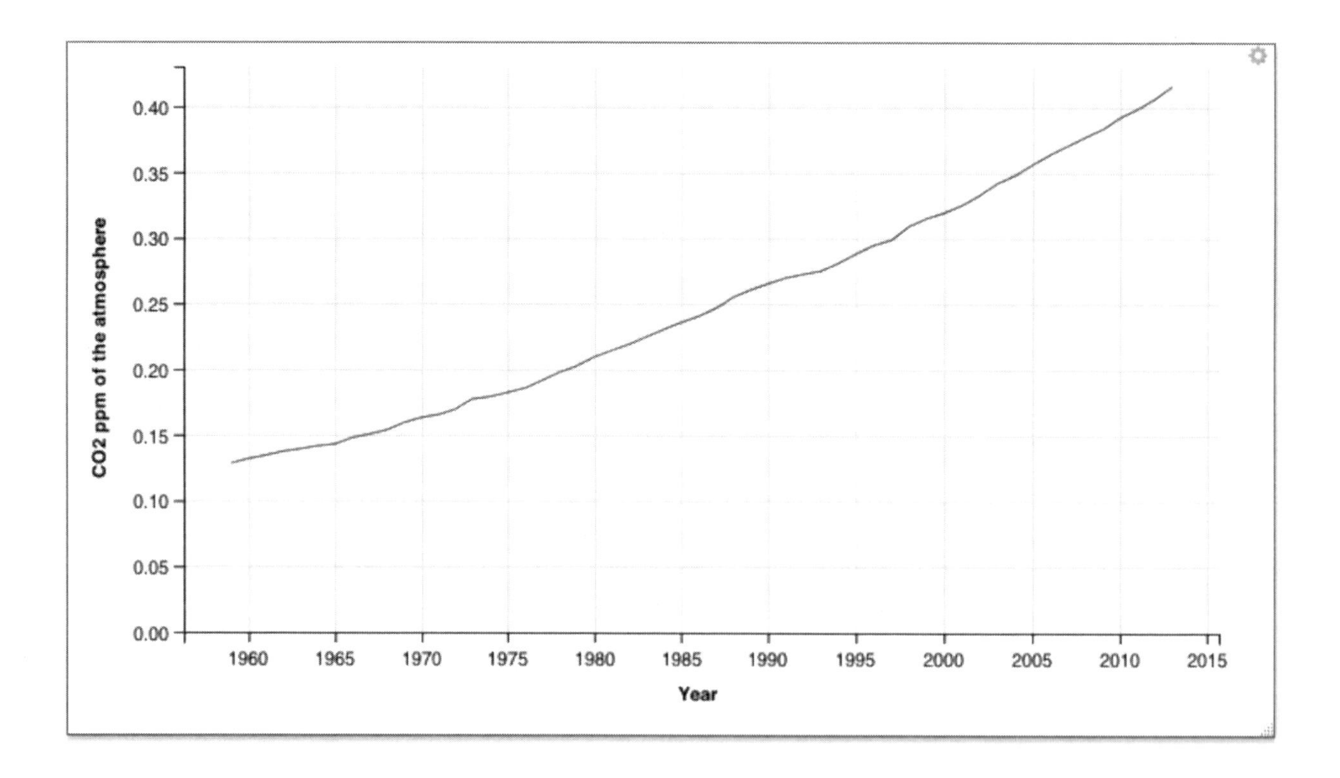

As the prior graphic shows, the present level of CO2 in the atmosphere is now above 40 percent higher than the pre-industrial period and is continuing to grow.

Assignment function

You will have seen three types of assignment symbols in the preceding examples. The notion of scaling helps in understanding the difference between a value and a property. A property is fixed. So not matter how you resize the preceding graph, the stroke will always be blue. Whereas, a value, such as the title, is scalable. The difference between a value and a property can seem confusing, so don't agonize over it.

Symbol		Example	
~	Data assignment	y ~ CO2	y is CO2 column
=	Set a value	title = 'year'	Title is scaled
:=	Set a property	stroke := 'blue'	Stroke is unscaled

Guides

Axes and legends are both forms of guides, which help the reader to understand a graphic. In ggvis, legends and axes are generated automatically based on the parameters specified in the ggvis command. You have the capability to override the defaults for axes, but the legend is quite fixed in its format.

In the following graphic, for each axis there is a label and there are tick marks with values so that you eyeball a point on the graphic to its x and y values. The legend enables you to determine which color, in this case, matches each year. A legend could also use shape (e.g., a diamond) and shape size to aid matching.

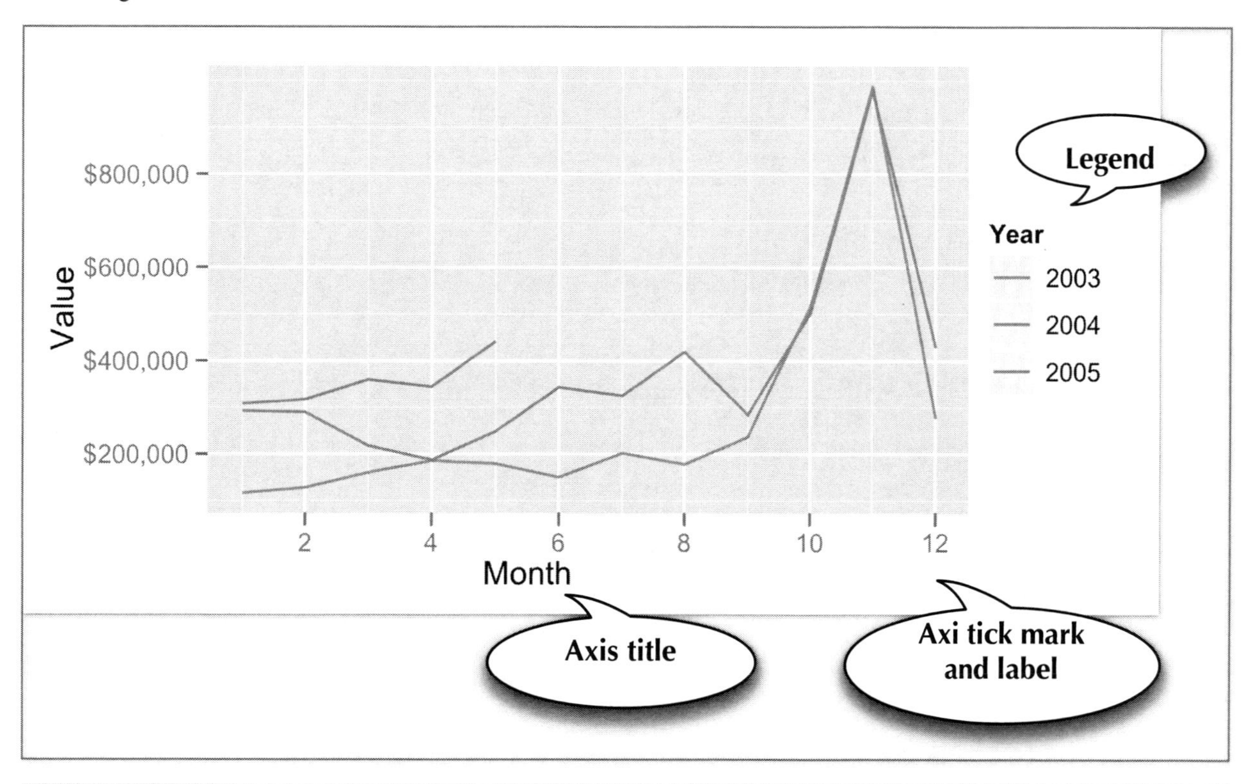

Skill builder

Create a line plot using the data in the following table.

Year	1804	1927	1960	1974	1987	1999	2012	2027	2046
Population (billions)	1	2	3	4	5	6	7	8	9

Some recipes

Learning the full power of ggvis is quite an undertaking, so here are a few recipes for visualizing data.

Histogram

Histograms are useful for showing the distribution of values in a single column. A histogram has a vertical orientation. The number of occurrences of a particular value in a column are automatically counted by the ggvis function. In the following sample code, we show the distribution of temperatures in Celsius for the Central Park temperature data. Notice that the Celsius conversion is rounded to an integer for plotting. Width specifies the size of the bin, which is two in this case. This means that the bin above the tick mark 10 contains all values in the range 9 to 11. There is online a list of names of colors you can use in ggvis.[47]

```
library(ggvis)
library(readr)
library(measurements)
url <-  'http://people.terry.uga.edu/rwatson/data/centralparktemps.txt'
t <- read_delim(url, delim=',')
t %>%
  mutate(Celsius = conv_unit(t$temperature,'F','C')) %>%
  ggvis(~Celsius) %>%
  layer_histograms(width = 2, fill:='cornflowerblue') %>%
add_axis('x',title='Celsius') %>%
  add_axis('y',title='Frequency')
```

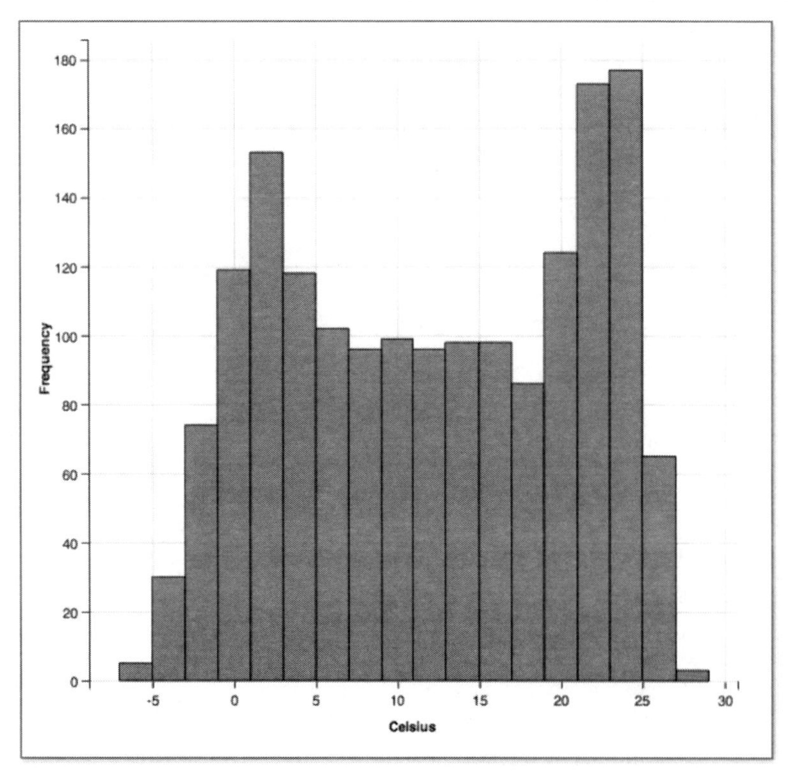

[47] www.w3schools.com/cssref/css_colornames.asp

Bar graph

In the following example, we query a database to get data for plotting. Because the column selected for graphing, productLine is categorical, we need to use a bar graph.

```
library(ggvis)
library(DBI)
conn <- dbConnect(RMySQL::MySQL(), "richardtwatson.com",
dbname="ClassicModels", user="db1", password="student")
# Query the database and create file for use with R
d <- dbGetQuery(conn,"SELECT * from Products;")
# Plot the number of product lines by specifying the appropriate column name
d %>%
  ggvis(~productLine) %>%
  layer_bars(fill:='chocolate') %>%
  add_axis('x',title='Product line') %>%
  add_axis('y',title='Count')
```

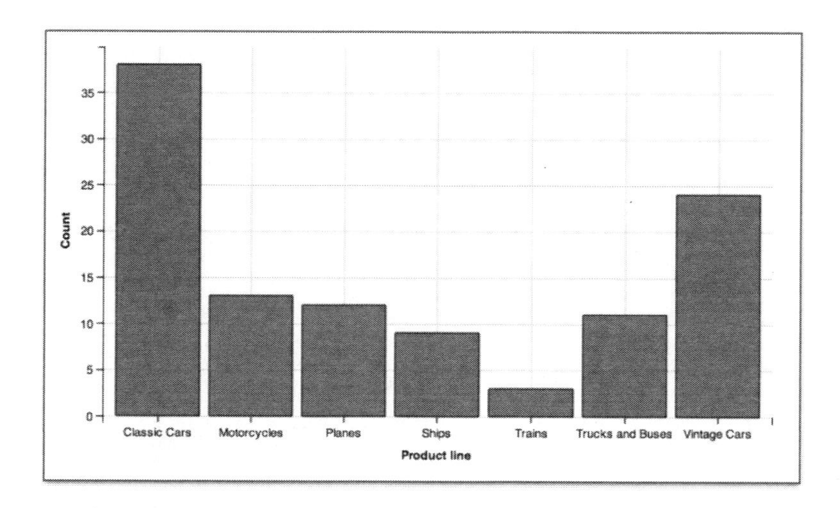

Skill builder

Create a bar graph using the data in the following table. Set population as the weight for each observation.

Year	1804	1927	1960	1974	1987	1999	2012	2027	2046
Population (billions)	1	2	3	4	5	6	7	8	9

Scatterplot

A scatterplot shows points on an x-y grid, which means you need to have an x and y with numeric values.

```
library(ggvis)
library(DBI)
```

```
library(dplyr)
library(lubridate)
conn <- dbConnect(RMySQL::MySQL(), "richardtwatson.com",
dbname="ClassicModels", user="db1", password="student")
o <- dbGetQuery(conn,"SELECT * FROM Orders")
od <- dbGetQuery(conn,"SELECT * FROM OrderDetails")
d <- inner_join(o,od)
# Get the monthly value of orders
d2 <- d %>%
  mutate(month = month(orderDate)) %>%
  group_by(month) %>% summarize(orderValue = sum(quantityOrdered*priceEach))
# Plot data orders by month
# Show the points and the line
d2 %>%
  ggvis(~month, ~orderValue/1000000) %>%
  layer_lines(stroke:='blue') %>%
  layer_points(fill:='red') %>%
  add_axis('x', title = 'Month') %>%
  add_axis('y',title='Order value (millions)', title_offset=30)
```

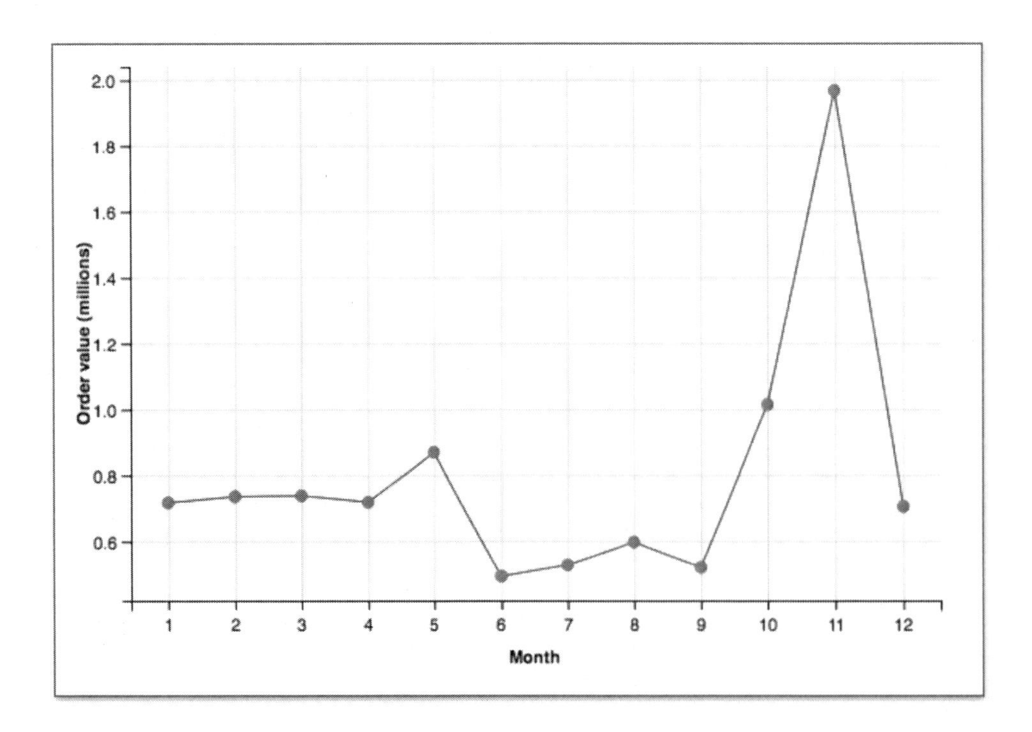

It is sometimes helpful to show multiple scatterplots on the one grid. In ggvis, you can create groups of points for plotting. Sometimes, you might find it convenient to recode data so you can use different colors or lines to distinguish values in set categories. Let's examine grouping by year.

```
library(ggvis)
library(DBI)
library(dplyr)
library(lubridate)
conn <- dbConnect(RMySQL::MySQL(), "richardtwatson.com",
dbname="ClassicModels", user="db1", password="student")
o <- dbGetQuery(conn,"SELECT * FROM Orders")
od <- dbGetQuery(conn,"SELECT * FROM OrderDetails")
d <- inner_join(o,od)
d2 <- d %>%
  mutate(month = month(orderDate)) %>%
  mutate(year = year(orderDate)) %>%
  group_by(year,month) %>% summarize(orderValue =
sum(quantityOrdered*priceEach))
# Plot data orders by month and display by year
# ggvis expects grouping variables to be a factor
d2 %>%
  ggvis(~month,~orderValue/1000, stroke = ~as.factor(year)) %>%
  layer_lines() %>%
  add_axis('x', title = 'Month') %>%
  add_axis('y',title='Order value (thousands)', title_offset=50)
```

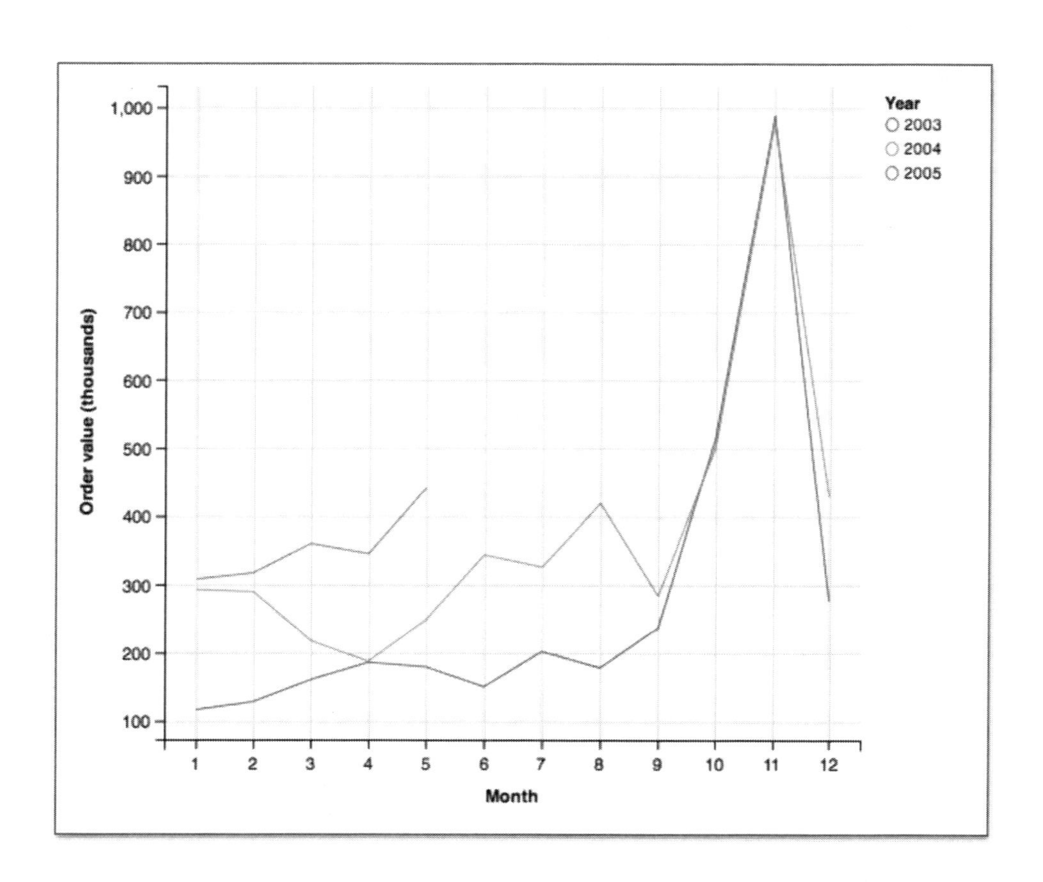

Because ggvis is based on a grammar of graphics, with a few changes, you can create a bar graph of the same data.

```
d2 %>%
  ggvis( ~month, ~orderValue/100000, fill = ~as.factor(year)) %>%
  layer_bars() %>%
  add_axis('x', title = 'Month') %>%
  add_axis('y',title='Order value (thousands)', title_offset=50)
```

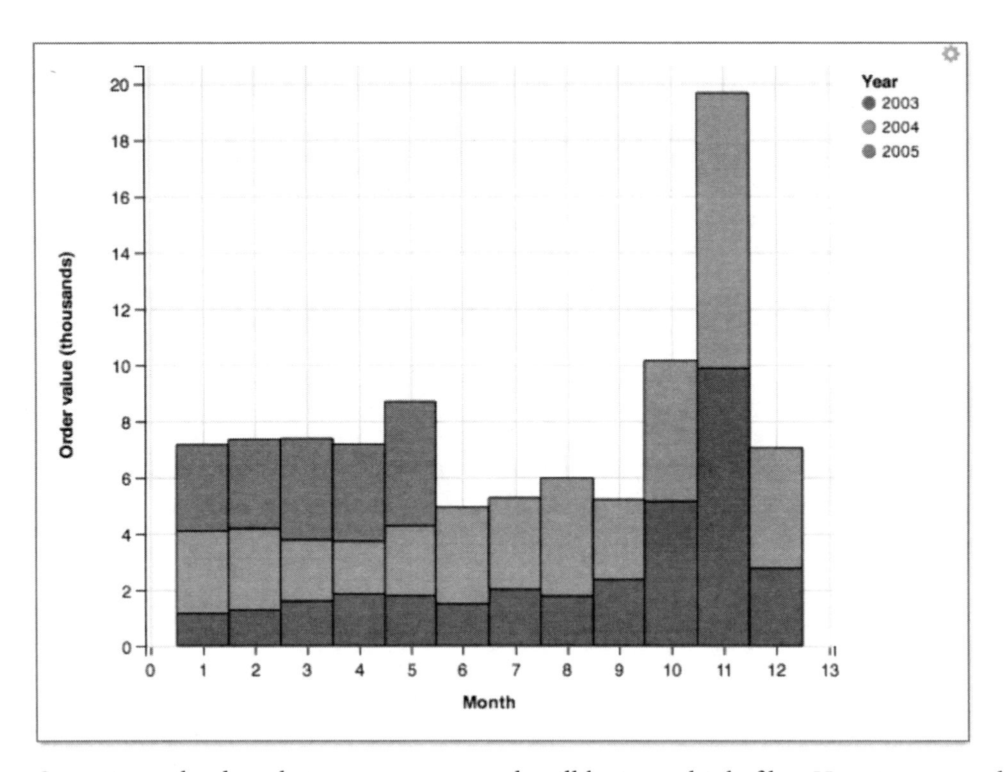

Sometimes the data that you want to graph will be in multiple files. Here is a case where we want to show ClassicCars orders and payments by month on the same plot.

```
library(ggvis)
library(DBI)
library(dplyr)
library(lubridate)
# Load the driver
conn <- dbConnect(RMySQL::MySQL(), "richardtwatson.com", dbname="ClassicModels",
user="db1", password="student")
o <- dbGetQuery(conn,"SELECT * FROM Orders")
od <- dbGetQuery(conn,"SELECT * FROM OrderDetails")
d <- inner_join(o,od)
d2 <-
  d %>%
  mutate(month = month(orderDate)) %>%
  mutate(year = year(orderDate)) %>%
  filter(year == 2004) %>%
```

```
  group_by(month) %>%
  summarize(value = sum(quantityOrdered*priceEach))
d2$category <- 'Orders'
p <-  dbGetQuery(conn,"SELECT * from Payments;")
p2 <- p %>%
  mutate(month = month(paymentDate)) %>%
  mutate(year = year(paymentDate)) %>%
  filter(year==2004) %>%
  group_by(month) %>%
  summarize(value = sum(amount))
p2$category <- 'Payments'
m <- rbind(d2,p2) # bind by rows
m %>%
  group_by(category) %>%
  ggvis(~month, ~value, stroke = ~ category) %>%
  layer_lines() %>%
  add_axis('x',title='Month') %>%
  add_axis('y',title='Value',title_offset=70)
```

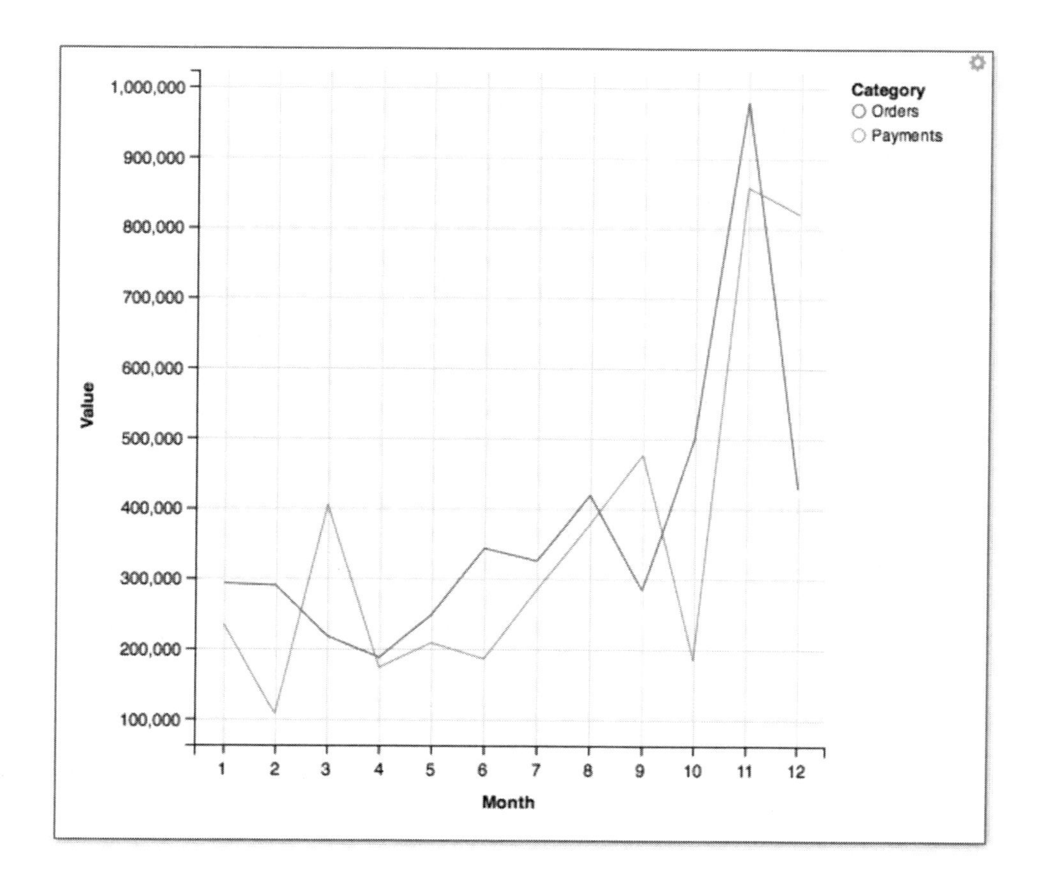

Smoothing

Smoothing helps to detect a trend in a line plot. The following example shows the average mean temperatures for August for Central Park.

```
library(ggvis)
library(readr)
library(dplyr)
url <-  "http://people.terry.uga.edu/rwatson/data/centralparktemps.txt"
t <- read_delim(url, delim=',')
t %>%
  filter(month == 8) %>%
  ggvis(~year,~temperature) %>%
  layer_lines(stroke:='red') %>%
  layer_smooths(se=T, stroke:='blue') %>%
  add_axis('x',title='Year',format = '####') %>%
  add_axis('y',title='Temperature (F)', title_offset=30)
```

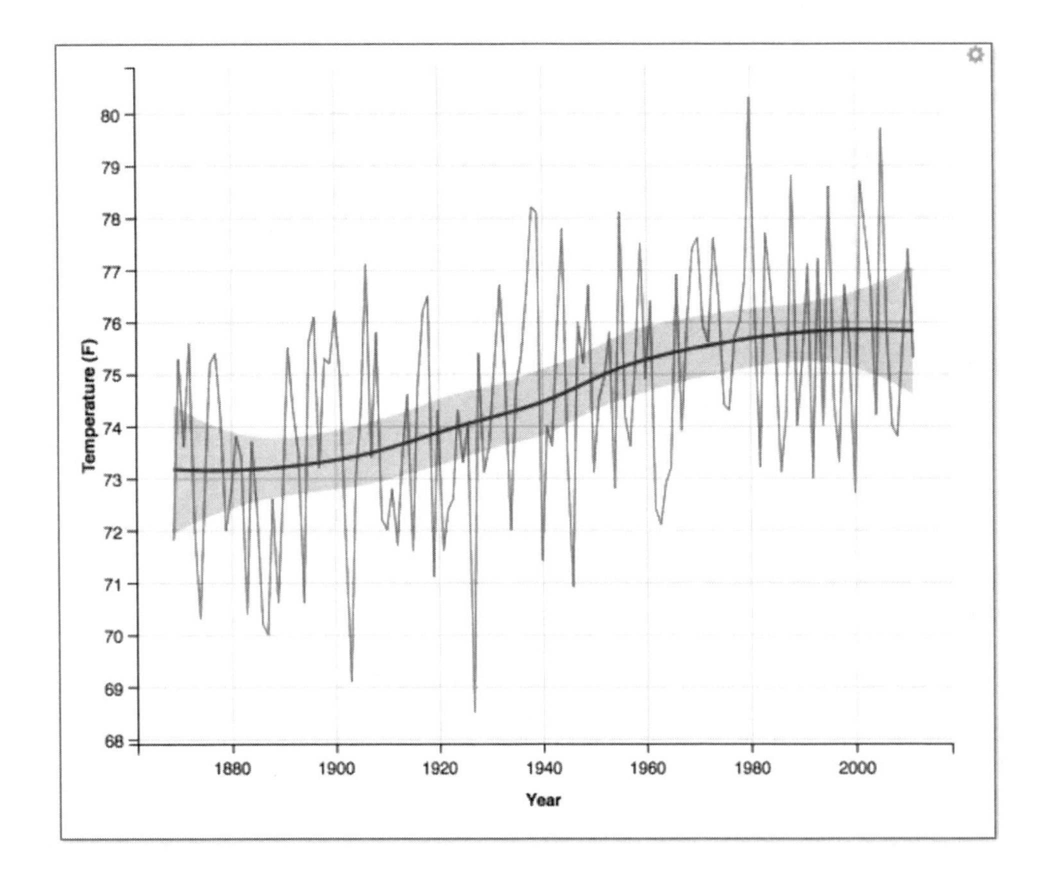

Skill builder

Using the Central Park data, plot the temperatures for February and fit a straight line to the points. Use layer_model_predictions(). What do you observe?

Box Plot

A box plot is an effective means of displaying information about one or more variables. It shows minimum and maximum, range, lower quartile, median, and upper quartile for each variable. The following code creates a box plot for a single variable. Notice that we use factor(0) to indicate there is no grouping variable.

```
library(ggvis)
library(DBI)
conn <- dbConnect(RMySQL::MySQL(), "richardtwatson.com", dbname="ClassicModels",
user="db1", password="student")
d <- dbGetQuery(conn,"SELECT * from Payments;")
# Boxplot of amounts paid
d %>%
  ggvis(~factor(0),~amount) %>%
  layer_boxplots() %>%
  add_axis('x',title='Checks') %>%
  add_axis('y',title='')
```

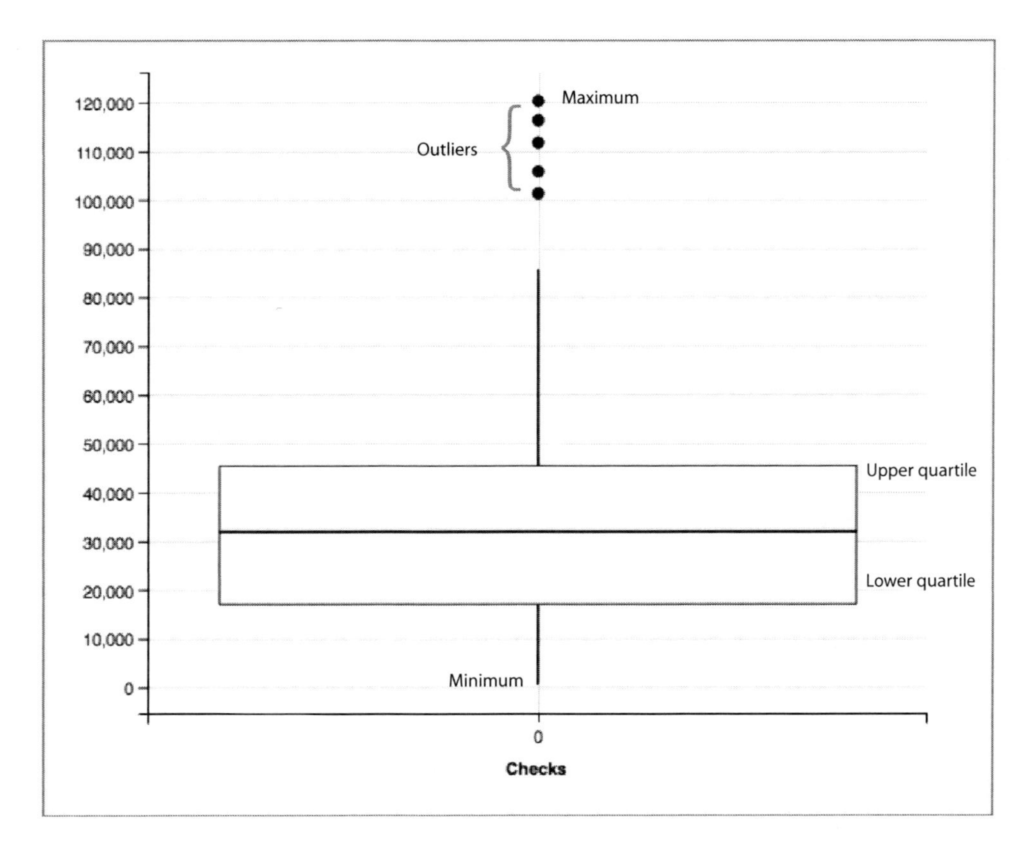

The following example shows a box plot for each month. Notice how to indicate the values for each of the ticks for the x-axis. Try running the code without the values specification.

```
library(ggvis)
library(DBI)
library(lubridate)
conn <- dbConnect(RMySQL::MySQL(), "richardtwatson.com",
dbname="ClassicModels", user="db1", password="student")
d <- dbGetQuery(conn,"SELECT * from Payments;")
# Boxplot of amounts paid
```

```
d %>%
  mutate(month = month(paymentDate)) %>%
  ggvis(~month,~amount) %>%
  layer_boxplots() %>%
  add_axis('x',title='Month', values=c(1:12)) %>%
  add_axis('y',title='Amount', title_offset=70)
```

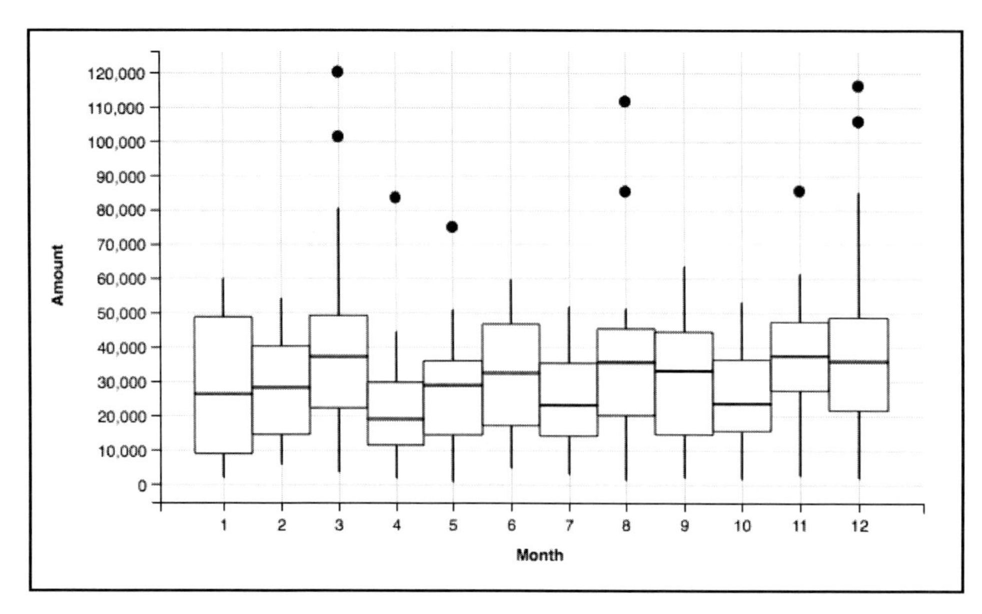

Heat map

A heat map visualizes tabular data. It is useful when you have two categorical variables cross tabulated. Consider the case for the ClassicModels database where we want to get an idea of the different number of model scales in each product line. We can get a quick picture with the following code.

```
library(ggvis)
library(DBI)
library(dplyr)
# Load the driver
conn <- dbConnect(RMySQL::MySQL(), "richardtwatson.com", dbname="ClassicModels",
user="db1", password="student")
d <- dbGetQuery(conn,'SELECT * FROM Products;')
d2 <-
  d %>%
  group_by(productLine, productScale) %>%
  summarize(frequency = n())
d2 %>%
  ggvis( ~productScale, ~productLine, fill= ~frequency) %>%
  layer_rects(width = band(), height = band()) %>%
  add_axis('y',title='Product Line', title_offset=70) %>%
  # add frequency to each cell
  layer_text(text:=~frequency, stroke:='white', align:='left', baseline:='top')
```

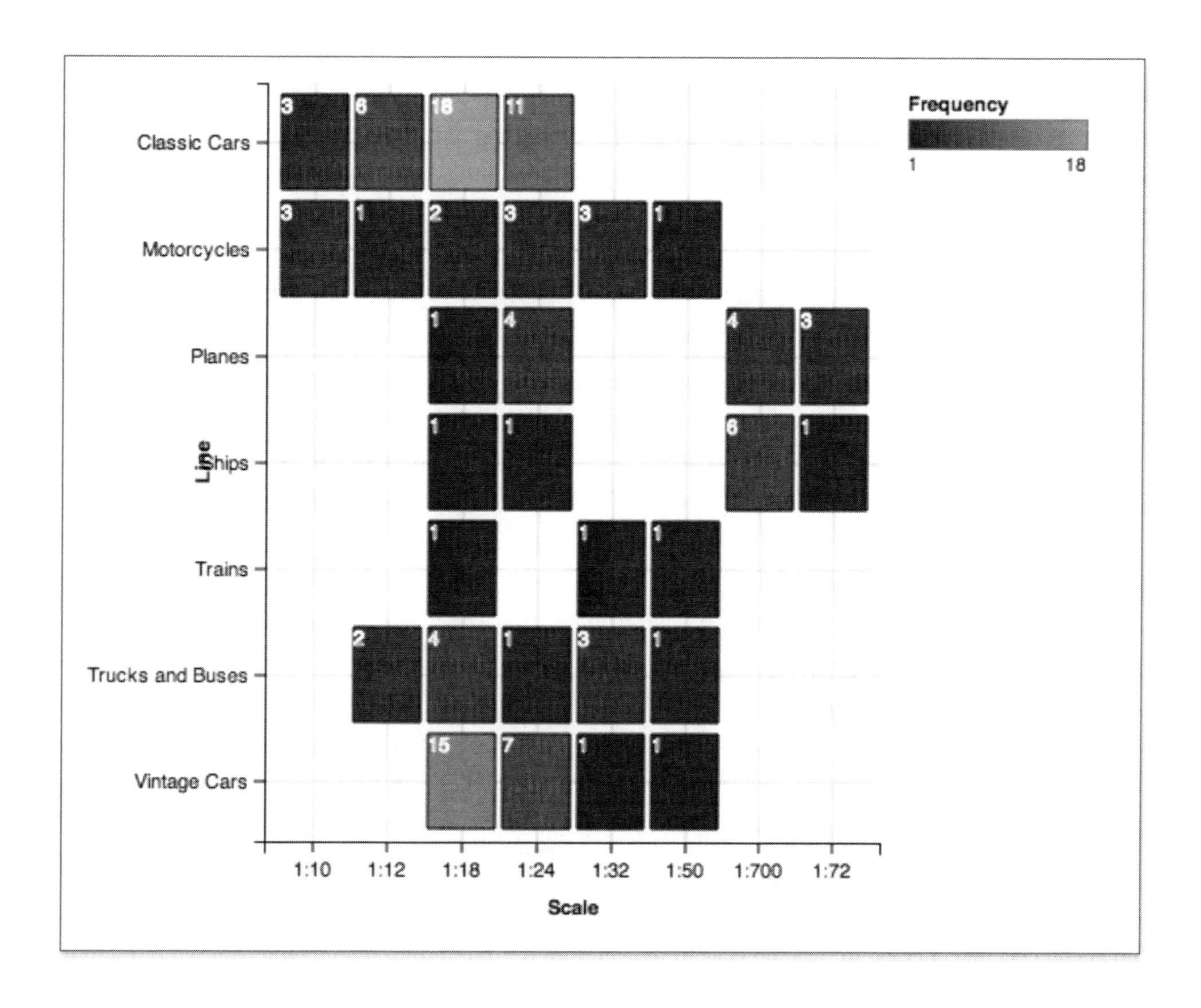

Interactive graphics

The ggvis package incorporates features of shiny, an R package, that enable you to create interactive graphics. For example, you might want to let the viewer select the color and width of a line on a graphic. Many of the interactive controls, shown in the following table, should be familiar as the the various options are often used for web forms.

Interactive controls for ggvis

Function	Purpose
input_checkbox()	Check one or more boxes
input_checkboxgroup()	A group of checkboxes
input_numeric()	A spin box
input_radiobuttons()	Pick one from a set of options

Function	Purpose
input_select()	Select from a drop-down text box
input_slider()	Select using a slider
input_text()	Input text

Selecting a property using a drop-down list

The following example uses the input_select function to present the viewer with an option of one of three colors (red, green, or blue) for the stroke of the graph. When you execute the code, you will see a selection list on the left bottom. Because the graph is interactive, you will need to click on the stop button (top right of the Viewer window) before running anymore R commands.

```
library(shiny)
library(ggvis)
library(dplyr)
carbon %>%
  mutate(relCO2 = (CO2-280)/280) %>%
  ggvis(~year,~relCO2) %>%
  layer_lines(stroke:=input_select(c("red", "green", "blue"))) %>%
  scale_numeric('y',zero=T) %>%
  add_axis('y', title = "CO2 ppm of the atmosphere", title_offset=50) %>%
  add_axis('x', title ='Year', format='####')
```

Skill builder

Create a point plot using the data in the following table. Give the viewer a choice of three colors and three shapes (square, cross, or diamond) for the points.

Year	1804	1927	1960	1974	1987	1999	2012	2027	2046
Population (billions)	1	2	3	4	5	6	7	8	9

Selecting a numeric value with a slider

A slider enables the viewer to select a numeric value in a range. In the following code, a slider is set up for values between 1 to 5, inclusively. Note in this case, the stroke width slider and color selector are defined outside the ggvis code. This is a useful technique for writing code that makes it possible for some chunks to be easily reused

```
slider <- input_slider(1, 5, label = "Width")
select_color <- input_select(label='Color',c("red", "green", "blue"))
carbon %>%
  mutate(relCO2 = (CO2-280)/280) %>%
  ggvis(~year,~relCO2) %>%
  layer_lines(stroke:=select_color, strokeWidth:=slider) %>%
  scale_numeric('y',zero=T) %>%
  add_axis('y', title = "CO2 ppm of the atmosphere", title_offset=50) %>%
```

```
add_axis('x', title ='Year', format='####')
```

Skill builder

Using the Central Park data, plot the temperatures for a selected year using a slider.

Geographic data

The ggmap package supports a variety of mapping systems, including Google maps. As you might expect, it offers many features, and we just touch on the basics in this example.

The Offices table in the Classic Models database includes the latitude and longitude of each office in officeLocation, which has a datatype of POINT. The following R code can be used to mark each office on a Google map. After loading the required packages, a database query is executed to return the longitude and latitude for each office. Then, a Google map of the United States is displayed. The marker parameter specifies the name of the table containing the values for longitude and latitude. Offices that are not in the U.S. (e.g., Sydney) are ignored. Adjust zoom, an integer, to get a suitably sized map. Zoom can range from 3 (a continent) to 20 (a building).

```
library(ggplot)
library(ggmap)
library(mapproj)
library(DBI)
conn <- dbConnect(RMySQL::MySQL(), "richardtwatson.com",
dbname="ClassicModels", user="db1", password="student")
# Google maps requires lon and lat, in that order, to create markers
d <- dbGetQuery(conn,"SELECT y(officeLocation) AS lon, x(officeLocation) AS
lat FROM Offices;")
# show offices in the United States
# vary zoom to change the size of the map
map <-  get_googlemap('united states',marker=d,zoom=4)
ggmap(map) + labs(x = 'Longitude', y = 'Latitude') + ggtitle('US offices')
```

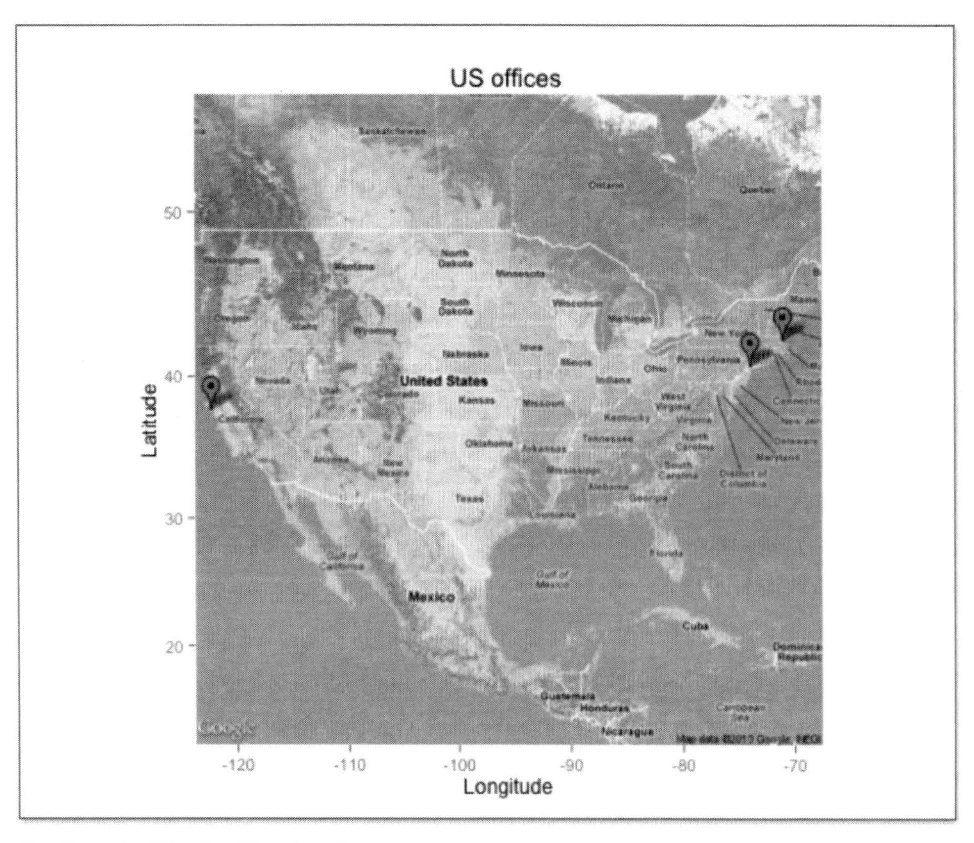

Skill builder

Create the following maps.

1. Offices in Europe
2. Office location in Paris
3. Customers in the US
4. Customers in Sydney

R resources

The developer of ggvis, Hadley Wickham, maintains a web site. Another useful web site for data visualization is FlowingData, which has an associated book, titled *Visualize This*. If you become heavily involved in visualization, we suggest you read some of the works of Edward Tufte. One of his books is listed in the references.

Summary

Because of evolutionary pressure, humans became strong visual processors. Consequently, graphics can be very effective for presenting information. The grammar of graphics provides a logical foundation for the construction of a graphic by adding successive layers of information. ggvis is a package implementing the grammar of graphics in R, the open source statistics platform. Data can be extracted from an MySQL database for processing and graphical presentation in R. Spatial data can be selected from a database and displayed on a Google map.

Key terms and concepts

Bar chart

Box plot

dplyr

ggvis

Grammar of graphics

Graph

Graphic

Heat map

Histogram

Scatterplot

Smooth

References

Kabacoff, R. I. (2009). *R in action: data analysis and graphics with R*. Greenwich, CT: Manning Publications.

Kahle, D., & Wickham, H. (forthcoming). ggmap : Spatial Visualization with ggplot. *The R Journal.*

Tufte, E. (1983). *The visual display of quantitative information*. Cheshire, CT: Graphics Press.

Yau, N. (2011). *Visualize this: the flowing data guide to design, visualization, and statistics.* Indianapolis, IN: Wiley.

Wilkinson, L. (2005). *The grammar of graphics* (2nd ed.). New York: Springer.

Exercises

1. Given the following data on the world's cargo carrying capacity in millions of dead weight tons (dwt)[48] for the given years, graph the relationship as a point plot. Add a linear prediction model to the graph. What is the approximate increase in the world capacity each year?

```
dwt <-  c(2566, 3704, 4008, 5984, 7700, 8034, 8229, 7858, 8408, 8939)
year <-  c(1970, 1980, 1990, 2000, 2006, 2007, 2008, 2009, 2010, 2011)
```

2. Create a bar graph showing the relative ocean shipping cost as a percentage of total cost for some common household items using the following data.

```
unitCost <-  c(700, 200, 150, 50, 15, 3, 1)
shipCost <-  c(10.00, 1.50, 1.00, 0.15, 0.15, 0.05, 0.01)
item <-  c('TV set', 'DVD player', 'Vacuum cleaner', 'Scotch whisky', 'Coffee',
'Biscuits', 'Beer')
```

SQL input

3. Visualize in blue the number of items for each product scale.

4. Prepare a line plot with appropriate labels for total payments for each month in 2004.

---

[48] Dead weight tonnage is a measure of how much weight, in tonnes, a ship can safely carry.

5. Create a histogram with appropriate labels for the value of orders received from the Nordic countries (Denmark, Finland, Norway, Sweden).

6. Create a heatmap for product lines and Norwegian cities.

7. Show on a Google map the customers in Japan.

8. Show on a Google map the European customers who have never placed an order.

File input

9. Access http://people.terry.uga.edu/rwatson/data/manheim.txt, which contains details of the sales of three car models: X, Y, and Z.

    a. Create a bar chart for sales of each model (X, Y , or Z)

    b. Create bar chart for sales by each form of sale (online or auction).

10. Use the 'Import Dataset' feature of RStudio to read http://people.terry.uga.edu/rwatson/data/electricityprices.csv, which contains details of electricity prices for a major city.[49] Do a Box plot of cost. What do you conclude about cost?

11. Read the table http://people.terry.uga.edu/rwatson/data/wealth.csv containing details of the wealth of various countries. Create histograms for each of the wealth measures. Consult the ggvis color chart[50] for a choice of colors.

12. Merge the data for weather <http://people.terry.uga.edu/rwatson/data/weather.csv> and electricity prices <http://people.terry.uga.edu/rwatson/data/electricityprices.csv> for a major city. The merged file should contain air temperature and electricity cost. Also, you need to convert air temperature from a factor to a numeric (hint: first convert to a character). As readr does not currently handle date and time stamps, use the following code to read the files.

```
wurl <-  'http://people.terry.uga.edu/rwatson/data/weather.csv'
w <-  read.csv(wurl,sep=',',header=T)
eurl <-  'http://people.terry.uga.edu/rwatson/data/electricityprices.csv'
e <-  read.csv(eurl,sep=',',header=T)
```

    a. Compute the correlation between temperature and electricity price. What do you conclude?

    b. Graph the relationship between temperature and electricity price.

    c. Graph the relationship between temperature and electricity price when the temperature is 95°F and above.

    d. Create a single graph showing the relationship between temperature and electricity price differentiating by color when the temperature is above or below 90°F. (Hint: Trying recoding temperature).

49 Note these prices have been transformed from the original values, but are still representative of the changes over time.

50 www.w3schools.com/cssref/css_colornames.asp

16. Text mining & natural language processing

From now on I will consider a language to be a set (finite or infinite) of sentences, each finite in length and constructed out of a finite set of elements. All natural languages in their spoken or written form are languages in this sense.

Noam Chomsky, *Syntactic Structures*

Learning objectives

Students completing this chapter will:

- Have a realistic understanding of the capabilities of current text mining and NLP software;
- Be able to use R and associated packages for text mining and NLP.

The nature of language

Language enables humans to cooperate through information exchange. We typically associate language with sound and writing, but gesturing, which is older than speech, is also a means of collaboration. The various dialects of sign languages are effective tools for visual communication. Some species, such as ants and bees, exchange information using chemical substances known as pheromones. Of all the species, humans have developed the most complex system for cooperation, starting with gestures and progressing to digital technology, with language being the core of our ability to work collectively.

Natural language processing (NLP) focuses on developing and implementing software that enables computers to handle large scale processing of language in a variety of forms, such as written and spoken. While it is a relatively easy task for computers to process numeric information, language is far more difficult because of the flexibility with which it is used, even when grammar and syntax are precisely obeyed. There is an inherent ambiguity of written and spoken speech. For example, the word "set" can be a noun, verb, or adjective, and the *Oxford English Dictionary* defines over 40 different meanings. Irregularities in language, both in its structure and use, and ambiguities in meaning make NLP a challenging task. Be forewarned. Don't expect NLP to provide the same level of exactness and starkness as numeric processing. NLP output can be messy, imprecise, and confusing – just like the language that goes into an NLP program. One of the well-known maxims of information processing is "garbage-in, garbage-out." While language is not garbage, we can certainly observe that "language-in, language-out" is a truism. You can't start with something that is marginally ambiguous and expect a computer to turn it into a precise statement. Legal and religious scholars can spend years learning how to interpret a text and still reach different conclusions as to meaning.

NLP, despite its limitations, enables humans to process large volumes of language data (e.g., text) quickly and to identify patterns and features that might be useful. A well-educated human with domain knowledge specific to the same data might make more sense of these data, but it might take months or years. For example, a firm might receive over a 1,000 tweets, 500 Facebook mentions, and 20 blog references in a day. It needs NLP to identify within minutes or hours which of these many messages might need human action.

Text mining and NLP overlap in their capabilities and goals. The ultimate objective is to extract useful and valuable information from text using analytical methods and NLP. Simply counting words in a document is a an example of text mining because it requires minimal NLP technology, other than separating text into words. Whereas, recognizing entities in a document requires prior extensive machine learning and more intensive NLP knowledge. Whether you call it text mining or NLP, you are processing natural language. We will use the terms somewhat interchangeably in this chapter.

The human brain has a special capability for learning and processing languages and reconciling ambiguities,[51] and it is a skill we have yet to transfer to computers. NLP can be a good servant, but enter its realm with realistic expectations of what is achievable with the current state-of-the-art.

Levels of processing

There are three levels to consider when processing language.

Semantics

Semantics focuses on the meaning of words and the interactions between words to form larger units of meaning (such as sentences). Words in isolation often provide little information. We normally need to read or hear a sentence to understand the sender's intent. One word can change the meaning of a sentence (e.g., "Help needed versus Help not needed"). It is typically an entire sentence that conveys meaning. Of course, elaborate ideas or commands can require many sentences.

Discourse

Building on semantic analysis, discourse analysis aims to determine the relationships between sentences in a communication, such as a conversation, consisting of multiple sentences in a particular order. Most human communications are a series of connected sentences that collectively disclose the sender's goals. Typically, interspersed in a conversation are one or more sentences from one or more receivers as they try to understand the sender's purpose and maybe interject their thoughts and goals into the discussion. The points and counterpoints of a blog are an example of such a discourse. As you might imagine, making sense of discourse is frequently more difficult, for both humans and machines, than comprehending a single sentence. However, the braiding of question and answer in a discourse, can sometimes help to reduce ambiguity.

Pragmatics

Finally, pragmatics studies how context, world knowledge, language conventions, and other abstract properties contribute to the meaning of human conversation. Our shared experiences and knowledge often help us to make sense of situations. We derive meaning from the manner of the discourse, where it takes place, its time and length, who else is involved, and so forth. Thus, we usually find it much easier to communicate with those with whom we share a common culture, history, and socioeconomic status because the great collection of knowledge we jointly share assists in overcoming ambiguity.

Tokenization

Tokenization is the process of breaking a document into chunks (e.g., words), which are called tokens. Whitespaces (e.g., spaces and tabs) are used to determine where a break occurs. Tokenization typically creates a *bag of words* for subsequent processing. Many text mining functions use words as the foundation for analysis.

51 Ambiguities are often the inspiration for puns. "You can tune a guitar, but you can't tuna fish. Unless of course, you play bass," by Douglas Adams

Counting words

To count the number of words in a string, simply count the number of times there are one or more consecutive spaces using the pattern " [[:space:]]+" and then add one, because the last word is not followed by a space.

```
library(stringr)
str_count("The dead batteries were given out free of charge", "[[:space:]]+")
+ 1
```

Sentiment analysis

Sentiment analysis is a popular and simple method of measuring aggregate feeling. In its simplest form, it is computed by giving a score of +1 to each "positive" word and -1 to each "negative" word and summing the total to get a sentiment score. A text is decomposed into words. Each word is then checked against a list to find its score (i.e., +1 or -1), and if the word is not in the list, it doesn't score.

A major shortcoming of sentiment analysis is that irony (e.g., "The name of Britain's biggest dog (until it died) was Tiny") and sarcasm (e.g., "I started out with nothing and still have most of it left") are usually misclassified. Also, a phrase such as "not happy" might be scored as +1 by a sentiment analysis program that simply examines each word and not those around it.

The **sentimentr package** offers an advanced implementation of sentiment analysis. It is based on a polarity table, in which a word and its polarity score (e.g., -1 for a negative word) are recorded. The default polarity table provided by the syuzhet package. You can create a polarity table suitable for your context, and you are not restricted to 1 or -1 for a word's polarity score. Here are the first few rows of the default polarity table.

```
> library(sentimentr)
> library(syuzhet)
> head(get_sentiment_dictionary())
         word value
1     abandon -0.75
2   abandoned -0.50
3   abandoner -0.25
4 abandonment -0.25
5    abandons -1.00
6    abducted -1.00
```

In addition, sentimentr supports valence shifters, which are words that alter or intensify the meaning of a polarizing word (i.e., a word appearing in the polarity table) appearing the text or document under examination. Each word has a value to indicate how to interpret its effect (negators (1), amplifiers(2), de-amplifiers (3), and conjunction (4).

Now, let's see how we use the **sentiment function**. We'll start with an example that does no use valence shifters, in which case we specify that the sentiment function should not look for valence words before or after any polarizing word. We indicate this by setting n.before and n.after to 0. Our sample text consists of several sentences, as shown in the following code, where polarizing words are shown in green (positive) and red (negative).

```
library(sentimentr)
```

```
sample = c("You're awesome and I love you", "I hate and hate and hate. So angry.
Die!", "Impressed and amazed: you are peerless in your achievement of unparalleled
mediocrity.")
sentiment(sample, n.before=0, n.after=0, amplifier.weight=0)
```

The results are:

```
   element_id sentence_id word_count  sentiment
1:          1           1          6  0.5511352
2:          2           1          6 -0.9185587
3:          2           2          2 -0.5303301
4:          2           3          1 -0.7500000
5:          3           1         12  0.3608439
```

Notice that the sentiment function breaks each element (the text between quotes in this case) into sentences, identifies each sentence in an element, and computes the word count for each of these sentences. The sentiment score is the sum of the polarity scores divided by the square root of the number of words in the associated sentence..

To get the overall sentiment for the sample text, -0.26 in this case, we can use:

```
y <-  sentiment(sample, n.before=0, n.after=0)
mean(y$sentiment)
```

When a **valence shift** is detected before or after a polarizing word, its effect is incorporated in the sentiment calculation. The size of the effect is indicated by the amplifier.weight, a sentiment function parameter with a value between 0 and 1. The weight amplifies or de-amplifies by multiplying the polarized terms by 1 + the amplifier weight. A ***negator*** flips the sign of a polarizing word. A ***conjunction*** amplifies the current clause and down weights the prior clause. Some examples in the following table illustrate the results of applying the function

```
sentiment(text, n.before=2, n.after=2, amplifier.weight=.8, but.weight = .9)
```

to a variety of input text. The polarities are -1 (crazy) and 1 (love). There is a negator (not), two amplifiers (very and much), and a conjunction (but). Contractions are treated as amplifiers and so get weights based on the contraction (.9 in this case) and amplification (.8) in this case.

Type	Code	Text	Sentiment
		You're crazy, and I love you.	0
Negator	1	You're not crazy, and I love you.	0.57
Amplifier	2	You're crazy, and I love you very much.	0.21
De-amplifier	3	You're slightly crazy, and I love you.	0.23
Conjunction	4	You're crazy, but I love you.	0.45

Skill builder

Run the following R code and comment on how sensitive sentiment analysis is to the n.before and n.after parameters.

```
sample = c("You're not crazy and I love you very much.")
sentiment(sample, n.before = 4, n.after=3, amplifier.weight=1)
sentiment(sample, n.before = Inf, n.after=Inf, amplifier.weight=1)
```

Sentiment analysis has given you an idea of some of the issues surrounding text mining. Let's now look at the topic in more depth and explore some of the tools available in tm, a general purpose text mining package for R. We will also use a few other R packages which support text mining and displaying the results.

Corpus

A collection of text is called a corpus. It is common to use N for the *corpus size*, the number of tokens, and V for the *vocabulary*, the number of distinct tokens.

In the following examples, our corpus consists of Warren Buffett's annual letters to the shareholders of Berkshire Hathaway[52] for the period 1998-2012.[53] The letters, available in html or pdf, were converted to separate text files using Abbyy Fine Reader. Tables, page numbers, and graphics were removed during the conversion.

The following R code sets up a loop to read each of the letters and add it to a data frame. When all the letters have been read, they are turned into a corpus.

```
require(stringr)
require(tm)
#set up a data frame to hold up to 100 letters
df <-  data.frame(num=100)
begin <-  1998 # date of first letter
i <-  begin
# read the letters
while (i < 2013) {
  y <- as.character(i)
# create the file name
  f <- str_c('http://www.richardtwatson.com/BuffettLetters/',y,
'ltr.txt',delim='')
# read the letter as on large string
  d <-  readChar(f,nchars=1e6)
# add letter to the data frame
  df[i-begin+1,] <-  d
  i <-  i + 1
```

52 http://www.berkshirehathaway.com/letters/letters.html

53 The converted letters are available at http://www.richardtwatson.com/BuffettLetters/ and will be extended to include earlier and more recent letters. The folder also contains the original letter in pdf.

```
}
# create the corpus
letters <-  Corpus(DataframeSource(as.data.frame(df), encoding = "UTF-8"))
```

Readability

There are several approaches to estimating the readability of a selection of text. They are usually based on counting the words in each sentence and the number of syllables in each word. For example, the Flesch-Kincaid method uses the formula:

```
(11.8 * SYLLABLES_PER_WORD) + (0.39 * WORDS_PER_SENTENCE) - 15.59
```

It estimates the grade-level or years of education required of the reader. The bands are:

13-16	Undergraduate
16-18	Masters
19-	PhD

The R package koRpus has a number of methods for calculating readability scores. You first need to tokenize the text using the package's tokenize function. Then complete the calculation.

```
library(koRpus)
#tokenize the first letter in the corpus after converting to character vector
txt <-  letters[[1]][1] # first element in the list
tagged.text <- tokenize(as.character(txt),format="obj",lang="en")
# score
readability(tagged.text, hyphen=NULL,index="FORCAST")
```

Preprocessing

Before commencing analysis, a text file typically needs to be prepared for processing. Several steps are usually taken.

Case conversion

For comparison purposes, all text should be of the same case. Conventionally, the choice is to convert to all lower case.

```
clean.letters <-  tm_map(letters,tolower)
```

Punctuation removal

Punctuation is usually removed when the focus is just on the words in a text and not on higher level elements such as sentences and paragraphs.

```
clean.letters <-  tm_map(clean.letters,removePunctuation)
```

Number removal

You might also want to remove all numbers.

```
clean.letters <-  tm_map(clean.letters,removeNumbers)
```

Stripping extra white spaces

Removing extra spaces, tabs, and such is another common preprocessing action.

```
clean.letters <-  tm_map(clean.letters,stripWhitespace)
```

Skill builder

Redo the readability calculation after executing the preprocessing steps described in the previous section. What do you observe?

Stop word filtering

Stop words are short common words that can be removed from a text without affecting the results of an analysis. Though there is no commonly agreed upon list of stop works, typically included are *the, is, be, and, but, to,* and *on*. Stop word lists are typically all lowercase, thus you should convert to lowercase before removing stop words. Each language has a set of stop words. In the following sample code, we use the SMART list of English stop words. 54

```
clean.letters <-  tm_map(clean.letters,removeWords,stopwords("SMART"))
```

Specific word removal

There can also specify particular words to be removed via a character vector. For instance, you might not be interested in tracking references to *Berkshire Hathaway* in Buffett's letters. You can set up a dictionary with words to be removed from the corpus.

```
dictionary <-  c("berkshire","hathaway", "million", "billion", "dollar")
clean.letters <-  tm_map(clean.letters,removeWords,dictionary)
```

Word length filtering

You can also apply a filter to remove all words less than or greater than a specified lengths. The tm package provides this option when generating a term frequency matrix, something you will read about shortly.

Parts of speech (POS) filtering

Another option is to remove particular types of words. For example, you might scrub all adverbs and adjectives.

Stemming

Stemming reduces inflected or derived words to their stem or root form. For example, *cats* and *catty* stem to *cat*. *Fishlike* and *fishy* stem to *fish*. What is the stem of *catfish*? As a stemmer generally work by suffix stripping, *catfish* should stem to *cat*. As you would expect, stemmers are available for different languages, and thus the language must be specified.

```
# stem the document -- might take a while to run
stem.letters <-  tm_map(clean.letters,stemDocument, language = "english")
```

54 See http://jmlr.csail.mit.edu/papers/volume5/lewis04a/a11-smart-stop-list/english.stop

Following stemming, you can apply **stem completion** to return stems to their original form to make the text more readable. The original document that was stemmed, in this case, is used as the dictionary to search for possible completions. Stem completion can apply several different rules for converting a stem to a word, including "prevalent" for the most frequent match, "first" for the first found completion, and "longest" and "shortest" for the longest and shortest, respectively, completion in terms of characters

```
# stem completion -- might take a while to run
stem.letters <-  tm_map(stem.letters,stemCompletion, dictionary=clean.letters,
type=c("prevalent"))
```

Regex filtering

The power of regex (regular expressions) can also be used for filtering text or searching and replacing text. You might recall that we covered regex when learning SQL.

Word frequency analysis

Word frequency analysis is a simple technique that can also be the foundation for other analyses. The method is based on creating a matrix in one of two forms.

- A **term-document matrix** contains one row for each term and one column for each document.

```
tdm <-  TermDocumentMatrix(stem.letters,control = list(minWordLength=3))
dim(tdm)
# report those words occurring more than 100 times
findFreqTerms(tdm, lowfreq = 100, highfreq = Inf)
```

- A **document-term matrix** contains one row for each document and one column for each term.

```
dtm <-  DocumentTermMatrix(stem.letters,control = list(minWordLength=3))
dim(dtm)
```

The function dtm() reports the number of distinct terms, the vocabulary, and the number of documents in the corpus.

Term frequency

Words that occur frequently within a document are usually a good indicator of the document's content. Term frequency (tf) measures word frequency.

tf_{td} = number of times term t occurs in document d.

Here is the R code for determining the frequency of words in a corpus.

```
tdm <-  TermDocumentMatrix(stem.letters,control = list(minWordLength=3))
# convert term document matrix to a regular matrix to get frequencies of words
m <-  as.matrix(tdm)
# sort on frequency of terms
```

```
v <- sort(rowSums(m), decreasing=TRUE)
# display the ten most frequent words
v[1:10]
```

A probability density plot shows the distribution of words in a document visually. As you can see, there is a very long and thin tail because a very small number of words occur frequently. Note that this plot shows the distribution of words after the removal of stop words.

```
require(ggplot2)
# get the names corresponding to the words
names <- names(v)
# create a data frame for plotting
d <- data.frame(word=names, freq=v)
ggplot(d,aes(freq)) + geom_density(fill="salmon") + xlab("Frequency")
```

Probability density plot of word frequency

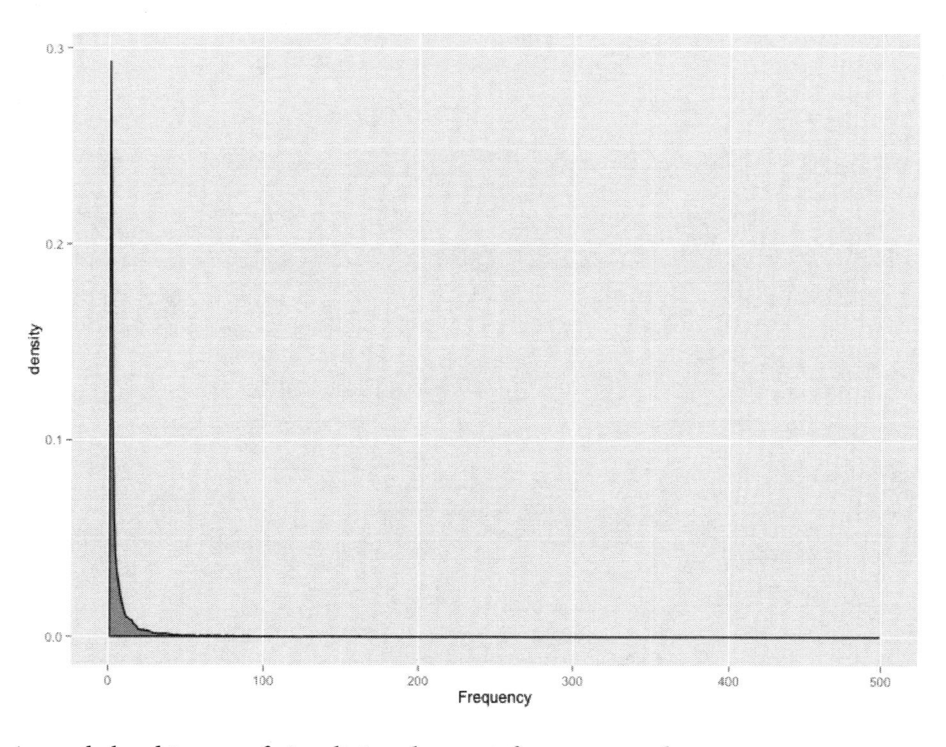

A word cloud is way of visualizing the most frequent words.

```
require(wordcloud)
# select the color palette
pal = brewer.pal(5,"Accent")
# generate the cloud based on the 30 most frequent words
wordcloud(d$word, d$freq, min.freq=d$freq[30],colors=pal)
```

A word cloud

Skill builder

Start with the original letters corpus (i.e., prior to preprocessing) and identify the 20 most common words and create a word cloud for these words.

Co-occurrence and association

Co-occurrence measures the frequency with which two words appear together. In the case of document level association, if the two words both appear or neither appears, then the correlation or association is 1. If two words never appear together in the same document, their association is -1.

A simple example illustrates the concept. The following code sets up a corpus of five elementary documents.

```
data <-  c("word1", "word1 word2","word1 word2 word3","word1 word2 word3
word4","word1 word2 word3 word4 word5")
frame <-  data.frame(data)
test <-  Corpus(DataframeSource(frame, encoding = "UTF-8"))
tdm <-  TermDocumentMatrix(test)
as.matrix(tdm)
```

```
       Docs
Terms    1 2 3 4 5
  word1 1 1 1 1 1
  word2 0 1 1 1 1
  word3 0 0 1 1 1
  word4 0 0 0 1 1
  word5 0 0 0 0 1
```

We compute the correlation of rows to get a measure of association across documents.

```
# Correlation between word2 and word3, word4, and word5
cor(c(0,1,1,1,1),c(0,0,1,1,1))
cor(c(0,1,1,1,1),c(0,0,0,1,1))
cor(c(0,1,1,1,1),c(0,0,0,0,1))
```

```
> cor(c(0,1,1,1,1),c(0,0,1,1,1))
[1] 0.6123724
> cor(c(0,1,1,1,1),c(0,0,0,1,1))
[1] 0.4082483
> cor(c(0,1,1,1,1),c(0,0,0,0,1))
[1] 0.25
```

Alternatively, use the findAssocs function, which computes all correlations between a given term and all terms in the term-document matrix and reports those higher than the correlation threshold.

```
# find associations greater than 0.1
findAssocs(tdm,"word2",0.1)
```

```
word3 word4 word5
 0.61  0.41  0.25
```

Now that you have an understanding of how association works across documents, here is an example for the corpus of Buffett letters.

```
# Select the first ten letters
tdm <-  TermDocumentMatrix(stem.letters[1:10])
# compute the associations
findAssocs(tdm, "invest",0.80)
```

```
   shooting  cigarettes    eyesight        feed moneymarket    pinpoint
       0.83        0.82        0.82        0.82        0.82        0.82
 ringmaster     suffice     tunnels     unnoted
       0.82        0.82        0.82        0.82
```

Cluster analysis

Cluster analysis is a statistical technique for grouping together sets of observations that share common characteristics. Objects assigned to the same group are more similar in some way than those allocated to another cluster. In the case of a corpus, cluster analysis groups documents based on their similarity. Google, for instance, uses clustering for its news site.

The general principle of cluster analysis is to map a set of observations in multidimensional space. For example, if you have seven measures for each observation, each will be mapped into seven-dimensional space. Observations that are close together in this space will be grouped together. In the case of a corpus, cluster analysis is based on mapping frequently occurring words into a multidimensional space. The frequency with which each word appears in a document is used as a weight, so that frequently occurring words have more influence than others.

There are multiple statistical techniques for clustering, and multiple methods for calculating the distance between points. Furthermore, the analyst has to decide how many clusters to create. Thus, cluster analysis requires some judgment and experimentation to develop a meaningful set of groups.

The following code computes all possible clusters using the Ward method of cluster analysis. A term-document matrix is sparse, which means it consists mainly of zeroes. In other words, many terms occur in only one or two documents, and the cell entries for the remaining documents are zero. In order to reduce the computations required, sparse terms are removed from the matrix. You can vary the sparse parameter to see how the clusters vary.

```
# Cluster analysis
# name the columns for the letter's year
colnames(tdm) <-  1998:2012
# Remove sparse terms
tdm1 <- removeSparseTerms(tdm, 0.5)
# transpose the matrix
tdmtranspose <-  t(tdm1)
cluster = hclust(dist(tdmtranspose))
# plot the tree
plot(cluster)
```

The cluster analysis is shown in the following figure as a dendrogram, a tree diagram, with a leaf for each year. Clusters seem to from around consecutive years. Can you think of an explanation?

Dendrogram for Buffett letters from 1998-2012

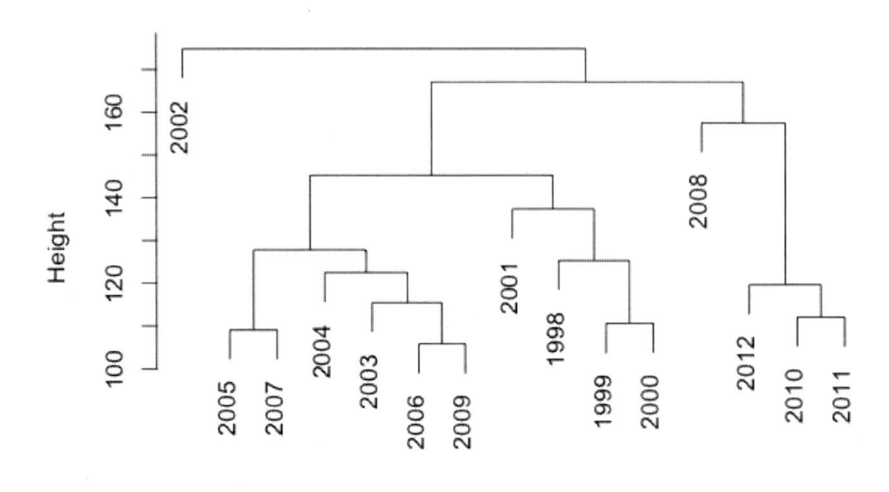

Topic modeling

Topic modeling is a set of statistical techniques for identifying the themes that occur in a document set. The key assumption is that a document on a particular topic will contain words that identify that topic. For example, a report on gold mining will likely contain words such as "gold" and "ore." Whereas, a document on France, would likely contain the terms "France," "French," and "Paris."

The package **topicmodels** implements topic modeling techniques within the R framework. It extends tm to provide support for topic modeling. It implements two methods: latent Dirichlet allocation (LDA), which assumes topics are uncorrelated; and correlated topic models (CTM), an extension of LDA that permits correlation between topics.[55] Both LDA and CTM require that the number of topics to be extracted is determined a priori. For example, you might decide in advance that five topics gives a reasonable spread and is sufficiently small for the diversity to be understood.[56]

Words that occur frequently in many documents are not good at distinguishing among documents. The weighted term frequency inverse document frequency (tf-idf) is a measure designed for determining which terms discriminate among documents. It is based on the term frequency (tf), defined earlier, and the inverse document frequency.

Inverse document frequency

The inverse document frequency (idf) measures the frequency of a term across documents.

$$idf_t = \log_2 \frac{m}{df_t}$$

Where

m = number of documents (i.e., rows in the case of a term-document matrix);

df_t = number of documents containing term *t*.

If a term occurs in every document, then its idf = 0, whereas if a term occurs in only one document out of 15, its idf = 3.91.

To calculate and display the idf for the letters corpus, we can use the following R script.

```
# calculate idf for each term
idf <-  log2(nDocs(dtm)/col_sums(dtm > 0))
# create dataframe for graphing
df.idf <-  data.frame(idf)
ggplot(df.idf,aes(idf)) + geom_histogram(fill="chocolate") + xlab("Inverse
document frequency")
```

55 The mathematics of LDA and CTM are beyond the scope of this text. For details, see Grün, B., & Hornik, K. (2011). Topicmodels: An R package for fitting topic models. *Journal of Statistical Software*, 40(13), 1-30.

56 Humans have a capacity to handle about 7±2 concepts at a time. Miller, G. A. (1956). The magical number seven, plus or minus two: some limits on our capacity for processing information. *The Psychological Review*, 63(2), 81-97.

Inverse document frequency (corpus has 15 documents)

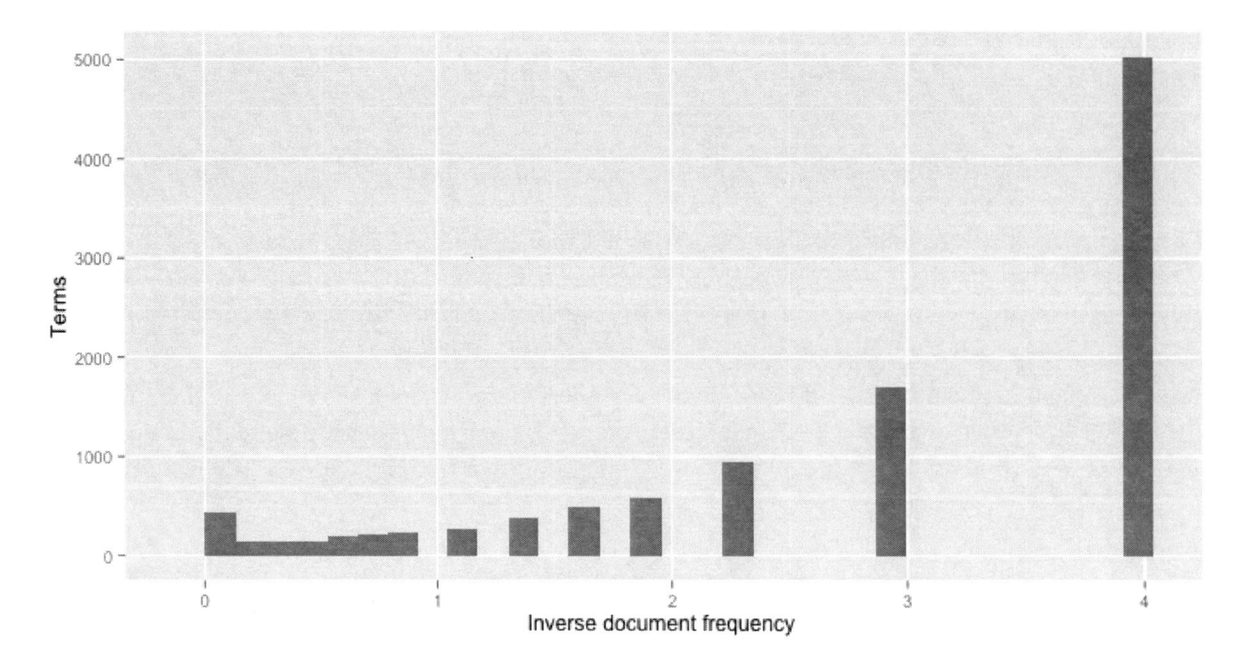

The preceding graphic shows that about 5,000 terms occur in only one document (i.e., the idf = 3.91) and less than 500 terms occur in every document. The terms with an idf in the range 1 to 2 are likely to be the most useful in determining the topic of each document.

Term frequency inverse document frequency (tf-idf)

The weighted term frequency inverse document frequency (tf-idf or ω_{td}) is calculated by multiplying a term's frequency (tf) by its inverse document frequency (idf).

$$\omega_{td} = tf_{td} \bullet \log_2 \frac{m}{df_t}$$

Where

tf_{td} = frequency of term t in document d.

Topic modeling with R

Prior to topic modeling, pre-process a text file in the usual fashion (e.g., convert to lower case, remove punctuation, and so forth). Then, create a document-term matrix.

The mean term frequency-inverse document frequency (tf-idf) is used to select the vocabulary for topic modeling. We use the median value of tf-idf for all terms as a threshold.

```
require(topicmodels)
require(slam)
# calculate tf-idf for each term
tfidf <-  tapply(dtm$v/row_sums(dtm)[dtm$i], dtm$j, mean) * log2(nDocs(dtm)/
col_sums(dtm > 0))
# report dimensions (terms)
```

```
dim(tfidf)
# report median to use as cut-off point
median(tfidf)
```

The goal of topic modeling is to find those terms that distinguish a document set. Thus, terms with low frequency should be omitted because they don't occur often enough to define a topic. Similarly, those terms occurring in many documents don't differentiate between documents.

A common heuristic is to select those terms with a tf-idf > median(tf-idf). As a result, we reduce the document-term matrix by keeping only those terms above the threshold and eliminating rows that have zero terms. Because the median is a measure of central tendency, this approach reduces the number of columns by roughly half.

```
# select columns with tf-idf > median
dtm <- dtm[,tfidf >= median(tfidf)]
#select rows with rowsum > 0
dtm <- dtm[row_sums(dtm) > 0,]
# report reduced dimension
dim(dtm)
```

As mentioned earlier, the topic modeling method assumes a set number of topics, and it is the responsibility of the analyst to estimate the correct number of topics to extract. It is common practice to fit models with a varying number of topics, and use the various results to establish a good choice for the number of topics. The analyst will typically review the output of several models and make a judgment on which model appears to provide a realistic set of distinct topics. Here is some code that starts with five topics.

```
# set number of topics to extract
k <- 5 # number of topics
SEED <- 2010 # seed for initializing the model rather than the default random
# try multiple methods - takes a while for a big corpus
TM <- list(VEM = LDA(dtm, k = k, control = list(seed = SEED)),
VEM_fixed = LDA(dtm, k = k, control = list(estimate.alpha = FALSE, seed =
SEED)),
Gibbs = LDA(dtm, k = k, method = "Gibbs", control = list(seed = SEED, burnin =
1000,  thin = 100, iter = 1000)), CTM = CTM(dtm, k = k,control = list(seed =
SEED, var = list(tol = 10^-3), em = list(tol = 10^-3))))
topics(TM[["VEM"]], 1)
terms(TM[["VEM"]], 5)
```

```
> topics(TM[["VEM"]], 1)
 1  2  3  4  5  6  7  8  9 10 11 12 13 14 15
 4  4  4  2  2  5  4  4  4  3  3  5  1  5  5
> terms(TM[["VEM"]], 5)
     Topic 1         Topic 2          Topic 3     Topic 4         Topic 5
[1,] "thats"         "independent"    "borrowers" "clayton"       "clayton"
```

```
[2,] "bnsf"         "audit"         "clayton"    "eja"           "bnsf"
[3,] "cant"         "contributions" "housing"    "contributions" "housing"
[4,] "blackscholes" "reserves"      "bhac"       "merger"        "papers"
[5,] "railroad"     "committee"     "derivative" "reserves"      "marmon"
```

The output indicates that the first three letter (1998-2000) are about topic 4, the fourth (2001) topic 2, and so on.

Topic 1 is defined by the following terms: thats, bnsf, cant, blacksholes, and railroad. As we have seen previously, some of these words (e.g., thats and cant, which we can infer as being that's and can't) are not useful differentiators, and the dictionary could be extended to remove them from consideration and topic modeling repeated. For this particular case, it might be that Buffett's letters don't vary much from year to year, and he returns to the same topics in each annual report.

Skill builder

Experiment with the topicmodels package to identify the topics in Buffett's letters. You might need to use the dictionary feature of text mining to remove selected words from the corpus to develop a meaningful distinction between topics.

Named-entity recognition (NER)

Named-entity recognition (NER) places terms in a corpus into predefined categories such as the names of persons, organizations, locations, expressions of times, quantities, monetary values, and percentages. It identifies some or all mentions of these categories, as shown in the following figure, where an organization, place, and date are recognized.

Named-entity recognition example

The Olympics were in London in 2012.

Tags are added to the corpus to denote the category of the terms identified.

```
The <organization>Olympics</organization> were in <place>London</place> in
<date>2012</date>.
```

There are two approaches to developing an NER capability. A *rules-based* approach works well for a well-understood domain, but it requires maintenance and is language dependent. *Statistical classifiers*, based on machine learning, look at each word in a sentence to decide whether it is the start of a named-entity, a continuation of an already identified named-entity, or not part of a named-entity. Once a named-entity is distinguished, its category (e.g., place) must be identified and surrounding tags inserted.

Statistical classifiers need to be trained on a large collection of human-annotated text that can be used as input to machine learning software. Human-annotation, while time-consuming, does not require a high level of skill. It is the sort of task that is easily parallelized and distributed to a large number of human coders who have a reasonable understanding of the corpus's context (e.g., able to recognize that London is a place and that the Olympics is an organization). The software classifier need to be trained on approximately 30,000 words.

The accuracy of NER is dependent on the corpus used for training and the domain of the documents to be classified. For example, NER based on a collection of news stories is unlikely to be very accurate for recognizing entities in medical or scientific literature. Thus, for some domains, you will likely need to annotate a set of sample documents to create a relevant model. Of course, as times change, it might be necessary to add new annotated text to the learning script to accommodate new organizations, place, people and so forth. A well-trained statistical classifier applied appropriately is usually capable of correctly recognizing entities with 90 percent accuracy.

NER software

At this point, R does not have a package for NER, so you will need to consider tools such as OpenNLP or KNIME. OpenNLP[57] is an Apache Java-based machine learning based toolkit for the processing of natural language in text format. It is a collection of natural language processing tools, including a sentence detector, tokenizer, parts-of-speech(POS)-tagger, syntactic parser, and named-entity detector. The NER tool can recognize people, locations, organizations, dates, times. percentages, and money. You will need to write a Java program to take advantage of the toolkit.

KNIME (Konstanz Information Miner)[58] is a general purpose data management and analysis package that supports NER. It is released as professional open-source software, which means you can freely download the application. You can opt to buy a support contract to gain access to additional features and get help with problem resolution. KNIME is available for Linux, OS X, and Windows. The following diagram illustrates how a string of actions can be sequenced to read the latest *New York Times* RSS feed, apply NER, and produce a word cloud.

KNIME process flow for NER

Future developments

Text mining and natural language processing are developing areas and you can expect new tools to emerge. Document summarization, relationship extraction, advanced sentiment analysis, and cross-language information retrieval (e.g., a Chinese speaker querying English documents and getting a Chinese translation of the search and selected documents) are all areas of research that will likely result in generally available software with possible R versions. If you work in this area, you will need to continually scan for new software that extends the power of existing methods and adds new text mining capabilities.

Summary

Language enables cooperation through information exchange. Natural language processing (NLP) focuses on developing and implementing software that enables computers to handle large scale processing of language in a variety of forms, such as written and spoken. The inherent ambiguity in written and spoken speech makes NLP challenging. Don't expect NLP to provide the same level of exactness and starkness as

57 http://opennlp.apache.org

58 http://knime.org

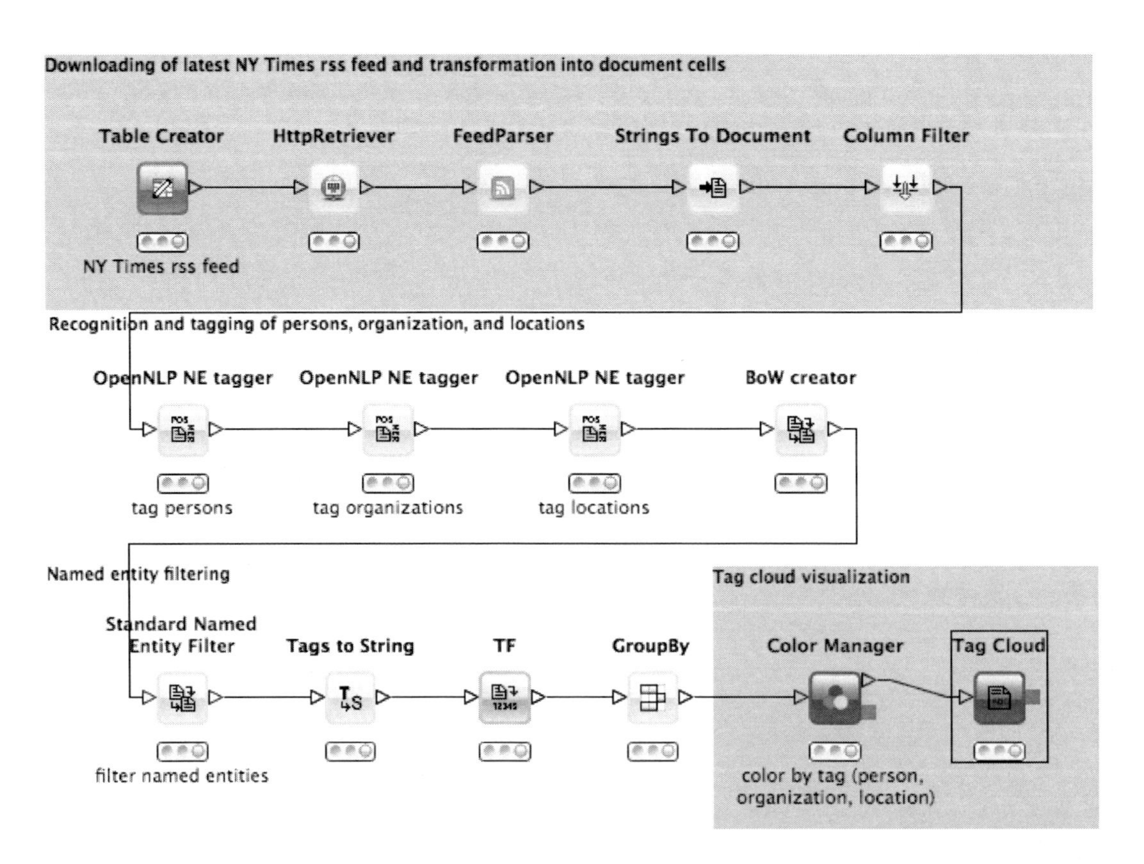

numeric processing. There are three levels to consider when processing language: semantics, discourse, and pragmatics.

Sentiment analysis is a popular and simple method of measuring aggregate feeling. Tokenization is the process of breaking a document into chunks. A collection of text is called a corpus. The Flesch-Kincaid formula is a common way of assessing readability. Preprocessing, which prepares a corpus for text mining, can include case conversion, punctuation removal, number removal, stripping extra white spaces, stop word filtering, specific word removal, word length filtering, parts of speech (POS) filtering, Stemming, and regex filtering.

Word frequency analysis is a simple technique that can also be the foundation for other analyses. A term-document matrix contains one row for each term and one column for each document. A document-term matrix contains one row for each document and one column for each term. Words that occur frequently within a document are usually a good indicator of the document's content. A word cloud is way of visualizing the most frequent words. Co-occurrence measures the frequency with which two words appear together. Cluster analysis is a statistical technique for grouping together sets of observations that share common characteristics. Topic modeling is a set of statistical techniques for identifying the topics that occur in a document set. The inverse document frequency (idf) measures the frequency of a term across documents. Named-entity recognition (NER) places terms in a corpus into predefined categories such as the names of persons, organizations, locations, expressions of times, quantities, monetary values, and percentages. Statistical classification is used for NER. OpenNLP is an Apache Java-based machine learning-based toolkit for the processing of natural language in text format. KNIME (Konstanz Information Miner) is a general purpose data management and analysis package that supports NER. Document summarization, relationship extraction, advanced sentiment analysis, and cross-language information retrieval are all areas of research.

Key terms and concepts

Association

Cluster analysis

Co-occurrence

Corpus

Dendrogram

Document-term matrix

Flesch-Kincaid formula

Inverse document frequency

KNIME

Named-entity recognition (NER)

Natural language processing (NLP)

Number removal

OpenNLP

Parts of speech (POS) filtering

Preprocessing

Punctuation removal

Readability

Regex filtering

Sentiment analysis

Statistical classification

Stemming

Stop word filtering

Stripping extra white spaces

Term-document matrix

Term frequency

Term frequency inverse document frequency

Text mining

Tokenization

Topic modeling

Word cloud

Word frequency analysis

Word length filtering

References

Feinerer, I. (2008). An introduction to text mining in R. *R News*, 8(2), 19-22.

Feinerer, I., Hornik, K., & Meyer, D. (2008). Text mining infrastructure in R. *Journal of Statistical Software*, 25(5), 1-54.

Grün, B., & Hornik, K. (2011). topicmodels: An R package for fitting topic models. *Journal of Statistical Software*, 40(13), 1-30.

Ingersoll, G., Morton, T., & Farris, L. (2012). *Taming Text: How to find, organize and manipulate it*. Greenwich, CT: Manning Publications.

Exercises

1. Take the recent annual reports for UPS[59] and convert them to text using an online service, such as http://www.fileformat.info/convert/doc/pdf2txt.htm. Complete the following tasks:
   a. Count the words in the most recent annual report.
   b. Compute the readability of the most recent annual report.
   c. Create a corpus.
   d. Preprocess the corpus.
   e. Create a term-document matrix and compute the frequency of words in the corpus.

59 http://www.investors.ups.com/phoenix.zhtml?c=62900&p=irol-reportsannual

f. Construct a word cloud for the 25 most common words.

g. Undertake a cluster analysis, identify which reports are similar in nature, and see if you can explain why some reports are in different clusters.

h. Build a topic model for the annual reports.

2. Download KNIME and run the *New York Times* RSS feed analyzer.[60] Note that you will need to add the community contributed plugin, Palladian, to run the analyzer. See Preferences > Install/Update > Available Software Sites and the community contribution page <http://tech.knime.org/community>.

3. Merge the annual reports for Berkshire Hathaway (i.e., Buffett's letters) and UPS into a single corpus.

   a. Undertake a cluster analysis and identify which reports are similar in nature.

   b. Build a topic model for the combined annual reports.

   c. Do the cluster analysis and topic model suggest considerable differences in the two sets of reports?

60 http://tech.knime.org/named-entity-recognizer-and-tag-cloud-example

17. Clustering computing

Let us change our traditional attitude to the construction of programs: Instead of imagining that our main task is to instruct a computer what to do, let us concentrate rather on explaining to humans what we want the computer to do

Donald E. Knuth, *Literate Programming*, 1984

Learning objectives

Students completing this chapter will:

- Understand the paradigm shift in decision-oriented data processing;
- Understand the principles of cluster computing

A paradigm shift

There is much talk of big data, and much of it is not very informative. Rather a lot of big talk but not much smart talk. Big data is not just about greater variety and volumes of data at higher velocity, which is certainly occurring. The more important issue is the paradigm shift in data processing so that large volumes of data can be handled in a timely manner to support decision making. The foundations of this new model is the shift to cluster computing, which means using large numbers of commodity processors for massively parallel computing.

We start by considering what is different between the old and new paradigms for decision data analysis. Note that we are not considering transaction processing, for which the relational model is a sound solution. Rather, we are interested in the processing of very large volumes of data at a time, and the relational model was not designed for this purpose. It is suited for handling transactions, which typically involve only a few records. The multidimensional database (MDDB) is the "old" approach for large datasets and cluster compute is the "new."

Another difference is the way data are handled. The old approach is to store data on a high speed disk drive and load it into computer memory for processing. To speed up processing, the data might be moved to multiple computers to enable parallel processing and the results merged into a single file. Because data files are typically much larger than programs, moving data from disks to computers is time consuming. Also, high performance disk storage devices are expensive. The new method is to spread a large data file across multiple commodity computers, possibly using HDFS, and then send each computer a copy of the program to run in parallel. The results from the individual jobs are then merged. While data still need to be moved to be processed, they are moved across a high speed data channel within a computer rather than the lower speed cables of a storage network.

Old	New
Data to the program	Program to the data
Mutable data	Immutable data
Special purpose hardware	Commodity hardware

The drivers

Exploring the drivers promoting the paradigm shift is a good starting point for understanding this important change in data management. First, you will recall that you learned in Chapter 1 that decision making is the central organizational activity. Furthermore, because data-driven decision making increases organizational performance, many executives are now demanding data analytics to support their decision making.

Second, as we also explained in Chapter 1, there is a societal shift in dominant logic as we collectively recognize that we need to focus on reducing environmental degradation and carbon emissions. Service and sustainability dominant logics are both data intensive. Customer service decisions are increasingly based on the analysis of large volumes of operational and social data. Sustainability oriented decisions also require large volumes of operational data, which are combined with environmental data collected by massive sensor networks to support decision making that reduces an organization's environmental impact.

Third, the world is in the midst of a massive data generating digital transformation. Large volumes of data are collected about the operation on an aircraft's jet engines, how gamers play massively online games, how people interact in social media space, and the operation of cell phone networks, for example. The digital transformation of life and work is creating a bits and bytes tsunami.

The bottleneck and its solution

In a highly connected and competitive world, speedy high quality decisions can be a competitive advantage. However, large data sets can take some time and expense to process, and so as more data are collected, there is a the danger that decision making will gradually slow down and its quality lowered. Data analytics becomes a bottleneck when the conversion of data to information is too slow. Second, decision quality is lowered when there is a dearth of skills for determining what data should be converted to information and interpreting the resulting conversion. We capture these problems in the elaboration of a diagram that was introduced in Chapter 1, which now illustrates the causes of the conversion, request, and interpretation bottlenecks.

Data analytics bottleneck

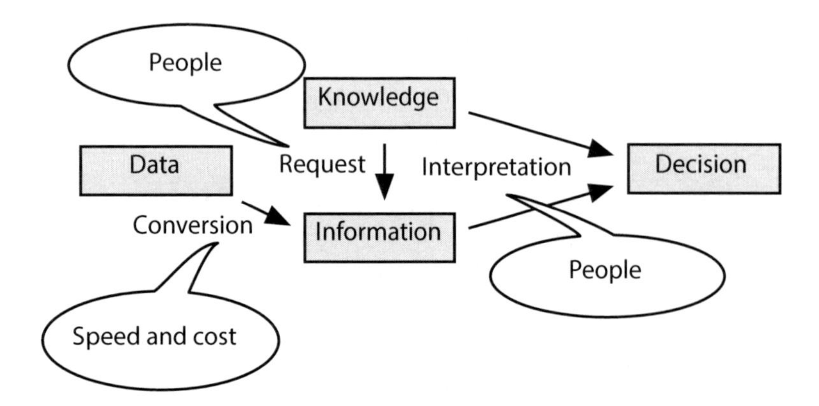

The people skills problem is being addressed by the many universities that have added graduate courses in data analytics. The **Lambda Architecture**[61] is a proposed general solution to the speed and cost problem.

61 Marz, N., & Warren, J. (2012). *Big Data*: Manning Publications.

Lambda Architecture

We will now consider the three layers of the Lambda Architecture: batch, speed, and serving.

The batch layer

Batch computing describes the situation where a computer works on one or more large tasks with minimal interruption. Because early computers were highly expensive and businesses operated at a different tempo, batch computing was common in the early days of information systems. The efficiency gains of batch computing mainly come from uninterrupted sequential file processing. The computer spends less time waiting for data to be retrieved from disks, particularly with HDFS where files are in 64Mb chunks. Batch computing is very efficient, though not timely, and the Lambda Architecture takes advantage of this efficiency.

The batch layer is used to precompute queries by running them with the most recent version of the dataset. The precomputed results are saved and can then be used as required. For example, a supermarket chain might want to know how often each pair of products appears in each shopper's basket for each day for each store. These data might help it to set cross-promotional activities within a store (e.g., a joint special on steak and mashed potatoes). The batch program could precompute the count of joint sales for each pair of items for each day for each store in a given date range. This highly summarized data could then be used for queries about customers' baskets (e.g., how many customers purchased both shrimp and grits in the last week in all Georgia stores?). The batch layer works with a dataset essentially consisting of every supermarket receipt because this is the record of a customer's basket. This dataset is also stored by the batch layer. Hadoop is well-suited for handling the batch layer, as you will see later, but it is not the only option.

New data are appended to the master dataset to preserve is immutability, so that it remains a complete record of transactions for a particular domain (e.g., all receipts). These incremental data are processed the next time the batch process is restarted.

The batch layer can be processing several batches simultaneously. It typically keeps recomputing batch views using the latest dataset every few hours or maybe overnight.

The serving layer

The serving layer processes views computed by the batch layer so they can be queried. Because the batch layer produces a flat file, the serving layer indexes it for random access. The serving layer also replaces the old batch output with the latest indexed batch view when it is received from the batch layer. In a typical Lambda Architecture system, there might be several or more hours between batch updates.

The combination of the batch and serving layers provides for efficient reporting, but it means that any queries on the files generated by the batch layer might be several or more hours old. We have efficiency but not timeliness, for which we need the speed layer.

Speed layer

Once a batch recompute has started running, all newly collected data cannot be part of the resulting batch report. The purpose of the speed layer is to process incremental data as they arrive so they can be merged with the latest batch data report to give current results. Because the speed layer modifies the results as each chunk of data (e.g., a transaction) is received, the merge of the batch and speed layer computations can be used to create real-time reports.

Merging speed and serving layer results to create a report (source: (Marz and Warren 2012))

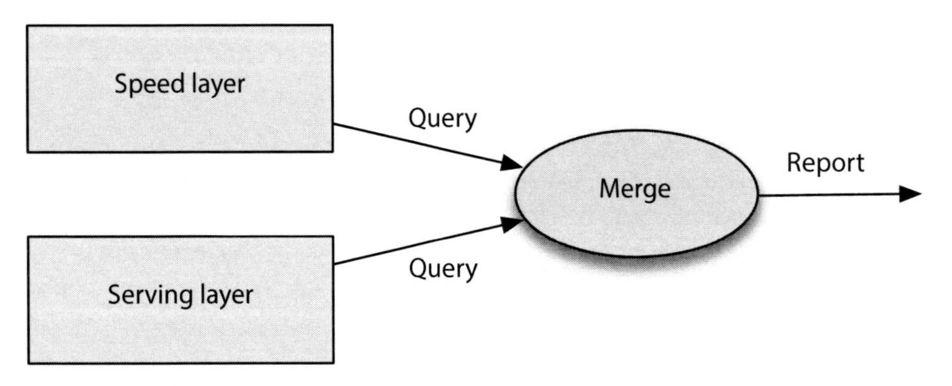

Putting the layers together

We now examine the process in detail.

Assume $batch_{n-1}$ has just been processed.

1. During the processing of $batch_{n-1}$, $increment_{n-1}$ was created from the records received. The combination of these two data sets creates $batch_n$.
2. As the data for $increment_{n-1}$ were received, speed layer ($Sresults_{n-1}$) were dynamically recomputed.
3. A current report can be created by combining speed layer and batch layer results (i.e., $Sresults_{n-1}$ and $Bresults_{n-1}$).

Now, assume batch computation resumes with $batch_n$.

1. $Sresults_n$ are computed from the data collected ($increment_n$) while $batch_n$ is being processing.
2. Current reports are based on $Bresults_{n-1}$, $Sresults_{n-1}$, and $Sresults_n$.
3. At the end of processing $batch_n$, $Sresults_{n-1}$ can be discarded because $Bresults_n$ includes all the data from $batch_{n-1}$ and $increment_{n-1}$.

The preparation of a real-time report using batch and speed layer results when processing $batch_n$

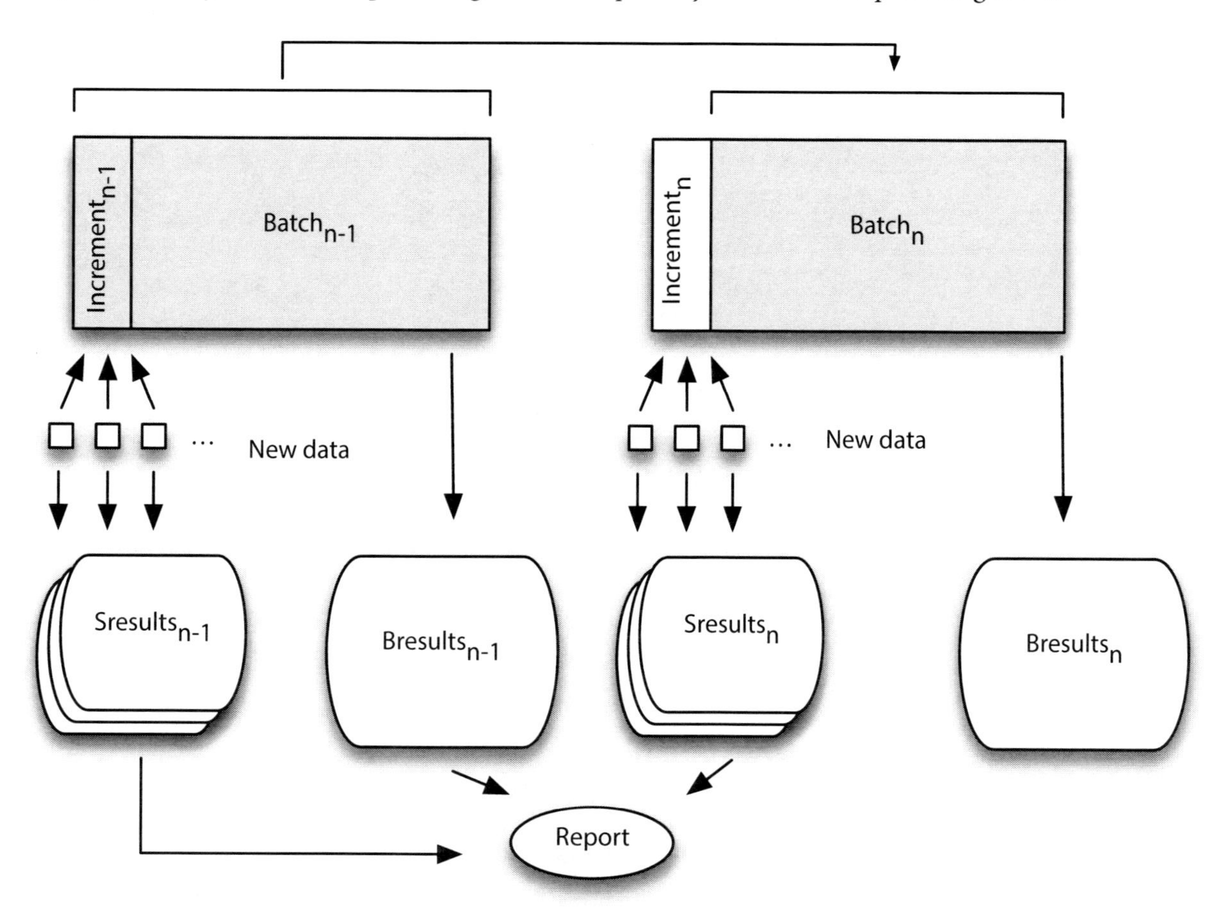

In summary, the batch layer pre-computes reports using all the currently available data. The serving layer indexes the results of the batch layers and creates views that are the foundation for rapid responses to queries. The speed layer does incremental updates as data are received. Queries are handled by merging data from the serving and speed layers.

Lambda Architecture (source: (Marz and Warren 2012))

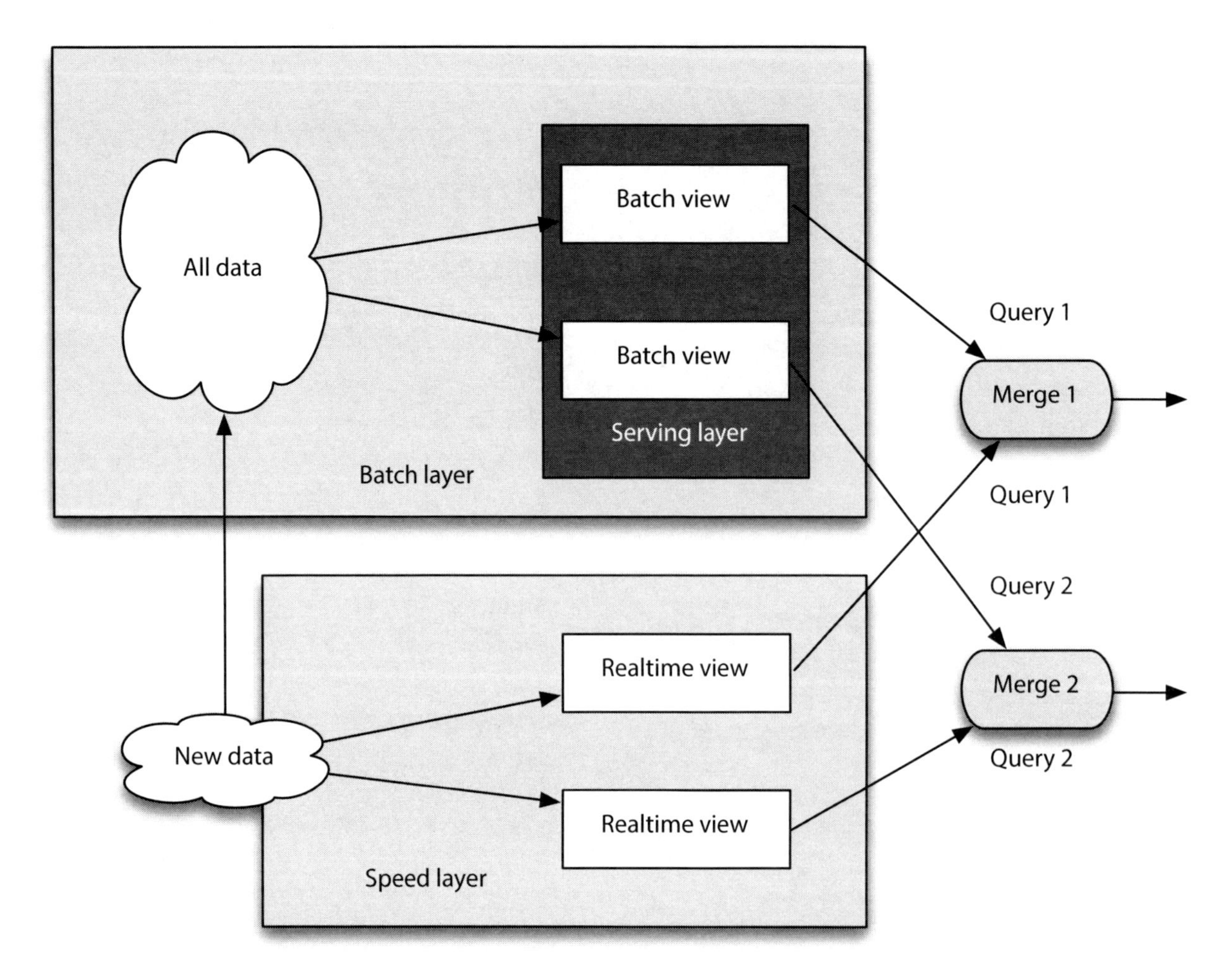

Benefits of the Lambda Architecture

The Lambda Architecture provides some important advantages for processing large datasets, and these are now considered.

Robust and fault-tolerant

Programming for batch processing is relatively simple and also it can easily be restarted if there is a problem. Replication of the batch layer dataset across computers increases fault tolerance. If a block is unreadable, the batch processor can shift to the identical block on another node in the cluster. Also, the redundancy intrinsic to a distributed file system and distributed processors provides fault-tolerance.

The speed layer is the complex component of the Lambda Architecture. Because complexity is isolated to this layer, it does not impact other layers. In addition, since the speed layer produced temporary results, these can be discarded in the event of an error. Eventually the system will right itself when the batch layer produces a new set of results, though intermediate reports might be a little out of date.

Low latency reads and updates

The speed layer overcomes the long delays associated with batch processing. Real-time reporting is possible through the combination of batch and speed layer outputs.

Scalable

Scalability is achieved using a distributed file system and distributed processors. To scale, new computers are added.

Support a wide variety of applications

The general architecture can support reporting for a wide variety of situations provided that a key, value pair can be extracted from a dataset.

Extensible

New data types can be added to the master dataset or new master datasets created. Furthermore, new computations can be added to the batch and speed layers to create new views.

Ad hoc queries

On the fly queries can be run on the output of the batch layer provided the required data are available in a view.

Minimal maintenance

The batch and serving layers are relatively simple programs because they don't deal with random updates or writes. Simple code requires less maintenance.

Debuggable

Batch programs are easier to debug because you can have a clear link between the input and output. Data immutability means that debugging is easier because no records have been overwritten during batch processing.

Relational and Lambda Architectures

Relational technology must support both transaction processing and data analytics. As a result, it needs to be more complex than the Lambda Architecture. Separating out data analytics from transaction processing simplifies the supporting technology and makes it suitable for handling large volumes of data efficiently. Relational systems can continue to support transaction processing and, as a byproduct, produce data that are fed to Lambda Architecture based business analytics.

Hadoop

Hadoop, an Apache project,[62] supports distributed processing of large data sets across a cluster of computers. A Hadoop cluster consists of many standard processors, nodes, with associated main memory and disks. They are connected by Ethernet or switches so they can pass data from node to node. Hadoop is highly scalable and reliable. It is a suitable technology for the batch layer of the Lambda architecture. Hadoop is used for data analytics, machine learning, search ranking, email anti-spam, ad optimization, and other areas of applications which are constantly emerging.

A market analysis projects that the Hadoop market is growing at 58% per year.[63] It also asserts that, "Hadoop is the only cost-sensible and scalable open source alternative to commercially available Big Data management packages. It also becomes an integral part of almost any commercially available Big Data solution and de-facto industry standard for business intelligence (BI)."

62 http://hadoop.apache.org

63 https://www.marketanalysis.com/?p=279 (Hadoop Market Forecast 2017-2022, May 2017)

Hadoop distributed file system (HDFS)

HDFS is a highly scalable, fault-toleration, distributed file system. When a file is uploaded to HDFS, it is split into fixed sized blocks of at least 64MB. Blocks are replicated across nodes to support parallel processing and provide fault tolerance. As the following diagram illustrates, an original file when written to HDFS is broken into multiple large blocks that are spread across multiple nodes. HDFS provides a set of functions for converting a file to and from HDFS format and handling HDFS.

Splitting of file across a HDFS cluster.

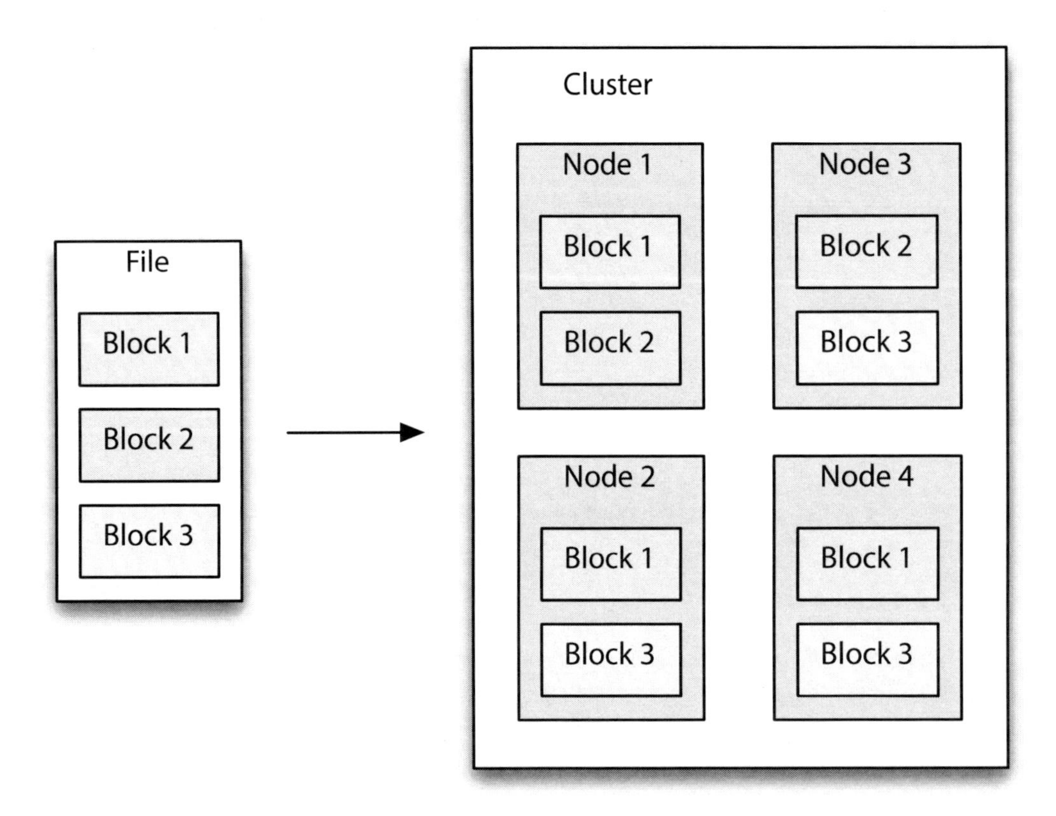

On each node, blocks are stored sequentially to minimize disk head movement. Blocks are grouped into files, and all files for a dataset are grouped into a single folder. As part of the simplification to support batch processing, there is no random access to records and new data are added as a new file.

Scalability is facilitated by the addition of new nodes, which means adding a few more pieces of inexpensive commodity hardware to the cluster. Appending new data as files on the cluster also supports scalability.

HDFS also supports partitioning of data into folders for processing at the folder level. For example, you might want all employment related data in a single folder.

Spark

Spark is an Apache project[64] for cluster computing that was initiated at the University of California Berkeley and later transferred to Apache as an open source project. Spark's distributed file system, resilient distributed dataset (RDD), has similar characteristics to HDFS. Spark can also interface with HDFS and

64 https://spark.apache.org

other distributed file systems. For testing and development, Spark has a local mode, that can work with a local file system.

Spark includes several component, including Spark SQL for SQL-type queries, Spark streaming for real-time analysis of event data as it is received, and a machine learning (ML) library. This library is designed for in-memory processing and is approximately10 times faster than disk-based equivalent approaches. Distributed graph processing is implemented using GraphX.

Computation with Spark

Spark applications can be written in Java, Scala, Python, and R. In our case, we will use the sparklyr, [65] an R interface to Spark. This package provides a simple, easy to use set of commands for exposing the distributed processing power of Spark to those familiar with R. In particular, it supports dplyr for data manipulation of Spark datasets and access to Sparks ML library.

Before starting with sparklyr, you need to check that you have latest version of Java on your machine.[66] Use RStudio to install sparklyr. For developing and testing on your computer, install a local version of Spark.

```
library(sparklyr)
spark_install(version='2.0.2')
```

Use the `spark_connect` function to connect to Spark either locally or on a remote Spark cluster. The following code shows how to specify local Spark connection (sc).

```
library(sparklyr)
sc <- spark_connect(master = "local")
```

Tabulation

In this example, we have a list of average monthly temperatures for New York's Central Park[67] and we want to determine how often each particular temperature occurred. R

Average monthly temperatures since 1869 are read, and the temperature rounded to an integer for the convenience of tabulation.

```
library(tidyverse)
url <-  "http://people.terry.uga.edu/rwatson/data/centralparktemps.txt"
t <- read_delim(url, delim=',')
# tabulate frequencies for temperature
t %>%
  mutate(Fahrenheit = round(temperature,0)) %>%
  group_by(Fahrenheit) %>%
  summarize(Frequency = n())
```

```
# A tibble: 61 x 2
   Fahrenheit Frequency
```

65 http://spark.rstudio.com

66 https://www.java.com/en/download/help/version_manual.xml

67 http://www.erh.noaa.gov/okx/climate/records/monthannualtemp.html

```
        <dbl>      <int>
 1         20          1
 2         22          2
 3         23          3
 4         24          5
 5         25         15
 6         26         10
 7         27          8
 8         28         19
 9         29         24
10         30         32
# ... with 51 more rows
```

Spark

By using dplyr in the prior R code, we can copy and paste and add a few commands for the Spark implementation. The major differences are the creation of a Spark connection (sc) and copying the R tibble to Spark with `copy-to`. Also, note that you need to sort the resulting tibble, which is not required in regular R.

```
library(sparklyr)
library(tidyverse)
spark_install(version='2.0.2')
sc <- spark_connect(master = "local", spark_home=spark_home_dir(version = "2.0.2"))
url <-  "http://people.terry.uga.edu/rwatson/data/centralparktemps.txt"
t <- read_delim(url, delim=',')
t_tbl <- copy_to(sc,t)
t_tbl %>%
  mutate(Fahrenheit = round(temperature,0)) %>%
  group_by(Fahrenheit) %>%
  summarize(Frequency = n()) %>%
  arrange(Fahrenheit)
```

```
# A tibble: 61 x 2
   Fahrenheit Frequency
        <dbl>      <dbl>
 1         20          1
 2         22          2
 3         23          3
 4         24          5
 5         25         15
 6         26          9
 7         27          9
 8         28         17
 9         29         26
10         30         30
# ... with 51 more rows
```

It you observe the two sets of output carefully, you will note that the results are not identical. It is because rounding can vary across systems. The IEEE Standard for Floating-Point Arithmetic (IEEE 754)[68] states on rounding, “if the number falls midway it is rounded to the nearest value with an even (zero) least significant bit.” Compare the results for `round(12.5,0)` and `round(13.5,0)`. R follows the IEEE standard, but Spark apparently does not.

Skill builder

Redo the tabulation example with temperatures in Celsius.

Basic statistics with Spark

We now take the same temperature dataset and calculate mean, min, and max monthly average temperatures for each year and put the results in a single file.

R

```
library(tidyverse)
url <-  "http://people.terry.uga.edu/rwatson/data/centralparktemps.txt"
t <- read_delim(url, delim=',')
# report minimum, mean, and maximum by year
t %>%
  group_by(year) %>%
  summarize(Min=min(temperature),
            Mean = round(mean(temperature),1),
            Max = max(temperature))
```

```
# A tibble: 148 x 4
    year   Min  Mean   Max
   <int> <dbl> <dbl> <dbl>
 1  1869  34.5  51.4  72.8
 2  1870  31.3  53.6  76.6
 3  1871  28.3  51.1  73.6
 4  1872  26.7  51.0  77.5
 5  1873  28.6  51.0  75.4
 6  1874  31.3  51.3  73.9
 7  1875  23.8  49.4  74.0
 8  1876  24.9  51.9  79.4
 9  1877  27.7  52.8  75.4
10  1878  30.3  53.5  77.8
# ... with 138 more rows
```

Spark

Again, the use of dplyr makes the conversion to Spark with simple.

```
library(sparklyr)
library(tidyverse)
spark_install(version='2.0.2')
sc <- spark_connect(master = "local", spark_home=spark_home_dir(version = "2.0.2"))
url <-  "http://people.terry.uga.edu/rwatson/data/centralparktemps.txt"
t <- read_delim(url, delim=',')
t_tbl <- copy_to(sc,t)
```

[68] https://en.wikipedia.org/wiki/IEEE_floating_point

```
# report minimum, mean, and maximum by year
t_tbl %>%
  group_by(year) %>%
  summarize(Min=min(temperature),
            Mean = round(mean(temperature),1),
            Max = max(temperature)) %>%
  arrange(year)
```

```
# A tibble: 148 x 4
    year   Min  Mean   Max
   <int> <dbl> <dbl> <dbl>
 1  1869  34.5  51.4  72.8
 2  1870  31.3  53.6  76.6
 3  1871  28.3  51.1  73.6
 4  1872  26.7  51.0  77.5
 5  1873  28.6  51.0  75.4
 6  1874  31.3  51.3  73.9
 7  1875  23.8  49.4  74.0
 8  1876  24.9  51.9  79.4
 9  1877  27.7  52.8  75.4
10  1878  30.3  53.5  77.8
# ... with 138 more rows
```

Skill builder

A file[69] of electricity costs for a major city contains a timestamp and cost separated by a comma. Compute the minimum, mean, and maximum costs.

Summary

Big data is a paradigm shift to new file structures, such as HDFS and RDD, and algorithms for the parallel processing of large volumes of data. The new file structure approach is to spread a large data file across multiple commodity computers and then send each computer a copy of the program to run in parallel. The drivers of the transformation are the need for high quality data-driven decisions, a societal shift in dominant logic, and digital transformation. The speed and cost of converting data to information is a critical bottleneck as is a dearth of skills for determining what data should be converted to information and interpreting the resulting conversion. The people skills problem is being addressed by universities' graduate courses in data analytics. The Lambda Architecture, a solution for handling the speed and cost problem, consists of three layers: speed, serving, and batch. The batch layer is used to precompute queries by running them with the most recent version of the dataset. The serving layer processes views computed by the batch layer so they can be queried. The purpose of the speed layer is to process incremental data as they arrive so they can be merged with the latest batch data report to give current results. The Lambda Architecture provides some important advantages for processing large datasets. Relational systems can continue to support transaction processing and, as a byproduct, produce data that are fed to Lambda Architecture based business analytics.

Hadoop supports distributed processing of large data sets across a cluster of computers. A Hadoop cluster consists of many standard processors, nodes, with associated main memory and disks. HDFS is a highly

69 http://people.terry.uga.edu/rwatson/data/electricityprices.csv

scalable, fault-toleration, distributed file system. Spark is a distributed computing method for scalable and fault-tolerant cluster computation.

Key terms and concepts

Batch layer

Bottleneck

Cluster computing

Hadoop

HDFS

Immutable data

Lambda Architecture

Parallelism

Serving layer

Spark

Speed layer

References and additional readings

Lam, C. (2010). *Hadoop in action*: Manning Publications Co.

Marz, N., & Warren, J. (2012). *Big Data*: Manning Publications.

Exercises

1. Write a MapReduce function for the following situations. In each case, write the corresponding R code to understand how MapReduce and conventional programming differ.
    a. Compute the square and cube of the numbers in the range 1 to 25. Display the results in a data frame.
    b. Using the average monthly temperatures for New York's Central Park, compute the maximum, mean, and average temperature in Celsius for each month.
    c. Using the average monthly temperatures for New York's Central Park, compute the max, mean, and min for August. You will need to use subsetting, as discussed in this chapter.
    d. Using the electricity price data, compute the average hourly cost.
    e. Read the national GDP file,[70] which records GDP in millions, and count how many countries have a GDP greater than or less than 10,000 million.

[70] http://people.terry.uga.edu/rwatson/data/GDP.csv

18. Dashboards

I think, aesthetically, car design is so interesting - the dashboards, the steering wheels, and the beauty of the mechanics. I don't know how any of it works, I don't want to know, but it's inspirational.

Paloma Picasso, designer, 2013[71]

Learning objectives

Students completing this chapter will:

- Understand the purpose of dashboards;
- Be able to use the R package shinydashboard to create a dashboard.

The value of dashboards

A dashboard is a web page or mobile app screen designed to present important information, primarily visual format, that can be quickly and easily comprehended. Dashboards are often used to show the current status, such as the weather app you have on your smart phone. Sometimes, a dashboard can also show historical data as a means of trying to identify long term trends. Key economic indicators for the last decade or so might be shown graphically to help strategic planners identify major shifts in economic activity. In a world overflowing with data, dashboards are an information system for summarizing and presenting key data. They can be very useful for maintaining situation awareness, by providing information about key environmental measures of the current situation and possible future developments.

A dashboard typically has a header, sidebar and body

Header

There is a header across the top of a page indicating the purpose of the dashboard, and additional facts about the content, such as the creator. A search engine is a common element of a header. A header can also contain tabs to various sources of information or reports (e.g., social media or Fortune 1000 companies).

Sidebar

A sidebar usually contains features that enable the body of the report to be customized to the viewer's needs. There might be filters to fine-tune what is displayed. Alternatively, there can be links to select information on a particular topic. In the following example, the sidebar contains some high level summary data, such as the number of documents in a category.

Body

The body of a dashboard contains the information selected via a sidebar. It can contained multiple panes, as shown in the following example, that display information in a variety of ways. The body of the following dashboard contains four types of visuals (a multiple time series graph, a donut style pie chart, a map, and a relationship network). It also shows a list of documents satisfying the specified criteria and a word cloud of these documents.

71 http://articles.latimes.com/2013/apr/18/news/la-paloma-picasso-discusses-new-olive-leaf-collection-for-tiffany-co-20130418

I encourage you to visit the ecoresearch.net dashboard. It is an exemplar and will give you some idea of the diversity of ways in which information can be presented and the power of a dashboard to inform.

A dashboard <http://www.ecoresearch.net/climate/>

Designing a dashboard

The purpose of a dashboard is to communicate, so the first step is to work with the client to determine the key performance indicators (KPIs), visuals, or text that should be on the dashboard. Your client will have some ideas as to what is required, and by asking good questions and prototyping, you can clarify needs.

You will need to establish that high quality data sources are available for conversion into the required information. If data are not available, you will have to work with other IS professionals to establish them, or you might conclude that it is infeasible to try to meet some requirements. You should keep the client informed as you work through the process of resolving data source issues. Importantly, you should make the client aware of any data quality problems. Sometimes your client might have to settle for less than desirable data quality in order to get some idea of directional swings.

Try several ways of visualizing data to learn what suits the client. The ggvis package works well with shinydashboard, the dashboard package we will learn. Also, other R packages, such as dygraphs, can be deployed for graphing time series, a frequent dashboard element for showing trends and turning points. Where possible, use interactivity to enable the client to customize as required (e.g., let the client select the period of a time series).

Design for ease of navigation and information retrieval. Simplicity should generally be preferred to complexity. Try to get chunks of information that should be viewed at the same time on the same page, and put other collections of cohesive information on separate tabs.

Use colors consistently and in accord with general practice. For example, red is the standard color for danger, so a red information box should contain only data that indicate a key problem (e.g., a machine not working or a major drop in a KPI). Consistency in color usage can support rapid scanning of a dashboard to identify the attention demanding situations.

Study dashboards that are considered as exhibiting leading business practices or are acknowledged as exemplars. Adopt or imitate the features that work well.

Build a prototype as soon as you have a reasonable idea of what you think is required and release it to your client. This will enable the client to learn what is possible and for you to get a deeper understanding of the needs. Learn from the response to the prototype, redesign, and release version 2. Continue this process for a couple of iterations until the dashboard is accepted.

Dashboards with R

R requires two packages, shiny and shinydashboard, to create a dashboard. Also, you must use RStudio to run your R script. Shiny is a R package for building interactive web applications in R. It was contributed to the R project by RStudio. It does not require knowledge of traditional web development tools such as HTML, CSS, or JavaScript. Shinydashboard uses Shiny to create dashboards, but you you don't need to learn Shiny. Some examples of dashboards built with shinydashboard will give you an idea of what is feasible.

The basics

In keeping with the style of this book, we will start by creating a minimalist dashboard without any content. There are three elements: header, sidebar, and body. It this case, they are all null. Together, these three elements create the UI (user interface) of a dashboard page. The dynamic side of the page, the server, is also null. A dashboard is a Shiny app, and the final line of code runs the application.

A UI script defines the layout and appearance of dashboard's web page. The server script contains commands to run the dashboard app and to make a page dynamic.

```
library(shiny)
library(shinydashboard)
header <- dashboardHeader()
sidebar <- dashboardSidebar()
body <- dashboardBody()
ui <- dashboardPage(header,sidebar,body)
server <- function(input, output) {}
shinyApp(ui, server)
```

When you select all the code and run it, a new browser window will open and display a blank dashboard.

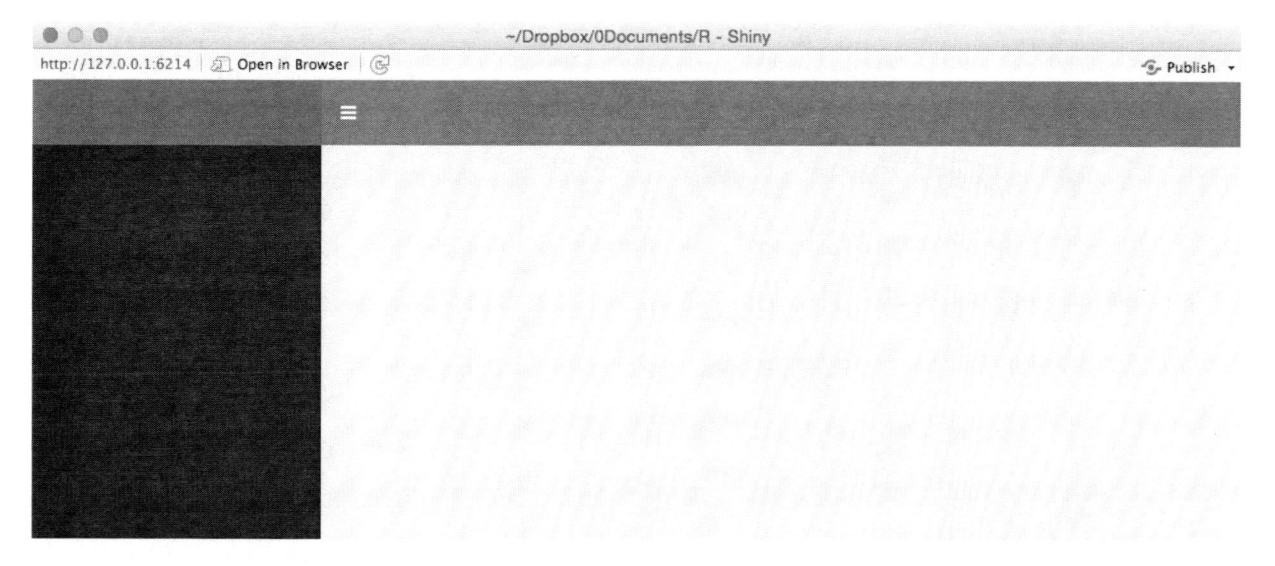

When you inspect some shinydashboard code, you will sometimes see a different format in which commands are deeply embedded in other commands. The following code illustrates this approach. I suggest you avoid it because it is harder to debug such code, especially when you are a novice.

```
library(shiny)
library(shinydashboard)
ui <- dashboardPage(
  dashboardHeader(),
  dashboardSidebar(),
  dashboardBody()
)
server <- function(input, output) {}
shinyApp(ui, server)
```

Terminating a dashboard

A dashboard is meant to be a dynamic web page that is updated when refreshed or when some element of the page is clicked on. It remains active until terminated. This means that when you want to test a new version of a dashboard or run another one, you must first stop the current one. You will need to click the stop icon on the console's top left to terminate a dashboard and close its web page.

A header and a body

We now add some content to a basic dashboard by giving it a header and a body. This page reports the share price of Apple, which has the stock exchange symbol of AAPL. The getQuote function[72] of the quantmod package returns the latest price, with about a two hour delay, every time the page is opened or refreshed. Notice the use of paste to concatenate a title and value.

```
library(shiny)
library(shinydashboard)
```

72 getQuote returns a dataframe, containing eight values :Trade Time, Last, Change, % Change, Open, High, Low, Volume. For example, getQuote('AAPL')$`% Change` returns the percentage change in price since the daily opening price.

```
library(quantmod)
header <- dashboardHeader(title = 'Apple stock watch')
sidebar <- dashboardSidebar()
body <- dashboardBody(paste('Latest price ',getQuote('AAPL')$Last))
ui <- dashboardPage(header,sidebar,body)
server <- function(input, output) {}
shinyApp(ui, server)
```

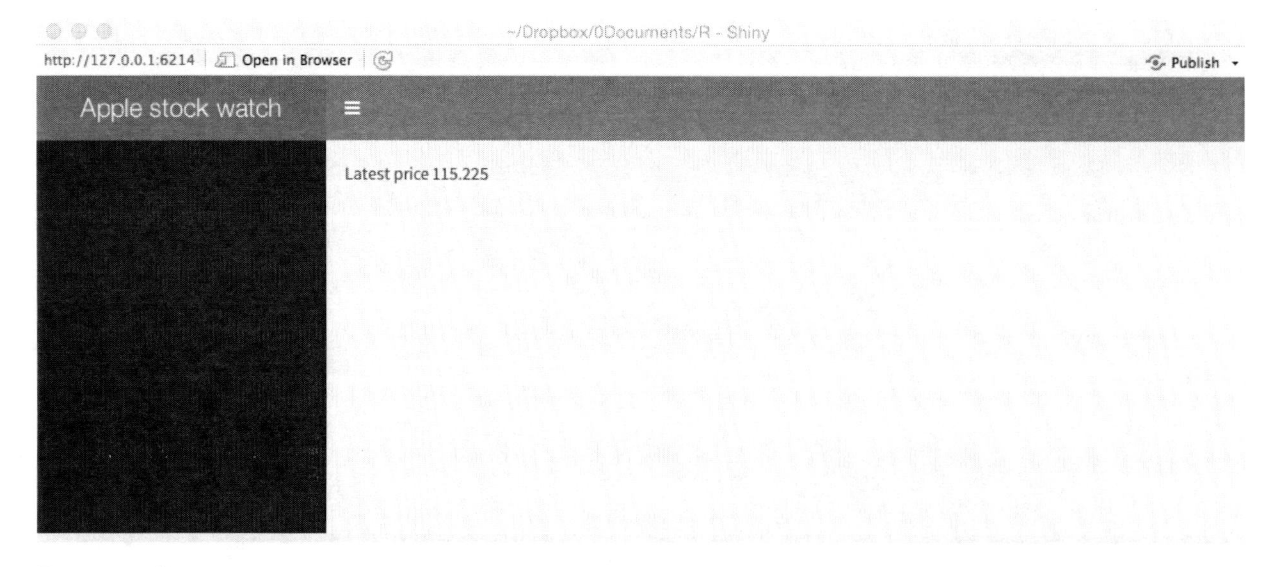

Boxes and rows

Boxes are the building blocks of a dashboard, and they can be assembled into rows or columns. The fluidRow function is used to place boxes into rows and columns. The following code illustrates the use of fluidRow.

```
library(shiny)
library(shinydashboard)
library(quantmod)
header <- dashboardHeader(title = 'Apple stock watch')
sidebar <- dashboardSidebar()
boxLatest <- box(title = 'Latest price: ',getQuote('AAPL')$Last, background =
'blue' )
boxChange <-  box(title = 'Change ',getQuote('AAPL')$Change, background =
'red' )
row <- fluidRow(boxLatest,boxChange)
body <- dashboardBody(row)
ui <- dashboardPage(header,sidebar,body)
server <- function(input, output) {}
shinyApp(ui, server)
```

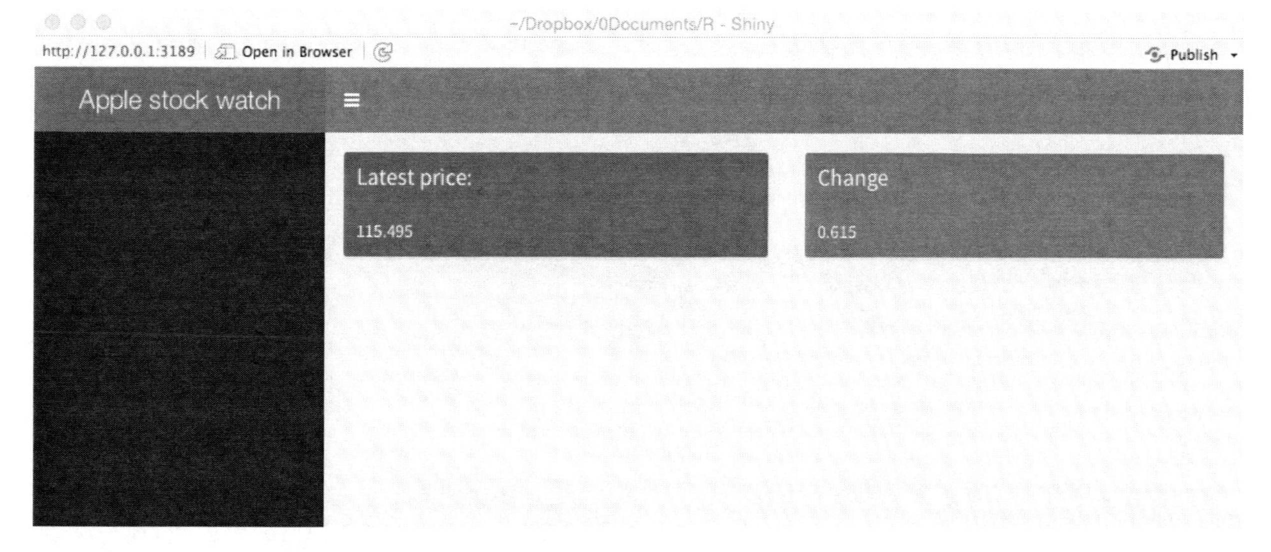

Skill builder

Add three more boxes (e.g., high price) to the dashboard just created.

Multicolumn layout

Shinydashboard is based on dividing up the breadth of a web page into 12 units, assuming it is wide enough. Thus, a box with width = 6 will take up half a page. The tops of the boxes in each row will be aligned, but the bottoms may not be because of the volume of data they contain. The fluidRows function ensures that a row's elements appear on the same line, if the browser has adequate width.

You can also specify that boxes are placed in columns. The column function defines how much horizontal space, within the 12-unit width grid, each element should occupy.

In the following code, note how:

- a background for a box is defined (e.g., background='navy');
- five boxes are organized into two columns (rows <- fluidRow(col1,col2));
- volume of shares traded is formatted with commas (formatC(getQuote('AAPL') $Volume,big.mark=',').

```
library(shiny)
library(shinydashboard)
library(quantmod)
header <- dashboardHeader(title = 'Apple stock watch')
sidebar <- dashboardSidebar()
boxLast <-  box(title = 'Latest', width=NULL, getQuote('AAPL')$Last,
background='navy')
boxHigh <-  box(title = 'High', width=NULL, getQuote('AAPL')$High ,
background='light-blue')
boxVolume <- box(title = 'Volume', width=NULL, formatC(getQuote('AAPL')
$Volume,big.mark=','), background='aqua')
boxChange <-  box(title = 'Change', width=NULL, getQuote('AAPL')$Change,
background='light-blue')
```

```
boxLow <-  box(title = 'Low', width=NULL, getQuote('AAPL')$Low,
background='light-blue')
col1 <-  column(width = 4,boxLast,boxHigh,boxVolume)
col2 <-  column(width = 4,boxChange,boxLow)
rows <- fluidRow(col1,col2)
body <- dashboardBody(rows)
ui <- dashboardPage(header,sidebar,body)
server <- function(input, output) {}
shinyApp(ui, server)
```

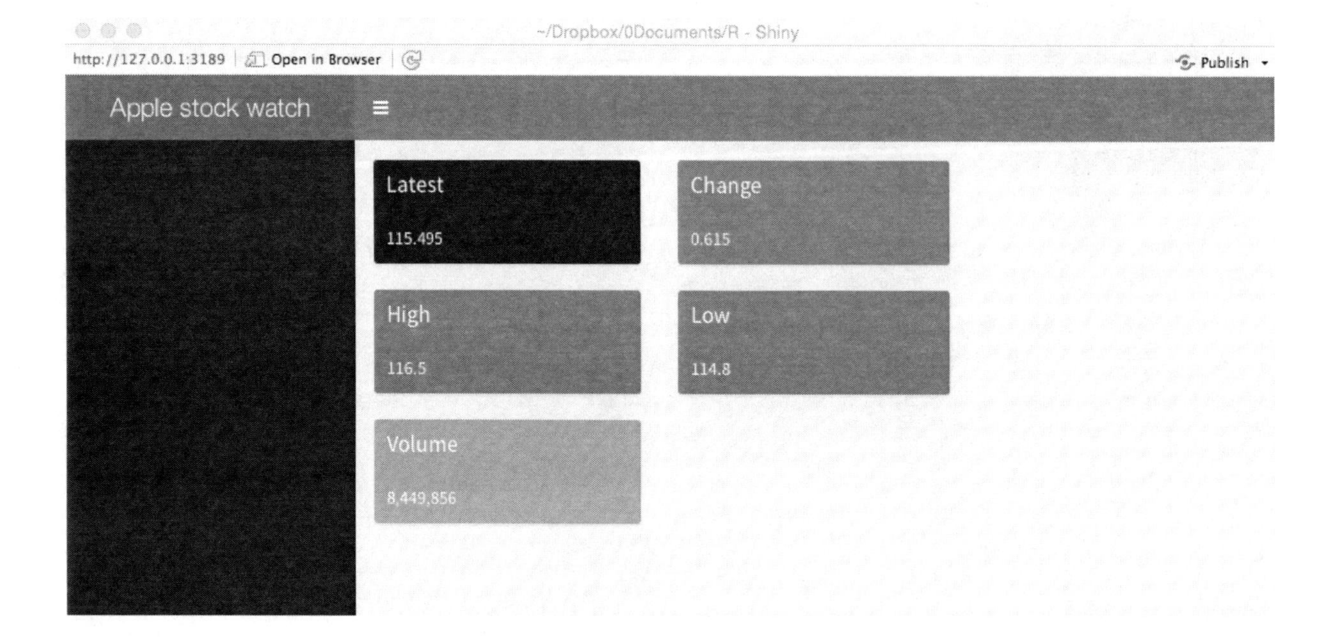

Sidebar

A sidebar is typically used to enable quick navigation of a dashboard's features. It can contain layers of menus, and by clicking on a menu link or icon, the dashboard can display different content in the body area.

A library of icons is available (Font-Awesome and Glyphicons) for use in the creation of a dashboard.

```
library(shiny)
library(shinydashboard)
library(quantmod)
header <-  dashboardHeader(title = 'Stock watch')
menuApple <-  menuItem("Apple", tabName = "Apple", icon = icon("dashboard"))
menuGoogle <- menuItem("Google", tabName = "Google", icon = icon("dashboard"))
sidebar <-  dashboardSidebar(sidebarMenu(menuApple, menuGoogle))
tabApple <-  tabItem(tabName = "Apple", getQuote('AAPL')$Last)
tabGoogle <-  tabItem(tabName = "Google", getQuote('GOOG')$Last)
tabs <-  tabItems(tabApple,tabGoogle)
```

```
body <-  dashboardBody(tabs)
ui <- dashboardPage(header, sidebar, body)
server <- function(input, output) {}
shinyApp(ui, server)
```

For the following dashboard, by clicking on Apple, you get its latest share price, and similarly for Google.

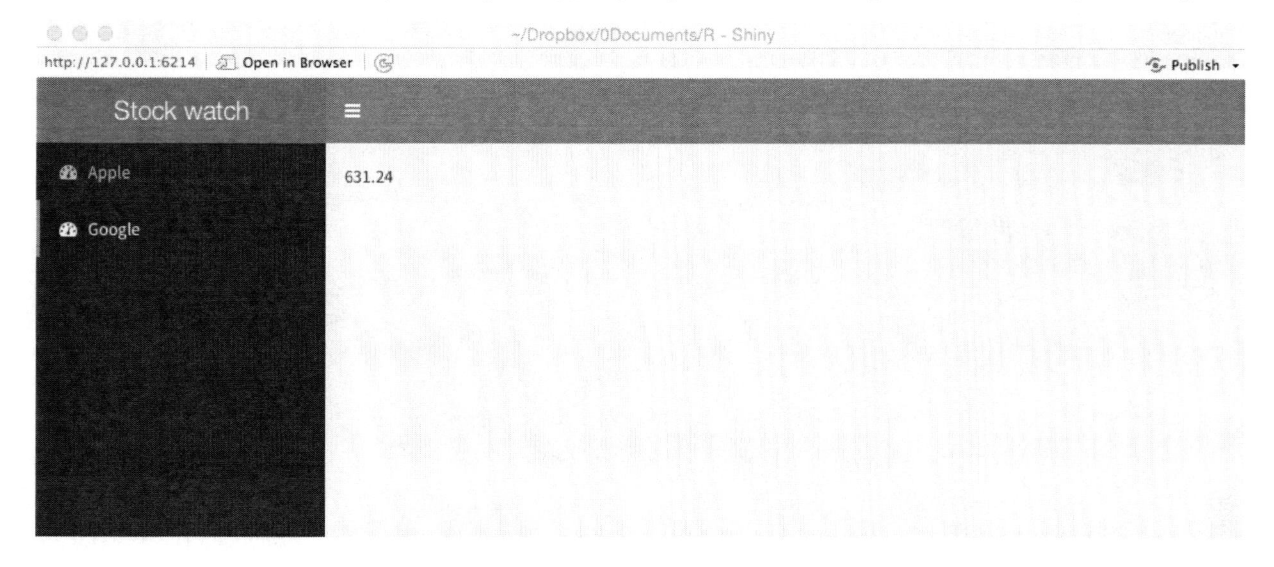

Infobox

An infobox is often used to display a single measure, such as a KPI.

```
library(shiny)
library(shinydashboard)
library(quantmod)
header <- dashboardHeader(title = 'Apple stock watch')
sidebar <- dashboardSidebar()
infoLatest <- infoBox(title = 'Latest', icon('dollar'), getQuote('AAPL')$Last,
color='red')
infoChange <- infoBox(title = 'Web site', icon('apple'),href='http://
investor.apple.com', color='purple')
row <- fluidRow(width=4,infoLatest,infoChange)
body <- dashboardBody(row)
ui <- dashboardPage(header,sidebar,body)
server <- function(input, output) {}
shinyApp(ui, server))
```

The following dashboard shows the latest price for Apple's shares. By clicking on the purple infobox, you access Apple investors' web site.

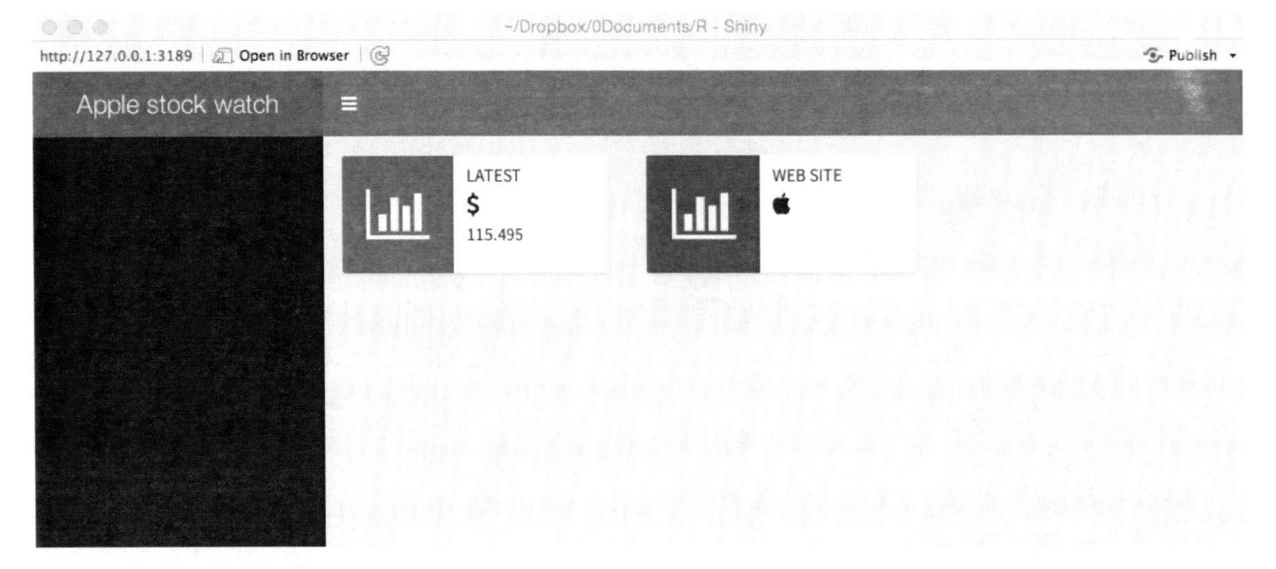

Dynamic dashboards

Dashboards are more useful when they give managers access to determine what is presented. The server function supports dynamic dashboards and executes when a dashboard is opened or refreshed.

The following basic dashboard illustrates how a server function is specified and how it communicates with the user interface. Using the time series graphing package, dygraphs, it creates a dashboard showing the closing price for Apple. Key points to note:

- The ui function indicates that it wants to create a graph with the code `dygraphOutput('apple')`.
- The server executes `output$apple <- renderDygraph({dygraph(Cl(get(getSymbols('AAPL'))))})` to produce the graph.
- The linkage between the UI and the server functions is through the highlighted code, as shown in the preceding two bullets and the following code block.
- The text parameter of dynagraphOutput() in the UI function must match the text following `output$` in the server function.
- The data to be graphed are retrieved with the Cl function of the quantmod package.[73]

```
library(shiny)
library(shinydashboard)
library(quantmod)
library(dygraphs) # graphic package for time series
header <-  dashboardHeader(title = 'Apple stock watch')
sidebar <- dashboardSidebar(NULL)
```

73 The quantmod function getSymbols('X') returns an time series object named X. The get function retrieves the name of the object and passes it to the Cl function and then the time series is graphed by dygraph. The code is a bit complicated, but necessary to generalize it for use with an interactive dashboard.

```
boxPrice <- box(title='Closing share price', width = 12, height = NULL,
dygraphOutput('apple'))
body <-   dashboardBody(fluidRow(boxPrice))
ui <- dashboardPage(header, sidebar, body)
server <- function(input, output) {
# quantmod retrieves closing price as a time series
output$apple <- renderDygraph({dygraph(Cl(get(getSymbols('AAPL'))))})
}
shinyApp(ui, server)
```

When you create the dashboard, mouse over the points on the graph and observe the data that are reported.

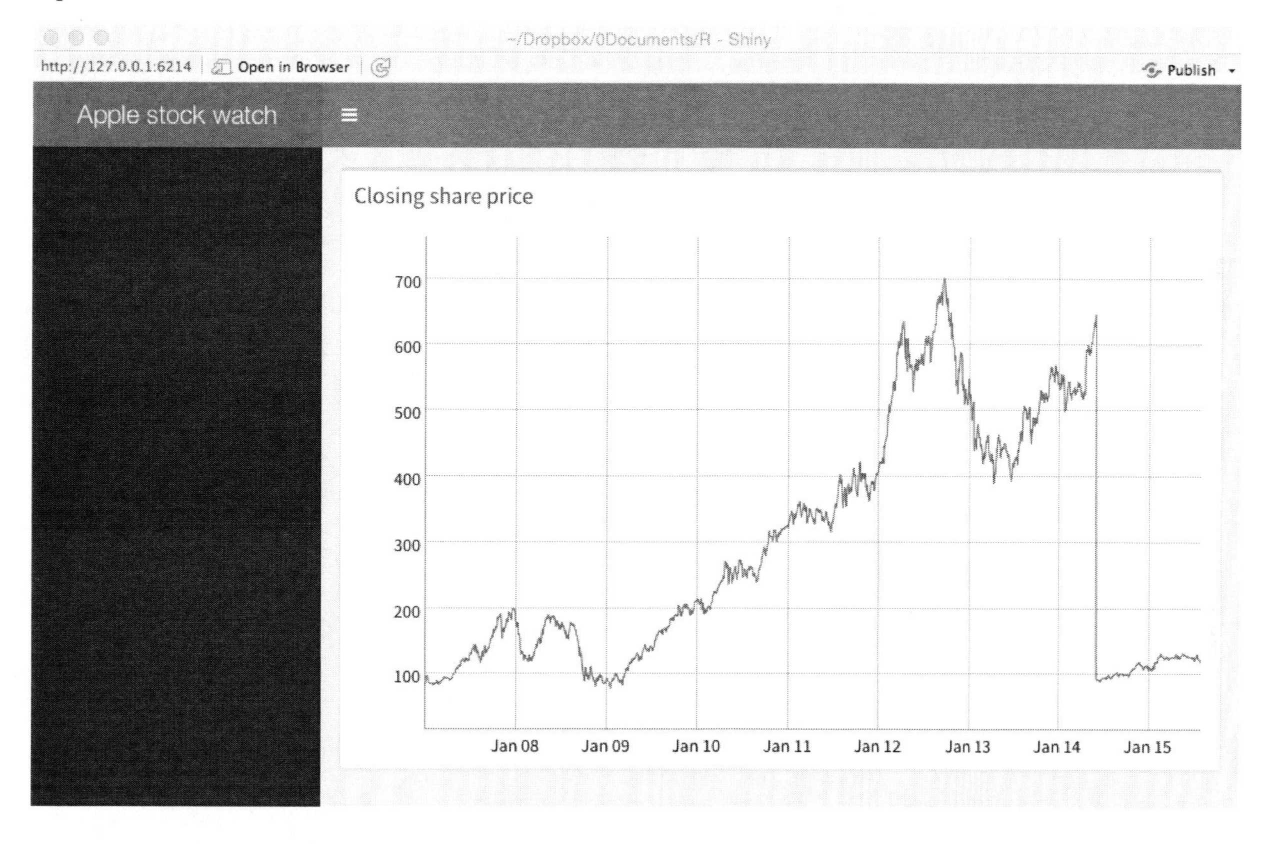

The following code illustrates how to create a dashboard that enables an analyst to graph the closing share price for Apple, Google, or Ford. Important variables in the code have been highlighted so you can easily see the correspondence between the UI and server functions. Note:

- The use of a selection list to pick one of three companies (`selectInput("symbol", "Equity:", choices = c("Apple" = "AAPL",  "Ford" = "F", "Google" = “GOOG")`).
- When an analyst selects one of the three firms, its stock exchange symbol (`symbol`) is passed to the server function.

- The value of symbol is used to retrieve the time series for the stock and to generate the graphic (`chart`) for display with `boxSymbol`. The symbol is also inserted into a text string (`text`) for display with `boxOutput`.

```
library(shiny)
library(shinydashboard)
library(quantmod)
library(dygraphs)
header <-  dashboardHeader(title = 'Stock watch')
sidebar <- dashboardSidebar(NULL)
boxSymbol <-  box(selectInput("symbol", "Equity:", choices = c("Apple" =
"AAPL",  "Ford" = "F", "Google" = "GOOG"), selected = 'AAPL'))
boxPrice <- box(title='Closing price', width = 12, height = NULL,
dygraphOutput("chart"))
boxOutput <-  box(textOutput("text"))
body <-   dashboardBody(fluidRow(boxSymbol, boxOutput, boxPrice))
ui <- dashboardPage(header, sidebar, body)
server <- function(input, output) {
  output$text <- renderText({
    paste("Symbol is:",input$symbol)
  })
# Cl in quantmod retrieves closing price as a time series
output$chart <- renderDygraph({dygraph(Cl(get(input$symbol)))}) # graph time
series
}
shinyApp(ui, server)
```

The following dashboard shows the pull down list in the top left for selecting an equity. The equity's symbol is then displayed in the top right box, and a time series of its closing price appears in the box occupying the entire second row.

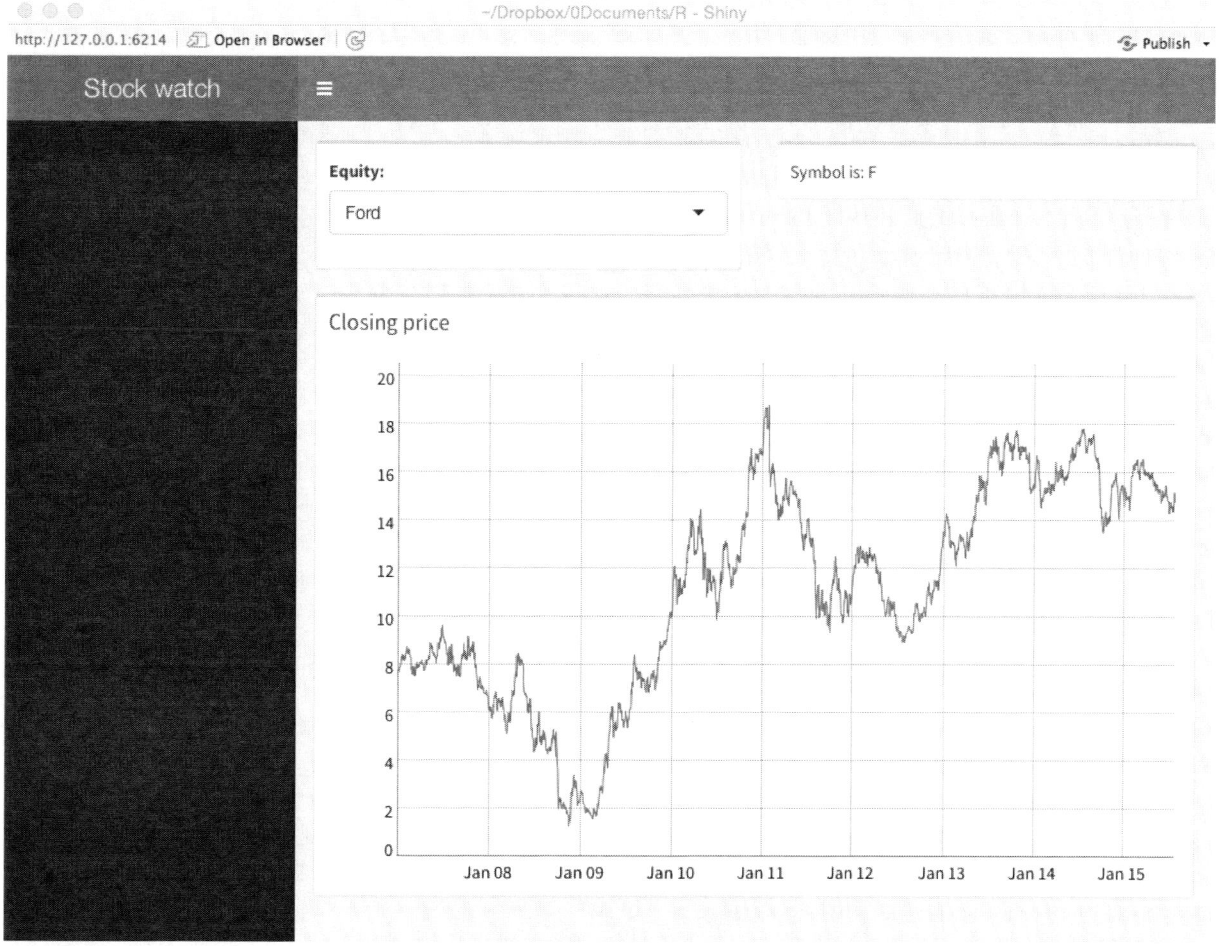

Input options

The preceding example illustrates the use of a selectInput function to select an equity from a list. There are other input options available, and these are listed in the following table.

Function	Purpose
checkboxInput()	Check one or more boxes
checkboxGroupInput()	A group of checkboxes
numericInput()	A spin box for numeric input
radioButtons()	Pick one from a set of options
selectInput()	Select from a drop-down text box
selectSlider()	Select using a slider
textInput()	Input text

Skill builder

Using ClassicModels, build a dashboard to report total orders by value and number for a given year and month.

.

Conclusion

This chapter has introduced you to the basic structure and use of shinydashboard. It has many options, and you will need to consult the online documentation and examples to learn more about creating dashboards.

Summary

A dashboard is a web page or mobile app screen that is designed to present important information in an easy to comprehend and primarily visual format. A dashboard consists of a header, sidebar, and body. Shinydashboard is an R package based on shiny that facilitates the creation of interactive real-time dashboards. It must be used in conjunction with RStudio.

Key terms and concepts

Body

Header

Server function

Sidebar

UI function

User-interface (ui)

References

Few, S. (2006). *Information dashboard design*: O'Reilly.

Exercises

1. Create a dashboard to show the current conditions and temperatures in both Fahrenheit and Celsius at a location. Your will need the rwunderground package and an API key.
2. Revise the dashboard created in the prior exercise to allow someone to select from up to five cities to get the weather details for that city.
3. Extend the previous dashboard. If the temperature is about 30C (86F), code the server function to give both temperature boxes a red background, and if it is below 10C (50F) give both a blue background. Otherwise the color should be yellow.
4. Use the WDI package to access World Bank Data and create a dashboard for a country of your choosing. Show three or more of the most current measures of the state of the selected country as an information box.
5. Use the WDI package to access World Bank Data for China, India, and the US for three variables, (1) CO2 emissions (metric tons per capita), (2) Electric power consumption (kWh per capita), and (3) forest area (% of land area). The corresponding WDI codes are: EN.ATM.CO2E.PC, EG.USE.ELEC.KH.PC, and AG.LND.FRST.ZS. Set up a dashboard so that a person can select the country from a pull down list and then the data for that country are shown in three infoboxes.
6. Use the WDI package to access World Bank Data for China, India, and the US for three variables, (1) CO2 emissions (metric tons per capita), (2) Electric power consumption (kWh per capita), and (3) forest area (% of land area). The corresponding WDI codes are: EN.ATM.CO2E.PC,

EG.USE.ELEC.KH.PC, and AG.LND.FRST.ZS. Set up a dashboard so that a person can select one of the three measures, and then the data for each country are shown in separate infoboxes.

7. Create a dashboard to:

   a. Show the conversion rate between two currencies using the quantmod package to retrieve exchange rates. Let a person select from one of five currencies using a drop down box;

   b. Show the value of input amount when converted one of the selected currencies to the other selected currency;

   c. Show the exchange rate between the two selected currencies over the last 100 days.

Section 4: Managing Organizational Memory

Every one complains of his memory, none of his judgment.

François Duc de La Rochefoucauld "Sentences et Maximes," Morales No. 89
1678

In order to manage data, you need a database architecture, a design for the storage and processing of data. Organizations strive to find an architecture that simultaneously achieves multiple goals:

1. Responds in a timely manner.
2. Minimizes the cost of

   processing data,

   storing data,

   data delivery,

   application development.
3. Highly secure.

The section deals with the approaches that can be used to achieve these goals. It covers **database structure and storage** alternatives. It provides the knowledge necessary to determine an appropriate **data storage structure** and device for a given situation. It also addresses the fundamental questions of where to store the data and where they should be processed.

Java has become a popular choice for writing programs that operate on multiple operating systems. Combining the interoperability of Java with standard SQL provides software developers with an opportunity to develop applications that can run on multiple operating systems and multiple DBMSs, and this typically reduces the cost of application development. Thus, for a comprehensive understanding of data management, it is important to learn how Java and SQL interface, which is the subject of the third chapter in this section.

As you now realize, organizational memory is an important resource requiring management. An inaccurate memory can result in bad decisions and poor customer service. Some aspects of organizational memory (e.g., chemical formulae, marketing strategy, and R&D plans) are critical to the well-being of an organization. The financial consequences can be extremely significant if these memories are lost or fall into the hands of competitors. Consequently, organizations must develop and implement procedures for maintaining data integrity. They need policies to protect the existence of data, maintain its quality, and ensure its confidentiality. Some of these procedures may be embedded in organizational memory technology, and others may be performed by data management staff. **Data integrity** is also addressed in this section.

When organizations recognize a resource as important to their long-term viability, they typically create a formal mechanism to manage it. For example, most companies have a human resources department responsible for activities such as compensation, recruiting, training, and employee counseling. People are the major resource of nearly every company, and the human resources department manages this resource. Similarly, the finance department manages a company's financial assets.

In the information age, data—the raw material of information—need to be managed. Consequently, **data administration** has become a formal organizational structure in many enterprises. It is designed to administer the data management and exploitation resources of an organization.

19. Data Structure and Storage

The modern age has a false sense of superiority because it relies on the mass of knowledge that it can use, but what is important is the extent to which knowledge is organized and mastered.

Goethe, 1810

Learning objectives

Students completing this chapter will, for a given situation, be able to recommend

- understand the implications of the data deluge;
- a data storage structure;
- a storage device.

Every quarter, The Expeditioner's IS group measures the quality of its service. It asks its clients to assess whether their hardware and software are adequate for their jobs, whether IS service is reliable and responsive, and whether they thought the IS staff were helpful and knowledgeable. The most recent survey revealed that some were experiencing unreasonably long delays for what were relatively simple queries. How could the IS group *tune* the database to reduce response time?

As though some grumpy clients were not enough, Alice dumped another problem on Ned's desk. The Marketing department had complained to her that the product database had been down for 30 minutes during a peak selling period. What was Ned going to do to prevent such an occurrence in the future?

Ned had just finished reading Alice's memo about the database problem when the Chief Accountant poked his head in the door. Somewhat agitated, he was waving an article from his favorite accounting journal that claimed that data stored on magnetic tape decayed with time. So what was he to do with all those financial records on magnetic tapes stored in the fireproof safe in his office? Were the magnetic bits likely to disappear tonight, tomorrow, or next week? "This business had lasted for centuries with paper ledgers. Why, you can still read the financial transactions for 1527," which he did whenever he had a few moments to spare. "But, if what I read is true, I soon won't be able to read the balance sheet from last year!"

It was just after 10 a.m. on a Monday, and Ned was faced with finding a way to improve response time, ensure that databases were continually available during business hours, and protect the long-term existence of financial records. It was going to be a long week.

The data deluge

With petabytes of new data being created daily, and the volume continuing to grow, many IS departments and storage vendors struggle to handle this data flood. "Big Data," as the deluge is colloquially known, arises from the flow of data created by Internet searches, Web site visits, social networking activity, streaming of videos, electronic health care records, sensor networks, large-scale simulations, and a host of other activities that are part of everyday business in today's world. The deluge requires a continuing investment in the management and storage of data.

A byte size table

Abbreviation	Prefix	Factor	Equivalent to
k	kilo	10^3	
M	mega	10^6	
G	giga	10^9	A digital audio recording of a symphony
T	tera	10^{12}	
P	peta	10^{15}	50 percent of all books in U.S. academic libraries
E	exa	10^{18}	5 times all the world's printed material
Z	zetta	10^{21}	
Y	yotta	10^{24}	

The following pages explore territory that is not normally the concern of application programmers or database users. Fortunately, the relational model keeps data structures and data access methods hidden. Nevertheless, an overview of what happens **under the hood** is part of a well-rounded education in data management and will equip you to work on some of the problems of the data deluge.

Data structures and access methods are the province of the person responsible for physically designing the database so that it responds in a timely manner to both queries and maintenance operations. Of course, there may be installations where application programmers have such responsibilities, and in these situations you will need to know physical database design.

Data structures

An in-depth consideration of the internal level of database architecture provides an understanding of the basic structures and access mechanisms underlying database technology. As you will see, the overriding concern of the internal level is to minimize disk access. In dealing with this level, we will speak in terms of files, records, and fields rather than the relational database terms of tables, rows, and columns. We do this because the discussion extends beyond the relational model to file structures in general.

The time required to access data on a magnetic disk, the storage device for many databases, is relatively long compared to that for main memory. Disk access times are measured in milliseconds (10^{-3}), and main memory access times are referred to in nanoseconds (10^{-9}). There are generally around five orders of magnitude difference between disk and main memory access—it takes about 10^5 times longer. This distinction is more meaningful if placed in an everyday context; it is like asking someone a question by phone or writing them a letter. The phone response takes seconds, and the written response takes days.

For many business applications, slow disk drives are a bottleneck. The computer often must wait for a disk to retrieve data before it can continue processing a request for information. This delay means that customers are also kept waiting. Appropriate selection of data structures and data access methods can considerably reduce delays. Database designers have two options: decrease disk read/write head movement or reduce disk accesses. Before considering these options, we need a general model of database access.

Database access

A three-layer model provides a framework for thinking about minimization of data access. This is a generic model, and a particular DBMS may implement the approach using a different number of layers. For simplicity, the discussion is based on retrieving a single record in a file, although the principles also apply to the retrieval of multiple records or an entire file.

Database access layers

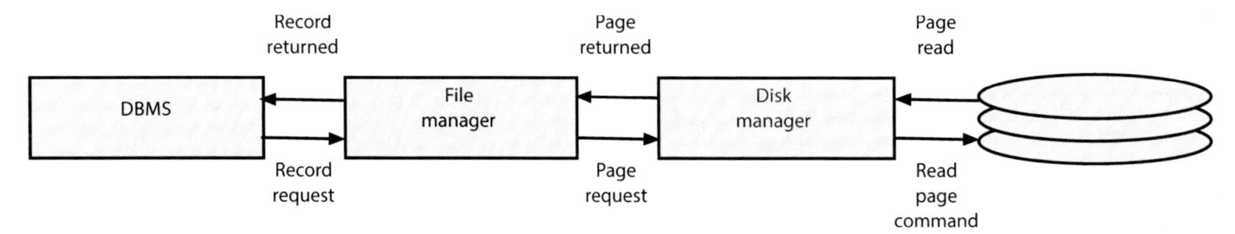

1. The DBMS determines which record is required and passes a request to the file manager to retrieve a particular record in a file.
2. The **file manager** converts this request into the address of the unit of storage (usually called a page) containing the specified record. A **page** is the minimum amount of storage accessed at one time and is typically around 1–4 kbytes. A page will often contain several short records (e.g., 200 bytes), but a long record (e.g., 10 kbytes) might be spread over several pages. In this example, we assume that records are shorter than a page.
3. The **disk manager** determines the physical location of the page, issues the retrieval instructions, and passes the page to the file manager.
4. The file manager extracts the requested record from the page and passes it to the DBMS.

The disk manager

The disk manager is that part of the operating system responsible for physical input and output (I/O). It maintains a directory of the location of each page on the disk with all pages identified by a unique page number. The disk manager's main functions are to retrieve pages, replace pages, and keep track of free pages.

Page retrieval requires the disk manager to convert the page number to a physical address and issue the command to read the physical location. Since a page can contain multiple records, when a record is updated, the disk manager must retrieve the relevant page, update the appropriate portion containing the record, and then replace the page without changing any of the other data on it.

The disk manager thinks of the disk as a collection of uniquely numbered pages. Some of these pages are allocated to the storage of data, and others are unused. When additional storage space is required, the disk manager allocates a page address from the set of unused page addresses. When a page becomes free because a file or some records are deleted, the disk manager moves that page's address to the unallocated set. Note, it does not erase the data, but simply indicates the page is available to be overwritten. This means that it is sometimes possible to read portions of a deleted file. If you want to ensure that an old file is truly deleted, you need to use an erase program that writes random data to the deleted page.

Disk manager's view of the world

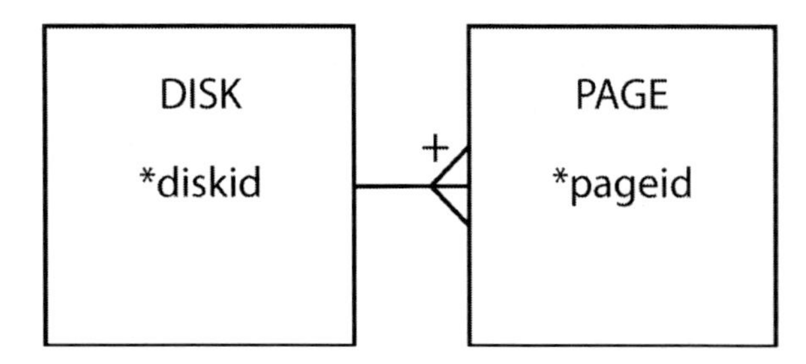

The file manager

The file manager, a level above the disk manager, is concerned with the storage of files. It thinks of the disk as a set of stored files. Each file has a unique file identifier, and each record within a file has a record identifier that is unique within that file.

File manager's view of the world

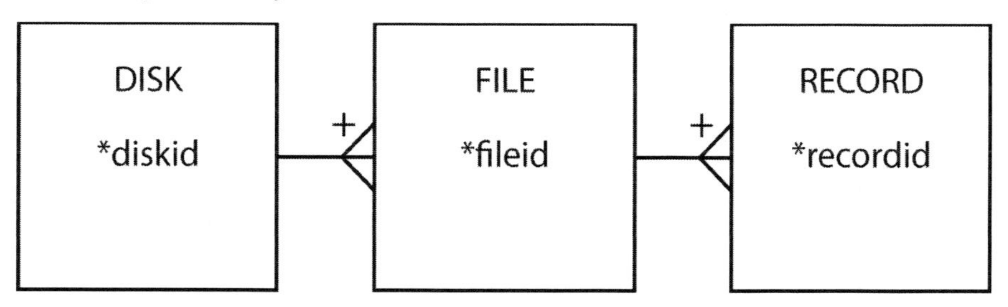

The file manager can

- Create a file
- Delete a file
- Retrieve a record from a file
- Update a record in a file
- Add a new record to a file
- Delete a record from a file

Techniques for reducing head movement

All disk storage devices have some common features. They have one or more recording surfaces. Typically, a magnetic disk drive has multiple recording surfaces, with data stored on tracks on each surface.

The key characteristics of disk storage devices that affect database access are **rotational speed** and **access arm speed**. The rotational speed of a magnetic disk is in the range of 3,000 to 15,000 rpm. Reading or writing a page to disk requires moving the read/write head to the destination track and waiting for the storage address to come under the head. Because moving the head usually takes more time (e.g., about 9 msec) than waiting for the storage address to appear under it (e.g., about 4 msec), data access times can be reduced by minimizing the movement of the read/write head or by rotating the disk faster. Since rotational speed is set by the disk manufacturer, minimizing read/write head movement is the only option available to database designers.

Cylinders

Head movement is reduced by storing data that are likely to be accessed at the same time, such as records in a file, on the same track on a single surface. When a file is too large to fit on one track, then it can be stored on the same track on different surfaces (i.e., one directly above or below the current track); such a collection of tracks is called a cylinder. The advantage of cylinder storage is that all tracks can be accessed without moving the read/write head. When a cylinder is full, remaining data are stored on adjacent cylinders. Adjacent cylinders are ideal for sequential file storage because the record retrieval pattern is predefined—the first record is read, then the second, and so on.

Clustering

Cylinder storage can also be used when the record retrieval pattern has some degree of regularity to it. Consider the following familiar data model of the following figure. Converting this data model to a relational database creates two tables. Conceptually, we may think of the data in each of the tables as being stored in adjacent cylinders. If, however, you frequently need to retrieve one row of nation and all the corresponding rows of stock, then `NATION` and `STOCK` rows should be intermingled to minimize access time.

Data model for `NATION` *and* `STOCK`

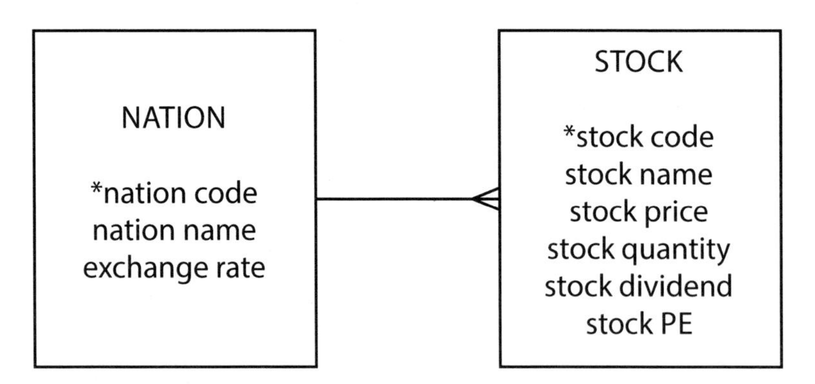

The term **clustering** denotes the concept that records that are frequently used together should be physically close together on a disk. Some DBMSs permit the database designer to specify clustering of different files to tune the database to reduce average access times. If usage patterns change, clustering specifications should be altered. Of course, clustering should be totally transparent to application programs and clients.

Techniques for reducing disk accesses

Several techniques are used to accelerate retrieval by reducing disk accesses. The ideal situation is when the required record is obtained with a single disk access. In special circumstances, it may be possible to create a file where the primary key can convert directly to a unique disk address (e.g., the record with primary key 1 is stored at address 1001, the record with primary key 2 is stored at address 1002, and so on). If possible, this method of **direct addressing** should be used because it is the fastest form of data access; however, it is most unusual to find such a direct mapping between the primary key and a disk address. For example, it would not be feasible to use direct addressing with a student file that has a Social Security number as the primary key because so many disk addresses would be wasted. Furthermore, direct addressing can work only for the primary key. What happens if there is a need for rapid retrieval on another field?

In most cases, database designers use features such as indexing, hashing, and linked lists. **Indexing**, a flexible and commonly used method of reducing disk accesses when searching for data, centers on the creation of a compact file containing the index field and the address of its corresponding record. The **B-tree** is a particular form of index structure that is widely used as the storage structure of relational DBMSs. **Hashing** is a direct access method based on using an arithmetic function to compute a disk address from a field within a file. A **linked list** is a data structure that accelerates data access by using pointers to show relationships existing between records.

Indexing

Consider the item file partially shown in the following table. Let's assume that this file has 10,000 records and each record requires 1 kbyte, which is a page on the particular disk system used. Suppose a common query is to request all the items of a particular type. Such a query might be

- *Find all items of type E.*

Regardless of the number of records with `itemtype = 'E'`, this query will require 10,000 disk accesses. Every record has to be retrieved and examined to determine the value of `itemtype`. For instance, if 20 percent of the items are of type E, then 8,000 of the disk accesses are wasted because they retrieve a record that is not required. The ideal situation would be to retrieve only those 2,000 records that contain an item of type E. We get closer to this ideal situation by creating a small file containing just the value of `itemtype` for each record and the address of the full record. This small file is called an index.

Portion of a 10,000 Record File

itemno	itemname	itemtype	itemcolor
1	Pocket knife — Nile	E	Brown
2	Pocket knife — Thames	E	Brown
3	Compass	N	—
4	Geopositioning system	N	—
5	Map measure	N	—
6	Hat — polar explorer	C	Red
7	Hat — polar explorer	C	White
8	Boots — snake proof	C	Green
9	Boots — snake proof	C	Black
10	Safari chair	F	Khaki

Part of the `ITEMTYPE` *index*

itemtype index

itemtype	
C	
C	
C	
C	
E	
E	
F	
N	
N	
N	

item

itemno	itemname	itemtype	itemcolor
1	Pocket knife-Nile	E	Brown
2	Pocket knife-Thames	E	Brown
3	Compass	N	-
4	Geo positioning system	N	-
5	Map measure	N	-
6	Hat-polar explorer	C	Red
7	Hat-polar explorer	C	White
8	Boots-snakeproof	C	Green
9	Boots-snakeproof	C	Black
10	Safari hat	F	Khaki

The `itemtype` index is a file. It contains 10,000 records and two fields. There is one record in the index for each record in `ITEM`. The first columns contains a value of `itemtype` and the second contains a pointer, an address, to the matching record of `item`. Notice that the index is in `itemtype` sequence. Storing the index in a particular order is another means of reducing disk accesses, as we will see shortly. The index is quite small. One byte is required for `itemtype` and four bytes for the pointer. So the total size of the index is 50 kbytes, which in this case is 50 pages of disk space.

Now consider finding all records with an item type of E. One approach is to read the entire index into memory, search it for type E items, and then use the pointers to retrieve the required records from `item`. This method requires 2,050 disk accesses—50 to load the index and 2,000 accesses of `item`, as there are 2,000 items of type E. Creating an index for item results in substantial savings in disk accesses for this example. Here, we assume that 20 percent of the records in item contained `itemtype = 'E'`. The number of disk accesses saved varies with the proportion of records meeting the query's criteria. If there are no records meeting the criteria, 9,950 disk accesses are avoided. At the other extreme, when all records meet the criteria, it takes 50 extra disk accesses to load the index.

The SQL for creating the index is

```
CREATE INDEX itemtypeindx ON item (itemtype);
```

The entire index need not be read into memory. As you will see when we discuss tree structures, we can take advantage of an index's ordering to reduce disk accesses further. Nevertheless, the clear advantage of an index is evident: it speeds up retrieval by reducing disk accesses. Like many aspects of database management, however, indexes have a drawback. Adding a record to a file without an index requires a single disk write. Adding a record to an indexed file requires at least two, and maybe more, disk writes, because an entry has to be added to both the file and its index. The trade-off is between faster retrievals and slower updates. If there are many retrievals and few updates, then opt for an index, especially if the indexed field can have a wide variety of values. If the file is very volatile and updates are frequent and retrievals few, then an index may cost more disk accesses than it saves.

Indexes can be used for both sequential and direct access. Sequential access means that records are retrieved in the sequence defined by the values in the index. In our example, this means retrieving records in `itemtype` sequence with a range query such as

- *Find all items with a type code in the range E through K.*

Direct access means records are retrieved according to one or more specified values. A sample query requiring direct access would be

- *Find all items with a type code of E or N.*

Indexes are also handy for existence testing. Remember, the `EXISTS` clause of SQL returns *true* or *false* and not a value. An index can be searched to check whether the indexed field takes a particular value, but there is no need to access the file because no data are returned. The following query can be answered by an index search:

- *Are there any items with a code of R?*

Multiple indexes

Multiple indexes can be created for a file. The `item` file could have an index defined on `itemcolor` or any other field. Multiple indexes may be used independently, as in this query:

- *List red items.*

or jointly, with a query such as

- *Find red items of type C.*

The preceding query can be solved by using the indexes for `itemtype` and `itemcolor`.

Indexes for fields itemtype and itemcolor

itemtype index	
itemtype	Disk address
C	d6
C	d7
C	d8
C	d9
E	d1
E	d2
F	d10
N	d3
N	d4
N	d5

itemcolor index	
itemcolor	Disk address
Black	d9
Brown	d1
Brown	d2
Green	d8
Khaki	d10
Red	d6
White	d7
–	d3
–	d4
–	d5

Examination of the `itemtype` index indicates that items of type C are stored at addresses d6, d7, d8, and d9. The only red item recorded in the `itemcolor` index is stored at address d6, and since it is the only record satisfying the query, it is the only record that needs to be retrieved.

Multiple indexes, as you would expect, involve a trade-off. Whenever a record is added or updated, each index must also be updated. Savings in disk accesses for retrieval are exchanged for additional disk accesses in maintenance. Again, you must consider the balance between retrieval and maintenance operations.

Indexes are not restricted to a single field. It is possible to specify an index that is a combination of several fields. For instance, if item type and color queries were very common, then an index based on the concatenation of both fields could be created. As a result, such queries could be answered with a search of a single index rather than scanning two indexes as in the preceding example. The SQL for creating the combined index is

```
CREATE INDEX typecolorindx ON item (itemtype, itemcolor);
```

Sparse indexes

Indexes are used to reduce disk accesses to accelerate retrieval. The simple model of an index introduced earlier suggests that the index contains an entry for each record of the file. If we can shrink the index to eliminate an entry for each record, we can save more disk accesses. Indeed, if an index is small enough, it, or key parts of it, can be retained continuously in primary memory.

There is a physical sequence to the records in a file. Records within a page are in a physical sequence, and pages on a disk are in a physical sequence. A file can also have a logical sequence, the ordering of the file on some field within a record. For instance, the item file could be ordered on `ITEMNO`. Making the physical and logical sequences correspond is a way to save disk accesses. Remember that the `ITEM` file was assumed to have a record size of 1,024 bytes, the same size as a page, and one record was stored per page. If we now assume the record size is 512 bytes, then two records are stored per page. Furthermore, suppose that item is physically stored in `itemno` sequence. The index can be compressed by storing `itemno` for the second record on each page and that page's address.

A sparse index for item

itemno index

itemno	
2	
4	
6	
8	
10	

item

itemno	itemname	itemtype	itemcolor	
1	Pocket knife–Nile	E	Brown	page p
2	Pocket knife–Thames	E	Brown	
3	Compass	N	–	page p + 1
4	Geo positioning system	N	–	
5	Map measure	N	–	page p + 2
6	Hat–polar explorer	C	Red	
7	Hat–polar explorer	C	White	page p + 3
8	Boots–snakeproof	C	Green	
9	Boots–snakeproof	C	Black	page p + 4
10	Safari chair	F	Khaki	

Consider the process of finding the record with `ITEMNO = 7`. First, the index is scanned to find the first value for `itemno` that is greater than or equal to 7, the entry for `ITEMNO = 8`. Second, the page on which this record is stored (page p + 3) is loaded into memory. Third, the required record is extracted from the page.

Indexes that take advantage of the physical sequencing of a file are known as sparse or non-dense because they do not contain an entry for every value of the indexed field. (A dense index is one that contains an entry for every value of the indexed field.)

As you would expect, a sparse index has pros and cons. One major advantage is that it takes less storage space and so requires fewer disk accesses for reading. One disadvantage is that it can no longer be used for existence tests because it does not contain a value for every record in the file.

A file can have only one sparse index because it can have only one physical sequence. This field on which a sparse index is based is often called the primary key. Other indexes, which must be dense, are called secondary indexes.

In SQL, a sparse index is created using the `CLUSTER` option. For example, to define a sparse index on `ITEM` the command is

```
CREATE INDEX itemnoindx ON item (itemno) CLUSTER;
```

B-trees

The B-tree is a particular form of index structure that is frequently the main storage structure for relational systems. It is also the basis for IBM's VSAM (Virtual Storage Access Method), the file structure underlying DB2, IBM's relational database. A B-tree is an efficient structure for both sequential and direct accessing of a file. It consists of two parts: the sequence set and the index set.

The **sequence set** is a single-level index to the file with pointers to the records (the vertical arrows in the lower part of the following figure). It can be sparse or dense, but is normally dense. Entries in the sequence set are grouped into pages, and these pages are linked together (the horizontal arrows) so that the logical ordering of the sequence set is the physical ordering of the file. Thus, the file can be processed sequentially by processing the records pointed to by the first page (records with identifiers 1, 4, and 5), the records pointed to by the next logical page (6, 19, and 20), and so on.

Structure of a simple B-tree

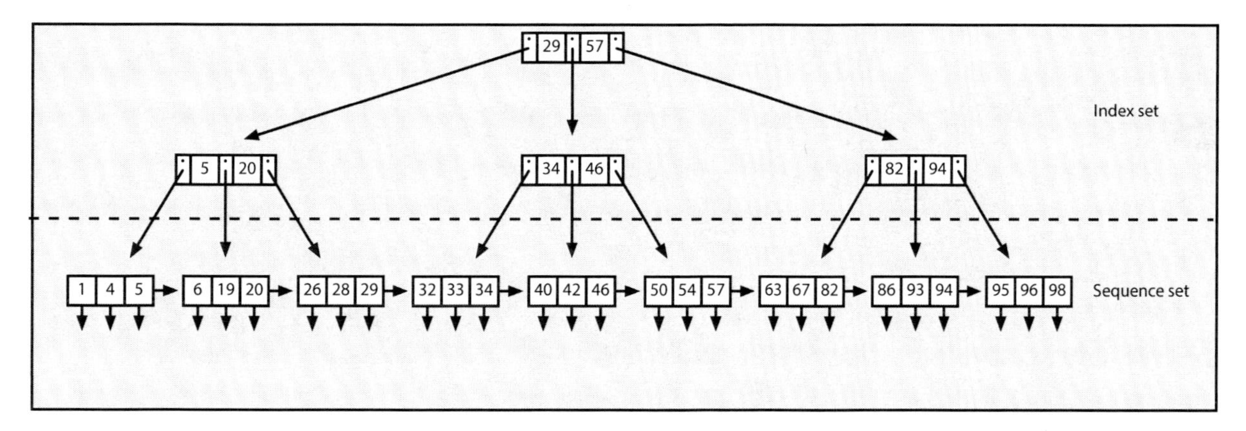

The **index set** is a tree-structured index to the sequence set. The top of the index set is a single node called the **root**. In this example, it contains two values (29 and 57) and three pointers. Records with an identifier less than or equal to 29 are found in the left branch, records with an identifier greater than 29 and less than or equal to 57 are found in the middle branch, and records with an identifier greater than 57 are found in the right branch. The three pointers are the page numbers of the left, middle, and right branches. The nodes at the next level have similar interpretations. The pointer to a particular record is found by moving down the tree until a node entry points to a value in the sequence set; this value can be used to retrieve the record. Thus, the index set provides direct access to the sequence set and then the data.

The index set is a B-tree. The combination of index set and sequence set is generally known as a B+ tree (**B-plus tree**). The B-tree simplifies the concept in two ways. *First,* the number of data values and pointers for any given node is not restricted to 2 and 3, respectively. In its general form, a B-tree of order n can have at least n and no more than $2n$ data values. If it has k values, the B-tree will have $k + 1$ pointers (in the example tree, nodes have two data values, $k = 2$, and there are three, $k + 1 = 3$, pointers). *Second,* B-trees typically have free space to permit rapid insertion of data values and possible updating of pointers when a new record is added.

As usual, there is a trade-off with B-trees. Retrieval will be fastest when each node occupies one page and is packed with data values and pointers. This lack of free space will slow down the addition of new records, however. Most implementations of the B+ tree permit a specified portion of free space to be defined for both the index and sequence set.

Hashing

Hashing reduces disk accesses by allowing direct access to a file. As you know, direct accessing via an index requires at least two or more accesses. The index must be loaded and searched, and then the record retrieved. For some applications, direct accessing via an index is too slow. Hashing can reduce the number of accesses to almost one by using the value in some field (the **hash field**, which is usually the primary key) to compute a record's address. A **hash function** converts the hash field into a **hash address**.

Consider the case of a university that uses the nine-digit Social Security number (SSN) as the student key. If the university has 10,000 students, it could simply use the last four digits of the SSN as the address. In effect, the file space is broken up into 10,000 slots with one student record in each slot. For example, the data for the student with SSN 417-03-4356 would be stored at address 4356. In this case, the hash field is SSN, and the hash function is

hash address = remainder after dividing SSN by 10,000.

What about the student with SSN 532-67-4356? Unfortunately, the hashing function will give the same address because most hashing schemes cannot guarantee a unique hash address for every record. When two hash fields have the same address, they are called **synonyms**, and a **collision** is said to have occurred.

There are techniques for handling synonyms. Essentially, you store the colliding record in an overflow area and point to it from the hash address. Of course, more than two records can have the same hash address, which in turn creates a synonym chain. The following figure shows an example of hashing with a synonym chain. Three SSNs hash to the same address. The first record (417-03-4356) is stored at the hash address. The second record (532-67-4356) is stored in the overflow area and is connected by a pointer to the first record. The third record (891-55-4356) is also stored in the overflow area and connected by a pointer from the second record. Because each record contains the full key (SSN in this case), during retrieval the system can determine whether it has the correct record or should follow the chain to the next record.

An example of hashing

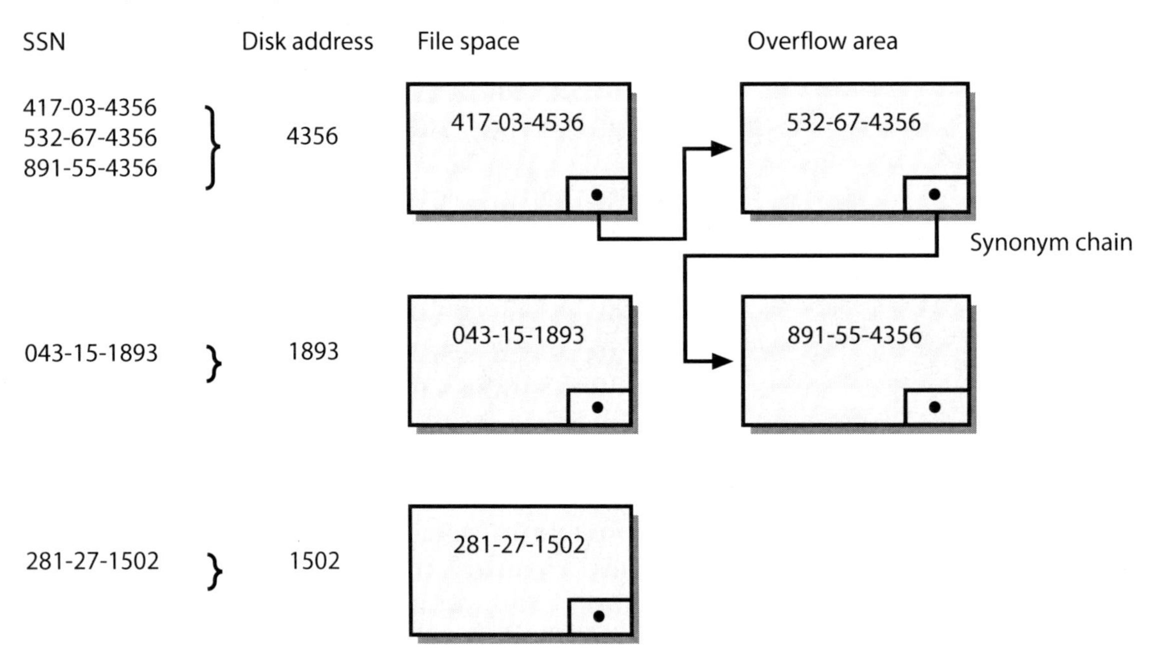

If there are no synonyms, hashing gives very fast direct retrieval, taking only one disk access to retrieve a record. Even with a small percentage of synonyms, retrieval via hashing is very fast. Access time degrades, however, if there are long synonym chains.

There are a number of different approaches to defining hashing functions. The most common method is to divide by a prime and use the remainder as the address. Before adopting a particular hashing function, test several functions on a representative sample of the hash field. Compute the percentage of synonyms and the length of synonym chains for each potential hashing function and compare the results.

Of course, hashing has trade-offs. There can be only one hashing field. In contrast, a file can have many indexed fields. The file can no longer be processed sequentially because its physical sequence loses any logical meaning if the records are not in primary key sequence or are sequenced on any other field.

Linked lists

A **linked list** is a useful data structure for interfile clustering. Suppose that the query, *Find all stocks of country X,* is a frequent request. Disk accesses can be reduced by storing a nation and its corresponding stocks together in a linked list.

A linked list

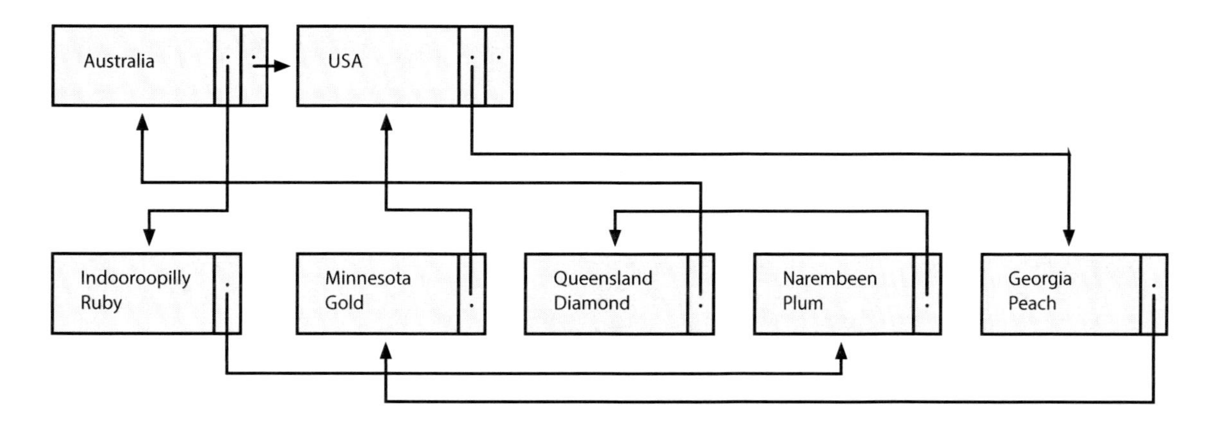

In this example, we have two files: `nation` and `stock`. Records in `nation` are connected by pointers (e.g., the horizontal arrow between Australia and USA). The pointers are used to maintain the `NATION` file in logical sequence (by nation, in this case). Each record in `nation` is linked to its `stock` records by a forward-pointing chain. The `nation` record for Australia points to the first `stock` record (Indooroopilly Ruby), which points to the next (Narembeen Plum), which points to the final record in the chain (Queensland Diamond). Notice that the last record in this `stock` chain points to the `nation` record to which the chain belongs (i.e., Australia). Similarly, there is a chain for the two USA stocks. In any chain, the records are maintained in logical sequence (by firm name, in this case).

This linked-list structure is also known as a parent/child structure. (The parent in this case is `nation` and the child is `stock`.) Although it is a suitable structure for representing a one-to-many (1:m) relationship, it is possible to depict more than one parent/child relationship. For example, stocks could be grouped into classes (e.g., high, medium, and low risk). A second chain linking all stocks of the same risk class can run through the file. Interfile clustering of a nation and its corresponding stocks will speed up access; so, the record for the parent Australia and its three children should be stored on one page. Similarly, all American stocks could be clustered with the USA record and so on for all other nations.

Of course, you expect some trade-offs. What happens with a query such as *Find the country in which Minnesota Gold is listed?* There is no quick way to find Minnesota Gold except by sequentially searching the `stock` file until the record is found and then following the chain to the parent nation record. This could take many disk accesses. One way to circumvent this is to build an index or hashing scheme for `stock` so that any record can be found directly. Building an index for the parent, `nation`, will speed up access as well.

Linked lists come in a variety of flavors:

- Lists that have two-way pointers, both forward and backward, speed up deletion.
- For some lists, every child record has a parent pointer. This helps prevent chain traversal when the query is concerned with finding the parent of a particular child.

Bitmap index

A **bitmap index** uses a single bit, rather than multiple bytes, to indicate the specific value of a field. For example, instead of using three bytes to represent *red* as an item's color, the color red is represented by a single bit. The relative position of the bit within a string of bits is then mapped to a record address.

Conceptually, you can think of a bitmap as a matrix. The following figure shows a bitmap containing details of an item's color and code. An item can have three possible colors; so, three bits are required, and two bits are needed for the two codes for the item. Thus, you can see that, in general, *n* bits are required if a field can have *n* possible values.

A bitmap index

itemcode	color			code		Disk address
	Red	Green	Blue	A	N	
1001	0	0	1	0	1	d1
1002	1	0	0	1	0	d2
1003	1	0	0	1	0	d3
1004	0	1	0	1	0	d4

When an item has a large number of values (i.e., *n* is large), the bitmap for that field will be very sparse, containing a large number of zeros. Thus, a bitmap is typically useful when the value of *n* is relatively small. When is *n* small or large? There is no simple answer; rather, database designers have to simulate alternative designs and evaluate the trade-offs.

In some situations, the bit string for a field can have multiple bits set to *on* (set to 1). A location field, for instance, may have two bits *on* to represent an item that can be found in Atlanta and New York.

The advantage of a bitmap is that it usually requires little storage. For example, we can recast the bitmap index as a conventional index. The core of the bitmap in the previous figure (i.e., the cells storing data about the color and code) requires 5 bits for each record, but the core of the traditional index requires 9 bytes, or 72 bits, for each record.

An index

itemcode	color char(8)	code char(1)	Disk address
1001	Blue	N	d1
1002	Red	A	d2
1003	Red	A	d3
1004	Green	A	d4

Bitmaps can accelerate query time for complex queries.

Join index

Many RDBMS queries frequently require two or more tables to be joined. Indeed, some people refer to the RDBMS as a join machine. A join index can be used to improve the execution speed of joins by creating indexes based on the matching columns of tables that are highly likely to be joined. For example, natcode is the common column used to join NATION and STOCK, and each of these tables can be indexed on natcode.

Indexes on natcode for nation and stock

nation index	
natcode	Disk address
UK	d1
USA	d2

stock index	
natcode	Disk address
UK	d101
UK	d102
UK	d103
USA	d104
USA	d105

A join index is a list of the disk addresses of the rows for matching columns. All indexes, including the join index, must be updated whenever a row is inserted into or deleted from the nation or stock tables. When a join of the two tables on the matching column is made, the join index is used to retrieve only those records that will be joined. This example demonstrates the advantage of a join index. If you think of a join as a product with a WHERE clause, then without a join index, 10 (2*5) rows have to be retrieved, but with the join index only 5 rows are retrieved. Join indexes can also be created for joins involving several tables. As usual, there is a trade-off. Joins will be faster, but insertions and deletions will be slower because of the need to update the indexes.

Join Index

join index	
nation disk address	stock disk address
d1	d101
d1	d102
d1	d103
d2	d104
d2	d105

Data coding standards

The successful exchange of data between two computers requires agreement on how information is coded. The American Standard Code for Information Interchange (ASCII) and Unicode are two widely used data coding standards.

ASCII

ASCII is the most common format for digital text files. In an ASCII file, each alphabetic, numeric, or special character is represented by a 7-bit code. Thus, 128 (2^7) possible characters are defined. Because most computers process data in eight-bit bytes or multiples thereof, an ASCII code usually occupies one byte.

Unicode

Unicode, officially the Unicode Worldwide Character Standard, is a system for the interchange, processing, and display of the written texts of the diverse languages of the modern world. Unicode provides a unique binary code for every character, no matter what the platform, program, or language. The Unicode standard currently contains 34,168 distinct coded characters derived from 24 supported language scripts. These characters cover the principal written languages of the world.

Unicode provides for two encoding forms: a default 16-bit form, and a byte (8-bit) form called UTF-8 that has been designed for ease of use with existing ASCII-based systems. As the default encoding of HTML and XML, Unicode is required for Internet protocols. It is implemented in all modern operating systems and computer languages such as Java. Unicode is the basis of software that must function globally.

Data storage devices

Many corporations double the amount of data they need to store every two to three years. Thus, the selection of data storage devices is a key consideration for data managers. When evaluating data storage options, data managers need to consider possible uses, which include:

1. Online data
2. Backup files
3. Archival storage

Many systems require data to be online — continually available. Here the prime concerns are usually access speed and capacity, because many firms require rapid response to large volumes of data. Backup files are required to provide security against data loss. Ideally, backup storage is high volume capacity at low cost. Archived data may need to be stored for many years; so the archival medium should be highly reliable, with no data decay over extended periods, and low-cost.

In deciding what data will be stored where, database designers need to consider a number of variables:

1. Volume of data
2. Volatility of data
3. Required speed of access to data
4. Cost of data storage
5. Reliability of the data storage medium
6. Legal standing of stored data

The design options are discussed and considered in terms of the variables just identified. First, we review some of the options.

Magnetic technology

Over USD 20 billion is spent annually on magnetic storage devices.[74] A significant proportion of many units IS hardware budgets is consumed by magnetic storage. Magnetic technology, the backbone of data storage for five decades, is based on magnetization and demagnetization of spots on a magnetic recording surface. The same spot can be magnetized and demagnetized repeatedly. Magnetic recording materials may be coated on rigid platters (hard disks), thin ribbons of material (magnetic tapes), or rectangular sheets (magnetic cards).

74 http://www.gartner.com/it/page.jsp?id=2149415

The main advantages of magnetic technology are its relative maturity, widespread use, and declining costs. A major disadvantage is susceptibility to strong magnetic fields, which can corrupt data stored on a disk. Another shortcoming is data storage life; magnetization decays with time.

As organizations continue to convert paper to images, they need a very long-term, unalterable storage medium for documents that could be demanded in legal proceedings (e.g., a customer's handwritten insurance claim or a client's completed form for a mutual fund investment). Because data resident on magnetic media can be readily changed and decay with time, magnetic storage is not an ideal medium for storing archival data or legal documents.

Fixed magnetic disk

A fixed magnetic disk, also known as a hard disk drive (HDD), containing one or more recording surfaces is permanently mounted in the disk drive and cannot be removed. The recording surfaces and access mechanism are assembled in a clean room and then sealed to eliminate contaminants, making the device more reliable and permitting higher recording density and transfer rates. Access time is typically between 4 and 10 ms, and transfer rates can be as high as 1,300 Mbytes per second. Disk unit capacities range from gigabytes to terabytes. Magnetic disk units are sometimes called direct-access storage devices or **DASD** (pronounced *dasdee*).

Fixed disk is the medium of choice for most systems, from personal computers to supercomputers. It gives rapid, direct access to large volumes of data and is ideal for highly volatile files. The major disadvantage of magnetic disk is the possibility of a head crash, which destroys the disk surface and data. With the read/write head of a disk just 15 millionths of an inch (40 millionths of a centimeter) above the surface of the disk, there is little margin for error. Hence, it is crucial to make backup copies of fixed disk files regularly.

RAID

RAID (redundant array of independent, or inexpensive, disks) takes advantage of the economies of scale in manufacturing disks for the personal computing market. The cost of drives increases with their capacity and speed. RAIDs use several cheaper drives whose total cost is less than one high-capacity drive but have the same capacity. In addition to lower cost, RAID offers greater data security. All RAID levels (except level 0) can reconstruct the data on any single disk from the data stored on the remaining disks in the array in a manner that is quite transparent to the user.

RAID uses a combination of mirroring and striping to provide greater data protection. When a file is written to a mirrored array, the disk controller writes identical copies of each record to each drive in the array. When a file is read from a mirrored array, the controller reads alternate pages simultaneously from each of the drives. It then puts these pages together in the correct sequence before delivering them to the computer. Mirroring reduces data access time by approximately the number of drives in the array because it interleaves the reading of records. During the time a conventional disk drive takes to read one page, a RAID system can read two or more pages (one from each drive). It is simply a case of moving from sequential to parallel retrieval of pages. Access times are halved for a two-drive array, quartered for a four-drive array, and so on.

If a read error occurs on a particular disk, the controller can always read the required page from another drive in the array because each drive has a full copy of the file. Mirroring, which requires at least two drives, improves response time and data security; however, it does take considerably more space to store a file because multiple copies are created.

Mirroring

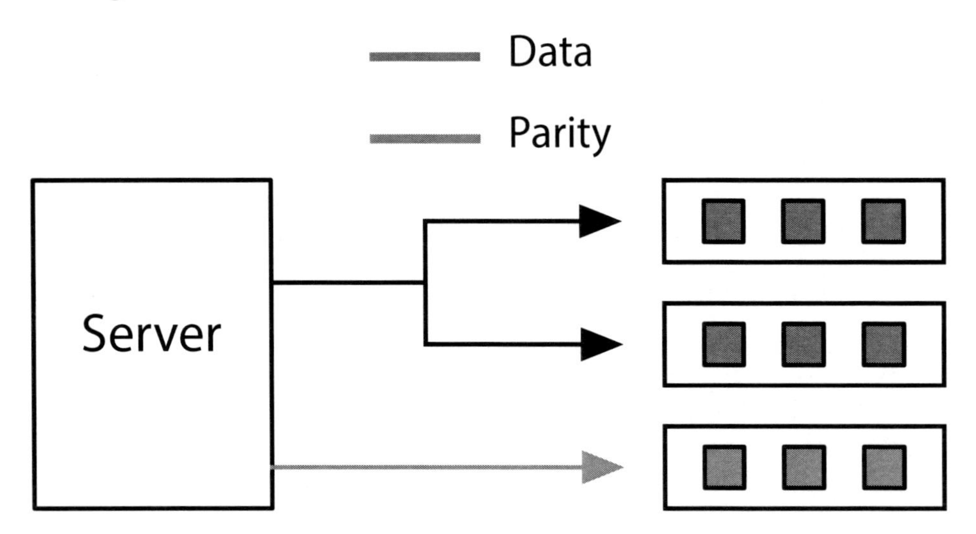

When a file is written to a striping array of three drives, for instance, one half of the file is written to the first drive and the second half to the second drive. The third drive is used for error correction. A parity bit is constructed for each corresponding pair of bits written to drives one and two, and this parity bit is written to the third drive. A parity bit is a bit added to a chunk of data (e.g., a byte) to ensure that the number of bits in the chunk with a value of one is even or odd. Parity bits can be checked when a file is read to detect data errors, and in many cases the data errors can be corrected. Parity bits demonstrate how redundancy supports recovery.

When a file is read by a striping array, portions are retrieved from each drive and assembled in the correct sequence by the controller. If a read error occurs, the lost bits can be reconstructed by using the parity data on the third drive. If a drive fails, it can be replaced and the missing data restored on the new drive. Striping requires at least three drives. Normally, data are written to every drive but one, and that remaining drive is used for the parity bit. Striping gives added data security without requiring considerably more storage, but it does not have the same response time increase as mirroring.

Striping

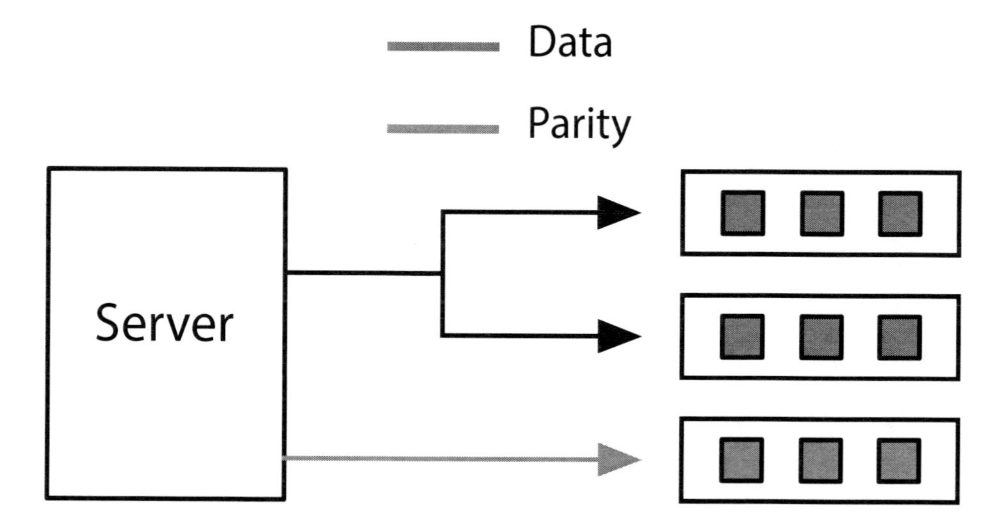

RAID subsystems are divided into seven levels, labeled 0 through 6. All RAID levels, except level 0, have common features:

- There is a set of physical disk drives viewed by the operating system as a single, logical drive.
- Data are distributed across corresponding physical drives.
- Parity bits are used to recover data in the event of a disk failure.

Level 0 has been in use for many years. Data are broken into blocks that are interleaved or *striped* across disks. By spreading data over multiple drives, read and write operations can occur in parallel. As a result, I/O rates are higher, which makes level 0 ideal for I/O intensive applications such as recording video. There is no additional parity information and thus no data recovery when a drive failure occurs.

Level 1 implements mirroring, as described previously. This is possibly the most popular form of RAID because of its effectiveness for critical nonstop applications, although high levels of data availability and I/O rates are counteracted by higher storage costs. Another disadvantage is that every write command must be executed twice (assuming a two-drive array), and thus level 1 is inappropriate for applications that have a high ratio of writes to reads.

Level 2 implements striping by interleaving blocks of data on each disk and maintaining parity on the check disk. This is a poor choice when an application has frequent, short random disk accesses, because every disk in the array is accessed for each read operation. However, RAID 2 provides excellent data transfer rates for large sequential data requests. It is therefore suitable for computer-aided drafting and computer-aided manufacturing (CAD/CAM) and multimedia applications, which typically use large sequential files. RAID 2 is rarely used, however, because the same effect can be achieved with level 3 at a lower cost.

Level 3 utilizes striping at the bit or byte level, so only one I/O operation can be executed at a time. Compared to level 1, RAID level 3 gives lower-cost data storage at lower I/O rates and tends to be most useful for the storage of large amounts of data, common with CAD/CAM and imaging applications.

Level 4 uses sector-level striping; thus, only a single disk needs to be accessed for a read request. Write requests are slower, however, because there is only one parity drive.

Level 5, a variation of striping, reads and writes data to separate disks independently and permits simultaneous reading and writing of data. Data and parity are written on the same drive. Spreading parity data evenly across several drives avoids the bottleneck that can occur when there is only one parity drive. RAID 5 is well designed for the high I/O rates required by transaction processing systems and servers, particularly e-mail servers. It is the most balanced implementation of the RAID concept in terms of price, reliability, and performance. It requires less capacity than mirroring with level 1 and higher I/O rates than striping with level 3, although performance can decrease with update-intensive operations. RAID 5 is frequently found in local-area network (LAN) environments.

RAID level 5

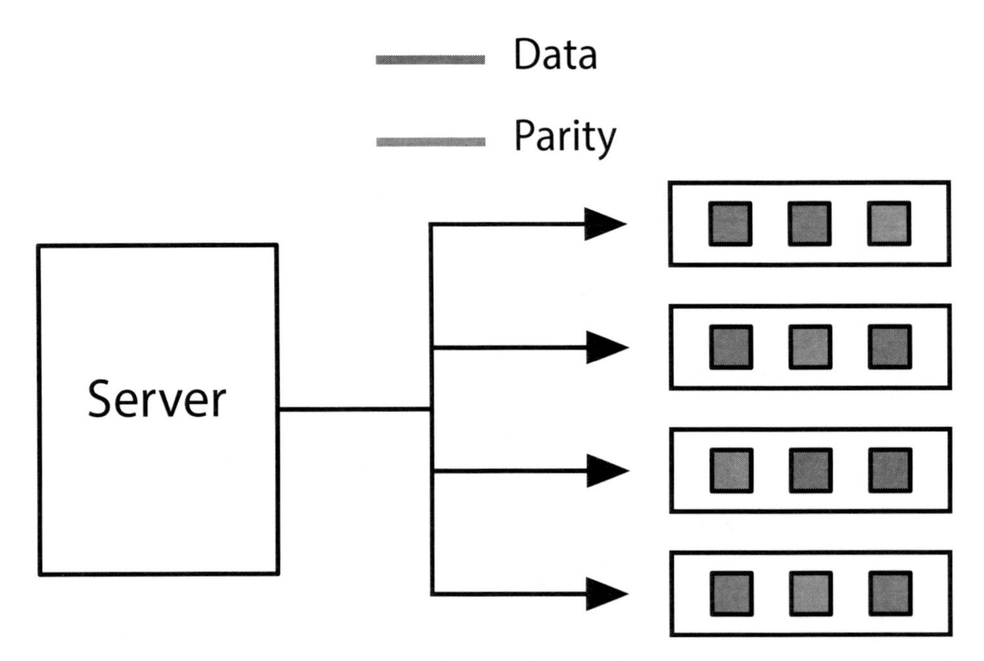

RAID 6 takes fault tolerance to another level by enabling a RAID system to still function when two drives fail. Using a method called double parity, it distributes parity information across multiple drives so that, on the fly, a disk array can be rebuild even if another drive fails before the rebuild is complete.

RAID does have drawbacks. It often lowers a system's performance in return for greater reliability and increased protection against data loss. Extra disk space is required to store parity information. Extra disk accesses are required to access and update parity information. You should remember that RAID is not a replacement for standard backup procedures; it is a technology for increasing the fault tolerance of critical online systems. RAID systems offer terabytes of storage.

Removable magnetic disk

Removable drives can be plugged into a system as required. The disk's removability is its primary advantage, making it ideal for backup and transport. For example, you might plug in a removable drive to the USB port of your laptop once a day to backup its hard drive. Removable disk is also useful when applications need not be continuously online. For instance, the monthly payroll system can be stored on a removable drive and mounted as required. Removable drives cost about USD 100 per terabyte.

Magnetic tape

A magnetic tape is a thin ribbon of plastic coated with ferric oxide. The once commonly used nine-track 2,400-foot (730 m) tape has a capacity of about 160 Mbytes and a data transfer rate of 2 Mbytes per second. The designation **nine-track** means nine bits are stored across the tape (8 bits plus one parity bit).

Magnetic tape was used extensively for archiving and backup in early database systems; however, its limited capacity and sequential nature have resulted in its replacement by other media, such as magnetic tape cartridge.

Magnetic tape cartridges

Tape cartridges, with a capacity measured in Gbytes and transfer rates of up to 6 Mbytes per second, have replaced magnetic tape. They are the most cost effective (from a purchase price $/GB perspective) magnetic recording device. Furthermore, a tape cartridge does not consume any power when unmounted and stored in a library. Thus, the operating costs for a tape storage archival system are much lower than an equivalent hard disk storage system, though hard disks provide faster access to archived data.

Mass storage

There exists a variety of mass storage devices that automate labor-intensive tape and cartridge handling. The storage medium, with a capacity of terabytes, is typically located and mounted by a robotic arm. Mass storage devices can currently handle hundreds of petabytes of data. They have slow access time because of the mechanical loading of tape storage, and it might take several minutes to load a file.

Solid-state memory

A solid-state disk (SSD) connects to a computer in the same way as regular magnetic disk drives do, but they store data on arrays of memory chips. SSDs require lower power consumption (longer battery life) and come in smaller sizes (smaller and lighter devices). As well, SSDs have faster access times. However, they cost around USD 1.50 per Gbyte, compared to hard disks which are around USD 0.10 per Gbyte. SSD is rapidly becoming the standard for laptops.

A **flash drive**, also known as a keydrive or jump drive, is a small, removable storage device with a Universal Serial Bus (USB) connector. It contains solid-state memory and is useful for transporting small amounts of data. Capacity is in the range 1 to 128 Gbytes. The price for low capacity drives is about about USD 1-2 per Gbytes.

While price is a factor, the current major drawback holding back the adoption of SSD is the lack of manufacturing capacity for the NAND chips used in SSDs.[75] The introduction of tablets, such as the iPad, and smartphones is exacerbating this problem.

Optical technology

Optical technology is a more recent development than magnetic. Its advantages are high-storage densities, low-cost media, and direct access. Optical storage systems work by reflecting beams of laser light off a rotating disk with a minutely pitted surface. As the disk rotates, the amount of light reflected back to a sensor varies, generating a stream of ones and zeros. A tiny change in the wavelength of the laser translates into as much as a tenfold increase in the amount of information that can be stored. Optical technology is highly reliable because it is not susceptible to head crashes.

There are three storage media based on optical technology: CD-ROM, DVD, and Blu-ray. Most optical disks can reliably store records for at least 10 years under prescribed conditions of humidity and temperature. Actual storage life may be in the region of 30 to 50 years. Optical media are compact medium for the storage of permanent data. Measuring 12 cm (~ 4.75 inches) in diameter, the size consistency often means later generation optical readers can read earlier generation media.

75 www.pcworld.com/businesscenter/article/217883/seagate_solidstate_disks_are_doomed_at_least_for_now.html

Optical technology

Medium	Introduced	Capacity
Compact disc (CD)	1982	.65 Gbytes
Digital versatile disc (DVD)	1995	1.5-17 Gbytes
Blu-ray disc (BD)	2006	25-50 Gbytes

Optical media come in three types. Read-only memory (ROM) is typically used for the large scale manufacturing of information to be distributed, such as a music album or movie, and it will also usually incorporate some form of digital rights management to prevent illegal copying. Recordable (R) for write-once media can be purchased for the storing of information that should not be alterable. Re-recordable media are re-usable, and are a general backup and distribution medium.

Optical media options

Format	Description
ROM	Read-only
R	Recordable or write-once
RW	Read/write or re-recordable

Optical media are commonly used to distribute various types of information. DVDs and Blu-ray have replaced CDs to distribute documents, music, and software, though the use of optical media for information distribution is declining because it is cheaper and faster to transmit via the Internet. Services such as Apple's App store and Netflix are pushing the distribution of software and movies, respectively, to the Internet.

Data stored on read-only or write-once optical media are generally reckoned to have the highest legal standing of any of the forms discussed because, once written, the data cannot be altered. Thus, they are ideal for storage of documents that potentially may be used in court. Consequently, we are likely to see archival storage emerge as the major use for optical technology.

Storage-area networks

A storage-area network (SAN) is a high-speed network for connecting and sharing different kinds of storage devices, such as tape libraries and disk arrays. In the typical LAN, the storage device (usually a disk) is closely coupled to a server and communicates through a bus connection. Communication among storage devices occurs through servers and over the LAN, which can lead to network congestion.

SANs support disk mirroring, backup and restore, archival and retrieval of archived data, data migration from one storage device to another, and the sharing of data among different servers in a network. Typically, a SAN is part of the overall network of computing resources for an enterprise. SANs are likely to be a critical element for supporting e-commerce and applications requiring large volumes of data. As Storage Area Networks (SANs) have become increasingly affordable, their use in enterprises and even smaller businesses has become widespread.

Long-term storage

Long-term data storage has always been of concern to societies and organizations. Some data, such as the location of toxic-waste sites, must be stored for thousands of years. Increasingly, governments are converting their records to electronic format. Unlike paper, magnetic media do not show degradation until it is too late to recover the data. Magnetic tapes can become so brittle that the magnetic coating separates from the backing. In addition, computer hardware and software rapidly become obsolete. The medium may be readable, but there could be no hardware to read it and no software to decode it.

Paper, it seems, is still the best medium for long-term storage, and research is being conducted to create extra-long-life paper that can store information for hundreds of years. This paper, resistant to damage from heat, cold, and magnetism, will store data in a highly compact format but, obviously, nowhere near optical disk densities.

The life expectancy of various media at 20°C (68°F) and 40 percent relative humidity (source: National Media Lab)

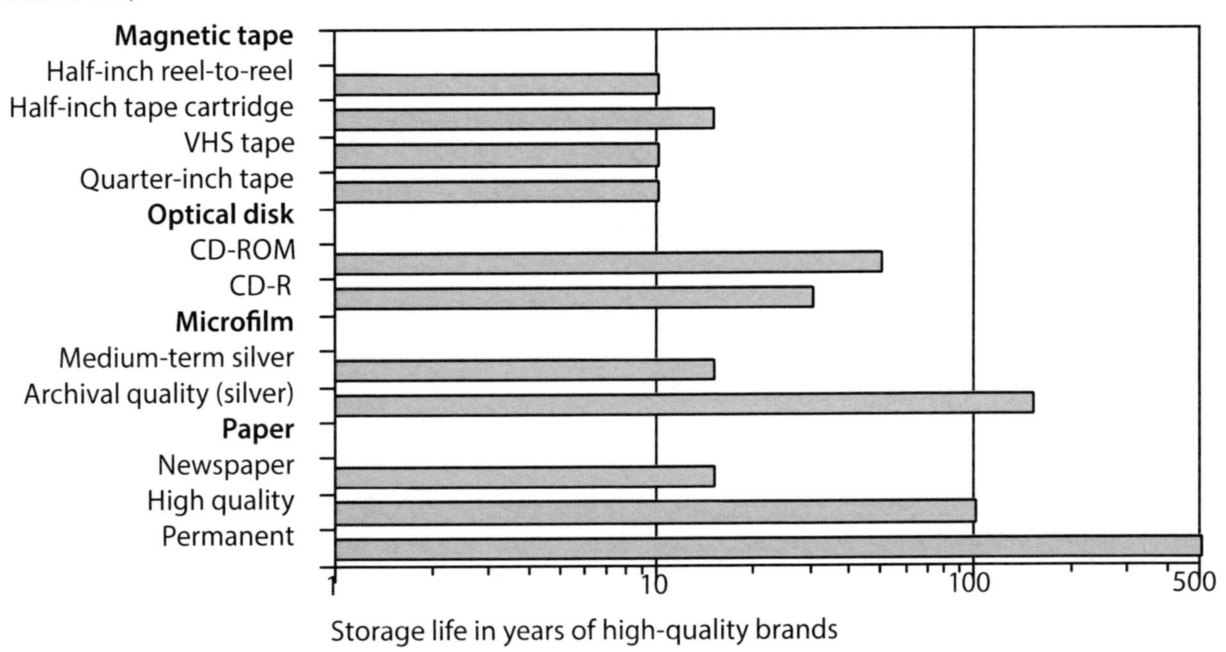

Data compression

Data compression is a method for encoding digital data so that they require less storage space and thus less communication bandwidth. There are two basic types of compression: lossless methods, in which no data are lost when the files are restored to their original format, and lossy methods, in which some data are lost when the files are decompressed.

Lossless compression

During lossless compression, the data-compression software searches for redundant or repetitive data and encodes it. For example, a string of 100 asterisks (*) can be stored more compactly by a compression system that records the repeated character (i.e., *) and the length of the repetition (i.e., 100). The same principle can be applied to a photo that contains a string of horizontal pixels of the same color. Clearly, you want to use lossless compression with text files (e.g., a business report or spreadsheet).

Lossy compression

Lossy compression is used for graphics, video, and audio files because humans are often unable to detect minor data losses in these formats. Audio files can often be compressed to 10 percent of their original size (e.g., an MP3 version of a CD recording).

Skill builder

An IS department in a major public university records the lectures for 10 of its classes for video streaming to its partner universities in India and China. Each twice-weekly lecture runs for 1.25 hours and a semester is 15 weeks long. A video streaming expert estimates that one minute of digital video requires 6.5 Mbytes using MPEG-4 and Apple's QuickTime software. What is MPEG-4? Calculate how much storage space will be required and recommend a storage device for the department.

Comparative analysis

Details of the various storage devices are summarized in the following table. A simple three-star rating system has been used for each device—the more stars, the better. In regard to access speed, RAID gets more stars than DVD-RAM because it retrieves a stored record more quickly. Similarly, optical rates three stars because it costs less per megabyte to store data on a magneto-optical disk than on a fixed disk. The scoring system is relative. The fact that removable disk gets two stars for reliability does not mean it is an unreliable storage medium; it simply means that it is not as reliable as some other media.

Relative merits of data storage devices

Device	Access speed	Volume	Volatility	Cost per megabyte	Reliability	Legal standing
Solid state	***	*	***	*	***	*
Fixed disk	***	***	***	***	**	*
RAID	***	***	***	***	***	*
Removable disk	**	**	***	**	**	*
Flash memory	**	*	***	*	***	*
Tape	*	**	*	***	**	*
Cartridge	**	***	*	***	**	*
Mass Storage	**	***	*	***	**	*
SAN	***	***	***	***	***	*
Optical-ROM	*	***	*	***	***	***
Optical-R	*	***	*	***	***	**
Optical-RW	*	***	**	***	***	*

Legend

Characteristic	More stars mean ...
Access speed	Faster access to data
Volume	Device more suitable for large files
Volatility	Device more suitable for files that change frequently
Cost per megabyte	Less costly form of storage
Reliability	Device less susceptible to an unrecoverable read error
Legal standing	Media more acceptable as evidence in court

Conclusion

The internal and physical aspects of database design are a key determinant of system performance. Selection of appropriate data structures can substantially curtail disk access and reduce response times. In making data structure decisions, the database administrator needs to weigh the pros and cons of each choice. Similarly, in selecting data storage devices, the designer needs to be aware of the trade-offs. Various devices and media have both strengths and weaknesses, and these need to be considered.

Summary

The data deluge is increasing the importance of data management for organizations. It takes considerably longer to retrieve data from a hard disk than from main memory. Appropriate selection of data structures and data access methods can considerably reduce delays by reducing disk accesses. The key characteristics of disk storage devices that affect database access are rotational speed and access arm speed. Access arm movement can be minimized by storing frequently used data on the same track on a single surface or on the same track on different surfaces. Records that are frequently used together should be clustered together. Intrafile clustering applies to the records within a single file. Interfile clustering applies to multiple files. The disk manager, the part of the operating system responsible for physical I/O, maintains a directory of pages. The file manager, a level above the disk manager, contains a directory of files.

Indexes are used to speed up retrieval by reducing disk accesses. An index is a file containing the value of the index field and the address of its full record. The use of indexes involves a trade-off between faster retrievals and slower updates. Indexes can be used for both sequential and direct access. A file can have multiple indexes. A sparse index does not contain an entry for every value of the indexed field. The B-tree, a particular form of index structure, consists of two parts: the sequence set and the index set. Hashing is a technique for reducing disk accesses that allows direct access to a file. There can be only one hashing field. A hashed file can no longer be processed sequentially because its physical sequence has lost any logical meaning. A linked list is a useful data structure for interfile clustering. It is a suitable structure for representing a 1:m relationship. Pointers between records are used to maintain a logical sequence. Lists can have forward, backward, and parent pointers.

Systems designers have to decide what data storage devices will be used for online data, backup files, and archival storage. In making this decision, they must consider the volume of data, volatility of data, required speed of access to data, cost of data storage, reliability of the data storage medium, and the legal standing of the stored data. Magnetic technology, the backbone of data storage for six decades, is based on magnetization and demagnetization of spots on a magnetic recording surface. Fixed disk, removable disk, magnetic tape, tape cartridge, and mass storage are examples of magnetic technology. RAID uses several cheaper drives whose total cost is less than one high-capacity drive. RAID uses a combination of mirroring or striping to provide greater data protection. RAID subsystems are divided into six levels labeled 0 through 6. A storage-area network (SAN) is a high-speed network for connecting and sharing different kinds of storage devices, such as tape libraries and disk arrays.

Optical technology offers high storage densities, low-cost medium, and direct access. CD, DVD, and Blu-ray are examples of optical technology. Optical disks can reliably store records for at least 10 years and possibly up to 30 years. Optical technology is not susceptible to head crashes.

Data compression techniques reduce the need for storage capacity and bandwidth. Lossless methods result in no data loss, whereas with lossy techniques, some data are lost during compression.

Key terms and concepts

Access time

Archival file

ASCII

B-tree

Backup file

Bitmap index

Blue-ray disc (BD)

Compact disc (CD)

Clustering
Conceptual schema
Cylinder
Data compression
Data deluge
Data storage device
Database architecture
Digital versatile disc (DVD)
Disk manager
External schema
File manager
Hash address
Hash field
Hash function
Hashing
Index
Index set
Interfile clustering
Internal schema
Intrafile clustering
Join index
Linked list
Lossless compression
Lossy compression
Magnetic disk
Magnetic tape
Mass storage
Mirroring
Page
Parity
Pointer
Redundant arrays of inexpensive or independent drives (RAID)
Sequence set
Solid-state disk (SSD)
Sparse index
Storage-area network (SAN)
Striping
Track
Unicode
VSAM

Exercises

1. Why is a disk drive considered a bottleneck?
2. What is the difference between a record and a page?
3. Describe the two types of delay that can occur prior to reading a record from a disk. What can be done to reduce these delays?
4. What is clustering? What is the difference between intrafile and interfile clustering?
5. Describe the differences between a file manager and a disk manager.
6. What is an index?
7. What are the advantages and disadvantages of indexing?
8. Write the SQL to create an index on the column `NATCODE` in the `NATION` table.
9. A Paris insurance firm keeps paper records of all policies and claims made on it. The firm now has a vault containing 100 filing cabinets full of forms. Because Paris rental costs are so high, the CEO has asked you to recommend a more compact medium for long-term storage of these documents.

Because some insurance claims are contested, she is very concerned with ensuring that documents, once stored, cannot be altered. What would you recommend and why?

10. A video producer has asked for your advice on a data storage device. She has specified that she must be able to record video at 5 to 7 Mbytes per second. What would you recommend and why?

11. A German consumer research company collects scanning data from supermarkets throughout central Europe. The scanned data include product code identifier, price, quantity purchased, time, date, supermarket location, and supermarket name, and in some cases where the supermarket has a frequent buyer plan, it collects a consumer identification code. It has also created a table containing details of the manufacturer of each product. The database contains over 500 Tbytes of data. The data are used by market researchers in consumer product companies. A researcher will typically request access to a slice of the database (e.g., sales of all detergents) and analyze these data for trends and patterns. The consumer research company promises rapid access to its data. Its goal is to give clients access to requested data within one to two minutes. Once clients have access to the data, they expect very rapid response to queries. What data storage and retrieval strategy would you recommend?

12. A magazine subscription service has a Web site for customers to place orders, inquire about existing orders, or check subscription rates. All customers are uniquely identified by an 11-digit numeric code. All magazines are identified by a 2- to 4-character code. The company has approximately 10 million customers who subscribe to an average of four magazines. Subscriptions are available to 126 magazines. Draw a data model for this situation. Decide what data structures you would recommend for the storage of the data. The management of the company prides itself on its customer service and strives to answer customer queries as rapidly as possible.

13. A firm offers a satellite-based digital radio service to the continental U.S. market. It broadcasts music, sports, and talk-radio programs from a library of 1.5 million digital audio files, which are sent to a satellite uplink and then beamed to car radios equipped to accept the service. Consumers pay $9.95 per month to access 100 channels.

    a. Assuming the average size of a digital audio file is 5 Mbytes (~4 minutes of music), how much storage space is required?

    b. What storage technology would you recommend?

14. Is MP3 a lossless or lossy compression standard?

15. What is the data deluge? What are the implications for data management?

16. According to Wikipedia, Apple has more than 28 million songs in the iTunes store.[76] What storage technology might be a good choice for this library?

---

[76] http://en.wikipedia.org/wiki/ITunes_Store

20. Data Processing Architectures

The difficulty in life is the choice.

George Moore, *The Bending of the Bough,* 1900

Learning Objectives

Students completing this chapter will be able to

- recommend a data architecture for a given situation;
- understand multi-tier client/server architecture;
- discuss the fundamental principles that a hybrid architecture should satisfy;
- demonstrate the general principles of distributed database design.

On a recent transatlantic flight, Sophie, the Personnel and PR manager, had been seated next to a very charming young man who had tried to impress her with his knowledge of computing. He worked for a large systems consulting firm, and his speech was sprinkled with words like "clouding computing," "XML," "server," "client," and "software as a service." In return, Sophie responded with some "ums," "aahs," and an occasional "that's interesting." Of course, this was before the third glass of champagne. After that, he was more intent on finding out how long Sophie would be in New York and what restaurants and shows piqued her curiosity.

After an extremely pleasant week of wining, dining, and seeing the town—amid, of course, many hours of valuable work for The Expeditioner—Sophie returned to London. Now, she must ask Ned what all these weird terms meant. After all, it was difficult to carry on a conversation when she did not understand the language. That's the problem with IS, she thought, new words and technologies were always being invented. It makes it very hard for the rest of us to make sense of what's being said.

Architectural choices

In general terms, data can be stored and processed locally or remotely. Combinations of these two options provide the basic information systems architectures.

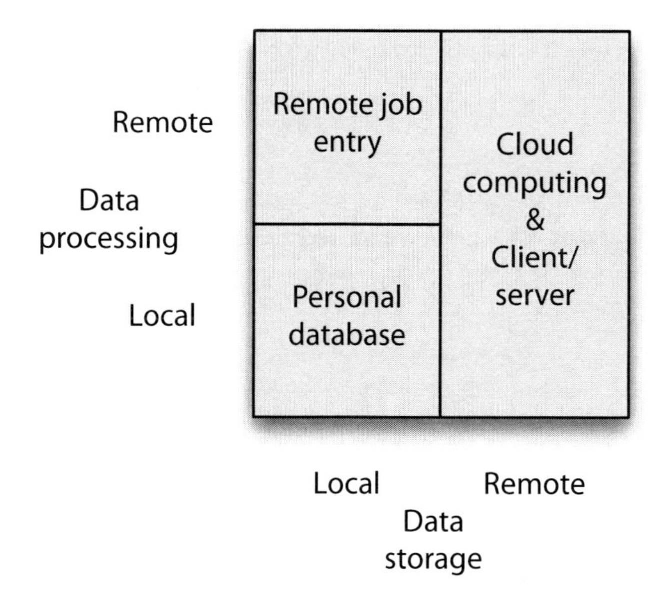

Remote job entry

In remote job entry, data are stored locally and processed remotely. Data are sent over a network to a remote computer for processing, and the output is returned the same way. This once fairly common form of data processing is still used today because remote job entry can overcome speed or memory shortcomings of a personal computer. Scientists and engineers typically need occasional access to a supercomputer for applications, such as simulating global climate change, that are data or computational intensive. Supercomputers typically have the main memory and processing power to handle such problems.

Local storage is used for several reasons. *First*, it may be cheaper than storing on a supercomputer. *Second*, the analyst might be able to do some local processing and preparation of the data before sending them to the supercomputer. Supercomputing time can be expensive. Where possible, local processing is used for tasks such as data entry, validation, and reporting. *Third*, local data storage might be deemed more secure for particularly sensitive data.

Personal database

People can store and process their data locally when they have their own computers. Many personal computer database systems (e.g., MS Access and FileMaker) permit people to develop their own applications, and many common desktop applications require local database facilities (e.g., a calendar).

Of course, there is a downside to personal databases. *First,* there is a great danger of repetition and redundancy. The same application might be developed in a slightly different way by many users. The same data get stored on many different systems. (It is not always the same, however, because data entry errors or maintenance inconsistencies result in discrepancies between personal databases.) *Second*, data are not readily shared because various users are not aware of what is available or find it inconvenient to share data. Personal databases are exactly that; but much of the data may be of corporate value and should be shared. *Third*, data integrity procedures are often quite lax for personal databases. People might not make regular backups, databases are often not secured from unauthorized access, and data validation procedures are often ignored. *Fourth*, often when the employee leaves the organization or moves to another role, the application and data are lost because they are not documented and the organization is unaware of their existence. *Fifth*, most personal databases are poorly designed and, in many cases, people use a spreadsheet as a poor substitute for a database. Personal databases are clearly very important for many organizations— when used appropriately. Data that are shared require a different processing architecture.

Client/server

The client/server architecture, in which multiple computers interact in a superior and subordinate mode, is the dominant architecture these days. A client process initiates a request and a server responds. The client is the dominant partner because it initiates the request to which the server responds. Client and server processes can run on the same computer, but generally they run on separate, linked computers. In the three-tier client/server model, the application and database are on separate servers.

A generic client/server architecture consists of several key components. It usually includes a mix of operating systems, data communications managers (usually abbreviated to DC manager), applications, clients (e.g., browser), and database management systems. The DC manager handles the transfer of data between clients and servers.

Three-tier client/server computing

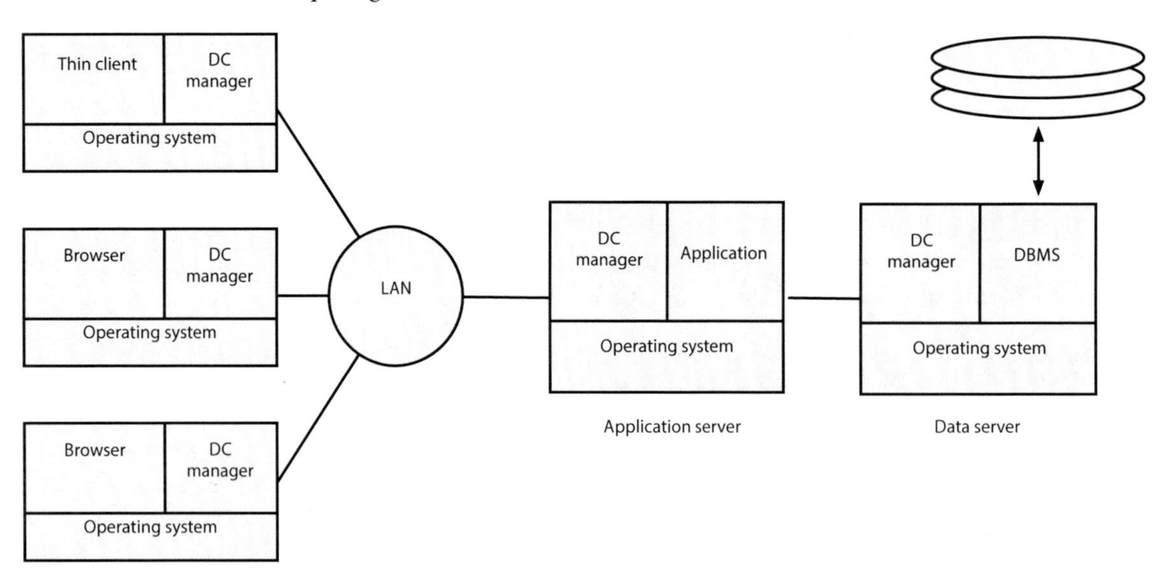

The three-tier model consists of three types of systems:

- **Clients** perform presentation functions, manage the graphical user interface (GUI), and execute communication software that provides network access. In most cases, the client is an Internet browser, though sometimes there might be a special program, a thin client, running on the client machine.
- **Application servers** are where the majority of the business and data logic are processed. They process commands sent by clients.
- **Data servers** run a DBMS. They respond to requests from an application, in which case the application is a client. They will also typically provide backup and recovery services and transaction management control.

Under the three-tier model, the client requests a service from an application server, and the application server requests data from a data server. The computing environment is a complex array of clients, application servers, and data servers. An organization can spread its processing load across many servers. This enables it to scale up the data processing workload more easily. For example, if a server running several applications is overloaded, another server can be purchased and some of the workload moved to the new server.

The client/server concept has evolved to describe a widely distributed, data-rich, cooperative environment. This is known as the n-tier client/server environment, which can consist of collections of servers dedicated to applications, data, transaction management, systems management, and other tasks. It extends the database side to incorporate nonrelational systems, such as multidimensional databases, multimedia databases, and legacy systems.

Evolution of client/server computing

Architecture	Description
Two-tier	Processing is split between client and server, which also runs the DBMS.
Three-tier	Client does presentation, processing is done by the server, and the DBMS is on a separate server.
N-tier	Client does presentation. Processing and DBMS can be spread across multiple servers. This is a distributed resources environment.

The rise of n-tier client/server can be attributed to the benefits of adopting a component-based architecture. The goal is to build quickly scalable systems by plugging together existing components. On the client side, the Web browser is a readily available component that makes deployment of new systems very simple. When everyone in the corporation has a browser installed on their personal computer, rolling out a new application is just a case of e-mailing the URL to the people concerned. Furthermore, the flexibility of the client/server model means that smartphones and tablets can easily be incorporated into the system. Thus, the client can be an app on an iPad.

On the data server side, many data management systems already exist, either in relational format or some other data model. Middle-tier server applications can make these data available to customers through a Web client. For example, UPS was able to make its parcel tracking system available via the Web, tablet, or smartphone because the database already existed. A middle-tier was written to connect customers using a range of devices to the database.

Skill builder

A European city plans to establish a fleet of two-person hybrid cars that can be rented for short periods (e.g., less than a day) for one-way trips. Potential renters first register online and then receive via the postal system a smart card that is used to gain entry to the car and pay automatically for its rental. The city also plans to have an online reservation system that renters can use to find the nearest car and reserve it. Reservations can be held for up to 30 minutes, after which time the car is released for use by other renters. Discuss the data processing architecture that you would recommend to support this venture. What technology would you need to install in each car? What technology would renters need? What features would the smart card require? Are there alternatives to a smart card?

Cloud computing

With the development of client/server computing, many organizations created data centers to contain the host of servers they needed to meet their information processing requirements. These so-called server farms can run to tens of thousands of servers. Because of security concerns, many firms are unwilling to reveal the location and size of their server farms. Google is reputed to have hundreds of thousands of servers. Many corporations run much smaller, but still significantly costly, data centers. Do they really need to run these centers?

The goal of the corporate IS unit should be to create and deploy systems that meet the operational needs of the business or give it a competitive advantage. It can gain little advantage from managing a data center. As a result, some organizations have turned to cloud computing, which is the use of shared hardware and software, to meet their processing needs. Instead of storing data and running applications on the corporate data center, applications and data are shifted to a third party data center, the cloud, which is accessed via the Internet. Companies can select from among several cloud providers. For example, it might run office applications (word processing, spreadsheet, etc.) on the Google cloud, and customer relationship management on Amazon's cloud.

There are usually economies of scale from switching from an in-house data center to the massive shared resources of a cloud provider, which lowers the cost of information processing. As well, the IS unit can turn its attention to the systems needs of the organization, without the distraction of managing a data center.

Cloud computing vendors specialize in managing large-scale data centers. As a result, they can lower costs in some areas that might be infeasible for the typical corporation. For example, it is much cheaper to move photons through fiber optics than electrons across the electricity grid.[77] This means that some cloud server farms are located where energy costs are low and minimal cooling is required. Iceland and parts of Canada are good locations for server farms because of inexpensive hydroelectric power and a cool climate. Information can be moved to and fro on a fiber optic network to be processed at a cloud site and then presented on the customer's computer.

Cloud computing can be more than a way to save some data center management costs. It has several features, or capabilities, that have strategic implications.[78] We start by looking at the levels or layers of clouds.

Cloud layers

Layer	Description	Example
Infrastructure	A virtual server over which the developer has complete control	Amazon Elastic Compute Cloud (EC2)
Platform as a service	A developer can build a new application with the provided tools and programming language	Salesforce.com
Application	Access to cloud applications	Google docs
Collaboration	A special case of an application cloud	Facebook
Service	Consulting and integration services	Appirio

As the preceding table illustrates, there a several options for the cloud customer. Someone looking to replace an office package installed on each personal computer could select an offering in the Application

77 The further electrons are moved, the more energy is lost. As much as a third of the initial energy generated can be lost during transmission across the grid before it gets to the final customer.

78 This section is based on Iyer, B., & Henderson, J. C. (2010). Preparing for the future: understanding the seven capabilities of cloud computing *MIS Executive*, 9(2), 117-131.

layer. A CIO looking to have complete control over a new application being developed from scratch could look to the infrastructure cloud, which also has the potential for high scalability.

Ownership is another way of looking at cloud offerings.[79]

Cloud ownership

Type	Description
Public	A cloud provided to the general public by its owner
Private	A cloud restricted to an organization. It has many of the same capabilities as a public cloud but provides greater control.
Community	A cloud shared by several organizations to support a community project
Hybrid	Multiple distinct clouds sharing a standardized or proprietary technology that enables data and application portability

Cloud capabilities

Understanding the strategic implications of cloud computing is of more interest than the layers and types because a cloud's seven capabilities offer opportunities to address the five strategic challenges facing all organizations.

Interface control

Some cloud vendors provide their customers with an opportunity to shape the form of interaction with the cloud. Under *open co-innovation*, a cloud vendor has a very open and community-driven approach to services. Customers and the vendor work together to determine the roadmap for future services. Amazon follows this approach. The *qualified retail* model requires developers to acquire permission and be certified before releasing any new application on the vendor's platform. This is the model Apple uses with its iTunes application store. *Qualified co-innovation* involves the community in the creation and maintenance of application programming interfaces (APIs). Third parties using the APIs, however, must be certified before they can be listed in the vendor's application store. This is the model salesforce.com uses. Finally, we have the *open retail* model. Application developers have some influence on APIs and new features, and they are completely free to write any program on top of the system. This model is favored by Microsoft.

Location independence

This capability means that data and applications can be accessed, with appropriate permission, without needing to know the location of the resources. This capability is particularly useful for serving customers and employees across a wide variety of regions. There are some challenges to location independence. Some countries restrict where data about their citizens can be stored and who can access it.

Ubiquitous access

Ubiquity means that customers and employees can access any information service from any platform or device via a Web browser from anywhere. Some clouds are not accessible in all parts of the globe at this stage because of the lack of connectivity, sufficient bandwidth, or local restrictions on cloud computing services.

---

79 Mell, P., & Grance, T. (2009). The NIST definition of cloud computing: National Institute of Standards and Technology.

Sourcing independence

One of the attractions of cloud computing is that computing processing power becomes a utility. As a result, a company could change cloud vendors easily at low cost. It is a goal rather than a reality at this stage of cloud development.

Virtual business environments

Under cloud computing, some special needs systems can be built quickly and later abandoned. For example, in 2009 the U.S. Government had a *Cash for Clunkers* program that ran for a few months to encourage people to trade-in their old car for a more fuel-efficient new car. This short-lived system whose processing needs are hard to estimate is well suited to a cloud environment. No resources need to be acquired, and processing power can be obtained as required.

Addressability and traceability

Traceability enables a company to track customers and employees who use an information service. Depending on the device accessing an information service, a company might be able to precisely identify the person and the location of the request. This is particularly the case with smartphones or tablets that have a GPS capability.

Rapid elasticity

Organizations cannot always judge exactly their processing needs, especially for new systems, such as the previously mentioned *Cash for Clunkers* program. Ideally, an organization should be able to pull on a pool of computer processing resources that it can scale up and down as required. It is often easier to scale up than down, as this is in the interests of the cloud vendor.

Strategic risk

Every firm faces five strategic risks: demand, innovation, inefficiency, scaling, and control risk.

Strategic risks

Risk	Description
Demand	Fluctuating demand or market collapse
Inefficiency	Inability to match competitors' unit costs
Innovation	Not innovating as well as competitors
Scaling	Not scaling fast and efficiently enough to meet market growth
Control	Inadequate procedures for acquiring or managing resources

We now consider how the various capabilities of cloud computing are related to these risks which are summarized in a table following the discussion of each risk.

Demand risk

Most companies are concerned about loss of demand, and make extensive use of marketing techniques (e.g., advertising) to maintain and grow demand. Cloud computing can boost demand by enabling customers to access a service wherever they might be. For example, Amazon combines ubiquitous access to its cloud and the Kindle, its book reader, to permit customers to download and start reading a book wherever they have access. Addressability and traceability enable a business to learn about customers and their needs. By mining the data collected by cloud applications, a firm can learn the connections between the when, where, and what of customers' wants. In the case of excessive demand, the elasticity of cloud resources might help a company to serve higher than expected demand.

Inefficiency risk

If a firm's cost structure is higher than its competitors, its profit margin will be less, and it will have less money to invest in new products and services. Cloud computing offers several opportunities to reduce cost. First, location independence means it can use a cloud to provide many services to customers across a large geographic region through a single electronic service facility. Second, sourcing independence should result in a competitive market for processing power. Cloud providers compete in a way that a corporate data center does not, and competition typically drives down costs. Third, the ability to quickly create and test new applications with minimal investment in resources avoids many of the costs of procurement, installation, and the danger of obsolete hardware and software. Finally, ubiquitous access means employees can work from a wide variety of locations. For example, ubiquity enables some to work conveniently and productively from home or on the road, and thus the organization needs smaller and fewer buildings.

Innovation risk

In most industries, enterprises need to continually introduce new products and services. Even in quite stable industries with simple products (e.g., coal mining), process innovation is often important for lowering costs and delivering high quality products. The ability to modify a cloud's interface could be of relevance to innovation. Apple's App store restrictions might push some to Android's more open model. As mentioned in the previous section, the capability to put up and tear down a virtual business environment quickly favors innovation because the costs of experimenting are low. Ubiquitous access makes it easier to engage customers and employees in product improvement. Customers use a firm's products every day, and some reinvent them or use them in ways the firm never anticipated. A firm that learns of these unintended uses might identify new markets and new features that will extend sales. Addressability and traceability also enhance a firm's ability to learn about how, when, and where customers use electronic services and network connected devices. Learning is the first step of innovation.

Scaling risk

Sometimes a new product will takeoff, and a company will struggle to meet unexpectedly high demand. If it can't satisfy the market, competitors will move in and capture the sales that the innovator was unable to satisfy. In this case of digital products, a firm can use the cloud's elasticity to quickly acquire new storage and processing resources. It might also take advantage of sourcing independence to use multiple clouds, another form of elasticity, to meet burgeoning demand.

Control risk

The 2008 recession clearly demonstrated the risk of inadequate controls. In some cases, an organization acquired financial resources whose value was uncertain. In others, companies lost track of what they owned because of the complexity of financial instruments and the multiple transactions involving them. A system's interface is a point of data capture, and a well-designed interface is a control mechanism. Although it is usually not effective at identifying falsified data, it can require that certain elements of data are entered. Addressability and traceability also means that the system can record who entered the data, from which device, and when.

Cloud capabilities and strategic risks

Risk/Capability	Demand	Inefficiency	Innovation	Scaling	Control
Interface control			✔		✔
Location independence		✔			
Sourcing independence		✔		✔	
Virtual business environment		✔	✔		
Ubiquitous access	✔	✔	✔		
Addressability and traceability	✔		✔		✔
Rapid elasticity	✔			✔	

Most people think of cloud computing as an opportunity to lower costs by shifting processing from the corporate data center to a third party. More imaginative thinkers will see cloud computing as an opportunity to gain a competitive advantage.

Distributed database

The client/server architecture is concerned with minimizing processing costs by distributing processing between multiple clients and servers. Another factor in the total processing cost equation is communication. The cost of transmitting data usually increases with distance, and there can be substantial savings by locating a database close to those most likely to use it. The trade-off for a distributed database is lowered communication costs versus increased complexity. While the Internet and fiber optics have lowered the costs of communication, for a globally distributed organization data transmission costs can still be an issue because many countries still lack the bandwidth found in advanced economies.

Distributed database architecture describes the situation where a database is in more than one location but still accessible as if it were centrally located. For example, a multinational organization might locate its Indonesian data in Jakarta, its Japanese data in Tokyo, and its Australian data in Sydney. If most queries deal with the local situation, communication costs are substantially lower than if the database were centrally located. Furthermore, since the database is still treated as one logical entity, queries that require access to different physical locations can be processed. For example, the query "Find total sales of red hats in Australia" is likely to originate in Australia and be processed using the Sydney part of the database. A query of the form "Find total sales of red hats" is more likely to come from a headquarters' user and is resolved by accessing each of the local databases, though the user need not be aware of where the data are stored because the database appears as a single logical entity.

A distributed database management system (DDBMS) is a federation of individual DBMSs. Each site has a local DBMS and DC manager. In many respects, each site acts semi-independently. Each site also contains additional software that enables it to be part of the federation and act as a single database. It is this additional software that creates the DDBMS and enables the multiple databases to appear as one.

A DDBMS introduces a need for a data store containing details of the entire system. Information must be kept about the structure and location of every database, table, row, and column and their possible replicas. The system catalog is extended to include such information. The system catalog must also be distributed; otherwise, every request for information would have to be routed through some central point. For instance, if the systems catalog were stored in Tokyo, a query on Sydney data would first have to access the Tokyo-based catalog. This would create a bottleneck.

A hybrid, distributed architecture

Any organization of a reasonable size is likely to have a mix of the data processing architectures. Databases will exist on stand-alone personal computers, multiple client/server networks, distributed mainframes, clouds, and so on. Architectures continue to evolve because information technology is dynamic. Today's best solutions for data processing can become obsolete very quickly. Yet, organizations have invested large sums in existing systems that meet their current needs and do not warrant replacement. As a result, organizations evolve a hybrid architecture — a mix of the various forms. The concern of the IS unit is how to patch this hybrid together so that clients see it as a seamless system that readily provides needed information. In creating this ideal system, there are some underlying key concepts that should be observed. These fundamental principles were initially stated in terms of a distributed database. However, they can be considered to apply broadly to the evolving, hybrid architecture that organizations must continually fashion.

The fundamental principles of a hybrid architecture

Principle
Transparency
No reliance on a central site
Local autonomy
Continuous operation
Distributed query processing
Distributed transaction processing
Fragmentation independence
Replication independence
Hardware independence
Operating system independence
Network independence
DBMS independence

Transparency

The user should not have to know where data are stored nor how they are processed. The location of data, its storage format, and access method should be invisible to the client. The system should accept queries and resolve them expeditiously. Of course, the system should check that the person is authorized to access the requested data. Transparency is also known as **location independence**—the system can be used independently of the location of data.

No reliance on a central site

Reliance on a central site for management of a hybrid architecture creates two major problems. *First,* because all requests are routed through the central site, bottlenecks develop during peak periods. *Second*, if the central site fails, the entire system fails. A controlling central site is too vulnerable, and control should be distributed throughout the system.

Local autonomy

A high degree of local autonomy avoids dependence on a central site. Data are locally owned and managed. The local site is responsible for the security, integrity, and storage of local data. There cannot be absolute local autonomy because the various sites must cooperate in order for transparency to be feasible. Cooperation always requires relinquishing some autonomy.

Continuous operation

The system must be accessible when required. Since business is increasingly global and clients are geographically dispersed, the system must be continuously available. Many data centers now describe their operations as "24/7" (24 hours a day and 7 days a week). Clouds must operate continuously.

Distributed query processing

The time taken to execute a query should be generally independent of the location from which it is submitted. Deciding the most efficient way to process the query is the system's responsibility, not the client's. For example, a Sydney analyst could submit the query, "Find sales of winter coats in Japan." The system is responsible for deciding which messages and data to send between the various sites where tables are located.

Distributed transaction processing

In a hybrid system, a single transaction can require updating of multiple files at multiple sites. The system must ensure that a transaction is successfully executed for all sites. Partial updating of files will cause inconsistencies.

Fragmentation independence

Fragmentation independence means that any table can be broken into fragments and then stored in separate physical locations. For example, the sales table could be fragmented so that Indonesian data are stored in Jakarta, Japanese data in Tokyo, and so on. A fragment is any piece of a table that can be created by applying restriction and projection operations. Using join and union, fragments can be assembled to create the full table. Fragmentation is the key to a distributed database. Without fragmentation independence, data cannot be distributed.

Replication independence

Fragmentation is good when local data are mainly processed locally, but there are some applications that also frequently process remote data. For example, the New York office of an international airline may need both American (local) and European (remote) data, and its London office may need American (remote) and European (local) data. In this case, fragmentation into American and European data may not substantially reduce communication costs.

Replication means that a fragment of data can be copied and physically stored at multiple sites; thus the European fragment could be replicated and stored in New York, and the American fragment replicated and stored in London. As a result, applications in both New York and London will reduce their communication costs. Of course, the trade-off is that when a replicated fragment is updated, all copies also must be updated. Reduced communication costs are exchanged for increased update complexity.

Replication independence implies that replication happens behind the scenes. The client is oblivious to replication and requires no knowledge of this activity.

There are two major approaches to replication: synchronous or asynchronous updates. **Synchronous replication** means all databases are updated at the same time. Although this is ideal, it is not a simple task and is resource intensive. **Asynchronous replication** occurs when changes made to one database are relayed to other databases within a certain period established by the database administrator. It does not provide real-time updating, but it takes fewer IS resources. Asynchronous replication is a compromise strategy for distributed DBMS replication. When real-time updating is not absolutely necessary, asynchronous replication can save IS resources.

Hardware independence

A hybrid architecture should support hardware from multiple suppliers without affecting the users' capacity to query files. Hardware independence is a long-term goal of many IS managers. With virtualization, under which a computer can run another operating system, hardware independence has become less of an issue. For example, a Macintosh can run OS X (the native system), a variety of Windows operating systems, and many variations of Linux.

Operating system independence

Operating system independence is another goal much sought after by IS executives. Ideally, the various DBMSs and applications of the hybrid system should work on a range of operating systems on a variety of hardware. Browser-based systems are a way of providing operating system independence.

Network independence

Clearly, network independence is desired by organizations that wish to avoid the electronic shackles of being committed to any single hardware or software supplier.

DBMS independence

The drive for independence is contagious and has been caught by the DBMS as well. Since SQL is a standard for relational databases, organizations may well be able to achieve DBMS independence. By settling on the relational model as the organizational standard, ensuring that all DBMSs installed conform to this model, and using only standard SQL, an organization may approach DBMS independence. Nevertheless, do not forget all those old systems from the pre-relational days—a legacy that must be supported in a hybrid architecture.

Organizations can gain considerable DBMS independence by using Open Database Connectivity (ODBC) technology, which was covered earlier in the chapter on SQL. An application that uses the ODBC interface can access any ODBC-compliant DBMS. In a distributed environment, such as an n-tier client/server, ODBC enables application servers to access a variety of vendors' databases on different data servers.

Conclusion—paradise postponed

For data managers, the 12 principles just outlined are ideal goals. In the hurly-burly of everyday business, incomplete information, and an uncertain future, data managers struggle valiantly to meet clients' needs with a hybrid architecture that is an imperfect interpretation of IS paradise. It is unlikely that these principles will ever be totally achieved. They are guidelines and something to reflect on when making the inevitable trade-offs that occur in data management.

Now that you understand the general goals of a distributed database architecture, we will consider the major aspects of the enabling technology. First, we will look at distributed data access methods and then distributed database design. In keeping with our focus on the relational model, illustrative SQL examples are used.

Distributed data access

When data are distributed across multiple locations, the data management logic must be adapted to recognize multiple locations. The various types of distributed data access methods are considered, and an example is used to illustrate the differences between the methods.

Remote request

A remote request occurs when an application issues a single data request to a single remote site. Consider the case of a branch bank requesting data for a specific customer from the bank's central server located in Atlanta. The SQL command specifies the name of the server (`atlserver`), the database (`bankdb`), and the name of the table (`customer`).

```
SELECT * FROM atlserver.bankdb.customer
   WHERE custcode = 12345;
```

A remote request can extract a table from the database for processing on the local database. For example, the Athens branch may download balance details of customers at the beginning of each day and handle queries locally rather than issuing a remote request. The SQL is

```
SELECT custcode, custbalance FROM atlserver.bankdb.customer
   WHERE custbranch = 'Athens';
```

Remote transaction

Multiple data requests are often necessary to execute a complete business transaction. For example, to add a new customer account might require inserting a row in two tables: one row for the account and another row in the table relating a customer to the new account. A remote transaction contains multiple data requests for a single remote location. The following example illustrates how a branch bank creates a new customer account on the central server:

```
BEGIN WORK;
INSERT INTO atlserver.bankdb.account
   (accnum, acctype)
   VALUES (789, 'C');
INSERT INTO atlserver.bankdb.cust_acct
   (custnum, accnum)
   VALUES (123, 789);
COMMIT WORK;
```

The commands `BEGIN WORK` and `COMMIT WORK` surround the SQL commands necessary to complete the transaction. The transaction is successful only if both SQL statements are successfully executed. If one of the SQL statements fails, the entire transaction fails.

Distributed transaction

A distributed transaction supports multiple data requests for data at multiple locations. Each request is for data on a single server. Support for distributed transactions permits a client to access tables on different servers.

Consider the case of a bank that operates in the United States and Norway and keeps details of employees on a server in the country in which they reside. The following example illustrates a revision of the database to record details of an employee who moves from the United States to Norway. The transaction copies the data for the employee from the Atlanta server to the Oslo server and then deletes the entry for that employee on the Atlanta server.

```
BEGIN WORK;
INSERT INTO osloserver.bankdb.employee
   (empcode, emplname, …)
   SELECT empcode, emplname, …
      FROM atlserver.bankdb.employee
         WHERE empcode = 123;
DELETE FROM atlserver.bankdb.employee
   WHERE empcode = 123;
COMMIT WORK;
```

As in the case of the remote transaction, the transaction is successful only if both SQL statements are successfully executed.

Distributed request

A distributed request is the most complicated form of distributed data access. It supports processing of multiple requests at multiple sites, and each request can access data on multiple sites. This means that a distributed request can handle data replicated or fragmented across multiple servers.

Let's assume the bank has had a good year and decided to give all employees a 15 percent bonus based on their annual salary and add USD 1,000 or NOK 7,500 to their retirement account, depending on whether the employee is based in the United States or Norway.

```
BEGIN WORK;
CREATE VIEW temp
   (empcode, empfname, emplname, empsalary)
AS
   SELECT empcode, empfname, emplname, empsalary
      FROM atlserver.bankdb.employee
   UNION
   SELECT empcode, empfname, emplname, empsalary
      FROM osloserver.bankdb.employee;
SELECT empcode, empfname, emplname, empsalary*.15 AS bonus
      FROM temp;
UPDATE atlserver.bankdb.employee
   SET empusdretfund = empusdretfund + 1000;
UPDATE osloserver.bankdb.employee
   SET empkrnretfund = empkrnretfund + 7500;
COMMIT WORK;
```

The transaction first creates a view containing all employees by a union on the employee tables for both locations. This view is then used to calculate the bonus. Two SQL update commands then update the respective retirement fund records of the U.S. and Norwegian employees. Notice that retirement funds are recorded in U.S. dollars or Norwegian kroner.

Ideally, a distributed request should not require the application to know where data are physically located. A DDBMS should not require the application to specify the name of the server. So, for example, it should be possible to write the following SQL:

```
SELECT empcode, empfname, emplname, empsalary*.15 AS bonus
      FROM bankdb.employee;
```

It is the responsibility of the DDBMS to determine where data are stored. In other words, the DDBMS is responsible for ensuring data location and fragmentation transparency.

Distributed database design

Designing a distributed database is a two-stage process. *First,* develop a data model using the principles discussed in Section 2. *Second,* decide how data and processing will be distributed by applying the concepts of partitioning and replication. **Partitioning** is the fragmentation of tables across servers. Tables can be fragmented horizontally, vertically, or by some combination of both. **Replication** is the duplication of tables across servers.

Horizontal fragmentation

A table is split into groups of rows when horizontally fragmented. For example, a firm may fragment its employee table into three because it has employees in Tokyo, Sydney, and Jakarta and store the fragment on the appropriate DBMS server for each city.

Horizontal fragmentation

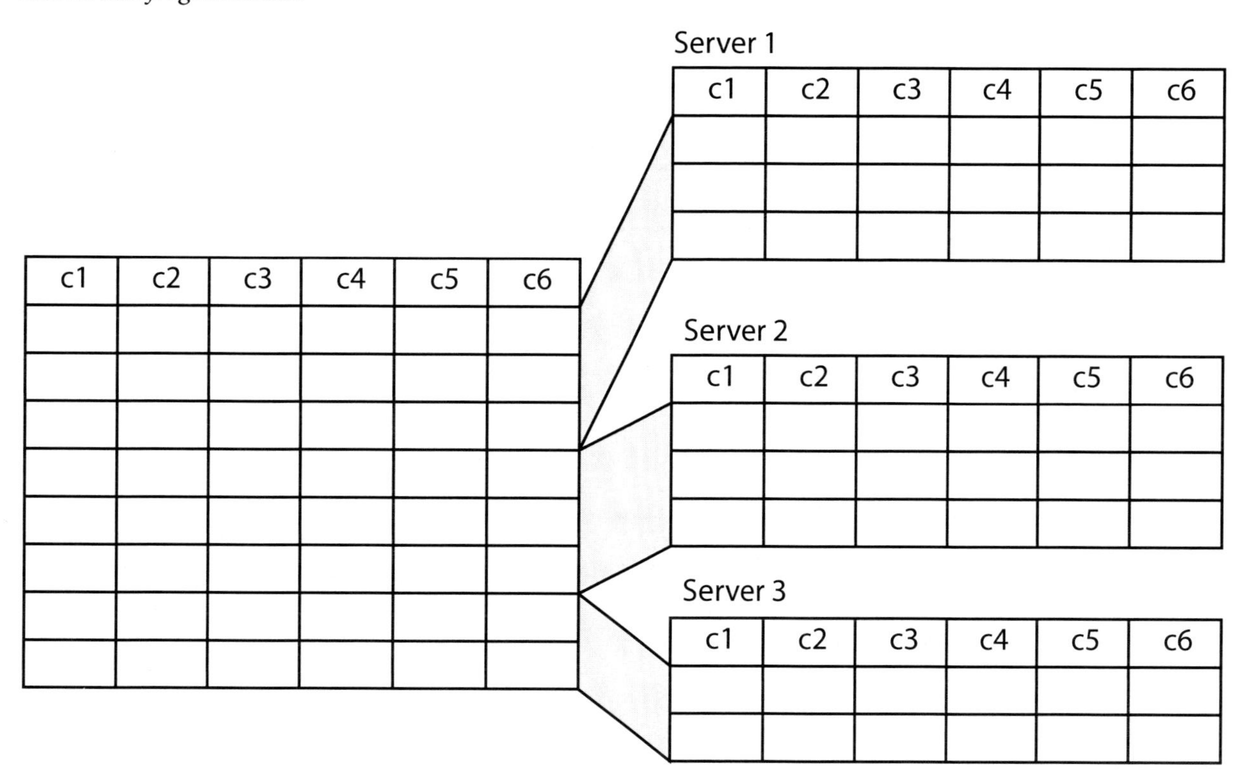

Three separate employee tables would be defined. Each would have a different name (e.g., `EMP-SYD`) but exactly the same columns. To insert a new employee, the SQL code is

```
INSERT INTO TABLE emp_syd
   SELECT * FROM new_emp
      WHERE emp_nation = 'Australia';
INSERT INTO table emp_tky
   SELECT * FROM new_emp
      WHERE emp_nation = 'Japan';
INSERT INTO table emp_jak
   SELECT * FROM new_emp
      WHERE emp_nation = 'Indonesia';
```

Vertical fragmentation

When vertically fragmented, a table is split into columns. For example, a firm may fragment its employee table vertically to spread the processing load across servers. There could be one server to handle address lookups and another to process payroll. In this case, the columns containing address information would be stored on one server and payroll columns on the other server. Notice that the primary key column (`C1`) must be stored on both servers; otherwise, the entity integrity rule is violated.

Vertical fragmentation

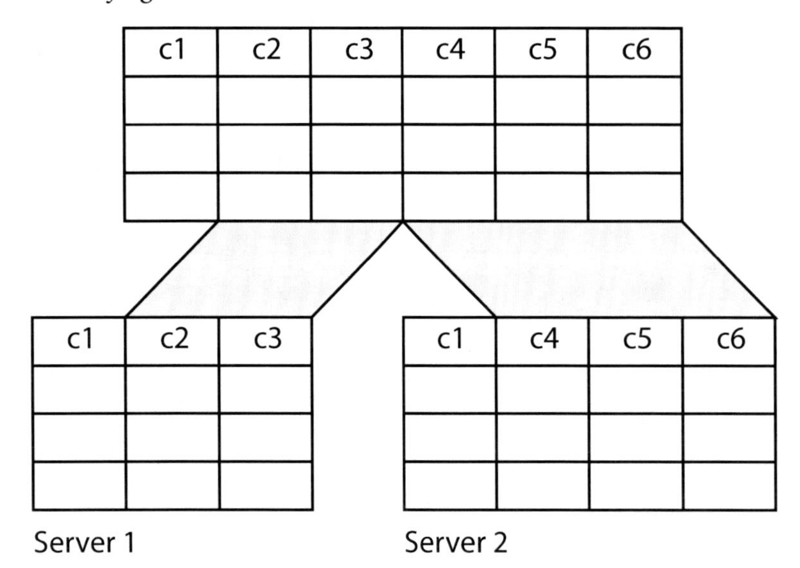

Hybrid fragmentation

Hybrid fragmentation is a mixture of horizontal and vertical. For example, the employee table could be first horizontally fragmented to distribute the data to where employees are located. Then, some of the horizontal fragments could be vertically fragmented to split processing across servers. Thus, if Tokyo is the corporate headquarters with many employees, the Tokyo horizontal fragment of the employee database could be vertically fragmented so that separate servers could handle address and payroll processing.

Horizontal fragmentation distributes data and thus can be used to reduce communication costs by storing data where they are most likely to be needed. Vertical fragmentation is used to distribute data across servers so that the processing load can be distributed. Hybrid fragmentation can be used to distribute both data and applications across servers.

Replication

Under replication, tables can be fully or partly replicated. **Full replication** means that tables are duplicated at each of the sites. The advantages of full replication are greater data integrity (because replication is essentially mirroring) and faster processing (because it can be done locally). However, replication is expensive because of the need to synchronize inserts, updates, and deletes across the replicated tables. When one table is altered, all the replicas must also be modified. A compromise is to use **partial replication** by duplicating the indexes only. This will increase the speed of query processing. The index can be processed locally, and then the required rows retrieved from the remote database.

Skill builder

A company supplies pharmaceuticals to sheep and cattle stations in outback Australia. Its agents often visit remote areas and are usually out of reach of a mobile phone network. To advise station owners, what data management strategy would you recommend for the company?

Conclusion

The two fundamental skills of data management, data modeling and data querying, are not changed by the development of a distributed data architecture such as client/server and the adoption of cloud computing. Data modeling remains unchanged. A high-fidelity data model is required regardless of where data are stored and whichever architecture is selected. SQL can be used to query local, remote, and distributed databases. Indeed, the adoption of client/server technology has seen a widespread increase in the demand for SQL skills.

Summary

Data can be stored and processed locally or remotely. Combinations of these two options provide the basic architecture options: remote job entry, personal database, and client/server. Many organizations are looking to cloud computing to reduce the processing costs and to allow them to focus on building systems. Cloud computing can also provide organizations with a competitive advantage if they exploit its seven capabilities to reduce one or more of the strategic risks.

Under distributed database architecture, a database is in more than one location but still accessible as if it were centrally located. The trade-off is lowered communication costs versus increased complexity. A hybrid architecture is a mix of data processing architectures. The concern of the IS department is to patch this hybrid together so that clients see a seamless system that readily provides needed information. The fundamental principles that a hybrid architecture should satisfy are transparency, no reliance on a central site, local autonomy, continuous operation, distributed query processing, distributed transaction processing, fragmentation independence, replication independence, hardware independence, operating system independence, network independence, and DBMS independence.

There are four types of distributed data access. In order of complexity, these are remote request, remote transaction, distributed transaction, and distributed request.

Distributed database design is based on the principles of fragmentation and replication. Horizontal fragmentation splits a table by rows and reduces communication costs by placing data where they are likely to be required. Vertical fragmentation splits a table by columns and spreads the processing load across servers. Hybrid fragmentation is a combination of horizontal and vertical fragmentation. Replication is the duplication of identical tables at different sites. Partial replication involves the replication of indexes. Replication speeds up local processing at the cost of maintaining the replicas.

Key terms and concepts

Application server
Client/server
Cloud computing
Continuous operation
Data communications manager (DC)
Data processing
Data server
Data storage
Database architecture
Database management system (DBMS)
DBMS independence
DBMS server
Distributed data access
Distributed database
Distributed query processing
Distributed request
Distributed transaction
Distributed transaction processing
Fragmentation independence
Hardware independence
Horizontal fragmentation
Hybrid architecture
Hybrid fragmentation
Local autonomy
Mainframe
N-tier architecture

Network independence

Personal database

Remote job entry

Remote request

Remote transaction

Replication

Replication independence

Server

Software independence

Strategic risk

Supercomputer

Three-tier architecture

Transaction processing monitor

Transparency

Two-tier architecture

Vertical fragmentation

References and additional readings

Child, J. (1987). Information technology, organizations, and the response to strategic challenges. *California Management Review, 30*(1), 33-50.

Iyer, B., & Henderson, J. C. (2010). Preparing for the future: understanding the seven capabilities of cloud computing. *MIS Executive, 9*(2), 117-131.

Morris, C. R., and C. H. Ferguson. 1993. How architecture wins technology wars. *Harvard Business Review* 71 (2):86–96.

Watson, R. T., Wynn, D., & Boudreau, M.-C. (2005). JBoss: The evolution of professional open source software. *MISQ Executive, 4*(3), 329-341.

Exercises

1. How does client/server differ from cloud computing?
2. In what situations are you likely to use remote job entry?
3. What are the disadvantages of personal databases?
4. What is a firm likely to gain when it moves from a centralized to a distributed database? What are the potential costs?
5. In terms of a hybrid architecture, what does transparency mean?
6. In terms of a hybrid architecture, what does fragmentation independence mean?
7. In terms of a hybrid architecture, what does DBMS independence mean?
8. How does ODBC support a hybrid architecture?
9. A university professor is about to develop a large simulation model for describing the global economy. The model uses data from 65 countries to simulate alternative economic policies and their possible outcomes. In terms of volume, the data requirements are quite modest, but the mathematical model is very complex, and there are many equations that must be solved for each quarter the model is run. What data processing/data storage architecture would you recommend?
10. A multinational company has operated relatively independent organizations in 15 countries. The new CEO wants greater coordination and believes that marketing, production, and purchasing should be globally managed. As a result, the corporate IS department must work with the separate national IS departments to integrate the various national applications and databases. What are the implications for the corporate data processing and database architecture? What are the key facts you would like to

know before developing an integration plan? What problems do you anticipate? What is your intuitive feeling about the key features of the new architecture?

11. A university wants to teach a specialized data management topic to its students every semester. It will take about two weeks to cover the topic, and during this period students will need access to a small high performance computing cluster on which the necessary software is installed. The software is Linux-based. Investigate three cloud computing offerings and make a recommendation as to which one the university should use.

21. SQL and Java

The vision for Java is to be the concrete and nails that people use to build this incredible network system that is happening all around us.

James Gosling, 2000

Learning objectives

Students completing this chapter will be able to

- write Java program to process a parameterized SQL query;
- read a CSV file and insert rows into tables;
- write a program using HMTL and JavaServer Pages (JSP) to insert data collected by a form data into a relational database;
- understand how SQL commands are used for transaction processing.

Introduction

Java is a platform-independent application development language. Its object-oriented nature makes it easy to create and maintain software and prototypes. These features make Java an attractive development language for many applications, including database systems.

This chapter assumes that you have some knowledge of Java programming, can use an integrated development environment (IDE) (e.g., BlueJ, Eclipse, or NetBeans), and know how to use a Java library. It is also requires some HTML skills because JavaServer Pages (JSP) are created by embedding Java code within an HTML document. MySQL is used for all the examples, but the fundamental principles are the same for all relational databases. With a few changes, your program will work with another implementation of the relational model.

JDBC

Java Database Connectivity (JDBC), a Java version of a portable SQL command line interface (CLI), is modeled on Open Database Connectivity (ODBC.) JDBC enables programmers to write Java software that is both operating system and DBMS independent. The JDBC core, which handles 90 percent of database programming, contains seven interfaces and two classes, which are part of the java.sql package. The purpose of each of these is summarized in the following table.

JDBC core interfaces and classes

Interfaces	Description
Connection	Connects an application to a database
Statement	A container for an SQL statement
PreparedStatement	Precompiles an SQL statement and then uses it multiple times
CallableStatement	Executes a stored procedure
ResultSet	The rows returned when a query is executed

ResultSetMetaData	The number, types, and properties of the result set
Classes	
DriverManager	Loads driver objects and creates database connections
DriverPropertyInfo	Used by specialized clients

For each DBMS, implementations of these interfaces are required because they are specific to the DBMS and not part of the Java package. For example, specific drivers are needed for MySQL, DB2, Oracle, and MS SQL. Before using JDBC, you will need to install these interfaces for the DBMS you plan to access. The standard practice appears to be to refer to this set of interfaces as the driver, but the 'driver' also includes implementations of all the interfaces.

Java EE

Java EE (Java Platform Enterprise Edition) is a platform for multi-tiered enterprise applications. In the typical three-tier model, a Java EE application server sits between the client's browser and the database server. A Java EE compliant server, of which there are a variety, is needed to process JSP.

Using SQL within Java

In this section, you will need a Java integrated development environment to execute the sample code. A good option is Eclipse,[80] a widely used open source IDE.

We now examine each of the major steps in processing an SQL query and, in the process, create Java methods to query a database.

Connect to the database

The getConnection method of the DriverManager specifies the URL of the database, the account identifier, and its password. You cannot assume that connecting to the database will be trouble-free. Thus, good coding practice requires that you detect and report any errors using Java's try-catch structure.

You connect to a DBMS by supplying its url and the account name and password.

```
try {
   conn = DriverManager.getConnection(url, account, password);
} catch (SQLException error) {
   System.out.println("Error connecting to database: " + error.toString());
   System.exit(1);
}
```

The format of the url parameter varies with the JDBC driver, and you will need to consult the documentation for the driver. In the case of MySQL, the possible formats are

```
jdbc:mysql:database
jdbc:mysql://host/database
jdbc:mysql://host:port/database
```

80 Install Eclipse IDE for Java EE Developers.

The default value for host is "localhost" and, for MySQL, the default port is "3306."

For example:

```
jdbc:mysql://www.richardtwatson.com:3306/Text
```

will enable connection to the database "Text" on the host "www.richardtwatson.com" on port 3306.

Create and execute an SQL statement

The prepareStatement method is invoked to produce a Statement object (see the following code). Note that the conn in conn.prepareStatement() refers to the connection created by the getConnection method. Parameters are used in prepareStatement to set execution time values, with each parameter indicated by a question mark (?). Parameters are identified by an integer for the order in which they appear in the SQL statement. In the following sample code, shrdiv is a parameter and is set to indiv by stmt.setInit(1,indiv), where 1 is its identifier (i.e., the first parameter in the SQL statement) and indiv is the parameter. The value of indiv is received as input at run time.

The SQL query is run by the executeQuery() method of the Statement object. The results are returned in a ResultSet object.

Create and execute an SQL statement

```
try {
   stmt = conn.prepareStatement("SELECT shrfirm, shrdiv FROM shr WHERE shrdiv > ?");
   // set the value for shrdiv to indiv
   stmt.setInt(1,indiv);
   rs = stmt.executeQuery();
   }
catch (SQLException error)
   {
      System.out.println("Error reporting from database: "
         + error.toString());
      System.exit(1);
   }
```

Report a SELECT

The rows in the table containing the results of the query are processed a row at a time using the next method of the ResultSet object (see the following code). Columns are retrieved one at a time using a get method within a loop, where the integer value specified in the method call refers to a column (e.g., `rs.getString(1)` returns the first column of the result set as a string). The get method used depends on the type of data returned by the query. For example, getString() gets a character string, getInt() gets an integer, and so on.

Reporting a SELECT

```
while (rs.next()) {
   String firm = rs.getString(1);
   int div = rs.getInt(2);
   System.out.println(firm + "  " + div);
}
```

Inserting a row

The prepareStatement() is also used for inserting a row in a similar manner, as the following example shows. The executeUpdate() method performs the insert. Notice the use of try and catch to detect and report an error during an insert.

Inserting a row

```
try {
    stmt = conn.prepareStatement("INSERT INTO alien (alnum, alname) VALUES (?,?)");
    stmt.setInt(1,10);
    stmt.setString(2, "ET");
    stmt.executeUpdate();
    }
catch (SQLException error)
{
    System.out.println("Error inserting row into database: "
     + error.toString());
    System.exit(1);
}
```

Release the Statement object

The resources associated with the various objects created are freed as follows:

```
stmt.close();
rs.close();
conn.close();
```

To illustrate the use of Java, a pair of programs is available. The first, DataTest.java, creates a database connection and then calls a method to execute and report an SQL. The second program, DatabaseAccess.java, contains the methods called by the first. Both programs are available on the book's web site for examination and use.

Skill builder

1. Get from this book's supporting Web site the code of DatabaseTest.java and DatabaseAccess.java.
2. Inspect the code to learn how you use SQL from within a Java application.
3. Run DatabaseTest.java with a few different values for indiv.
4. Make modifications to query a different table.

Loading a text file into a database

Java is useful when you have a dataset that you need to load into a database. For example, you might have set up a spreadsheet and later realized that it would be more efficient to use a database. In this case, you can export the spreadsheet and load the data into one or more tables.

We will illustrate loading data from a text file into the ArtCollection database with the following design. Notice that there different tables for different types of art, paintings, and sculptures in this case.[81]

ArtCollection data model

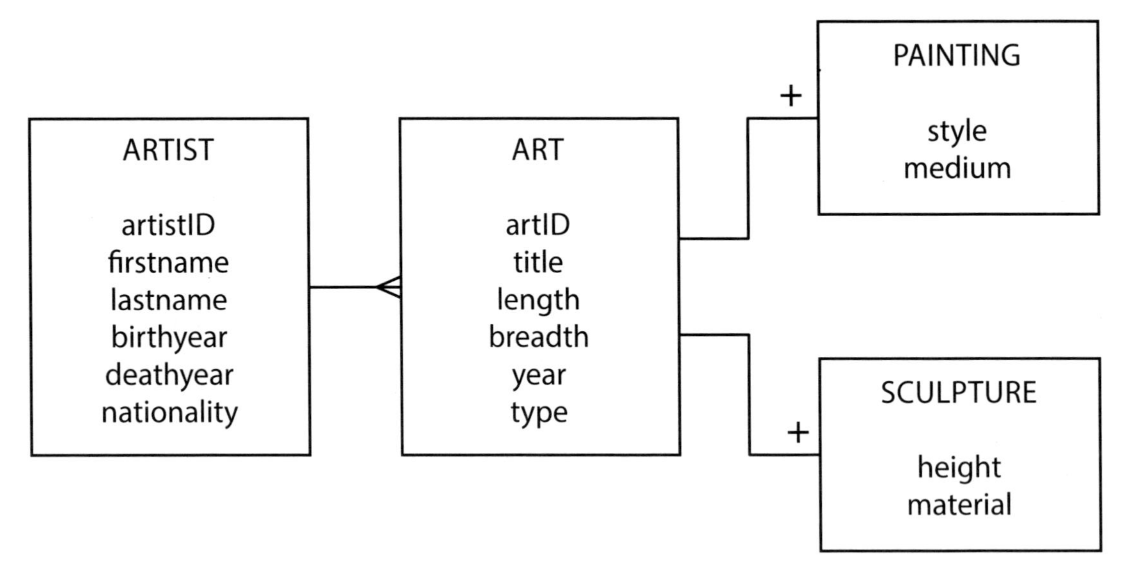

CSV file

A comma-separated values (CSV) file[82] is a common form of text file because most spreadsheet programs offer it as an export option. There are a variety of options available in terms of the separator for fields (usually a comma or tab) and the use of quotation marks to denote text fields. Usually, each record has the same number of fields. Frequently, the first record contains names for each of the fields. Notice in the example file that follows that a record contains details of an artist and one piece of that person's art. As the data model specifies, these two sets of data go into separate tables.

A CSV file

```
firstName,lastName,birthyear,deathyear,nationality,title,length,breadth,year,style,medium
Edvard,Munch,1863,1944,Norwegian,The Scream,36,29,1893,Cubist,Oil
Claude,Monet,1840,1926,French,Bridge over a Pond of Water Lilies,36,29,1899,Impressionist,Oil
Toulouse,Lautrec,1864,1901,French,Avril,55,33,1893,Impressionist,Oil
Vincent,Van Gogh,1853,1890,French,The Starry Night,29,36,1889,Impressionist,Oil
Johannes,Vermeer,1632,1675,Dutch,Girl with a Pearl Earring,17,15,1665,Impressionist,Oil
Salvador,Dali,1904,1989,Spanish,The Persistence of Memory,9.5,13,1931,Surreal,Oil
Andrew,Wyeth,1917,2009,American,Christina's World ,32,47,1948,Surreal,Oil
William,Turner,1789,1862,English,The Battle of Trafalgar,103,145,1822,Surreal,Oil
Tom,Roberts,1856,1931,Australian,Shearing the Rams,48,72,1890,Surreal,Oil
Paul,Gauguin,1848,1903,French,Spirit of the Dead Watching,28,36,1892,Surreal,Watercolor
```

81 The code for creating the database is online at http://richardtwatson.com/dm6e/Reader/java.html.

82 http://en.wikipedia.org/wiki/Comma-separated_values

Because CSV files are so widely used, there are Java libraries for processing them. We have opted for CsvReader.[83] Here is some code for connecting to a CSV file defined by its URL. You can also connect to text files on your personal computer.

Connecting to a CSV file

```
try {
   csvurl = new URL("http://people.terry.uga.edu/rwatson/data/painting.csv");
} catch (IOException error) {
   System.out.println("Error accessing url: " + error.toString());
   System.exit(1);
}
```

A CSV file is read a record at a time, and the required fields are extracted for further processing. The following code illustrates how a loop is used to read each record in a CSV file and extract some fields. Note the following:

- The first record, the header, contains the names of each field and is read before the loop starts.
- Field names can be used to extract each field with a get() method.
- You often need to convert the input string data to another format, such as integer.
- Methods, addArtist, and addArt are included in the loop to add details of an artist to the artist table and a piece of art to the art table.

Processing a CSV file

```
try {
   input = new CsvReader(new InputStreamReader(csvurl.openStream()));
   input.readHeaders();
   while (input.readRecord())
   {
   // Artist
      String firstName = input.get("firstName");
      String lastName = input.get("lastName");
      String nationality = input.get("nationality");
      int birthYear = Integer.parseInt(input.get("birthyear"));
      int deathYear = Integer.parseInt(input.get("deathyear"));
      artistPK = addArtist(conn, firstName, lastName, nationality, birthYear,
deathYear);
      // Painting
      String title = input.get("title");
      double length = Double.parseDouble(input.get("length"));
      double breadth = Double.parseDouble(input.get("breadth"));
      int year = Integer.parseInt(input.get("year"));
      addArt(conn, title, length, breadth, year,artistPK);
   }
input.close();
} catch (IOException error) {
```

83 Get the Java archive from http://sourceforge.net/projects/javacsv/

```
    System.out.println("Error reading CSV file: " + error.toString());
    System.exit(1);
}
```

We now need to consider the addArtist method. Its purpose is to add a row to the artist table. The primary key, artistID, is specified as AUTO_INCREMENT, which means it is set by the DBMS. Because of the 1:m relationship between artist and art, we need to know the value of artistID, the primary key of art, to set the foreign key for the corresponding piece of art. We can use the following code to determine the primary key of the last insert.

Determining the value of the primary key of the last insert

```
rs = stmt.executeQuery("SELECT LAST_INSERT_ID()");
    if (rs.next()) {
        autoIncKey = rs.getInt(1);
    }
```

The addArtist method returns the value of the primary key of the most recent insert so that we can then use this value as the foreign key for the related piece of art.

```
artistPK = addArtist(conn, firstName, lastName, nationality, birthYear, deathYear);
```

The call to the addArt method includes the returned value as a parameter. That is, it passes the primary key of artist to be used as foreign key of art.

```
addArt(conn, title, length, breadth, year,artistPK);
```

The Java code for the complete application is available on the book's web site.[84]

Skill builder

1. Add a few sculptors and a piece of sculpture to the CSV file. Also, decide how you will differentiate between a painting and a sculpture.
2. Add a method to ArtCollector.java to insert an artist and that person's sculpture.
3. Run the revised program so that it adds both paintings and sculptures for various artists.

JavaServer Pages (JSP)

Standard HTML pages are static. They contain text and graphics that don't change unless someone recodes the page. JSP and other technologies, such as PHP and ASP, enable you to create dynamic pages whose content can change based to suit the goals of the person accessing the page, the browser used, or the device on which the page is displayed.

A JSP contains standard HTML code, the static, and JSP elements, the dynamic. When someone requests a JSP, the server combines HTML and JSP elements to deliver a dynamic page. JSP is also useful for server side processing. For example, when someone submits a form, JSP elements can be used to process the form on the server.

84 http://richardtwatson.com/dm6e/Reader/java.html

Map collection case

The Expeditioner has added a new product line to meet the needs of its changing clientele. It now stocks a range of maps. The firm's data modeler has created a data model describing the situation . A map has a scale; for example, a scale of 1:1 000 000 means that 1 unit on the map is 1,000,000 units on the ground (or 1 cm is 10 km, and 1 inch is ~16 miles). There are three types of maps: road, rail, and canal. Initially, The Expeditioner decided to stock maps for only a few European countries.

The Expeditioner's map collection data model

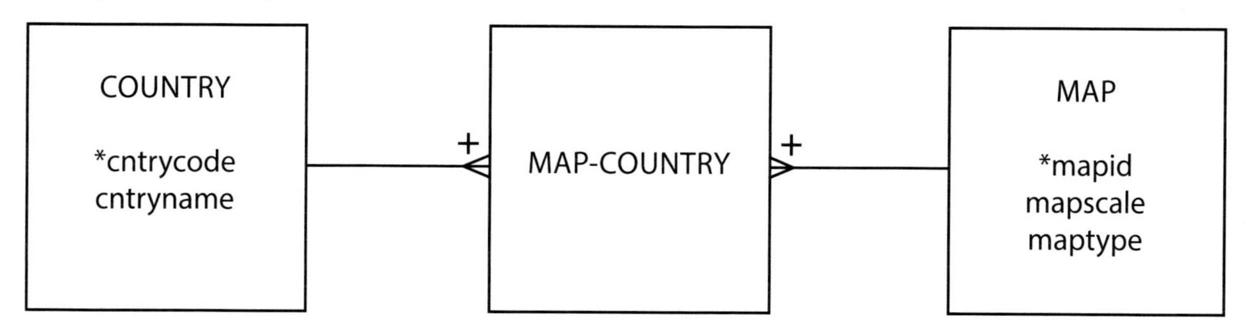

Data entry

Once the tables have been defined, we want to insert some records. This could be done using the insertSQL method of the Java code just created, but it would be very tedious, as we would type insert statements for each row (e.g., `INSERT INTO MAP VALUES (1, 1000000, 'RAIL');`). A better approach is to use a HTML data entry form. An appropriate form and its associated code follow.

Map collection data entry form

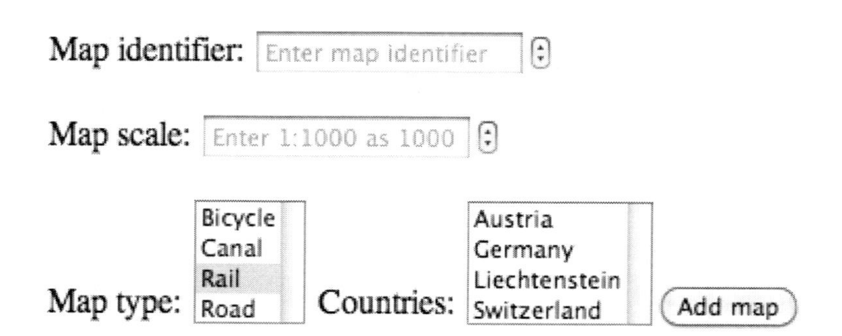

Map collection data entry HTML code (index.html)

```
<!DOCTYPE html PUBLIC "-//W3C//DTD HTML 4.01 Transitional//EN" "http://www.w3.org/TR/html4/loose.dtd">
<html>
<head>
<meta http-equiv="Content-Type" content="text/html; charset=ISO-8859-1">
<title>Map collection</title>
</head>
</body>
<form name="MapInsert" action="mapinsert.jsp" method="post">
    <p>
        <label>Map identifier: <input type="number" required
```

```
        pattern="[0-9]{3}" name="mapid" size="24" value=""
        placeholder="Enter map identifier"></label>
  <p>
    <label>Map scale: <input type="number" required
        pattern="[0-9]+" min="1000" max="100000" step="1000" name="mapscale"
        size="24" value="" placeholder="Enter 1:1000 as 1000"></label>
  <p>
    <label>Map type: <select name="maptype" size="3"
        placeholder="Select type of map"></label>
  </p>
  <option value="Bicycle">Bicycle</option>
  <option value="Canal">Canal</option>
  <option value="Rail" selected>Rail</option>
  <option value="Road">Road</option>
  </select> <label>Countries: <select name="countries" size="4" multiple
    placeholder="Select countries"></label>
  <option value="at">Austria</option>
  <option value="de">Germany</option>
  <option value="li">Liechtenstein</option>
  <option value="ch">Switzerland</option>
  <input type="submit" name="submitbutton" value="Add map">
</form>
</body>
</html>
```

The data entry form captures a map's identifier, scale, type, and the number of countries covered by the map. A map identifier is of the form M001 to M999, and a regular expression is used to validate the input.[85] A map scale must be numeric with a minimum of 1000 and maximum of 100000 in increments of 1000. The type of map is selected from the list, which is defined so that only one type can be selected. Multiple countries can be selected by holding down an appropriate key combination when selected from the displayed list of countries. When the "Add map" button is clicked, the page mapinsert.jsp is called, which is specified in the first line of the HTML body code.

Passing data from a form to a JSP

In a form, the various entry fields are each identified by a name specification (e.g., `NAME="MAPID"` and `NAME="MAPSCALE"`). In the corresponding JSP, which is the one defined in the form's action statement (i.e., `action="mapinsert.jsp"`) these same fields are accessed using the getParameter method when there is a single value or getParameterValues when there are multiple values. The following chunks of code show corresponding form and JSP code for maptype.

```
<label>Map type: <select name="maptype" size="3"
        placeholder="Select type of map"></label>
```

```
maptype = request.getParameter("maptype");
```

85 This is a feature of HTML5 and is not supported by older browsers.

Transaction processing

A transaction is a logical unit of work. In the case of adding a map, it means inserting a row in map for the map and inserting one row in mapCountry for each country on the map. All of these inserts must be executed without failure for the entire transaction to be processed. SQL has two commands for supporting transaction processing: `COMMIT` and `ROLLBACK`. If no errors are detected, then the various changes made by a transaction can be *committed*. In the event of any errors during the processing of a transaction, the database should be *rolled back* to the state prior to the transaction.

Autocommit

Before processing a transaction, you need to turn off autocommit to avoid committing each database change separately before the entire transaction is complete. It is a good idea to set the value for autocommit immediately after a successful database connection, which is what you will see when you inspect the code for mapinsert.jsp. The following code sets autocommit to false in the case where conn is the connection object.

```
conn.setAutoCommit(false);
```

Commit

The `COMMIT` command is executed when all parts of a transaction have been successfully executed. It makes the changes to the database permanent.

```
conn.commit(); // all inserts successful
```

Rollback

The `ROLLBACK` command is executed when any part of a transaction fails. All changes to the database since the beginning of the transaction are reversed, and the database is restored to its state before the transaction commenced.

```
conn.rollback(); // at least one insert failed
```

Completing the transaction

The final task, to commit or roll back the transaction depending on whether errors were detected during any of the inserts, is determined by examining `TRANSOK`, which is a boolean variable set to false when any errors are detected during a transaction.

Completing the transaction

```
if (transOK) {
   conn.commit(); // all inserts successful
} else {
   conn.rollback(); // at least one insert failed
}
conn.close(); // close database
```

Putting it all together

You have now seen some of the key pieces for the event handler that processes a transaction to add a map and the countries on that map. The code can be downloaded from the book's Web site.[86]

mapinsert.jsp

```
<%@ page language="java" contentType="text/html; charset=ISO-8859-1"
   pageEncoding="ISO-8859-1"%>
<!DOCTYPE html PUBLIC "-//W3C//DTD HTML 4.01 Transitional//EN" "http://www.w3.org/TR/
html4/loose.dtd">
<%@ page import="java.util.*"%>
<%@ page import="java.lang.*"%>
<%@ page import="java.sql.*"%>
<html>
<head>
<meta http-equiv="Content-Type" content="text/html; charset=ISO-8859-1">
<title>Map insert page</title>
</head>
<body>
   <%
   String url;
   String jdbc = "jdbc:mysql:";
   String database = "//localhost:3306/MapCollection";
   String username = "student", password = "student";
   String mapid, maptype, countries;
   String[] country;
   int mapscale = 0;
   boolean transOK = true;
   PreparedStatement insertMap;
   PreparedStatement insertMapCountry;
   Connection conn = null;
   // make connection
   url = jdbc + database;
   try {
      conn = DriverManager.getConnection(url, username, password);
   } catch (SQLException error) {
      System.out.println("Error connecting to database: "
               + error.toString());
      System.exit(1);
   }
   try {
      conn.setAutoCommit(false);
   } catch (SQLException error) {
      System.out.println("Error turning off autocommit"
               + error.toString());
      System.exit(2);
   }
```

86 http://richardtwatson.com/dm6e/Reader/java.html

```
    //form data
    mapid = request.getParameter("mapid");
    mapscale = Integer.parseInt(request.getParameter("mapscale"));
    maptype = request.getParameter("maptype");
    country = request.getParameterValues("countries");
    transOK = true;
    // insert the map
    try {
        insertMap = conn.prepareStatement("INSERT INTO map (mapid, mapscale, maptype) 
VALUES (?,?,?)");
        insertMap.setString(1, mapid);
        insertMap.setInt(2, mapscale);
        insertMap.setString(3, maptype);
        System.out.println(insertMap);
        insertMap.executeUpdate();
        // insert the countries
        for (int loopInx = 0; loopInx < country.length; loopInx++) {
            insertMapCountry = conn.prepareStatement("INSERT INTO mapCountry 
(mapid ,cntrycode ) VALUES (?,?)");
            insertMapCountry.setString(1, mapid);
            insertMapCountry.setString(2, country[loopInx]);
            System.out.println(insertMapCountry);
            insertMapCountry.executeUpdate();
        }
    } catch (SQLException error) {
        System.out.println("Error inserting row: " + error.toString());
        transOK = false;
    }
    if (transOK) {
        conn.commit(); // all inserts successful
        System.out.println("Transaction commit");
    } else {
        conn.rollback(); // at least one insert failed
        System.out.println("Transaction rollback");
    }
    conn.close();
    System.out.println("Database close"); // close database
    %>
</body>
</html>
```

The code in mapinsert.jsp does the following:

1. Creates a connection to a database
2. Gets the data from the map entry form
3. Inserts a row for a new map
4. Uses a loop to insert a row for each country on the map

5. Shows how to handle a transaction failure.

Conclusion

Java is a widely used object-oriented programming language for developing distributed multi-tier applications. JDBC is a key technology in this environment because it enables a Java application to interact with various implementations of the relational database model (e.g., Oracle, SQL Server, MySQL). As this chapter has demonstrated, with the help of a few examples, JDBC can be readily understood and applied.

Summary

Java is a platform-independent application development language. JDBC enables programs that are DBMS independent. SQL statements can be embedded within Java programs. JSP is used to support server side processing. `COMMIT` and `ROLLBACK` are used for transaction processing.

Key terms and concepts

Autocommit

Comma-separated values

IDE

COMMIT

Java

Java Database Connectivity (JDBC)

Java Server Pages (JSP)

ROLLBACK

Transaction processing

References and additional readings

Barnes, D. J., and M. Kölling. 2005. Objects first with Java : a practical introduction using Blue J. 2nd ed. Upper Saddle River, NJ: Prentice Hall.

Exercises

1. Write a Java program to run a parameterized query against the ClassicModels database. For example, you might report the sum of payments for a given customer's name.

Extend ArtCollection.java so that it can handle inserting rows for multiple pieces of art for a single artist. You will have to rethink the structure of the CSV file and how to record multiple pieces of art from a single artist.

2. Write a JSP application to maintain the database defined by the following data model. The database keeps track of the cars sold by a salesperson in an automotive dealership. Your application should be able to add a person and the cars a person has sold. These should be separate transactions.

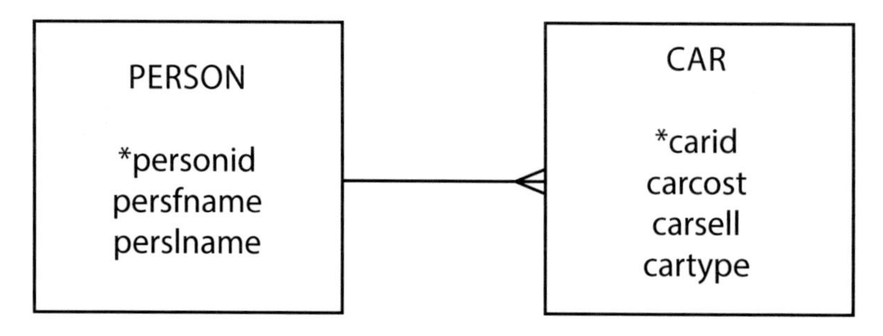

22. Data Integrity

Integrity without knowledge is weak and useless, and knowledge without integrity is dangerous and dreadful.

Samuel Johnson, ***Rasselas***, 1759

Learning objectives

After completing this chapter, you will

- understand the three major data integrity outcomes;
- understand the strategies for achieving each of the data integrity outcomes;
- understand the possible threats to data integrity and how to deal with them;
- understand the principles of transaction management;
- realize that successful data management requires making data available and maintaining data integrity.

The Expeditioner has become very dependent on its databases. The day-to-day operations of the company would be adversely affected if the major operational databases were lost. Indeed, The Expeditioner may not be able to survive a major data loss. Recently, there have also been a few minor problems with the quality and confidentiality of some of the databases. A part-time salesperson was discovered making a query about staff salaries. A major order was nearly lost when it was shipped to the wrong address because the complete shipping address had not been entered when the order was taken. The sales database had been offline for 30 minutes last Monday morning because of a disk sector read error.

The Expeditioner had spent much time and money creating an extremely effective and efficient management system. It became clear, however, that more attention needed to be paid to maintaining the system and ensuring that high-quality data were continuously available to authorized users.

Introduction

The management of data is driven by two goals: availability and integrity. **Availability** deals with making data available to whomever needs it, whenever and wherever he or she needs it, and in a meaningful form. As illustrated in following figure, availability concerns the creation, interrogation, and update of data stores. Although most of the book, thus far, has dealt with making data available, a database is of little use to anyone unless it has integrity. Maintaining data integrity implies three goals:[87]

3. Protecting existence: Data are available when needed.
4. Maintaining quality: Data are accurate, complete, and current.
5. Ensuring confidentiality: Data are accessed only by those authorized to do so.

87 To the author's knowledge, Gordon C. Everest was the first to define data integrity in terms of these three goals. See Everest, G. (1986). Database management: objectives, systems functions, and administration. New York, NY: McGraw-Hill.

Goals of managing organizational memory

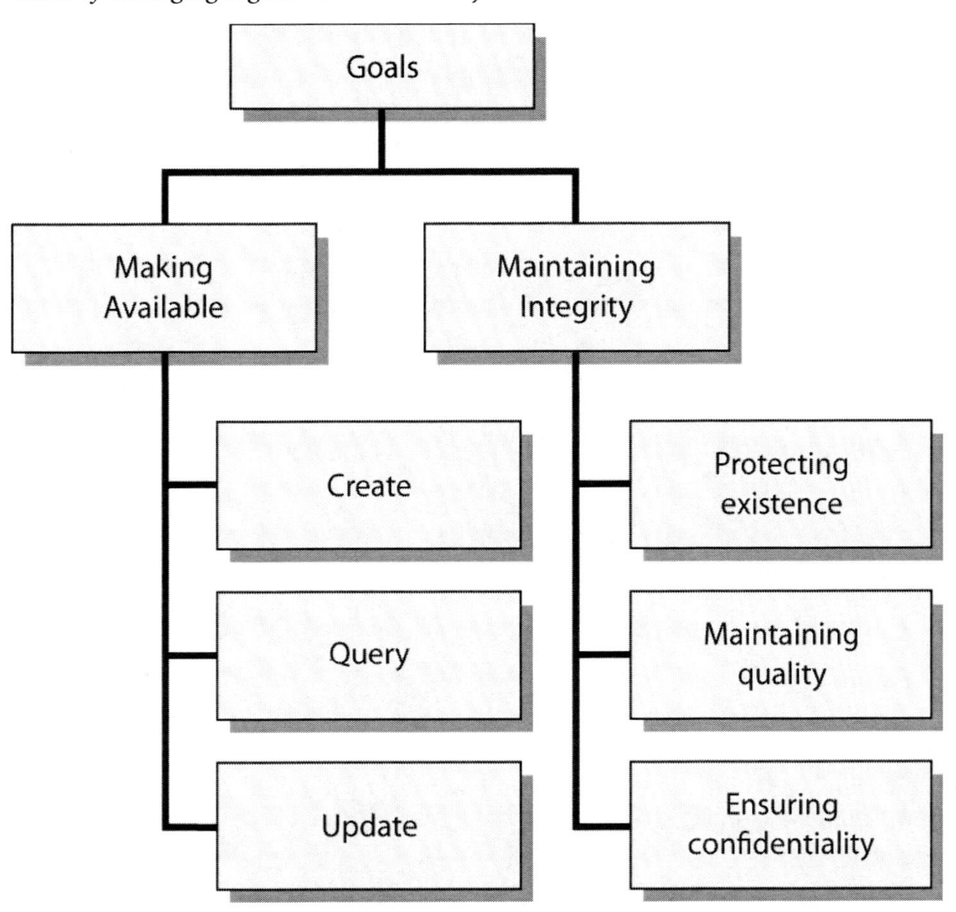

This chapter considers the three types of strategies for maintaining data integrity:

1. **Legal** strategies are externally imposed laws, rules, and regulations. Privacy laws are an example.
2. **Administrative** strategies are organizational policies and procedures. An example is a standard operating procedure of storing all backup files in a locked vault.
3. **Technical** strategies are those incorporated in the computer system (e.g., database management system [DBMS], application program, or operating system). An example is the inclusion of validation rules (e.g., `NOT NULL`) in a database definition that are used to ensure data quality when a database is updated.

A **consistent database** is one in which all data integrity constraints are satisfied.

Our focus is on data stored in multiuser computer databases and technical and administrative strategies for maintaining integrity in computer system environments., commonly referred to as **database integrity**. More and more, organizational memories are being captured in computerized databases. From an integrity perspective, this is a very positive development. Computers offer some excellent mechanisms for controlling data integrity, but this does not eliminate the need for administrative strategies. Both administrative and technical strategies are still needed.

Who is responsible for database integrity? Some would say the data users, others the database administrator. Both groups are right; database integrity is a shared responsibility, and the way it is managed may differ across organizations. Our focus is on the tools and strategies for maintaining data integrity, regardless of who is responsible.

The strategies for achieving the three integrity outcomes are summarized in the following table. We will cover procedures for protecting existence, followed by those for maintaining integrity, and finally those used to ensure confidentiality. Before considering each of these goals, we need to examine the general issue of transaction management.

Strategies for maintaining database integrity

Database integrity outcome	Strategies for achieving the outcome
Protecting existence	Isolation (preventive) Database backup and recovery (curative)
Maintaining quality	Update authorization Integrity constraints/data validation Concurrent update control
Ensuring confidentiality	Access control Encryption

Transaction management

Transaction management focuses on ensuring that transactions are correctly recorded in the database. The **transaction manager** is the element of a DBMS that processes transactions. A **transaction** is a series of actions to be taken on the database such that they must be entirely completed or entirely aborted. A transaction is a **logical unit of work**. All its elements must be processed; otherwise, the database will be incorrect. For example, with a sale of a product, the transaction consists of at least two parts: an update to the inventory on hand, and an update to the customer information for the items sold in order to bill the customer later. Updating only the inventory or only the customer information would create a database without integrity and an inconsistent database.

Transaction managers are designed to accomplish the **ACID** (atomicity, consistency, isolation, and durability) concept. These attributes are:

1. **Atomicity:** If a transaction has two or more discrete pieces of information, either all of the pieces are committed or none are.
2. **Consistency:** Either a transaction creates a valid new database state or, if any failure occurs, the transaction manager returns the database to its prior state.
3. **Isolation:** A transaction in process and not yet committed must remain isolated from any other transaction.
4. **Durability:** Committed data are saved by the DBMS so that, in the event of a failure and system recovery, these data are available in their correct state.

Transaction atomicity requires that all transactions are processed on an **all-or-nothing** basis and that any collection of transactions is **serializable**. When a transaction is executed, either all its changes to the database are completed or none of the changes are performed. In other words, the entire unit of work must be completed. If a transaction is terminated before it is completed, the transaction manager must undo the executed actions to restore the database to its state before the transaction commenced (the consistency concept). Once a transaction is successfully completed, it does not need to be undone. For efficiency reasons, transactions should be no larger than necessary to ensure the integrity of the database. For example, in an accounting system, a debit and credit would be an appropriate transaction, because this is the minimum amount of work needed to keep the books in balance.

Serializability relates to the execution of a set of transactions. An interleaved execution schedule (i.e., the elements of different transactions are intermixed) is serializable if its outcome is equivalent to a non-interleaved (i.e., serial) schedule. Interleaved operations are often used to increase the efficiency of computing resources, so it is not unusual for the components of multiple transactions to be interleaved. Interleaved transactions cause problems when they interfere with each other and, as a result, compromise the correctness of the database.

The ACID concept is critical to concurrent update control and recovery after a transaction failure.

Concurrent update control

When updating a database, most applications implicitly assume that their actions do not interfere with any other's actions. If the DBMS is a single-user system, then a lack of interference is guaranteed. Most DBMSs, however, are multiuser systems, where many analysts and applications can be accessing a given database at the same time. When two or more transactions are allowed to update a database concurrently, the integrity of the database is threatened. For example, multiple agents selling airline tickets should not be able to sell the same seat twice. Similarly, inconsistent results can be obtained by a retrieval transaction when retrievals are being made simultaneously with updates. This gives the appearance of a loss of database integrity. We will first discuss the integrity problems caused by concurrent updates and then show how to control them to ensure database quality.

Lost update

Uncontrolled concurrent updates can result in the lost-update or phantom-record problem. To illustrate the lost-update problem, suppose two concurrent update transactions simultaneously want to update the same record in an inventory file. Both want to update the quantity-on-hand field (quantity). Assume quantity has a current value of 40. One update transaction wants to add 80 units (a delivery) to quantity, while the other transaction wants to subtract 20 units (a sale).

Suppose the transactions have concurrent access to the record; that is, each transaction is able to read the record from the database before a previous transaction has been committed. This sequence is depicted in the following figure. Note that the first transaction, A, has not updated the database when the second transaction, B, reads the same record. Thus, both A and B read in a value of 40 for quantity. Both make their calculations, then A writes the value of 120 to disk, followed by B promptly overwriting the 120 with 20. The result is that the delivery of 80 units, transaction A, is lost during the update process.

Lost update when concurrent accessing is allowed

Time	*Action*		*Database record*	
			Part#	*Quantity*
			P10	40
T1	App A receives message for delivery of 80 units of P10			
T2	App A reads the record for P10	←	P10	40
T3	App B receives message for sale of 20 units of P10			
T4	App B reads the record for P10	←	P10	40

Time	Action		Database record	
			Part#	Quantit y
T5	App A processes delivery (40 + 80 = 120)			
T6	App A updates the record for P10	→	P10	120
T7	App B processes the sale of (40 - 20 = 20)			
T8	App B updates the record P10	→	P10	20

Inconsistent retrievals occur when a transaction calculates some aggregate function (e.g., sum) over a set of data while other transactions are updating the same data. The problem is that the retrieval may read some data before they are changed and other data after they are changed, thereby yielding inconsistent results.

The solution: locking

To prevent lost updates and inconsistent retrieval results, the DBMS must incorporate a resource **lock**, a basic tool of transaction management to ensure correct transaction behavior. Any data retrieved by one application with the intent of updating must be locked out or denied access by other applications until the update is completed (or aborted).

There are two types of locks: **Slocks** (shared or read locks) and **Xlocks** (exclusive or write locks). Here are some key points to understand about these types of locks:

1. When a transaction has a Slock on a database item, other transactions can issue Slocks on the same item, but there can be no Xlocks on that item.
2. Before a transaction can read a database item, it must be granted a Slock or Xlock on that item.
3. When a transaction has an Xlock on a database item, no other transaction can issue either a Slock or Xlock on that item.
4. Before a transaction can write to a database item, it must be granted an Xlock on that item.

Consider the example used previously. When A accesses the record for update, the DBMS must refuse all further accesses to that record until transaction A is complete (i.e., an Xlock). As the following figure shows, B's first attempt to access the record is denied until transaction A is finished. As a result, database integrity is maintained. Unless the DBMS controls concurrent access, a multiuser database environment can create both data and retrieval integrity problems.

Valid update when concurrent accessing is not allowed

Time	*Action*		*Database record*	
			Part#	*Quantity*
			P10	40
T1	App A receives message for delivery of 80 units of P10			
T2	App A reads the record for P10	←	P10	40
T3	App B receives message for sale of 20 units of P10			
T4	App B attempts to read the record for P10	deny	P10	40
T5	App A process delivery (40 + 80 = 120)			
T6	App A updates the record for P10	→	P10	120
T7	App B reads the record for P10	←	P10	120
T8	App B processes the sale of (120 - 20 = 100)			
T9	App B updates the record P10	→	P10	100

To administer locking procedures, a DBMS requires two pieces of information:

1. Whether a particular transaction will update the database;
2. Where the transaction begins and ends.

Usually, the data required by an update transaction are locked when the transaction begins and then released when the transaction is completed (i.e., committed to the database) or aborted. Locking mechanisms can operate at different levels of **locking granularity**: database, table, page, row, or data element. At the most precise level, a DBMS can lock individual data elements so that different update transactions can update different items in the same record concurrently. This approach increases processing overhead but provides the fewest resource conflicts. At the other end of the continuum, the DBMS can lock the entire database for each update. If there were many update transactions to process, this would be very unacceptable because of the long waiting times. Locking at the record level is the most common approach taken by DBMSs.

In most situations, applications are not concerned with locking, because it is handled entirely by the DBMS. But in some DBMSs, choices are provided to the programmer. These are primarily limited to programming language interfaces.

Resource locking solves some data and retrieval integrity problems, but it may lead to another problem, referred to as **deadlock** or the **deadly embrace**. Deadlock is an impasse that occurs because two applications lock certain resources, then request resources locked by each other. The following figure illustrates a deadlock situation. Both transactions require records 1 and 2. Transaction A first accesses record 1 and locks it. Then transaction B accesses record 2 and locks it. Next, B's attempt to access record 1 is denied, so the application waits for the record to be released. Finally, A's attempt to access record 2 is denied, so the application waits for the record to be released. Thus, application A's update transaction is waiting for record 2 (locked by application B), and application B is waiting for record 1 (locked by application A). Unless the DBMS intervenes, both applications will wait indefinitely.

There are two ways to resolve deadlock: prevention and resolution. **Deadlock prevention** requires applications to lock in advance all records they will require. Application B would have to lock both records 1 and 2 before processing the transaction. If these records are locked, B would have to wait.

An example of deadlock

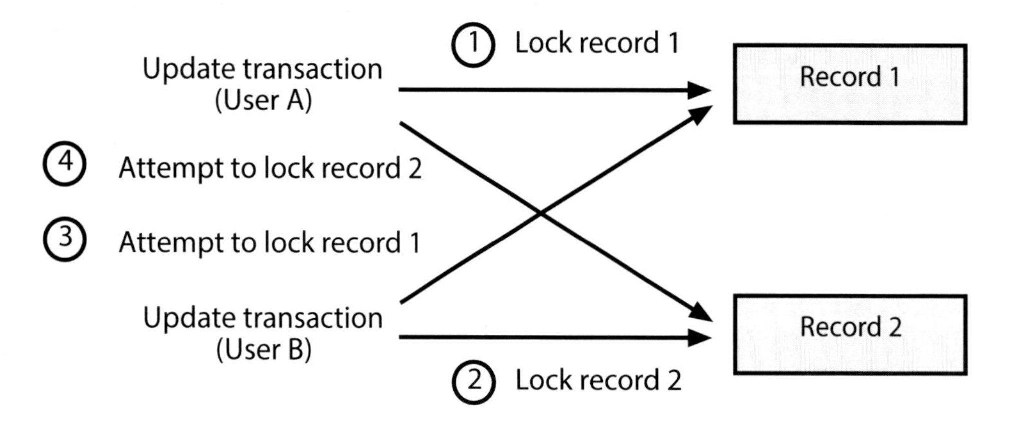

The **two-phase locking protocol** is a simple approach to preventing deadlocks. It operates on the notion that a transaction has a growing phase followed by a shrinking phase. During the growing phase, locks can be requested. The shrinking phase is initiated by a release statement, which means no additional locks can be requested. A release statement enables the program to signal to the DBMS the transition from requesting locks to releasing locks.

Another approach to deadlock prevention is **deadlock resolution,** whereby the DBMS detects and breaks deadlocks. The DBMS usually keeps a resource usage matrix, which instantly reflects which applications (e.g., update transactions) are using which resources (e.g., rows). By scanning the matrix, the DBMS can detect deadlocks as they occur. The DBMS then resolves the deadlock by backing out one of the deadlocked transactions. For example, the DBMS might release application A's lock on record 1, thus allowing application B to proceed. Any changes made by application A up to that time (e.g., updates to record 1) would be rolled back. Application A's transaction would be restarted when the required resources became available.

Transaction failure and recovery

When a transaction fails, there is the danger of an inconsistent database. Transactions can fail for a variety of reasons, including

- Program error (e.g., a logic error in the code)
- Action by the transaction manager (e.g., resolution of a deadlock)
- Self-abort (e.g., an error in the transaction data means it cannot be completed)
- System failure (e.g., an operating-system bug)

If a transaction fails for any reason, then the DBMS must be able to restore the database to a correct state. To do this, two statements are required: an **end of transaction (EOT)** and **commit**. EOT indicates the physical end of the transaction, the last statement. Commit, an explicit statement, must occur in the transaction code before the EOT statement. The only statements that should occur between commit and EOT are database writes and lock releases.

When a transaction issues a commit, the transaction manager checks that all the necessary write-record locks for statements following the commit have been established. If these locks are not in place, the transaction is terminated; otherwise, the transaction is committed, and it proceeds to execute the database writes and release locks. Once a transaction is committed, a system problem is the only failure to which it is susceptible.

When a transaction fails, the transaction manager must take one of two corrective actions:

- If the transaction has not been committed, the transaction manager must return the database to its state prior to the transaction. That is, it must perform a **rollback** of the database to its most recent valid state.
- If the transaction has been committed, the transaction manager must ensure that the database is established at the correct post-transaction state. It must check that all write statements executed by the transaction and those appearing between commit and EOT have been applied. The DBMS may have to redo some writes.

Protecting existence

One of the three database integrity outcomes is protecting the existence of the database—ensuring data are available when needed. Two strategies for protecting existence are isolation and database backup and recovery. **Isolation** is a preventive strategy that involves administrative procedures to insulate the physical database from destruction. Some mechanisms for doing this are keeping data in safe places, such as in vaults or underground; having multiple installations; or having security systems. For example, one organization keeps backup copies of important databases on removable magnetic disks. These are stored in a vault that is always guarded. To gain access to the vault, employees need a badge with an encoded personal voice print. Many companies have total-backup computer centers, which contain duplicate databases and documentation for system operation. If something should happen at the main center (e.g., a flood), they can be up and running at their backup center in a few hours, or even minutes in some highly critical situations. What isolation strategies do you use to protect the backup medium of your personal computer? Do you make backup files?

A study of 429 disaster declarations reported to a major international provider of disaster recovery services provides some insights as to the frequency and effects of different IT disasters. These data cover the period 1981–2000 and identify the most frequent disasters and statistics on the length of the disruption.[88]

Most frequent IT disasters

Category	Description
Disruptive act	Strikes and other intentional human acts, such as bombs or civil unrest, that are designed to interrupt normal organizational processes
Fire	Electrical or natural fires
IT failure	Hardware, software, or network problems
IT move/upgrade	Data center moves and CPU upgrades
Natural event	Earthquakes, hurricanes, severe weather
Power outage	Loss of power
Water leakage	Unintended escape of contained water (e.g., pipe leaks, main breaks)

Days of disruption per year

Category	Number	Minimum	Maximum	Mean
Natural event	122	0	85	6.38
IT failure	98	1	61	6.89

88 Source: Lewis Jr, W., Watson, R. T., & Pickren, A. (2003). An empirical assessment of IT disaster probabilities. Communications of the ACM, 46(9), 201-206.

Category	Number	Minimum	Maximum	Mean
Power outage	67	1	20	4.94
Disruptive act	29	1	97	23.93
Water leakage	28	0	17	6.07
Fire	22	1	124	13.31
IT move/upgrade	14	1	204	20.93
Environmental	6	1	183	65.67
Miscellaneous	5	1	416	92.8
IT capacity	2	4	8	6
Theft	2	1	3	2
Construction	1	2	2	2
Flood	1	13	13	13
IT user error	1	1	1	1

Backup and recovery

Database backup and recovery is a curative strategy to protect the existence of a physical database and to recreate or recover the data whenever loss or destruction occurs. The possibility of loss always exists. The use of, and choice among, backup and recovery procedures depends upon an assessment of the risk of loss and the cost of applying recovery procedures. The procedures in this case are carried out by the computer system, usually the DBMS. Data loss and damage should be anticipated. No matter how small the probability of such events, there should be a detailed plan for data recovery.

There are several possible causes for data loss or damage, which can be grouped into three categories.

Storage-medium destruction

In this situation, a portion or all of the database is unreadable as a result of catastrophes such as power or air-conditioning failure, fire, flood, theft, sabotage, and the overwriting of disks or tapes by mistake. A more frequent cause is a disk failure. Some of the disk blocks may be unreadable as a consequence of a read or write malfunction, such as a head crash.

Abnormal termination of an update transaction

In this case, a transaction fails part way through execution, leaving the database partially updated. The database will be inconsistent because it does not accurately reflect the current state of the business. The primary causes of an abnormal termination are a transaction error or system failure. Some operation within transaction processing, such as division by zero, may cause it to fail. A hardware or software failure will usually result in one or more active programs being aborted. If these programs were updating the database, integrity problems could result.

Incorrect data discovered

In this situation, an update program or transaction incorrectly updated the database. This usually happens because a logic error was not detected during program testing.

Because most organizations rely heavily on their databases, a DBMS must provide the following three mechanisms for restoring a database quickly and accurately after loss or damage:

1. **Backup facilities**, which create duplicate copies of the database
2. **Journaling facilities**, which provide backup copies or an audit trail of transactions or database changes
3. A **recovery facility** within the DBMS that restores the database to a consistent state and restarts the processing of transactions

Before discussing each of these mechanisms in more depth, let us review the steps involved in updating a database and how backup and journaling facilities might be integrated into this process.

An overview of the database update process is captured in the following figure. The process can be viewed as a series of database state changes. The initial database, state 1, is modified by an update transaction, such as deleting customer Jones, creating a new state (state 2). State 1 reflects the state of the organization with Jones as a customer, while state 2 reflects the organization without this customer. Each update transaction changes the state of the database to reflect changes in organizational data. Periodically, the database is copied or backed up, possibly onto a different storage medium and stored in a secure location. The backup is made when the database is in state 2.

Database state changes as transactions are processed

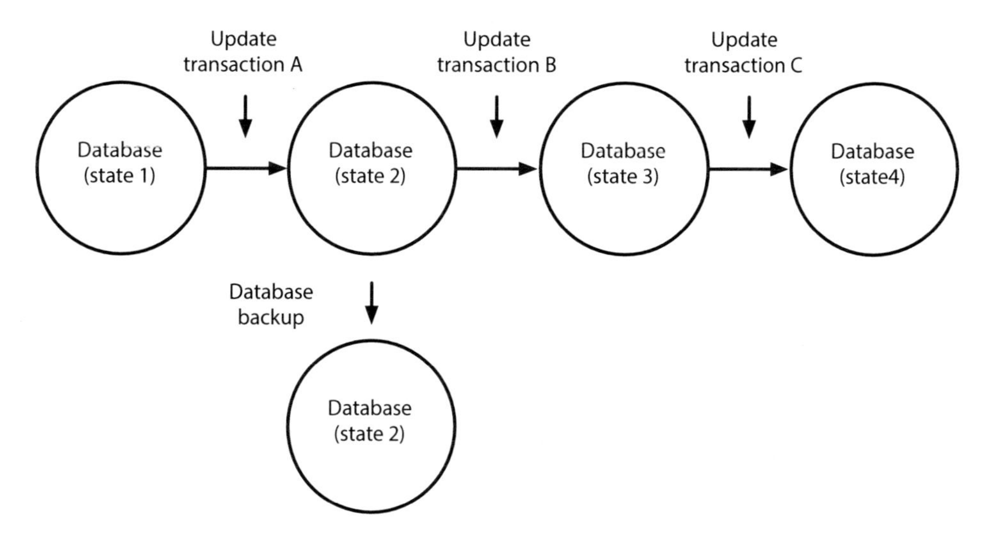

A more detailed illustration of database update procedures and the incorporation of backup facilities is shown in the following figure.

Possible procedures for a database update

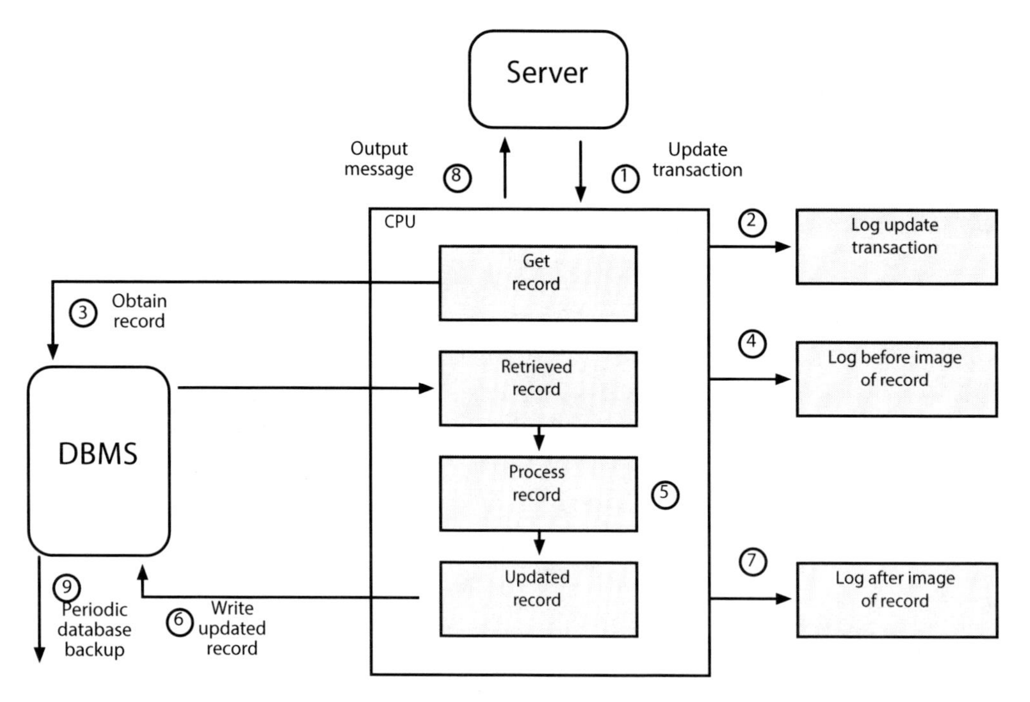

The updating of a single record is described below.

1. The client submits an update transaction from a workstation.
2. The transaction is edited and validated by an application program or the DBMS. If it is valid, it is logged or stored in the transaction log or journal. A journal or log is a special database or file that stores information for backup and recovery.
3. The DBMS obtains the record to be updated.
4. A copy of the retrieved record is logged to the journal file. This copy is referred to as the *before image*, because it is a copy of the database record before the transaction changes it.
5. The transaction is processed by changing the affected data items in the record.
6. The DBMS writes the updated record to the database.
7. A copy of the updated record is written to the journal. This copy is referred to as the *after image*, because it is a copy of the database record after the transaction has updated it.
8. An output message tells the application that the update has been successfully completed.
9. The database is copied periodically. This backup copy reflects all updates to the database up to the time when the copy was made. An alternative strategy to periodic copying is to maintain multiple (usually two) complete copies of the database online and update them simultaneously. This technique, known as *mirroring*, is discussed in Chapter11.

In order to recover from data loss or damage, it is necessary to store backup data, such as a complete copy of the database or the necessary data to restore the accuracy of a database. Preferably, backup data should be stored on another medium and kept separate from the primary database. As the description of the update process indicates, there are several options for backup data, depending on the objective of the backup procedure.

Backup option

Objective	Action
Complete copy of database	Dual recording of data (mirroring)
Past states of the database (also known as database dumps)	Database backup
Changes to the database	Before-image log or journal After-image log or journal
Transactions that caused a change in the state of the database	Transaction log or journal

Data stored for backup and recovery are generally some combination of periodic database backups, transaction logs, and before- and after-image logs. Different recovery strategies use different combinations of backup data to recover a database.

The recovery method is highly dependent on the backup strategy. The database administrator selects a strategy based on a trade-off between ease of recovery from data loss or damage and the cost of performing backup operations. For example, keeping a mirror database is more expensive then keeping periodic database backups, but a mirroring strategy is useful when recovery is needed very quickly, say in seconds or minutes. An airline reservations system could use mirroring to ensure fast and reliable recovery. In general, the cost of keeping backup data is measured in terms of interruption of database availability (e.g., time the system is out of operation when a database is being restored), storage of redundant data, and degradation of update efficiency (e.g., extra time taken in update processing to save before or after images).

Recovery strategies

The type of recovery strategy or procedure used in a given situation depends on the nature of the data loss, the type of backup data available, and the sophistication of the DBMS's recovery facilities. The following discussion outlines four major recovery strategies: switch to a duplicate database, backward recovery or rollback, forward recovery or roll forward, and reprocessing transactions.

The recovery procedure of switching to a duplicate database requires the maintenance of the mirror copy. The other three strategies assume a periodic dumping or backing up of the database. Periodic dumps may be made on a regular schedule, triggered automatically by the DBMS, or triggered externally by personnel. The schedule may be determined by time (hourly, daily, weekly) or by event (the number of transactions since the last backup).

Switching to a duplicate database

This recovery procedure requires maintaining at least two copies of the database and updating both simultaneously. When there is a problem with one database, access is switched to the duplicate. This strategy is particularly useful when recovery must be accomplished in seconds or minutes. This procedure offers good protection against certain storage-medium destruction problems, such as disk failures, but none against events that damage or make both databases unavailable, such as a power failure or a faulty update program. This strategy entails additional costs in terms of doubling online storage capacity. It can also be implemented with dual computer processors, where each computer updates its copy of the database. This duplexed configuration offers greater backup protection at a greater cost.

Backward recovery or rollback

Backward recovery (also called rollback or rolling back) is used to back out or undo unwanted changes to the database. For example, the database update procedures figure shows three updates (A, B, and C) to a database. Let's say that update B terminated abnormally, leaving the database, now in state 3, inconsistent. What we need to do is return the database to state 2 by applying the before images to the database. Thus, we would perform a rollback by changing the database to state 2 with the before images of the records updated by transaction B.

Backward recovery reverses the changes made when a transaction abnormally terminates or produces erroneous results. To illustrate the need for rollback, consider the example of a budget transfer of $1,000 between two departments.

1. The program reads the account record for Department X and subtracts $1,000 from the account balance and updates the database.
2. The program then reads the record for Department Y and adds $1,000 to the account balance, but while attempting to update the database, the program encounters a disk error and cannot write the record.

Now the database is inconsistent. Department X has been updated, but Department Y has not. Thus, the transaction must be aborted and the database recovered. The DBMS would apply the before image to Department X to restore the account balance to its original value. The DBMS may then restart the transaction and make another attempt to update the database.

Forward recovery or roll forward

Forward recovery (also called roll forward or bringing forward) involves recreating a database using a prior database state. Returning to the example in the database update procedures figure, suppose that state 4 of the database was destroyed and that we need to recover it. We would take the last database dump or backup (state 2) and then apply the after-image records created by update transactions B and C. This would return the database to state 4. Thus, roll forward starts with an earlier copy of the database, and by applying after images (the results of good transactions), the backup copy of the database is moved forward to a later state.

Reprocessing transactions

Although similar to forward recovery, this procedure uses update transactions instead of after images. Taking the same example shown above, assume the database is destroyed in state 4. We would take the last database backup (state 2) and then reprocess update transactions B and C to return the database to state 4. The main advantage of using this method is its simplicity. The DBMS does not need to create an after-image journal, and there are no special restart procedures. The one major disadvantage, however, is the length of time to reprocess transactions. Depending on the frequency of database backups and the time needed to get transactions into the identical sequence as previous updates, several hours of reprocessing may be required. Processing new transactions must be delayed until the database recovery is complete.

The following table reviews the three types of data losses and the corresponding recovery strategies one could use. The major problem is to recreate a database using a backup copy, a previous state of organizational memory. Recovery is done through forward recovery, reprocessing, or switching to a duplicate database if one is available. With abnormal termination or incorrect data, the preferred strategy is backward recovery, but other procedures could be used.

What to do when data loss occurs

Problem	Recovery procedures
Storage medium destruction (database is unreadable)	*Switch to a duplicate database—this can be transparent with RAID Forward recovery Reprocessing transactions
Abnormal termination of an update transaction (transaction error or system failure)	*Backward recovery Forward recovery or reprocessing transactions—bring forward to the state just before termination of the transaction
Incorrect data detected (database has been incorrectly updated)	*Backward recovery Reprocessing transactions (excluding those from the update program that created the incorrect data)

* Preferred strategy

Use of recovery procedures

Usually the person doing a query or an update is not concerned with backup and recovery. Database administration personnel often implement strategies that are automatically carried out by the DBMS. ANSI has defined standards governing SQL processing of database transactions that relate to recovery.

Transaction support is provided through the use of the two SQL statements `COMMIT` and `ROLLBACK`. These commands are employed when a procedural programming language such as Java is used to update a database. They are illustrated in the chapter on SQL and Java.

Skill builder

An Internet bank with more than 10 million customers has asked for your advice on developing procedures for protecting the existence of its data. What would you recommend?

Maintaining data quality

The second integrity goal is to maintain data quality, which typically means keeping data accurate, complete, and current. *Data are high-quality if they fit their intended uses in operations, decision making, and planning. They are fit for use if they are free of defects and possess desired features.*[89] The preceding definition implicitly recognizes that data quality is determined by the customer. It also implies that data quality is relative to a task. Data could be high-quality for one task and low-quality for another. The data provided by a flight-tracking system are very useful when you are planning to meet someone at the airport, but not particularly helpful for selecting a vacation spot. Defect-free data are accessible, accurate, timely, complete, and consistent with other sources. Desirable features include relevance, comprehensiveness, appropriate level of detail, easy-to-navigate source, high readability, and absence of ambiguity.

Poor-quality data have several detrimental effects. Customer service decreases when there is dissatisfaction with poor and inaccurate information or a lack of appropriate information. For many customers, information is the heart of customer service, and they lose confidence in firms that can't or don't provide relevant information. Bad data interrupt information processing because they need to be corrected before

89 Redman, T. C. (2001). *Data quality : the field guide*. Boston: Digital Press.

processing can continue. Poor-quality data can lead to a wrong decision because inaccurate inferences and conclusions are made.

As we have stated previously, data quality varies with circumstances, and the model in the following figure will help you to understand this linkage. By considering variations in a customer's uncertainty about a firm's products and a firm's ability to deliver consistently, we arrive at four fundamental strategies for customer-oriented data quality.

Customer-oriented data quality strategies

Firm performance variation	Customer uncertainty: Low	Customer uncertainty: High
High	**Tracking** Performance deviation	**Knowledge management** Advice
Low	**Transaction processing** Confirmation	**Expert system** Recommendation

Transaction processing: When customers know what they want and firms can deliver consistently, customers simply want fast and accurate transactions and data confirming details of the transaction. Most banking services fall into this category. Customers know they can withdraw and deposit money, and banks can perform reliably.

Expert system: In some circumstances, customers are uncertain of their needs. For instance, Vanguard offers personal investors a choice from more than 150 mutual funds. Most prospective investors are confused by such a range of choices, and Vanguard, by asking a series of questions, helps prospective investors narrow their choices and recommends a small subset of its funds. A firm's recommendation will vary little over time because the characteristics of a mutual fund (e.g., intermediate tax-exempt bond) do not change.

Tracking: Some firms operate in environments where they don't have a lot of control over all the factors that affect performance. Handling more than 2,200 take-offs and landings and over 250,000 passengers per day,[90] Atlanta's Hartsfield Airport becomes congested when bad weather, such as a summer thunderstorm, slows down operations. Passengers clearly know what they want—data on flight delays and their consequences. They assess data quality in terms of how well data tracks delays and notifies them of alternative travel arrangements.

Knowledge management: When customers are uncertain about their needs for products delivered by firms that don't perform consistently, they seek advice from knowledgeable people and organizations. Data quality is judged by the soundness of the advice received. Thus, a woman wanting a custom-built house would likely seek the advice of an architect to select the site, design the house, and supervise its

90 November 2012. http://www.atlanta-airport.com/docs/Traffic/201211.pdf

construction, because architects are skilled in eliciting clients' needs and knowledgeable about the building industry.

An organization's first step toward improving data quality is to determine in which quadrant it operates so it can identify the critical information customers expect. Of course, data quality in many situations will be a mix of expectations. The mutual fund will be expected to confirm fund addition and deletion transactions. However, a firm must meet its dominant goal to attract and retain customers.

A firm will also need to consider its dominant data quality strategy for its internal customers, and the same general principles illustrated by the preceding figure can be applied. In the case of internal customers, there can be varying degrees of uncertainty as to what other organizational units can do for them, and these units will vary in their ability to perform consistently for internal customers. For example, consulting firms develop centers of excellence as repositories of knowledge on a particular topic to provide their employees with a source of expertise. These are the internal equivalent of knowledge centers for external customers.

Once they have settled on a dominant data quality strategy, organizations need a corporate-wide approach to data quality, just like product and service quality. There are three generations of data quality:

- **First generation**: Errors in existing data stores are found and corrected. This data cleansing is necessary when much of the data is captured by manual systems.
- **Second generation**: Errors are prevented at the source. Procedures are put in place to capture high-quality data so that there is no need to cleanse it later. As more data are born digital, this approach becomes more feasible. Thus, when customers enter their own data or barcodes are scanned, the data should be higher-quality than when entered by a data-processing operator.
- **Third generation**: Defects are highly unlikely. Data capture systems meet six-sigma standards (3.4 defects per million), and data quality is no longer an issue.

Skill builder

1. A consumer electronics company with a well-respected brand offers a wide range of products. For example, it offers nine world-band radios and seven televisions. What data quality strategy would you recommend?
2. What data quality focus would you recommend for a regulated natural gas utility?
3. A telephone company has problems in estimating how long it takes to install DSL in homes. Sometimes it takes less than an hour and other times much longer. Customers are given a scheduled appointment and many have to make special arrangements so that they are home when the installation person arrives. What data might these customers expect from the telephone company, and how would they judge data quality?

Dimensions

The many writers on quality all agree on one thing—quality is multidimensional. Data quality also has many facets, and these are presented in the following table. The list also provides data managers with a checklist for evaluating overall high-quality data. Organizations should aim for a balanced performance across all dimensions because failure in one area is likely to diminish overall data quality.

The dimensions of data quality

Dimension	Conditions for high-quality data
Accuracy	Data values agree with known correct values.

Dimension	Conditions for high-quality data
Completeness	Values for all reasonably expected attributes are available.
Representation consistency	Values for a particular attribute have the same representation across all tables (e.g., dates).
Organizational consistency	There is one organization wide table for each entity and one organization wide domain for each attribute.
Row consistency	The values in a row are internally consistent (e.g., a home phone number's area code is consistent with a city's location).
Timeliness	A value's recentness matches the needs of the most time-critical application requiring it.
Stewardship	Responsibility has been assigned for managing data.
Sharing	Data sharing is widespread across organizational units.
Fitness	The format and presentation of data fit each task for which they are required.
Interpretation	Clients correctly interpret the meaning of data elements.
Flexibility	The content and format of presentations can be readily altered to meet changing circumstances.
Precision	Data values can be conveniently formatted to the required degree of accuracy.
International	Data values can be displayed in the measurement unit of choice (e.g., kilometers or miles).
Accessibility	Authorized users can readily access data values through a variety of devices from a variety of locations.
Security and privacy	Data are appropriately protected from unauthorized access.
Continuity	The organization continues to operate in spite of major disruptive events.
Granularity	Data are represented at the lowest level necessary to support all uses (e.g., hourly sales).
Metadata	There is ready access to accurate data about data.

Skill builder

1. What level of granularity of sales data would you recommend for an online retailer?
2. What level of data quality completeness might you expect for a university's student DBMS?

DBMS and data quality

To assist data quality, functions are needed within the DBMS to ensure that update and insert actions are performed by authorized persons in accordance with stated rules or integrity constraints, and that the results are properly recorded. These functions are accomplished by update authorization, data validation using integrity constraints, and concurrent update control. Each of these functions is discussed in turn.

Update authorization

Without proper controls, update transactions can diminish the quality of a database. Unauthorized users could sabotage a database by entering erroneous values. The first step is to ensure that anyone who wants to update a database is authorized to do so. Some responsible person—usually the database owner or database administrator—must tell the DBMS who is permitted to initiate particular database operations. The DBMS must then check every transaction to ensure that it is authorized. Unauthorized access to a database exposes an organization to many risks, including fraud and sabotage.

Update authorization is accomplished through the same access mechanism used to protect confidentiality. We will discuss access control more thoroughly later in this chapter. In SQL, access control is implemented through `GRANT`, which gives a user a privilege, and `REVOKE`, which removes a privilege. (These commands are discussed in Chapter 10.) A control mechanism may lump all update actions into a single privilege or separate them for greater control. In SQL, they are separated as follows:

- `UPDATE` (privilege to change field values using `UPDATE`; this can be column specific)
- `DELETE` (privilege to delete records from a table)
- `INSERT` (privilege to insert records into a table)

Separate privileges for each of the update commands allow tighter controls on certain update actions, such as updating a salary field or deleting records.

Data validation using integrity constraints

Once the update process has been authorized, the DBMS must ensure that a database is accurate and complete before any updates are applied. Consequently, the DBMS needs to be aware of any integrity constraints or rules that apply to the data. For example, the `qdel` table in the relational database described previously would have constraints such as

- Delivery number (`delno`) must be unique, numeric, and in the range 1–99999.
- Delivered quantity (`delqty`) must be nonzero.
- Item name (`itemname`) must appear in the `qitem` table.
- Supplier code (`splno`) must appear in the `qspl` table.

Once integrity constraints are specified, the DBMS must monitor or validate all insert and update operations to ensure that they do not violate any of the constraints or rules.

The key to updating data validation is a clear definition of valid and invalid data. Data validation cannot be performed without integrity constraints or rules. A person or the DBMS must know acceptable data format, valid values, and procedures to invoke to determine validity. All data validation is based on a prior expression of integrity constraints.

Data validation may not always produce error-free data, however. Sometimes integrity constraints are unknown or are not well defined. In other cases, the DBMS does provide a convenient means for expressing and performing validation checks.

Based on how integrity constraints have been defined, data validation can be performed outside the DBMS by people or within the DBMS itself. External validation is usually done by reviewing input documents before they are entered into the system and by checking system outputs to ensure that the database was updated correctly. Maintaining data quality is of paramount importance, and data validation preferably should be handled by the DBMS as much as possible rather than by the application, which should handle the exceptions and respond to any failed data validation checks.

Integrity constraints are usually specified as part of the database definition supplied to the DBMS. For example, the primary-key uniqueness and referential integrity constraints can be specified within the SQL

CREATE statement. DBMSs generally permit some constraints to be stored as part of the database schema and are used by the DBMS to monitor all update operations and perform appropriate data validation checks. Any given database is likely to be subject to a very large number of constraints, but not all of these can be automatically enforced by the DBMS. Some will need to be handled by application programs.

The general types of constraints applied to a data item are outlined in the following table. Not all of these necessarily would be supported by a DBMS, and a particular database may not use all types.

Types of data items in integrity constraint

Type of integrity constraint	Explanation	Example
type	Validating a data item value against a specified data type	Supplier number is numeric.
size	Defining and validating the minimum and maximum size of a data item	Delivery number must be at least 3 digits, and at most 5.
values	Providing a list of acceptable values for a data item	Item colors must match the list provided.
range	Providing one or more ranges within which the data item must lie	Employee numbers must be in the range 1–100.
pattern	Providing a pattern of allowable characters that define permissible formats for data values	Department phone number must be of the form 542-nnnn (stands for exactly four decimal digits).
procedure	Providing a procedure to be invoked to validate data items	A delivery must have valid item name, department, and supplier values before it can be added to the database (tables are checked for valid entries).
Conditional	Providing one or more conditions to apply against data values	If item type is "Y," then color is null.
Not null (mandatory)	Indicating whether the data item value is mandatory (not null) or optional; the not-null option is required for primary keys	Employee number is mandatory.
Unique	Indicating whether stored values for this data item must be compared to other values of the item within the same table	Supplier number is unique.

As mentioned, integrity constraints are usually specified as part of the database definition supplied to the DBMS. The following table contains some typical specifications of integrity constraints for a relational DBMS.

Examples of integrity constraints

Examples	Explanation

`CREATE TABLE stock (`
`stkcode CHAR(3),`
`…,`
	Column `stkcode` must always have 3 or fewer alphanumeric characters, and `stkcode` must be unique because it is a primary key.

`natcode CHAR(3),`
`PRIMARY KEY(stkcode),`
`CONSTRAINT fk_stock_nation`
`FOREIGN KEY(natcode)`
	Column `natcode` must be assigned a value of 3 or less alphanumeric characters and must exist as the primary key of the `nation`.

`REFERENCES nation(natcode)` `ON DELETE RESTRICT);`	Do not allow the deletion of a row in `nation` while there still exist rows in `stock` containing the corresponding value of `natcode`.

Data quality control does not end with the application of integrity constraints. Whenever an error or unusual situation is detected by the DBMS, some form of response is required. Response rules need to be given to the DBMS along with the integrity constraints. The responses can take many different forms, such as abort the entire program, reject entire update transaction, display a message and continue processing, or let the DBMS attempt to correct the error. The response may vary depending on the type of integrity constraint violated. If the DBMS does not allow the specification of response rules, then it must take a default action when an error is detected. For example, if alphabetic data are entered in a numeric field, most DBMSs will have a default response and message (e.g., nonnumeric data entered in numeric field). In the case of application programs, an error code is passed to the program from the DBMS. The program would then use this error code to execute an error-handling procedure.

Ensuring confidentiality

Thus far, we have discussed how the first two goals of data integrity can be accomplished: Data are available when needed (protecting existence); data are accurate, complete, and current (maintaining quality). This section deals with the final goal: ensuring confidentiality or data security. Two DBMS functions—access control and encryption—are the primary means of ensuring that the data are accessed only by those authorized to do so. We begin by discussing an overall model of data security.

General model of data security

The following figure depicts the two functions for ensuring data confidentiality: access control and encryption. Access control consists of two basic steps—identification and authorization. Once past access control, the user is permitted to access the database. Access control is applied only to the established avenues of entry to a database. Clever people, however, may be able to circumvent the controls and gain unauthorized access. To counteract this possibility, it is often desirable to hide the meaning of stored data by encrypting them, so that it is impossible to interpret their meaning. Encrypted data are stored in a transformed or coded format that can be decrypted and read only by those with the appropriate key.

A general model of data security

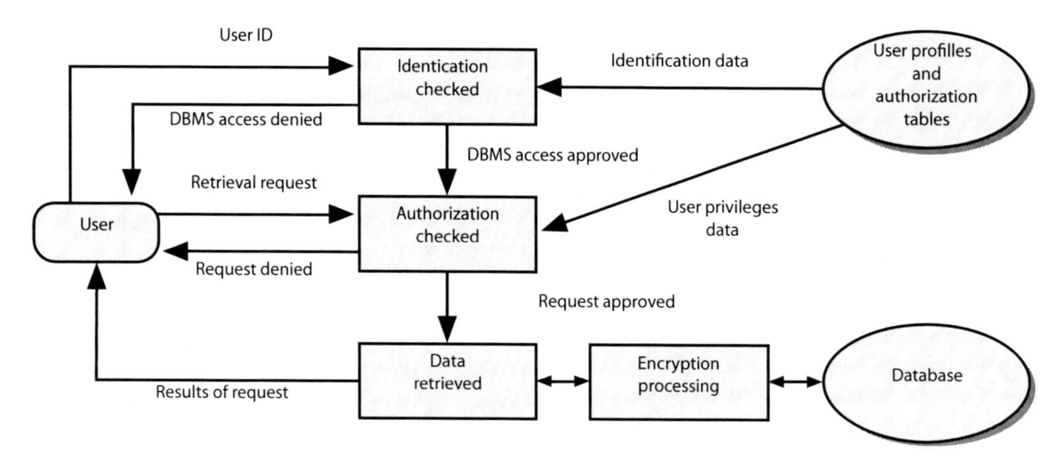

Now let us walk through the figure in detail.

1. A client must be identified and provide additional information required to authenticate this identification (e.g., an account name and password). Client profile information (e.g., a password or a voice print) is used to verify or authenticate a connection.
2. Having authenticated the client, the authorization step is initiated by a request (retrieve or update database). The previously stored client authorization rules (what data each client can access and the authorized actions on those data) are checked to determine whether the client has the right or privilege to access the requested data. (The previously stored client's privileges are created and maintained by an authorized person, database owner, or administrator.) A decision is made to permit or deny the execution of the request. If access is permitted, the transaction is processed against the database.
3. Data are encrypted before storage, and retrieved data are decrypted before presentation.

Data access control

Data access control begins with identification of an organizational entity that can access the database. Examples are individuals, departments, groups of people, transactions, terminals, and application programs. Valid combinations may be required, for example, a particular person entering a certain transaction at a particular terminal. A user identification (often called ***userid***) is the first piece of data the DBMS receives from the subject. It may be a name or number. The user identification enables the DBMS to locate the corresponding entry in the stored user profiles and authorization tables (see preceding figure).

Taking this information, the DBMS goes through the process of authentication. The system attempts to match additional information supplied by the client with the information previously stored in the client's profile. The system may perform multiple matches to ensure the identity of the client (see the following table for the different types). If all tests are successful, the DBMS assumes that the subject is an authenticated client.

Authenticating mechanisms[1]

Class	Examples
Something a person knows: remembered information	Name, account number, password
Something the person has: possessed object	Badge, plastic card, key

Class	Examples
Something the person is: personal characteristic	Fingerprint, voiceprint, signature, hand size

Many systems use remembered information to control access. The problem with such information is that it does not positively identify the client. Passwords have been the most widely used form of access control. If used correctly, they can be very effective. Unfortunately, people leave them around where others can pick them up, allowing unauthorized people to gain access to databases.

To deal with this problem, organizations are moving toward using personal characteristics and combinations of authenticating mechanisms to protect sensitive data. Collectively, these mechanisms can provide even greater security. For example, access to a large firm's very valuable marketing database requires a smart card and a fingerprint, a combination of personal characteristic and a possessed object. The database can be accessed through only a few terminals in specific locations, an isolation strategy. Once the smart card test is passed, the DBMS requests entry of other remembered information—password and account number—before granting access.

Data access authorization is the process of permitting clients whose identity has been authenticated to perform certain operations on certain data objects in a shared database. The authorization process is driven by rules incorporated into the DBMS. Authorization rules are in a table that includes subjects, objects, actions, and constraints for a given database. An example of such a table is shown in the following table. Each row of the table indicates that a particular subject is authorized to take a certain action on a database object, perhaps depending on some constraint. For example, the last entry of the table indicates that Brier is authorized to delete supplier records with no restrictions.

Sample authorization table

Subject/Client	Action	Object	Constraint
Accounting department	Insert	Supplier table	None
Purchase department clerk	Insert	Supplier table	If quantity < 200
Purchase department supervisor	Insert	Delivery table	If quantity >= 200
Production department	Read	Delivery table	None
Todd	Modify	Item table	Type and color only
Order-processing program	Modify	Sale table	None
Brier	Delete	Supplier table	None

We have already discussed subjects, but not objects, actions, and constraints. Objects are database entities protected by the DBMS. Examples are databases, views, files, tables, and data items. In the preceding table, the objects are all tables. A view is another form of security. It restricts the client's access to a database. Any data not included in a view are unknown to the client. Although views promote security, several persons may share a view or unauthorized persons may gain access. Thus, a view is another object to be included in the authorization process. Typical actions on objects are shown in the table: read, insert, modify, and delete. Constraints are particular rules that apply to a subject - action - object relationship.

Implementing authorization rules

Most contemporary DBMSs do not implement the complete authorization table shown in the table. Usually, they implement a simplified version. The most common form is an authorization table for subjects with limited applications of the constraint column. Let us take the granting of table privileges, which are needed in order to authorize subjects to perform operations on both tables and views .

Authorization commands

SQL Command	Result
SELECT	Permission to retrieve data
UPDATE	Permission to change data; can be column specific
DELETE	Permission to delete records or tables
INSERT	Permission to add records or tables

The GRANT and REVOKE SQL commands discussed in Chapter 10 are used to define and delete authorization rules. Some examples:

```
GRANT SELECT ON qspl TO vikki;
GRANT SELECT, UPDATE (splname) ON qspl TO huang;
GRANT ALL PRIVILEGES ON qitem TO vikki;
GRANT SELECT ON qitem TO huang;
```

The GRANT commands have essentially created two authorization tables, one for Huang and the other for Vikki. These tables illustrate how most current systems create authorization tables for subjects using a limited set of objects (e.g., tables) and constraints.

A sample authorization table

Client	Object (table)	Action	Constraint
vikki	qspl	SELECT	None
vikki	qitem	UPDATE	None
vikki	qitem	INSERT	None
vikki	qitem	DELETE	None
vikki	qitem	SELECT	None
huang	qspl	SELECT	None
huang	qspl	UPDATE	splname only
huang	qitem	SELECT	None

Because authorization tables contain highly sensitive data, they must be protected by stringent security rules and encryption. Normally, only selected persons in data administration have authority to access and modify them.

Encryption

Encryption techniques complement access control. As the preceding figure illustrates, access control applies only to established avenues of access to a database. There is always the possibility that people will circumvent these controls and gain unauthorized access to a database. To counteract this possibility, encryption can be used to obscure or hide the meaning of data. Encrypted data cannot be read by an intruder unless that person knows the method of encryption and has the key. Encryption is any transformation applied to data that makes it difficult to extract meaning. Encryption transforms data into cipher text or disguised information, and decryption reconstructs the original data from cipher text.

Public-key encryption is based on a pair of private and public keys. A person's public key can be freely distributed because it is quite separate from his or her private key. To send and receive messages, communicators first need to create private and public keys and then exchange their public keys. The sender encodes a message with the intended receiver's public key, and upon receiving the message, the receiver applies her private key. The receiver's private key, the only one that can decode the message, must be kept secret to provide secure message exchanging.

Public-key encryption

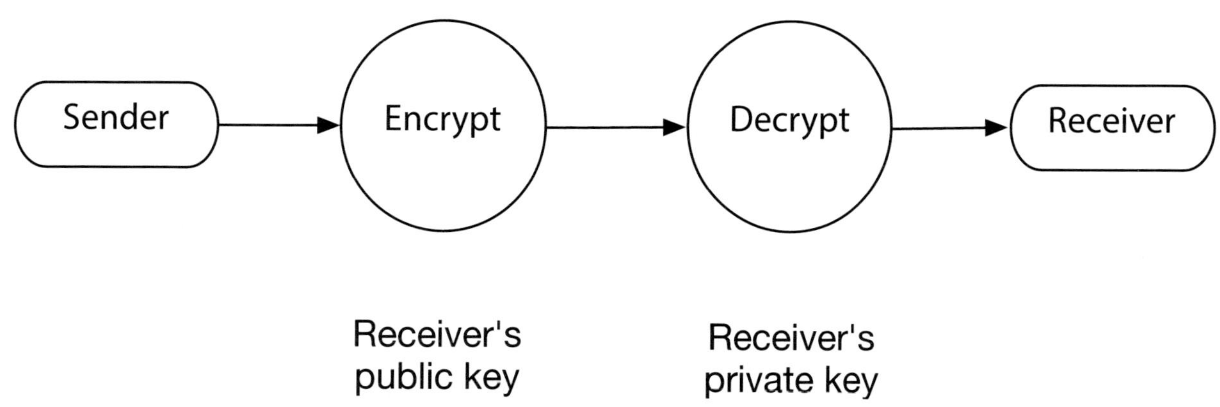

In a DBMS environment, encryption techniques can be applied to transmitted data sent over communication lines to and from devices, or between computers, and to all highly sensitive stored data in active databases or their backup versions. Some DBMS products include routines that automatically encrypt sensitive data when they are stored or transmitted over communication channels. Other DBMS products provide exits that allow users to code their own encryption routines. Encrypted data may also take less storage space because they are often compressed.

Skill builder

A university has decided that it will e-mail students their results at the end of each semester. What procedures would you establish to ensure that only the designated student opened and viewed the e-mail?

Monitoring activity

Sometimes no single activity will be detected as an example of misuse of a system. However, examination of a pattern of behavior may reveal undesirable behavior (e.g., persistent attempts to log into a system with a variety of userids and passwords). Many systems now monitor all activity using **audit trail analysis**. A time- and date-stamped audit trail of all system actions (e.g., database accesses) is maintained. This audit log is dynamically analyzed to detect unusual behavior patterns and alert security personnel to possible misuse.

A form of misuse can occur when an authorized user violates privacy rules by using a series of authorized queries to gain access to private data. For example, some systems aim to protect individual privacy by restricting authorized queries to aggregate functions (e.g., `AVG` and `COUNT`). Since it is impossible to do non-aggregate queries, this approach should prevent access to individual-level data, which it does at the single-query level. However, multiple queries can be constructed to circumvent this restriction.

Assume we know that a professor in the IS department is aged 40 to 50, is single, and attended the University of Minnesota. Consider the results of the following set of queries.[91]

```
SELECT COUNT(*) from faculty
   WHERE dept = 'MIS'
   AND age >= 40 AND age <= 50;
```

```
        10
```

```
SELECT COUNT(*) FROM faculty
   WHERE dept = 'MIS'
   AND age >= 40 AND age <= 50
   AND degree_from = 'Minnesota';
```

```
         2
```

```
SELECT COUNT(*) FROM faculty
   WHERE = 'MIS'
   AND age >= 40 AND age <= 50;
   AND degree_from = 'Minnesota'
   AND marital_status = 'S';
```

```
         1
```

```
SELECT AVG(salary) FROM faculty
   WHERE dept = 'MIS'
   AND age >= 40 AND age <= 50
   AND degree_from = 'Minnesota'
   AND marital_status = 'S';
```

```
     85000
```

The preceding set of queries, while all at the aggregate level, enables one to deduce the salary of the professor. This is an invasion of privacy and counter to the spirit of the restriction queries to aggregate functions. An audit trail should detect such **tracker queries**, one or more authorized queries that collectively violate privacy. One approach to preventing tracker queries is to set a lower bound on the number of rows on which a query can report.

Summary

The management of organizational data is driven by the joint goals of availability and integrity. Availability deals with making data available to whoever needs them, whenever and wherever they need them, and in a meaningful form. Maintaining integrity implies protecting existence, maintaining quality, and ensuring confidentiality. There are three strategies for maintaining data integrity: legal, administrative, and technical. A consistent database is one in which all data integrity constraints are satisfied.

A transaction must be entirely completed or aborted before there is any effect on the database. Transactions are processed as logical units of work to ensure data integrity. The transaction manager is responsible for ensuring that transactions are correctly recorded.

91 Adapted from Helman, P. (1994). The science of database management. Burr Ridge, IL: Richard D. Irwin, Inc. p. 434

Concurrent update control focuses on making sure updated results are correctly recorded in a database. To prevent loss of updates and inconsistent retrieval results, a DBMS must incorporate a resource-locking mechanism. Deadlock is an impasse that occurs because two users lock certain resources, then request resources locked by each other. Deadlock prevention requires applications to lock all required records at the beginning of the transaction. Deadlock resolution uses the DBMS to detect and break deadlocks.

Isolation is a preventive strategy that involves administrative procedures to insulate the physical database from destruction. Database backup and recovery is a curative strategy that protects an existing database and recreates or recovers the data whenever loss or destruction occurs. A DBMS needs to provide backup, journaling, and recovery facilities to restore a database to a consistent state and restart the processing of transactions. A journal or log is a special database or file that stores information for backup and recovery. A before image is a copy of a database record before a transaction changes the record. An after image is a copy of a database record after a transaction has updated the record.

In order to recover from data loss or damage, it is necessary to store redundant, backup data. The recovery method is highly dependent on the backup strategy. The cost of keeping backup data is measured in terms of interruption of database availability, storage of redundant data, and degradation of update efficiency. The four major recovery strategies are switching to a duplicate database, backward recovery or rollback, forward recovery or roll forward, and reprocessing transactions. Database administration personnel often implement recovery strategies automatically carried out by the DBMS. The SQL statements `COMMIT` and `ROLLBACK` are used with a procedural programming language for implementing recovery procedures.

Maintaining quality implies keeping data accurate, complete, and current. The first step is to ensure that anyone wanting to update a database is required to have authorization. In SQL, access control is implemented through `GRANT` and `REVOKE`. Data validation cannot be performed without integrity constraints or rules. Data validation can be performed external to the DBMS by personnel or within the DBMS based on defined integrity constraints. Because maintaining data quality is of paramount importance, it is desirable that the DBMS handles data validation rather than the application. Error response rules need to be given to the DBMS along with the integrity constraints.

Two DBMS functions, access control and encryption, are the primary mechanisms for ensuring that the data are accessed only by authorized persons. Access control consists of identification and authorization. Data access authorization is the process of permitting users whose identities have been authenticated to perform certain operations on certain data objects. Encrypted data cannot be read by an intruder unless that person knows the method of encryption and has the key. In a DBMS environment, encryption can be applied to data sent over communication lines between computers and data storage devices.

Database activity is monitored to detect patterns of activity indicating misuse of the system. An audit trail is maintained of all system actions. A tracker query is a series of aggregate function queries designed to reveal individual-level data.

Key terms and concepts

ACID

Administrative strategies

After image

All-or-nothing rule

Atomicity

Audit trail analysis

Authentication

Authorization

Backup

Before image

COMMIT

Concurrent update control

Consistency
Data access control
Data availability
Data quality
Data security
Database integrity
Deadlock prevention
Deadlock resolution
Deadly embrace
Decryption
Durability
Encryption
Ensuring confidentiality
GRANT
Integrity constraint
Isolation
Journal
Legal strategies
Locking
Maintaining quality
Private key
Protecting existence
Public-key encryption
Recovery
Reprocessing
REVOKE
Roll forward
ROLLBACK
Rollback
Serializability
Slock
Technical strategies
Tracker query
Transaction
Transaction atomicity
Transaction manager
Two-phase locking protocol
Validation
Xlock

Exercises

1. What are the three goals of maintaining organizational memory integrity?
2. What strategies are available for maintaining data integrity?
3. A large corporation needs to operate its computer system continuously to remain viable. It currently has data centers in Miami and San Francisco. Do you have any advice for the CIO?
4. An investment company operates out of a single office in Boston. Its business is based on many years of high-quality service, honesty, and reliability. The CEO is concerned that the firm has become too dependent on its computer system. If some disaster should occur and the firm's databases were lost, its reputation for reliability would disappear overnight and so would many of its customers in this highly competitive business. What should the firm do?
5. What mechanisms should a DBMS provide to support backup and recovery?
6. What is the difference between a before image and an after image?
7. A large organization has asked you to advise on backup and recovery procedures for its weekly, batch payroll system. They want reliable recovery at the lowest cost. What would you recommend?
8. An online information service operates globally and prides itself on its uptime of 99.98 percent. What sort of backup and recovery scheme is this firm likely to use? Describe some of the levels of redundancy you would expect to find.

9. The information systems manager of a small manufacturing company is considering the backup strategy for a new production planning database. The database is used every evening to create a plan for the next day's production. As long as the production plan is prepared before 6 a.m. the next day, there is no impact upon plant efficiency. The database is currently 200 Mbytes and growing about 2 percent per year. What backup strategy would you recommend and why?
10. How do backward recovery and forward recovery differ?
11. What are the advantages and disadvantages of reprocessing transactions?
12. When would you use `ROLLBACK` in an application program?
13. When would you use `COMMIT` in an application program?
14. Give three examples of data integrity constraints.
15. What is the purpose of locking?
16. What is the likely effect on performance between locking at a row compared to locking at a page?
17. What is a deadly embrace? How can it be avoided?
18. What are three types of authenticating mechanisms?
19. Assume that you want to discover the grade point average of a fellow student. You know the following details of this person. She is a Norwegian citizen who is majoring in IS and minoring in philosophy. Write one or more aggregate queries that should enable you to determine her GPA.
20. What is encryption?
21. What are the disadvantages of the data encryption standard (DES)?
22. What are the advantages of public-key encryption?
23. A national stock exchange requires listed companies to transmit quarterly reports to its computer center electronically. Recently, a hacker intercepted some of the transmissions and made several hundred thousand dollars because of advance knowledge of one firm's unexpectedly high quarterly profits. How could the stock exchange reduce the likelihood of this event?

23. Data Administration

Bad administration, to be sure, can destroy good policy; but good administration can never save bad policy.

Adlai Stevenson, speech given in Los Angeles, September 11, 1952

Learning objectives

Students completing this chapter will

- understand the role of the Chief Data Officer (CDO);
- understand the importance and role of data administration;
- understand how system-level data administration functions are used to manage a database environment successfully;
- understand how project-level data administration activities support the development of a database system;
- understand what skills data administration requires and why it needs a balance of people, technology, and business skills to carry out its roles effectively;
- understand how computer-based tools can be used to support data administration activities;
- understand the management issues involved in initiating, staffing, and locating data administration organizationally.

The Tahiti tourist resort had been an outstanding success for The Expeditioner. Located 40 minutes from Papeete, the capital city, on a stretch of tropical forest and golden sand, the resort had soon become the favorite meeting place for The Expeditioner's board. No one objected to the long flight to French Polynesia. The destination was well worth the journey, and The Expeditioner had historic ties to that part of the South Pacific. Early visitors to Tahiti, James Cook and William Bligh, had been famous customers of The Expeditioner in the eighteenth century. Before Paul Gauguin embarked on his journey to paint scenes of Tahiti, he had purchased supplies from L'Explorateur, now the French division of The Expeditioner.

Although The Expeditioner is very successful, there are always problems for the board to address. A number of board members are very concerned by the seeming lack of control over the various database systems that are vital to the firm's profitability. Recently, there had been a number of incidents that had underscored the problem. Purchasing had made several poor decisions. For example, it had ordered too many parkas for the North American stores and had to discount them heavily to sell all the stock. The problem was traced to poor data standards and policies within Sales. In another case, Personnel and Marketing had been squabbling for some time over access to the personnel database. Personnel claimed ownership of the data and was reluctant to share data with Marketing, which wanted access to some of the data to support its new incentive program. In yet another incident, a new database project for the Travel Division had been seriously delayed when it was discovered that the Travel Division's development team was planning to implement a system incompatible with The Expeditioner's existing hardware and software.

After the usual exchange of greetings and a presentation of the monthly financial report, Alice forthrightly raised the database problem. "We all know that we depend on information technology to manage The Expeditioner," she began as she glanced at her notes on her tablet. "The Information Systems department does a great job running the computers, building new systems, and providing us with excellent service, but," she stressed, "we seem to be focusing on managing the wrong things. We should be managing what really matters: the data we need to run the business. Data errors, internecine fighting over data, and project delays are costly. Our present system for managing data is fragmented. We don't have anyone or any group who manages data centrally. It is critical that we develop an action plan for the organizational management of data." Pointing to Bob, she continued, "I have invited Bob to brief us on data administration and present his proposal for solving our data management problem. It's all yours, Bob."

Introduction

In the information age, data are the lifeblood of every organization and need to be properly managed to retain their value to the organization. The importance of data as a key organizational resource has been emphasized throughout this book. Data administration is the management of organizational data stores.

Information technology permits organizations to capture, organize, and maintain a greater variety of data. These data can be hard (e.g., financial or production figures) or soft (e.g., management reports, correspondence, voice conversations, and video). If these data are to be used in the organization, they must be managed just as diligently as accounting information. *Data administration* is the common term applied to the task of managing organizational memory. Although common, basic management principles apply to most kinds of organizational data stores, the discussion in this chapter refers primarily to databases.

Why manage data?

Data are constantly generated in every act and utterance of every stakeholder (employee, shareholder, customer, or supplier) in relation to the organization. Some of these data are formal and structured, such as invoices, grade sheets, or bank withdrawals. A large amount of relatively unstructured data is generated too, such as tweets, blogs, and Facebook comments from customers. Much of the unstructured data generated within and outside the organizations are frequently captured but rarely analyzed deeply.

Organization units typically begin maintaining systematic records for data most likely to impinge on their performance. Often, different departments or individuals maintain records for the same data. As a result, the same data may be used in different ways by each department, and so each of them may adopt a different system of organizing its data. Over time, an organization accumulates a great deal of redundant data which demands considerable, needless administrative overhead for its maintenance. Inconsistencies may begin to emerge between the various forms of the same data. A department may incorrectly enter some data, which could result in embarrassment at best or a serious financial loss for the organization at worst.

When data are fragmented across several departments or individuals, and especially when there is personnel turnover, data may not be accessible when most needed. This is nearly as serious a problem as not having any data. Yet another motivation is that effective data management can greatly simplify and assist in the identification of new information system application development opportunities. Also, poor data management can result in breaches of security. Valuable information may be revealed to competitors or antagonists.

In summary, the problems arising from poor data management are:

- The same data may be represented through multiple, inconsistent definitions.
- There may be inconsistencies among different representations.
- Essential data may be missing from the database.
- Data may be inaccurate or incomplete.
- Some data may never be captured for the database and thus are effectively lost to the organization.
- There may be no way of knowing how to locate data when they are needed.

The overall goal of data administration is to prevent the occurrence of these problems by enabling users to access the data they need in the format most suitable for achieving organizational goals and by ensuring the integrity of organizational databases.

The Chief Data Officer

Firms are increasingly recognizing the importance of data to organizational performance, particularly with the attention given to big data in the years following 2010. As a result, some have created the new C-level position of Chief Data Officer (CDO), who is responsible for the strategic management of data systems and ensuring that the organization fully seizes data-driven opportunities to create new business, reduce costs, and increase revenues. The CDO assists the top management team in gaining full value from data, a key strategic asset.[92] In 2003, Capital One was perhaps to first firm to appoint a CDO. Other early creators of this new executive role were Yahoo! and Microsoft Germany. The US Federal Communications Commission (FCC) has appointed a CDO for each of its 11 major divisions. Many firms report plans to create data stewards and CDOs. Given the strategic nature of data for business, it is not surprising that one study reports that 30 percent of CDOs report to the CEO, with another 20% reporting to the COO.

CDO role dimensions

Three dimensions of the CDO role have been identified and described (see the following figure), namely, collaboration direction (inward or outward), data management focus (traditional transaction or big data, and value orientation (service or strategy). We now discuss each of these dimensions.

[92] This section is based on Lee, Y., Madnick, S., Wang, R., Forea, W., & Zhang, H. (2014). A cubic framework for the Chief Data Officer (CDO): Succeeding in a world of Big Data emergence of Chief Data Officers. *MISQ Executive.*

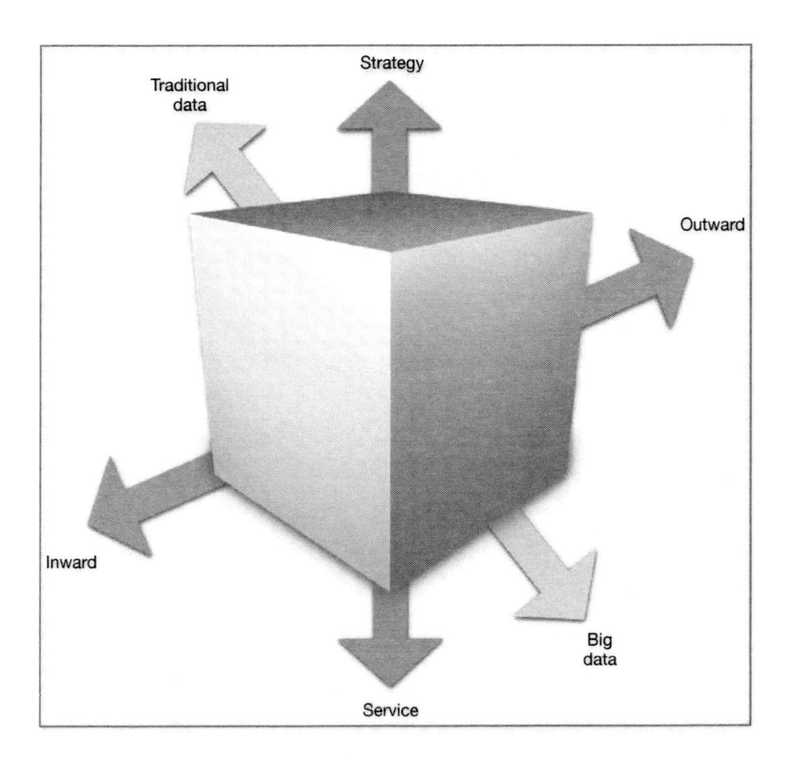

The three dimensions of the CDO role

Inward vs. outward collaboration

A CDO can focus collaborative efforts inward or outward. An inward emphasis might mean working with production to improve the processes for capturing manufacturing data. An outward oriented CDO, in contrast, might work with customers to improve the data flow between the firm and the customer.

Inward oriented initiatives might include developing data quality assessment methods, establishing data products standards, creating procedures for managing metadata, and establishing data governance. The goal is to ensure consistent data delivery and quality inside the organization.

An outwardly-focused CDO will strive to cooperate with an organization's external stakeholders. For example, one CDO led a program for "global unique product identification" to improve collaboration with external global partners. Another might pay attention to improving the quality of data supplied to external partners.

Traditional vs. big data management focus

A CDO can stress managing traditional transactional data, which is typically managed with a relational databases, or shift direction towards handling expanding data volumes with new files structures, such as Hadoop data file structure (HDFS).

Traditional data are still the foundation of many organization's operations, and there remains in many firms a need for a CDO with a transactional data orientation. Big data promises opportunities for improving operations or developing new business strategies based on analyses and insights not available from traditional data. A CDO attending to big data can provide leadership in helping a business gain deeper knowledge of its customers, suppliers, and so forth based on mining large volumes of data.

Service vs. strategy orientation

A CDO can stress improving services or exploring an organization's strategic opportunities. This dimension should reflect the organization's goals for the CDO position. If the top management team is mainly concerned with oversight and accountability, then the CDO should pay attention to improving existing data-related processes. Alternatively, if the senior team actively seeks new data-driven strategic value, then the CDO needs to be similarly aligned and might look at how to exploit digit data streams, for example. One strategy-directed CDO, for instance, led an initiative to identify new information products for advancing the firm's position in the financial industry.

CDO archetypes

Based on the three dimensions just discussed, eight different CDO roles can be identified, as shown in the following table.

Dimensions of CDO archetypes

	Inward	Outward	Traditional data	Big data	Strategy	Service
Coordinator						
Reporter						
Architect						
Ambassador						
Analyst						
Marketer						
Developer						
Experimenter						

While eight different roles are possible, it is important to note that a CDO might take on several of these roles as the situation changes and the position evolves. Also, treat an archetype as a broad indicator of a role rather than a confining specification.

CDO archetypes

Archetype	Definition
Coordinator	Fosters internal collaboration using transactional data to support business services.
Reporter	Provides high quality enterprise data delivery services for external reporting.
Architect	Designs databases and internal business processes to create new opportunities for the organization.

Archetype	Definition
Ambassador	Develops internal data policies to support business strategy and external collaboration using traditional data sources.
Analyst	Improves internal business performance by exploiting big data to provide new services.
Marketer	Develops relationships with external data partners and stakeholders to improve externally provided data services using big data.
Developer	Navigates and negotiates with internal enterprise divisions in order to create new services by exploiting big data.
Experimente r	Engages with external parties, such as suppliers and industry peers, to explore new, unidentified markets and products based on insights derived from big data.

Management of the database environment

In many large organizations, there is a formal data administration function to manage corporate data. For those companies with a CDO, data administration is subsumed within this function. For those without, data administration is typically the responsibility of the CIO. The relationship among the various components of data administration is shown in the following figure.

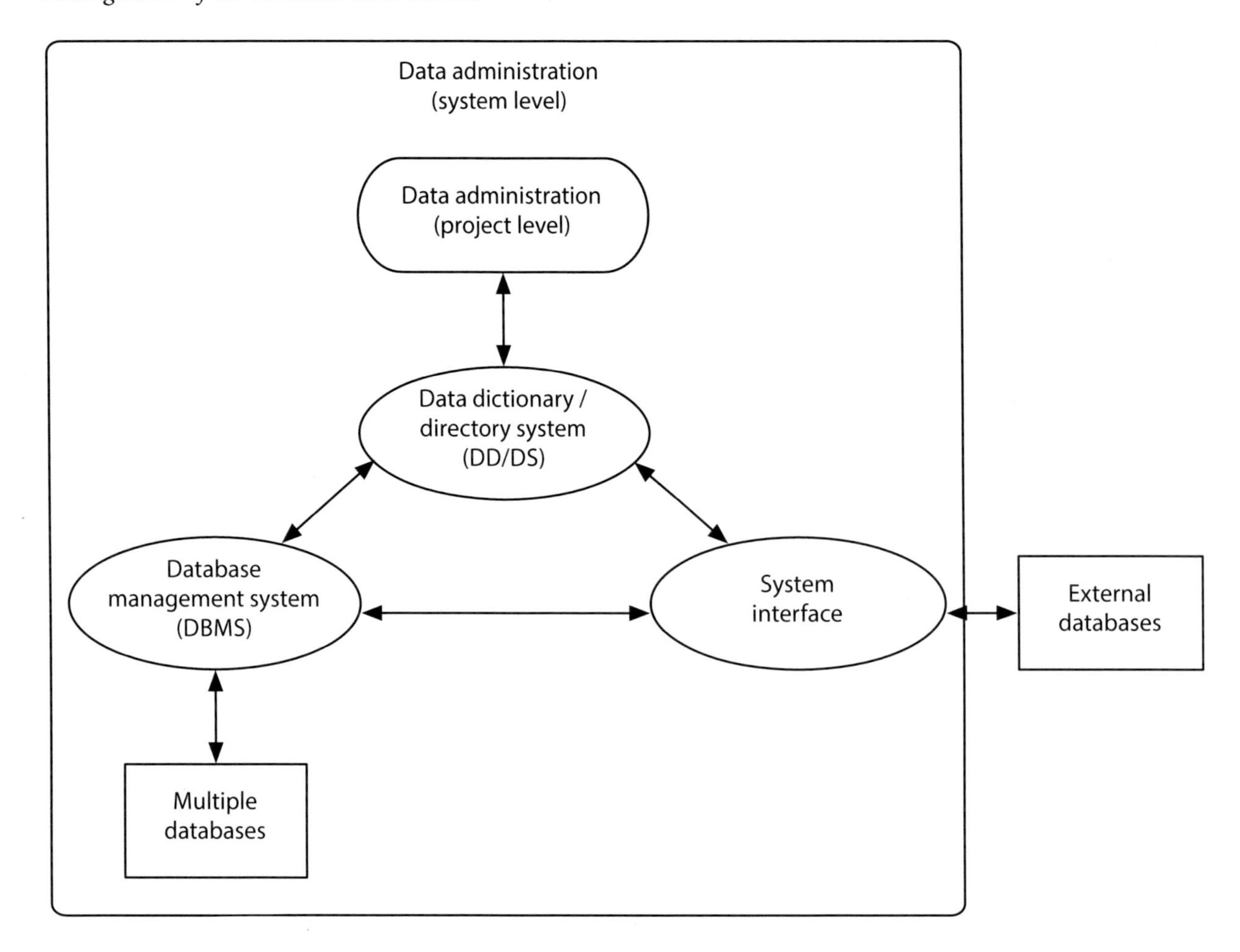

Databases

A database management system (DBMS) can manage multiple databases covering different aspects of an organization's activities. When a database has multiple clients, it may be designed to meet all their requirements, even though a specific person may need only a portion of the data contained within the database. For instance, a finished-goods database may be accessed by Production and Finance, as well as Marketing. It may contain cost information that is accessible by Production and Finance but not by Marketing.

System interface

The interface consists of windows, menus, icons, and commands that enable clients to direct the system to manipulate data. Clients may range from casual novices, who need to be insulated from the underlying complexity of the data, to expert application developers, who manipulate the data using programming languages or other data-handling tools.

Data dictionary

A data dictionary is a reference repository containing *metadata* (i.e., *data about data*) that is stored in the database. Among other things, the data dictionary contains a list of all the databases; their component parts and detailed descriptions such as field sizes, data types, and data validation information for data capture purposes; authorized clients and their access privileges; and ownership details.

A data dictionary is a map of the data in organizational data stores. It permits the data administration staff and users to document the database, design new applications, and redesign the database if necessary. The data dictionary/directory system (DD/DS), itself a DBMS, is software for managing the data dictionary.

External databases

For organizations to remain competitive in a rapidly changing marketplace, access to data from external sources is becoming increasingly critical. Research and development groups need access to the latest developments in their technical fields and need to track information such as patents filed, research reports, and new product releases. Marketing departments need access to data on market conditions and competitive situations as reported in various surveys and the media in general. Financial data regarding competitors and customers are important to senior executives. Monitoring political situations may be critical to many business decisions, especially in the international business arena.

Extensive external data are available electronically through information services such as Bloomberg (financial data), Reuters (news), and LexisNexis (legal data), various government sites,such as data.gov, or from various Web sites. Tools are available to download external data into internal databases, from which they may be accessed through the same interface as for internal data.

Data administration

Data administration is responsible for the management of data-related activities. There are two levels of data administration activities: system and project.

System-level administration is concerned with establishing overall policies and procedures for the management and use of data in the organization. Formulating a data strategy and specifying an information architecture for the organization are also system-level data administration functions.

Project-level administration deals more with the specifics, such as optimizing specific databases for operational efficiency, establishing and implementing database access rights, creating new databases, and monitoring database use.

In general, the system-level function takes a broader perspective of the role played by data in achieving business objectives, while the project-level function is more concerned with the actual mechanics of database implementation and operation. We use the term *data administration* to refer to both functional levels.

Data administration functions and roles

The functions of data administration may be accomplished through multiple roles or job titles such as database administrator, database developer, database consultant, and database analyst, collectively referred to as the data administration staff. A single role could be responsible for both system and project levels of data administration, or responsibility may be distributed among several persons, depending on the size of the organization, the number of database applications, and the number of clients.

In addition, data administration could be carried out entirely in a client department. For instance, a client could be the data steward responsible for managing all corporate data for some critical business-related entity or activity (e.g., a customer, a production facility, a supplier, a division, a project, or a product) regardless of the purpose for which the data are used.

Data stewards coordinate planning of the data for which they are responsible. Tasks include data definition, quality control and improvement, security, and access authorization. The data steward's role is especially important today because of the emphasis on customer satisfaction and cross-functional teams. Data stewardship seeks to align data management with organizational strategy.

Database levels

Databases may be maintained at several levels of use: personal, workgroup (e.g., project team or department), and organizational. More clients usually results in greater complexity of both the database and its management.

Personal databases in the form of calendars, planners, and name and address books have existed for a long time. The availability of laptops, tablets, and smartphones have made it convenient to maintain synchronized electronic personal databases across multiple devices. Behind the interface of many of these apps is a lightweight relational database, such as SQLite.[93]

Workgroup databases cannot be as idiosyncratic because they are shared by many people. Managing them requires more planning and coordination to ensure that all the various clients' needs are addressed and data integrity is maintained. Organizational databases are the most complex in terms of both structure and need for administration. All databases, regardless of scope or level, require administration.

Managing a personal database is relatively simple. Typically, the owner of the database is also its developer and administrator. Issues such as access rights and security are settled quite easily, perhaps by locking the computer when away from the desk. Managing workgroup databases is more complex. Controls almost certainly will be needed to restrict access to certain data. On the other hand, some data will need to be available to many group members. Also, responsibility for backup and recovery must be established. Small workgroups may jointly perform both system- and project-level data administration activities. Meetings may be a way to coordinate system-level data administration activities, and project-level activities may be distributed among different workgroup members. Larger groups may have a designated data administrator, who is also a group member.

Managing organizational databases is typically a full-time job requiring special skills to work with complex database environments. In large organizations, several persons may handle data administration, each carrying out different data administration activities. System-level data administration activities may be carried out by a committee led by a senior IS executive (who may be a full- or part-time data administrator), while project-level data administration activities may be delegated to individual data administration staff members.

System-level data administration functions

System-level data administration functions, which may be performed by one or more persons, are summarized in the following table.

Function
Planning
Developing data standards and policies
Defining XML data schemas
Maintaining data integrity
Resolving data conflict
Managing the DBMS
Establishing and maintaining the data dictionary
Selecting hardware and software
Managing external databases
Benchmarking

93 http://www.sqlite.org

Function
Internal marketing

Planning

Because data are a strategic corporate resource, planning is perhaps the most critical data administration function. A key planning activity is creating an organization's information architecture, which includes all the major data entities and the relationships between them. It indicates which business functions and applications access which entities. An information architecture also may address issues such as how data will be transmitted and where they will be stored. Since an information architecture is an organization's overall strategy for data and applications, it should dovetail with the organization's long-term plans and objectives.[94]

Developing data standards and policies

Whenever data are used by more than one person, there must be standards to govern their use. Data standards become especially critical in organizations using heterogeneous hardware and software environments. Why could this become a problem? For historical reasons, different departments may use different names and field sizes for the same data item. These differences can cause confusion and misunderstanding. For example, "sale date" may have different meanings for the legal department (e.g., the date the contract was signed) and the sales department (e.g., the date of the sales call). Furthermore, the legal department may store data in the form *yyyy-mm-dd* and the sales department as *dd-mm-yy*. Data administration's task is to develop and publish data standards so that field names are clearly defined, and a field's size and format are consistent across the enterprise.

Furthermore, some data items may be more important to certain departments or divisions. For instance, customer data are often critical to the marketing department. It is useful in such cases to appoint a data steward from the appropriate functional area as custodian for these data items.

Policies need to be established regarding who can access and manipulate which data, when, and from where. For instance, should employees using their home computers be allowed to access corporate data? If such access is permitted, then data security and risk exposure must be considered and adequate data safeguards implemented.

Defining XML data schemas

Data administration is also often responsible for defining data schemas for data exchange within the organization and among business partners. Data administration is also responsible for keeping abreast of industry schema standards so that the organization is in conformance with common practice. Some data administrators may even work on defining a common data schema for an industry.

Maintaining data integrity

Data must be made available when needed, but only to authorized users. The data management aspects of data integrity are discussed at length in the Data Integrity chapter.

Resolving data conflict

Data administration involves the custodianship of data *owned* or originating in various organizational departments or functions, and conflicts are bound to arise at some point. For instance, one department may be concerned about a loss of security when another department is allowed access to its data. In another instance, one group may feel that another is contaminating a commonly used data pool because of inadequate data validation practices. Incidents like these, and many others, require management

94 For more on information architecture, see Smith, H. A., Watson, R. T., & Sullivan, P. (2012). Delivering Effective Enterprise Architecture at Chubb Insurance. *MISQ Executive*. 11(2)

intervention and settlement through a formal or informal process of discussion and negotiation in which all parties are assured of an outcome that best meets the needs of the organization. Data administration facilitates negotiation and mediates dispute resolution.

Managing the DBMS

While project-level data administration is concerned more directly with the DBMS, the performance and characteristics of the DBMS ultimately impinge on the effectiveness of the system-level data administration function. It is, therefore, important to monitor characteristics of the DBMS. Over a period, benchmark statistics for different projects or applications will need to be compiled. These statistics are especially useful for addressing complaints regarding the performance of the DBMS, which may then lead to design changes, tuning of the DBMS, or additional hardware.

Database technology is rapidly advancing. For example, relational DBMSs are continually being extended, and Hadoop offers a new approach to handle large data processing tasks. Keeping track of developments, evaluating their benefits, and deciding on converting to new database environments are critical system-level data administration functions that can have strategic implications for the corporation.

Establishing and maintaining the data dictionary

A data dictionary is a key data administration tool that provides details of data in the organizational database and how they are used (e.g., by various application programs). If modifications are planned for the database (e.g., changing the size of a column in a table), the data dictionary helps to determine which applications will be affected by the proposed changes.

More sophisticated data dictionary systems are closely integrated with specific database products. They are updated automatically whenever the structure of the underlying database is changed.

Selecting hardware and software

Evaluating and selecting the appropriate hardware and software for an organizational database are critical responsibilities with strategic organizational implications. These are not easy tasks because of the dynamic nature of the database industry, the continually changing variety of available hardware and software products, and the rapid pace of change within many organizations. Today's excellent choice might become tomorrow's nightmare if, for instance, the vendor of a key database component goes out of business or ceases product development.

Extensive experience and knowledge of the database software business and technological progress in the field are essential to making effective database hardware and software decisions. The current and future needs of the organization need to be assessed in terms of capacity as well as features. Relevant questions include the following:

- How many people and apps will simultaneously access the database?
- Will the database need to be geographically distributed? If so, what is the degree to which the database will be replicated, and what is the nature of database replication that is supported?
- What is the maximum size of the database?
- How many transactions per second can the DBMS handle?
- What kind of support for online transaction processing is available?
- What are the initial and ongoing costs of using the product?
- Can the database be extended to include new data types?

- What is the extent of training required, who can provide it, and what are the associated costs?

DBMS selection should cover technical, operational, and financial considerations. An organization's selection criteria are often specified in a request for proposal (RFP). This document is sent to a short list of potential vendors, who are invited to respond with a software or hardware/software proposal outlining how their product or service meets each criterion in the RFP. Discussions with current customers are usually desirable to gain confirming evidence of a vendor's claims. The final decision should be based on the manner and degree to which each vendor's proposal satisfies these criteria.

Skill builder

A small nonprofit organization has asked for your help in selecting a relational database management system (RDBMS) for general-purpose management tasks. Because of budget limitations, it is very keen to adopt an open source RDBMS. Search the Web to find at least two open source RDBMSs, compare the two systems, and make a recommendation to the organization.

Benchmarking

Benchmarking, the comparison of alternative hardware and software combinations, is an important step in the selection phase. Because benchmarking is an activity performed by many IS units, the IS community gains if there is one group that specializes in rigorous benchmarking of a wide range of systems. The Transaction Processing Council[95](TPC) is the IS profession's Consumer Union. TPC has established benchmarks for a variety of business situations.

Managing external databases

Providing access to external databases has increased the level of complexity of data administration, which now has the additional responsibility of identifying information services that meet existing or potential managerial needs. Data administration must determine the quality of such data and the means by which they can be channeled into the organization's existing information delivery system. Costs of data may vary among vendors. Some may charge a flat monthly or annual fee, while others may have a usage charge. Data may arrive in a variety of formats, and data administration may need to make them adhere to corporate standards. Data from different vendors and sources may need to be integrated and presented in a unified format and on common screens.

Monitoring external data sources is critical because data quality may vary over time. Data administration must determine whether organizational needs are continuing to be met and data quality is being maintained. If they are not, a subscription may be canceled and an alternative vendor sought. Security is another critical problem. When corporate databases are connected to external communication links, there is a threat of hackers breaking into the system and gaining unauthorized access to confidential internal data. Also, corporate data may be contaminated by spurious data or even by viruses entering from external sources. Data administration must be cautious when incorporating external data into the database.

Internal marketing

Because IS applications can have a major impact on organizational performance, the IS function is becoming more proactive in initiating the development of new applications. Many clients are not aware of what is possible with newly emergent technologies, and hence do not see opportunities to exploit these developments. Also, as custodian of organizational data, data administration needs to communicate its goals and responsibilities throughout the organization. People and departments need to be persuaded to

95 http://www.tpc.org

share data that may be of value to other parts of the organization. There may be resistance to change when people are asked to switch to newly set data standards. In all these instances, data administration must be presented in a positive light to lessen resistance to change. Data administration needs to market internally its products and services to its customers.

Project-level data administration

At the project level, data administration focuses on the detailed needs of individual clients and applications. It supports the development and use of a specific database system.

Systems development life cycle (SDLC)

Database development follows a fairly predictable sequence of steps or phases similar to the systems development life cycle (SDLC) for applications. This sequence is called the database development life cycle (DDLC). The database and application development life cycles together constitute the systems development life cycle (SDLC).

Systems development life cycles

Application development life cycle (ADLC)	Database development life cycle (DDLC)
Project planning	Project planning
Requirements definition	Requirements definition
Application design	Database design
Application construction	
Application testing	Database testing
Application implementation	Database implementation
Operations	Database usage
Maintenance	Database evolution

Application development involves the eight phases shown in the preceding table. It commences with project planning which, among other things, involves determining project feasibility and allocating the necessary personnel and material resources for the project. This is followed by requirements definition, which involves considerable interaction with clients to clearly specify the system. These specifications become the basis for a conceptual application design, which is then constructed through program coding and tested. Once the system is thoroughly tested, it is installed and user operations begin. Over time, changes may be needed to upgrade or repair the system, and this is called system maintenance.

The database development phases parallel application development. Data administration is responsible for the DDLC. Data are the focus of database development, rather than procedures or processes. Database construction is folded into the testing phase because database testing typically involves minimal effort. In systems with integrated data dictionaries, the process of constructing the data dictionary also creates the database shell (i.e., tables without data). While the sequence of phases in the cycle as presented is generally followed, there is often a number of iterations within and between steps. Data modeling is iterative, and the final database design evolves from many data modeling sessions. A previously unforeseen requirement may surface during the database design phase, and this may prompt a revision of the specifications completed in the earlier phase.

System development may proceed in three different ways:

1. The database may be developed independently of applications, following only the DDLC steps.
2. Applications may be developed for existing databases, following only the ADLC steps.

3. Application and database development may proceed in parallel, with both simultaneously stepping through the ADLC and DDLC.

Consider each of these possibilities.

Database development may proceed independently of application development for a number of reasons. The database may be created and later used by an application or for ad hoc queries using SQL. In another case, an existing database may undergo changes to meet changed business requirements. In such situations, the developer goes through the appropriate stages of the DDLC.

Application development may proceed based on an existing database. For instance, a personnel database may already exist to serve a set of applications, such as payroll. This database could be used as the basis for a new personnel benefits application, which must go through all the phases of the ADLC.

A new system requires both application and database development. Frequently, a new system will require creation of both a new database and applications. For instance, a computer manufacturer may start a new Web-based order sales division and wish to monitor its performance. The vice-president in charge of the division may be interested in receiving daily sales reports by product and by customer as well as a weekly moving sales trend analysis for the prior 10 weeks. This requires both the development of a new sales database as well as a new application for sales reporting. Here, the ADLC and DDLC are both used to manage development of the new system.

Database development roles

Database development involves several roles, chiefly those of developer, client, and data administrator. The roles and their responsibilities are outlined in the following table.

Database development roles

Database development phase	Database developer	Data administrator	Client
Project planning	Does	Consults	Provides information
Requirements definition	Does	Consults	Provides requirements
Database design	Does	Consults Data integrity	Validates data models
Database testing	System and client testing	Consults Data integrity	Testing
Database implementation	System-related activities	Consults Data integrity	Client activities
Database usage	Consults	Data integrity monitoring	Uses
Database evolution	Does	Change control	Provides additional requirements

The database developer shoulders the bulk of the responsibility for developing data models and implementing the database. This can be seen in the table, where most of the cells in the database developer column are labeled "Does." The database developer does project planning, requirements definition, database design, database testing, and database implementation, and in addition, is responsible for database evolution.

The client's role is to establish the goals of a specific database project, provide the database developers with access to all information needed for project development, and review and regularly scrutinize the developer's work.

The data administrator's prime responsibilities are implementing and controlling, but the person also may be required to perform activities and consult. In some situations, the database developer is not part of the data administration staff and may be located in a client department, or may be an analyst from an IS project team. In these cases, the data administrator advises the developer on organizational standards and policies as well as provides specific technical guidelines for successful construction of a database. When the database developer is part of the data administration staff, developer and data administration activities may be carried out by the same person, or by the person(s) occupying the data administration role. In all cases, the data administrator should understand the larger business context in which the database will be used and should be able to relate business needs to specific technical capabilities and requirements.

Database development life cycle (DDLC)

Previously, we discussed the various roles involved in database development and how they may be assigned to different persons. In this section, we will assume that administration and development are carried out by the data administration staff, since this is the typical situation encountered in many organizations. The activities of developer and administrator, shown in the first two columns of the table, are assumed to be performed by data administration staff. Now, let us consider data administration project-level support activities in detail . These activities are discussed in terms of the DDLC phase they support.

Database development life cycle

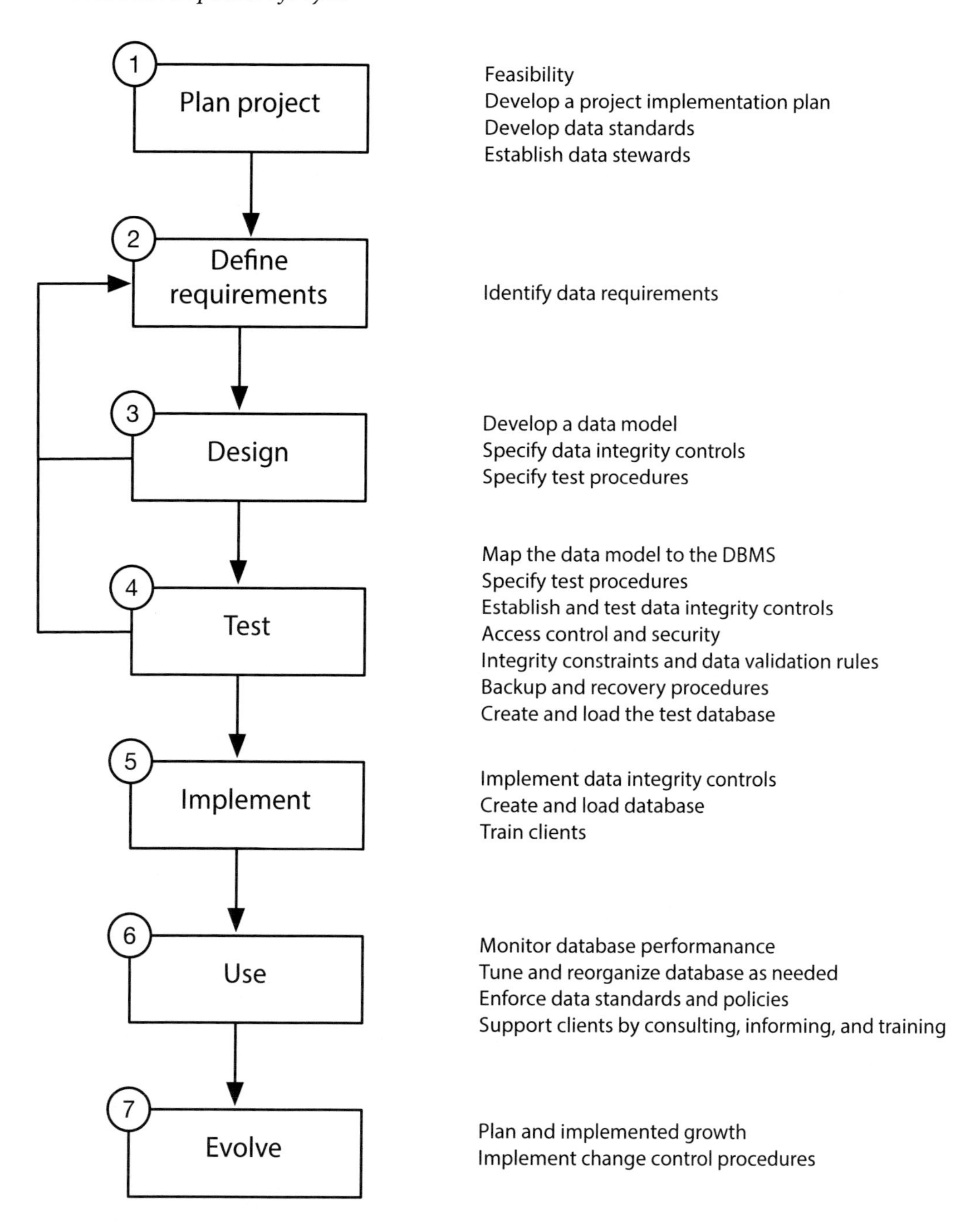

Database project planning

Database project planning includes establishing project goals, determining project feasibility (financial, technical, and operational), creating an implementation plan and schedule, assigning project responsibilities (including data stewards), and establishing standards. All project stakeholders, including clients, senior management, and developers, are involved in planning. They are included for their knowledge as well as to gain their commitment to the project.

Requirements definition

During requirements definition, clients and developers establish requirements and develop a mutual understanding of what the new system will deliver. Data are defined and the resulting definitions stored in the data dictionary. Requirements definition generates documentation that should serve as an unambiguous reference for database development. Although in theory the clients are expected to *sign-off* on the specifications and accept the developed database as is, their needs may actually change in practice. Clients may gain a greater understanding of their requirements and business conditions may change. Consequently, the original specifications may require revision. In the preceding figure, the arrows connecting phase 4 (testing) and phase 3 (design) to phase 2 (requirements definition) indicate that modeling and testing may identify revisions to the database specification, and these amendments are then incorporated into the design.

Database design

Conceptual and internal models of the database are developed during database design. Conceptual design, or data modeling, is discussed extensively in Section 2 of this book. Database design should also include specification of procedures for testing the database. Any additional controls for ensuring data integrity are also specified. The external model should be checked and validated by the user.

Database testing

Database testing requires previously developed specifications and models to be tested using the intended DBMS. Clients are often asked to provide operational data to support testing the database with realistic transactions. Testing should address a number of key questions.

- Does the DBMS support all the operational and security requirements?
- Is the system able to handle the expected number of transactions per second?
- How long does it take to process a realistic mix of queries?

Testing assists in making early decisions regarding the suitability of the selected DBMS. Another critical aspect of database testing is verifying data integrity controls. Testing may include checking backup and recovery procedures, access control, and data validation rules.

Database implementation

Testing is complete when the clients and developers are extremely confident the system meets specified needs. Data integrity controls are implemented, operational data are loaded (including historical data, if necessary), and database documentation is finalized. Clients are then trained to operate the system.

Database use

Clients may need considerable support as they learn and adapt to the system. Monitoring database performance is critical to keeping them satisfied; enables the data administrator to anticipate problems even before the clients begin to notice and complain about them, and tune the system to meet organizational needs; and also helps to enforce data standards and policies during the initial stages of database implementation.

Database evolution

Since organizations cannot afford to stand still in today's dynamic business environment, business needs are bound to change over time, perhaps even after a few months. Data administration should be prepared to meet the challenge of change. Minor changes, such as changes in display formats, or performance improvements, may be continually requested. These have to be attended to on an ongoing basis. Other evolutionary changes may emerge from constant monitoring of database use by the data administration staff. Implementing these evolutionary changes involves repeating phases 3 to 6. Significant business changes may merit a radical redesign of the database. Major redesign may require repeating all phases of the DDLC.

Data administration interfaces

Data administration is increasingly a key corporate function, and it requires the existence of established channels of communication with various organizational groups. The key data administration interfaces are with clients, management, development staff, and computer operations. The central position of the data administration staff, shown in the following figure, reflects the liaison role that it plays in managing databases. Each of the groups has a different focus and different terminology and jargon. Data administration should be able to communicate effectively with all participants. For instance, operations staff will tend to focus on technical, day-to-day issues to which management is likely to pay less attention. These different focuses can, and frequently do, lead to conflicting views and expectations among the different groups. Good interpersonal skills are a must for data administration staff in order to deal with a variety of conflict-laden situations. Data administration, therefore, needs a balance of people, technical, and business skills for effective execution of its tasks.

Major data administration interfaces

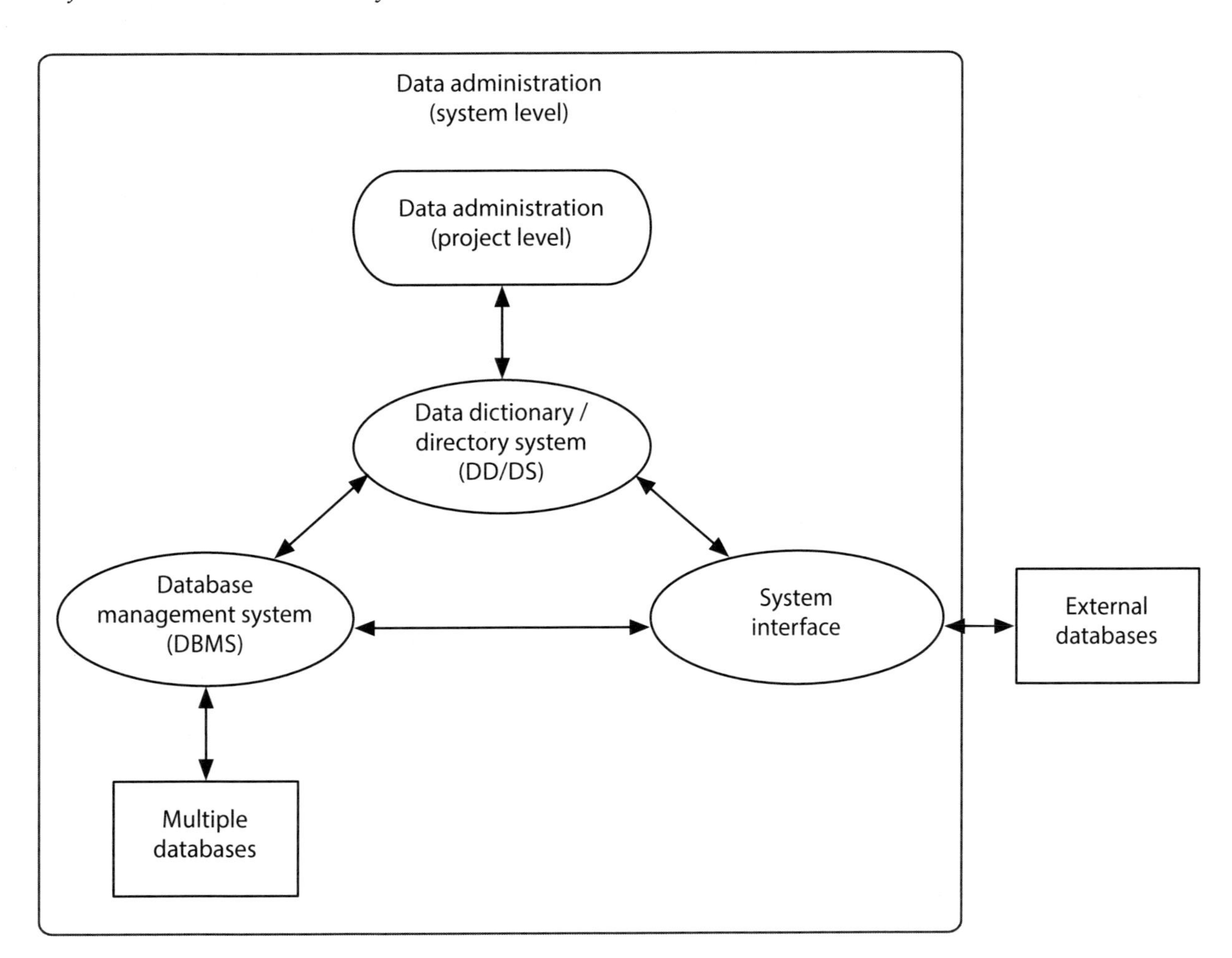

Data administration probably will communicate most frequently with computer operations and development staff, somewhat less frequently with clients, and least frequently with management. These differences, however, have little to do with the relative importance of communicating with each group. The interactions between the data administration staff and each of the four groups are discussed next.

Management

Management sets the overall business agenda for data administration, which must ensure that its actions directly contribute to the achievement of organizational goals. In particular, management establishes overall policy guidelines, approves data administration budgets, evaluates proposals, and champions major changes. For instance, if the data administration staff is interested in introducing new technology, management may request a formal report on the anticipated benefits of the proposed expenditure.

Interactions between the data administration staff and management may focus on establishing and evolving the information architecture for the organization. In some instances, these changes may involve the introduction of a new technology that could fundamentally transform roles or the organization.

Clients

On an ongoing basis, most clients will be less concerned with architectural issues and will focus on their personal needs. Data administration must determine what data should be collected and stored, how they should be validated to ensure integrity, and in what form and frequency they should be made available.

Typically, the data administration staff is responsible for managing the database, while the data supplier is responsible for ensuring accuracy. This may cause conflict, however, if there are multiple clients from separate departments. If conflict does arise, data administration has to arbitrate.

Development staff

Having received strategic directions from management, and after determining clients' needs, data administration works with application and database developers, in order to fulfill the organization's goals for the new system. On an ongoing basis, this may consist of developing specifications for implementation. Data administration works on an advisory basis with systems development, providing inputs on the database aspects. Data administration is typically responsible for establishing standards for program and database interfaces and making developers aware of these standards. Developers may need to be told which commands may be used in their programs and which databases they can access.

In many organizations, database development is part of data administration, and it has a very direct role in database design and implementation. In such instances, communication between data administration and database development is within the group. In other organizations, database development is not part of data administration, and communication is between groups.

Computer operations

The focus of computer operations is on the physical hardware, procedures, schedules and shifts, staff assignments, physical security of data, and execution of programs. Data administration responsibilities include establishing and monitoring procedures for operating the database. The data administration staff needs to establish and communicate database backup, recovery, and archiving procedures to computer operations. Also, the scheduling of new database and application installations needs to be coordinated with computer operations personnel.

Computer operations provide data administration with operational statistics and exception reports. These data are used by data administration to ensure that corporate database objectives are being fulfilled.

Communication

The diverse parties with which data administration communicates often see things differently. This can lead to misunderstandings and results in systems that fail to meet requirements. Part of the problem arises from a difference in perspective and approaches to viewing database technology.

Management is interested in understanding how implementing database technology will contribute to strategic goals. In contrast, clients are interested in how the proposed database and accompanying applications will affect their daily work. Developers are concerned with translating management and client needs into conceptual models and converting these into tables and applications. Operations staff are concerned primarily with efficient daily management of DBMS, computer hardware, and software.

Data models can serve as a common language for bridging the varying goals of clients, developers, management, and operational staff. A data model can reduce the ambiguity inherent in verbal communications and thereby ensure that overall needs are more closely met and all parties are satisfied with the results. A data model provides a common meeting point and language for understanding the needs of each group.

Data administration does not work in isolation. It must communicate successfully with all its constituents in order to be successful. The capacity to understand and correctly translate the needs of each stakeholder group is the key to competent data administration.

Data administration tools

Several computer-based tools have emerged to support data administration. There are five major classes of tools: data dictionary, DBMS, performance monitoring, computer-aided software engineering (CASE), and groupware tools. Each of these tools is now examined and its role in supporting data administration considered. We focus on how these tools support the DDLC. Note, however, that groupware is not shown in the following table because it is useful in all phases of the life cycle.

Data administration tool use during the DDLC

	Database development phase	**Data**	**DBMS**	**Performance monitoring**	**Case tools**
1	Project planning tools	Dictionary (DD)			Estimation
2	Requirements definition	Document Design aid			Document Design aid
3	Database design	Document Data map Design aid Schema generator			Document Design aid Data map
4	Database testing	Data map Design aid Schema generator	Define, create, test, data integrity	Impact analysis	Data generator Design aid
5	Database implementation	Document Change control	Data integrity Implement Design	Monitor Tune	
6	Database usage	Document Data map Schema generator Change control	Provide tools for retrieval and update Enforce integrity controls and procedures	Monitor Tune	

	Database development phase	Data	DBMS	Performance monitoring	Case tools
7	Database evolution	Document Data map Change control	Redefine	Impact analysis	

Data administration staff, clients, and computer operations all require information about organizational databases. Ideally, such information should be stored in one central repository. This is the role of the DD/DS, perhaps the main data administration tool. Thus, we start our discussion with this tool.

Data dictionary/directory system (DD/DS)

The DD/DS, a database application that manages the data dictionary, is an essential tool for data administration. The DD/DS is the repository for organizational metadata, such as data definitions, relationships, and privileges. The DBMS manages data, and the DD/DS manages data about data. The DD/DS also uses the data dictionary to generate table definitions or schema required by the DBMS and application programs to access databases.

Clients and data administration can utilize the DD/DS to ask questions about characteristics of data stored in organizational databases such as

- What are the names of all tables for which the user Todd has delete privileges?
- Where does the data item *customer number* appear or where is it used?

The report for the second query could include the names and tables, transactions, reports, display screens, account names, and application programs.

In some systems, such as a relational DBMS, the catalog, discussed in the SQL chapter, performs some of the functions of a DD/DS, although the catalog does not contain the same level of detail. The catalog essentially contains data about tables, columns, and owners of tables, whereas the DD/DS can include data about applications, forms, transactions, and many other aspects of the system. Consequently, a DD/DS is of greater value to data administration.

Although there is no standard format for data stored in a data dictionary, several features are common across systems. For example, a data dictionary for a typical relational database environment would contain descriptions of the following:

- All columns that are defined in all tables of all databases. The data dictionary stores specific data characteristics such as name, data type, display format, internal storage format, validation rules, and integrity constraints. It indicates to which table a column belongs.
- All relationships among data elements, what elements are involved, and characteristics of relationships, such as cardinality and degree.
- All defined databases, including who created each database, the date of creation, and where the database is located.
- All tables defined in all databases. The data dictionary is likely to store details of who created the table, the date of creation, primary key, and the number of columns.
- All indexes defined for each of the database tables. For each of the indexes, the DBMS stores data such as the index name, location, specific index characteristics, and creation date.
- All clients and their access authorizations for various databases.

- All programs that access the database, including screen formats, report formats, application programs, and SQL queries.

A data dictionary can be useful for both systems and project level data administration activities. The five major uses of a data dictionary are the following:

1. Documentation support: recording, classifying, and reporting metadata.
2. Data maps: a map of available data for data administration staff and users. A data map allows users to discover what data exist, what they mean, where they are stored, and how they are accessed.
3. Design aid: documenting the relationships between data entities and performing impact analysis.
4. Schema generation: automatic generation of data definition statements needed by software systems such as the DBMS and application programs.
5. Change control: setting and enforcing standards, evaluating the impact of proposed changes, and implementing amendments, such as adding new data items.

Database management systems (DBMSs)

The DBMS is the primary tool for maintaining database integrity and making data available to users. Availability means making data accessible to whoever needs them, when and where they need them, and in a meaningful form. Maintaining database integrity implies the implementation of control procedures to achieve the three goals discussed in a prior chapter: protecting existence, maintaining quality, and ensuring confidentiality. In terms of the DDLC life cycle (see the following figure), data administration uses, or helps others to use, the DBMS to create and test new databases, define data integrity controls, modify existing database definitions, and provide tools for clients to retrieve and update databases. Since much of this book has been devoted to DBMS functions, we limit our discussion to reviewing its role as a data administration tool.

Performance monitoring tools

Monitoring the performance of DBMS and database operations by gathering usage statistics is essential to improving performance, enhancing availability, and promoting database evolution. Monitoring tools are used to collect statistics and improve database performance during the implementation and use stages of the DDLC. Monitoring tools can also be used to collect data to evaluate design choices during testing.

Many database factors can be monitored and a variety of statistics gathered. Monitoring growth in the number of rows in each table can reveal trends that are helpful in projecting future needs for physical storage space. Database access patterns can be scrutinized to record data such as:

- Type of function requested: query, insert, update, or delete
- Response time (elapsed time from query to response)
- Number of disk accesses
- Identification of client
- Identification of error conditions

The observed patterns help to determine performance enhancements. For example, these statistics could be used to determine which tables or files should be indexed. Since gathering statistics can result in some degradation of overall performance, it should be possible to turn the monitoring function on or off with regard to selected statistics.

CASE tools

A CASE tool, as broadly defined, provides automated assistance for systems development, maintenance, and project management activities. Data administration may use project management tools to coordinate all phases within the DDLC. The dictionary provided by CASE systems can be used to supplement the DD/DS, especially where the database development effort is part of a systems development project.

One of the most important components of a CASE tool is an extensive dictionary that tracks all objects created by systems designers. Database and application developers can use the CASE dictionary to store descriptions of data elements, application processes, screens, reports, and other relevant information. Thus, during the first three phases of the life cycle, the CASE dictionary performs functions similar to a DD/DS. During stages 4 and 5, data from the CASE dictionary would be transferred, usually automatically, to the DD/DS.

Groupware

Groupware can be applied by data administration to support any of the DDLC phases shown in the figure. Groupware supports communication among people and thus enhances access to organizational memory residing within humans. As pointed out, data administration interfaces with four major groups during the DDLC: management, clients, developers, and computer operations. Groupware supports interactions with all of these groups.

Data administration is a complex task involving a variety of technologies and the need to interact with, and satisfy the needs of, a diverse range of clients. Managing such a complex environment demands the use of computer-based tools, which make data administration more manageable and effective. Software tools, such as CASE and groupware, can improve data administration.

Data integration

A common problem for many organizations is a lack of data integration, which can manifest in a number of ways:

- Different identifiers for the same instance of an entity (e.g., the same product with different codes in different divisions)
- The same, or what should be the same, data stored in multiple systems (e.g., a customer's name and address)
- Data for, or related to, a key entity stored in different databases (e.g., a customer's transaction history and profile stored in different databases)
- Different rules for computing the same business indicator (e.g., the Australian office computes net profit differently from the U.S. office)

In the following table, we see an example of a firm that practices data integration. Different divisions use the same numbers for parts, the same identifiers for customers, and have a common definition of sales date. In contrast, the table after that shows the case of a firm where there is a lack of data integration. The different divisions have different identifiers for the same part and different codes for the same customer, as well as different definitions for the sales date. Imagine the problems this nonintegrated firm would have in trying to determine how much it sold to each of its customers in the last six months.

Firm with data integration

	Red division	Blue division
partnumber (code for green widget)	27	27
customerid (code for UPS)	53	53
Definition of salesdate	The date the customer signs the order	The date the customer signs the order

Firm without data integration

	Red division	Blue division
partnumber (code for green widget)	27	10056
customerid (code for UPS)	53	613
Definition of salesdate	The date the customer signs the order	The date the customer receives the order

Skill builder

Complete the data integration lab exercise described on the book's Web site (Lab exercises > Data integration).

Not surprisingly, many organizations seek to increase their degree of data integration so that they can improve the accuracy of managerial reporting, reduce the cost of managing data, and improve customer service by having a single view of the customer.

There are several goals of data integration:

1. A standard meaning for all data elements within the organization (e.g., customer acquisition date is the date on which the customer first purchased a product)
2. A standard format for each and every data element (e.g., all dates are stored in the format yyyymmdd and reported in the format yyyy-mm-dd)
3. A standard coding system (e.g., female is coded "f" and male is coded "m")
4. A standard measurement system (e.g., all measurements are stored in metric format and reported in the client's preferred system)
5. A single corporate data model, or a least a single data model for each major business entity

These are challenging goals for many organizations and sometimes take years to achieve. Many organizations are still striving to achieve the fifth, and most difficult, goal of a single corporate data model. Sometimes, however, data integration might not be a goal worth pursuing if the costs outweigh the benefits.

The two major factors that determine the desirable degree of data integration between organizational units are unit interdependence and environmental turbulence. There is a high level of interdependence between organizational units when they affect each other's success (for example, the output of one unit is used by the other). As a result of the commonality of some of their goals, these units will gain from sharing standardized information. Data integration will make it easier for them to coordinate their activities and manage their operations. Essentially, data integration means that they will speak a common information language. When there is low interdependence between two organizational units, the gains from data integration are usually outweighed by the bureaucratic costs and delays of trying to enforce standards. If two units have different approaches to marketing and manufacturing, then a high level of data integration is unlikely to be beneficial. They gain little from sharing data because they have so little in common.

When organizational units operate in highly turbulent environments, they need flexibility to be able to handle rapid change. They will often need to change their information systems quickly to respond to new competitive challenges. Forcing such units to comply with organizational data integration standards will slow down their ability to create new systems and thus threaten their ability to respond in a timely fashion.

Firms have three basic data integration strategies based on the level of organizational unit interdependence and environmental turbulence. When unit interdependence is low and environmental turbulence high, a unit should settle for a low level of data integration, such as common financial reporting and human resources systems. Moderate data integration might mean going beyond the standard financial reporting and human resources system to include a common customer database. If unit independence is high and turbulence high, then moderate data integration would further extend to those areas where the units overlap (e.g., if they share a manufacturing system, this would be a target for data integration). A high level of data integration, a desirable target when unit interdependence is high and environmental turbulence low, means targeting common systems for both units.

Target level of data integration between organizational units

		Unit interdependence	
		Low	High
Environmental turbulence	High	Low	Moderate
	Low	Moderate	High

Skill builder

1. Global Electronics has nine factories in the United States and Asia producing components for the computer industry, and each has its own information systems. Although there is some specialization, production of any item can be moved to another factory, if required. What level of data integration should the company seek, and what systems should be targeted for integration?
2. European Radio, the owner of 15 FM radio stations throughout Europe, has just purchased an electronics retailing chain of 50 stores in Brazil. What level of data integration should the company seek, and what systems should be targeted for integration?
3. Australian Leather operates several tanneries in Thailand, two leather goods manufacturing plants in Vietnam, and a chain of leather retailers in Australia and New Zealand. Recently, it purchased an entertainment park in Singapore. The various units have been assembled over the last five years, and many still operate their original information systems. What level of data integration should the company seek, and what systems should be targeted for integration?

Conclusion

Data administration has become increasingly important for most organizations, particularly for those for whom data-driven decision making can significantly enhance performance. The emergence of the Chief Data Officer is a strong indicator of the growing impact of data on the ability of an organization to fulfill its mission. For many organizations, it is no longer sufficient to just focus on administering data. Rather, it is now critical for many to insert to insert data management and exploitation into the strategic thinking of the top management team.

Summary

Companies are finding that data management is so critical to their future that they need a Chief Data Officer (CDO). Eight different types of CDO roles have been identified: coordinator, reporter, architect, ambassador, analyst, marketer, developer, and experimenter. In practice, a CDO is likely to shift among these roles depending on the enterprise's needs.

Data administration is the task of managing that part of organizational memory that involves electronically available data. Managing electronic data stores is important because key organizational decisions are based on data drawn from them, and it is necessary to ensure that reliable data are available when needed. Data administration is carried out at both the system level, which involves overall policies and alignment with organizational goals, and the project level, where the specific details of each database are handled. Key modules in data administration are the DBMS, the DD/DS, user interfaces, and external databases.

Data administration is a function performed by those with assigned organizational roles. Data administration may be carried out by a variety of persons either within the IS department or in user departments. Also, this function may occur at the personal, workgroup, or organizational level.

Data administration involves communication with management, clients, developers, and computer operations staff. It needs the cooperation of all four groups to perform its functions effectively. Since each group may hold very different perspectives, which could lead to conflicts and misunderstandings, it is important for data administration staff to possess superior communication skills. Successful data administration requires a combination of interpersonal, technical, and business skills.

Data administration is complex, and its success partly depends on a range of computer-based tools. Available tools include DD/DS, DBMS, performance monitoring tools, CASE tools, and groupware.

Key terms and concepts

Application development life cycle (ADLC)

Benchmark

Change agent

Computer-aided software engineering (CASE)

Data administration

Data dictionary

Data dictionary/directory system (DD/DS)

Data integrity

Data steward

Database administrator

Database developer

Database development life cycle (DDLC)

Database management system (DBMS)

External database

Groupware

Matrix organization

Performance monitoring

Project-level data administration

Request for proposal (RFP)

System-level data administration

Systems development life cycle (SDLC)

Transaction Processing Council (TPC)

User interface

References and additional readings

Bostrom, R. P. 1989. Successful application of communication techniques to improve the systems development process. *Information & Management* 16:279–295.

Goodhue, D. L., J. A. Quillard, and J. F. Rockart. 1988. Managing the data resource: A contingency perspective. *MIS Quarterly* 12 (3):373–391.

Goodhue, D. L., M. D. Wybo, and L. J. Kirsch. 1992. The impact of data integration on the costs and benefits of information systems. *MIS Quarterly* 16 (3):293–311.

Redman, T. C. 2001. Data quality: *The field guide*. Boston: Digital Press.

Exercises

1. Why do organizations need to manage data?
2. What problems can arise because of poor data administration?
3. What is the purpose of a data dictionary?
4. Do you think a data dictionary should be part of a DBMS or a separate package?
5. How does the management of external databases differ from internal databases?
6. What is the difference between system- and project-level data administration?
7. What is a data steward? What is the purpose of this role?
8. What is the difference between workgroups and organizational databases? What are the implications for data administration?
9. What is an information architecture?
10. Why do organizations need data standards? Give some examples of typical data items that may require standardization.
11. You have been asked to advise a firm on the capacity of its database system. Describe the procedures you would use to estimate the size of the database and the number of transactions per second it will have to handle.
12. Why would a company issue an RFP?
13. How do the roles of database developer and data administrator differ?
14. What do you think is the most critical task for the user during database development?
15. A medium-sized manufacturing company is about to establish a data administration group within the IS department. What software tools would you recommend that group acquire?
16. What is a stakeholder? Why should stakeholders be involved in database project planning?
17. What support do CASE tools provide for the DDLC?
18. How can groupware support the DDLC?
19. A large international corporation has typically operated in a very decentralized manner with regional managers having considerable autonomy. How would you recommend the corporation establish its data administration function?
20. Describe the personality of a successful data administration manager. Compare your assessment to the personality appropriate for a database technical adviser.
21. Write a job advertisement for a data administrator for your university.
22. What types of organizations are likely to have data administration reporting directly to the CIO?

23. What do you think are the most critical phases of the DDLC? Justify your decision.
24. When might application development and database development proceed independently?
25. Why is database monitoring important? What data would you ask for in a database monitoring report?